T0075396

Sound Identification of Terrestrial Mammals of Britain & Ireland

Sound Identification of Terrestrial Mammals of Britain & Ireland

Neil Middleton
Stuart Newson
Huma Pearce

PELAGIC PUBLISHING

First published in 2024 by
Pelagic Publishing
20–22 Wenlock Road
London N1 7GU, UK

www.pelagicpublishing.com

Sound Identification of Terrestrial Mammals of Britain & Ireland

British Library Cataloguing in Publication Data
A catalogue record for this book is available from the British Library

https://doi.org/10.53061/VHUN1535

ISBN 978-1-78427-381-1 *Hardback*
ISBN 978-1-78427-382-8 *ePub*
ISBN 978-1-78427-383-5 *PDF*

Cover image: Red deer stag (*Cervus elaphus*) calling in the mist © Philippe Clement/naturepl.com

Typeset by BBR Design, Sheffield

Contents

Foreword

by Professor Robbie McDonald (University of Exeter)

Mammals can be such frustrating subjects. Whether our engagement with them stems from research, professional practice, natural history or the happy, chance encounter, it is this element of chance that so often restricts and confounds our ability to observe, analyse and learn. In studying the biology of mammals, it is so important to rely on the indirect approach, of finding signs, however ephemeral, and interpreting these through lenses clouded with bias and uncertainty.

Whilst at least some mammals are near ubiquitous, with our depauperate sensory systems, they remain steadfastly inconspicuous to us. Nevertheless, mammals clearly spend a good deal of time and energy making themselves conspicuous to one another, through their relatively heightened and diversified senses. Some mammals are readily seen, at least at times and in some places. Most mammals are certainly smelly, at least to other mammals, if not to our stunted noses. But naturalists and researchers have almost been taken unawares that many mammals are also rather noisy. Clearly, eavesdropping on the ultrasonic world of bats has been a primary means of learning about their nocturnal lives for decades. But keen observers of these bat recordings spotted that smaller, earth-bound voices of other mammals were contributing to the ultrasonic cacophony.

With the tremendous initiative of this book, three of the keenest observers have taken further action, with a truly innovative text and accompanying sound files, priming our dull ears to the sonic world of these non-flying, but no less vocal, mammals. Neil, Stuart and Huma have comprehensively laid out, in terms accessible and valuable to all, the audible and ultrasonic world of the terrestrial mammals with which we share these islands.

The best ideas are invariably the ones that seem most obvious after-the-fact and their book realises an excellent idea. This work will, I have no doubt, inspire the development of new perspectives on how mammals live their lives, as well as form a new touchstone on how we might survey, research and understand their populations and behaviours. More questions can now be asked of mammals. By listening attentively, as the authors have and as they here encourage and enable us to do, we will hear more of their answer.

Preface

It's May 2021, and Neil Middleton is working frantically to complete the second edition of *The Social Calls of Bats of Britain & Ireland*, having in 2020 seen *Is That a Bat?* hit the shelves, alerting bat workers to the fact that there are many things they may record that look and sound like bats, but are not (e.g. small land-based terrestrial mammals). Simultaneously and independently, Stuart Newson is carrying out research within the British Trust for Ornithology (BTO) relating to the identification of small mammal sound to species through the use of automated classifiers, a project that would feed into what we all now recognise as the BTO Acoustic Pipeline. And, whilst these guys were sitting relatively comfortably in their offices, spending the bulk of their lives on PCs, Huma Pearce was separately on a mission to record any creature that made a sound, any sound! Huma's relentlessness and resilience to keep going and find ways to get data was feeding into Stuart's work, and assisting Neil with the bat social calls book, and pushing, to the limit, the storage space available on all of their respective computers.

Unbeknown to us all at the time, during our separate, but sometimes intertwined journeys, in late 2020 we would together work on an important article published in *British Wildlife* relating to acoustic identification of small terrestrial mammals in Britain. This paper, which was rooted in the information that Stuart had been working on, expanded greatly upon and added new information to that which was made available via the *Is That a Bat?* book. It was great to see us all together in one place, on the same page, but at the same time we were beginning to realise that there were not that many other people studying the subject matter. Further, despite articles published, TV and radio appearances, chapters of books on bigger subjects, workshops to the bat community and the like, it was going to be a long and relatively lonely journey for each of us. We felt that this was a huge shame, as those interested in natural history were missing a massive opportunity to expand their horizon to include an acoustic landscape. Just think where the world of ornithology would be if bird calls and songs were never understood or considered. That seems a ridiculous thing to say, but it is by and large the situation when it comes to mammals, especially the smaller species that are so often difficult to notice or engage with at any level.

So, it's difficult, and this seems to be a major barrier to entry. It's just *too* difficult, and even if someone was interested – where do they find out about our mammal acoustic behaviour, in a relatively accessible manner, and in a single place? Goodness, there are so many resources available for bird- and bat-related sound, but why has no one done it for our land-based terrestrial mammals in Britain and Ireland? It's difficult!

This takes us to May 2021, when the phone rings. It's Nigel Massen of Pelagic Publishing, who has worked with Neil on other book projects, and he is asking, *'do you want to do a book on terrestrial mammal acoustics?'* Neil's initial reaction at the time was probably unprintable, but after the emotional fog of *'oh no, not another book – when will it all end?'*, the answer had to be, *'absolutely, yes, let's do it.'* It needed to be done, and the three authors working together, each approaching things from a slightly different perspective and with different experiences, would hopefully prove to be a good blend, and produce something that was accessible, making it easier for people to understand how to go about gathering data, and how to interpret data gathered or sounds heard in the field.

And here we are. Is it perfect? Of course not – a subject like this is always developing with new information coming to light. Will it answer the question of absolutely every terrestrial mammal sound you hear? Nope – there are some omissions that we were unable to source, as well as which, each species produces a complex range of sounds, some of which we may not yet have experienced. Will it help you to understand the subject better, and make you a better field naturalist? Will it highlight opportunities for research? Will it help support ecological consultancy-driven studies (survey methods, mitigation, compensation, habitat enhancement)? Will it support conservation? We really hope so, and the more that we and others add to the subject matter in the future, the more it will offer.

More about why the authors feel this book is important

Sound produced by terrestrial mammals is an area about which relatively little is known, with more research still required in order to verify what is already thought to be understood, and develop further our knowledge and understanding.

We have been asked to explain the value of studying the sounds of terrestrial mammals. In short, acoustic information is becoming increasingly recognised as a useful aid to identification, as in many situations sounds produced can be species-specific, as well as being indicative of behaviour (e.g. territorial behaviour of males during the mating season). In this day and age, when the gathering and interpretation of bird and bat sound, for example, is used as a matter of course during hobby, self-study, research, conservation and consultancy work, it seems a logical progression to build momentum within the mainstream mammal world for land-based terrestrial mammals. Indeed, a better understanding of this subject could mean that non-intrusive survey methods can be developed. This would reduce the need for direct disturbance or interference with the study subject, meaning that what is being observed or recorded is, as far as can be, reflecting natural behaviour. All of this could therefore contribute to better interpretation, and consequently, if required, more suitable mitigation, compensation and/or habitat enhancement solutions.

This book summarises much of what is understood so far relating to the species occurring within Britain and Ireland. The material provided will be of benefit to people interested in or carrying out studies, at whatever level or for whatever purpose, and will also encourage others to carry out further research. Sound is a fascinating subject to consider, in that quite often we do not see what we can hear, and therefore either intrigue kicks in, or alternatively, we end up ignoring it altogether, hoping that we have not missed out on the opportunity to document something important or interesting. Add to this that for some occasions, not only do we not see the animal, but due to its ultrasonic acoustic behaviour we do not hear it either. At least not without the presence of an ultrasonic recording device (e.g. a bat detector). But these ultrasonic sounds, when they are produced, can be quite common, and once noticed can be found far more often than you may at first consider.

To produce this work without the reader also having the opportunity to become a listener, and therefore fully appreciate the sounds produced, would be falling well short of what could be achieved. Hence one of the reasons for accompanying the book with a downloadable sound library. Another reason for providing you with these recordings is that it gives you an opportunity to put them through a software package of your choice, thus enabling you to explore those of interest as you wish in your own way.

The material produced here will not give all the definitive answers relating to the subject. It is a complex area of study, which still, relatively speaking, requires much more work. What we are seeking to do, at this stage, is pull together a lot of the information that currently exists, with the anticipation that with time, further research will increase knowledge and confidence. So, please do not feel that you should read this book with a view to knowing everything there is to know by the end of it. In many respects there are still far more questions than there are answers. For the time being, therefore, please regard this piece of work as a source of reference and a collection of material, blended with our own experiences and thoughts, with the aspiration of creating intrigue and inspiring others to further the understanding of this fascinating subject.

Acknowledgements

Since having announced in early 2022 that we were working on this book, we have been truly humbled by the positive reaction received from so many supporters and contributors to this work.

The idea for this book was born out of the frustration of not having anything that was easily accessible to reference, along with the fortunate coincidence of the three authors being known to each other and having previously been working, sometimes separately, sometimes together on the subject matter. All of this in conjunction with the publisher (Pelagic Publishing) suggesting the idea for a book devoted to the subject matter.

At the start of the writing process, for some of the species groups we already had lots of data. But in other areas we definitely felt that we would need to work hard to improve upon our own knowledge, so that we could help others to help themselves. Any initial feelings of inadequacy were dispelled by the support we have received on this journey. To everyone who has contributed to, enquired about, or promoted what we are doing, we owe a huge debt of thanks.

In many instances material obtained by ourselves or contributed by others has not been included; however we feel it is important to acknowledge that all the material received has greatly assisted in our overall understanding. There are many more examples that we could have used. There comes a point, however, when the time and space available need to be considered, and the work has to go to print. What gets produced in its final format may indeed not be perfect, but let's not make *'the perfect the enemy of the (hopefully) good'*. And if it is good, then it is only so because of the huge number of supporters and contributors. To that end, we acknowledge everyone here who contributed, in whatever way. Without you, we would not have the material to create this work. As such, we would like to thank and acknowledge the following people and organisations:

A special thanks to Professor Robbie McDonald (University of Exeter) for providing the foreword and for being a strong supporter and ambassador of what we are seeking to achieve, and also to Alexandra (Sandra) Graham for allowing us to use her excellent photographs accompanying the start of most chapters. Laurie Campbell (lauriecampbell.com) and Matt Errington for the use of their pictures. Simon Gillings for providing support, especially with distribution maps. Dr Philip J. Baker, Dr Johnny Birks, Melanie Findlay (Findlay Ecology Services) and Dr Gavin Wilson for providing technical support in their respective specialisms. Sara Magness for her copy-editing expertise and suggestions for improvements. Aileen Hendry, who has supported the project throughout, and given valued encouragement and feedback. Nigel Massen and David Hawkins of Pelagic Publishing, for their continued support and professional expertise throughout the project, and for believing in us and the message we sought to deliver.

Lars Pettersson (Pettersson Elektronik AB, Sweden) for his support in our use of 'BatSound' software which is used to produce most of the figures contained within the work. The following organisations who assisted and supported the project: BatAbility Courses & Tuition; The British Trust for Ornithology; City of London Corporation (Epping Forest); Echoes Ecology Ltd; The Mammal Society; Natural England; NatureScot (previously Scottish Natural Heritage).

Then we have those people who went well beyond what anyone in our position would ever dare to hope for, and helped in the gathering of lots of data, sharing experiences and supporting the cause in so many essential ways. So, for really going 'above and beyond', special thanks are due to: Dan Bagur, Kari Bettoney (Mid Devon Bat Rescue Centre), Matt Binstead (British Wildlife Centre), Laura Carter-Davis (Echoes Ecology Ltd), Simon Elliott, Mark Ferguson (Mark Ferguson Audio), Lorna Griffiths (Nottinghamshire Dormouse Group), Paul Howden-Leach, David Mellor, Aaron Middleton (WSP), Philip Mill, Gillian Prince (Hedgehog Bottom, Berkshire), Russell Savory (wildlifeonfilm.com) and the Wildlife Sound Recording Society (WSRS).

For providing assistance in so many ways, including unearthing research papers, making recordings available, arranging site access, suggesting ideas, or for general support and encouragement of the project, we would also like to acknowledge all of the following: Stewart Abbott (derbyshirebirdtours.com), Mick A'Court, Danny Alder, Marc Anderson (wildambience.com), Argaty Red Kites (Tom Bowser), Martin Bailey (Wildlife & Countryside Services), Marc Baldwin (wildlifeonline.me.uk), Bat Conservation Trust (in particular Lisa Worledge and Naomi Webster), BBC WinterWatch, Claudia Bieber (University of Camerino), Ian Bond, Simon Bowers, Caroline Boxall, Jonathan Bramley (Bramley Associates), Adele Brand, Andrew Brennan, British Deer Society, Simone Bullion, Katie Burrough (WSP), Heather Campbell, Andrea Catorci (University of Camerino), Paul Childerley, Clyde Muirshiel Regional Park, Arnold Cooke, Lizzie Croose (Vincent Wildlife Trust), David Darrell-Lambert, David Archer Associates, Adrian Dexter, Daan Drukker, Laurent Duvergé (Kestrel Wildlife Consultants Ltd), Bengt Edqvist, Elekon AG, Russell Elliott, Andrena Ellis (WSP), Pat Emslie, Falkirk Community Trust, FCC Environment, Anders Forsman, FPCR Environment and Design Ltd, Alasdair Fraser, Heidi French, Tony Fulford, Alison Fure, Leonida Fusani (University of Camerino), David (Davy) Galbraith, Geckoella Ltd, Leif Gjerde, Jean-François Godeau, Angela Goodwin, Brian Goodwin, Lauren Graham, Angel Perez Grandi (Arney Aquatic Acoustics), Sian Green (MammalWeb.org), Amy Hall, John Harrison-Bryant (HB Bat Surveys), Darren Hart (Isles of Scilly Wildlife Trust), Rosanna Hignett (Echoes Ecology Ltd), Alison Hitchens, Adam Hough, Jon Horn (Nurture Ecology Ltd), Morgan Hughes, Nick Hull (Two Owls Blog), Malcolm Ingham, Wim Jacobs, Bill Jackson, Nicola Jee, Nancy Vaughan Jennings, K. Lisa Yang Center for Conservation Bioacoustics, Inger Kaergaard, Kelly Kilfeather, Donna Kisielewski, Stefanie Kruse, Alex Lack, Landguard NR (Suffolk), Harry Lehto (University of Turku), Liza Lipscombe (British Wildlife Centre), Ben Locke, Alison Looser, Craig Macdonald (CSM Ecology and Environmental Assessment Ltd), Kate MacRae (wildlifekate.co.uk), Hugh Manistre, Barbara McInerney, Emiliano Mori, John Muddeman, Muiravonside Country Park (Falkirk), Angie Nash (panashadventures.com), Chris Nason, Lucy Newson, Samuel Olney, Mark Osborne (OS Ecology Ltd), Lotty Packman, Claire Parnwell (Greenwillows Associates Ltd), Hilary Phillips, Michael Pitzrick (K. Lisa Yang Center for Conservation Bioacoustics), Brian Power, Paul Pratley, Bob Reed, Pedro Ribeiro, Marius Ruchon, Paola Scocco (University of Camerino), the Royal Society for the Protection of Birds (RSPB), RSPB Buckenham Marshes, RSPB Strumpshaw Fen, Sharon Scott, Scottish Natural Heritage (Licensing Team), Peter Scrimshaw, Steve Senior, Kelly Sheldrick, Phil Sime (BBC Radio Scotland), Heather Simpson, Andrew Skinner, Greg Slack, Mr M. Smith (Banchory, Aberdeenshire), Charles Smith-Jones, Rune Sørås, Paul Stancliffe, Tim Strudwick (RSPB), Suffolk Wildlife Trust, Emily Sullivan, Jo Sutherland, Laura Torrent, Roger Trout, Goedele Verbeylen (Natuurpunt), Michael Walker (Nottinghamshire Bat Group), Liz Walsh, William Walton, Leonie Washington, Harriet Webb, Dan Wildsmith, Wildwood Trust (Kent), James Wilson, Wiltshire Bat Group, Bill Winters, Christopher Wren (trogtrogblog.blogspot.com) and WSP UK Limited.

About the photographer

Sandra Graham is a keen and respected photographer, with most of her work centred in Scotland, focusing on wildlife and the dramatic Scottish landscapes that inspire her. Many of the photographs within this book were taken by Sandra, and the authors are delighted and grateful of her passion for the natural world, her support and assistance in this respect.

Sandra shares much of her work on her Instagram site, which can be found at www.instagram.com/sandra_photostuff/

About the Authors

Neil Middleton is a licensed bat worker and trainer, and is the owner of **BatAbility Courses & Tuition**, an organisation that delivers ecology-related skills development to customers throughout the UK and beyond. He has a constant appetite for self-development, as well as seeking to develop those around him, and to this end he has designed and delivered in excess of 400 training events covering a wide range of business- and ecology-related subjects. Neil has had a strong interest in the natural world since childhood, particularly in relation to birds and mammals. He has studied bats for over 25 years, with a particular focus on their acoustic behaviour (echolocation and social calls).

Picture credit: © Laurie Campbell, 2011

Other books also available by Neil Middleton:

Social Calls of the Bats of Britain & Ireland (2nd Edition).
Pelagic Publishing, Exeter, 2022.

Is That a Bat? A Guide to Non-Bat Sounds Encountered During Bat Surveys.
Pelagic Publishing, Exeter, 2020.

The Effective Ecologist – Succeed in the Office Environment.
Pelagic Publishing, Exeter, 2016.

Email: neil.middleton@batability.co.uk
Website: www.batability.co.uk

Stuart Newson is a Senior Research Ecologist at the **British Trust for Ornithology**, where he is involved in survey design and data analysis from national 'citizen science' surveys. While birds have been the core of his work, Stuart has an interest in bats and acoustic monitoring – in particular, how technology can deliver opportunities for conservation and provide ways to engage with large audiences. Stuart's work on bioacoustics has included creating tools to identify European bats, bush-crickets and small mammal species from their 'calls'. This resulted in the BTO Acoustic Pipeline, which integrates online tools for coordinating fieldworkers, processing recordings, and returning feedback to encourage large-scale participation in acoustic surveys. Stuart is also a member of Natural England's Bat Expert Panel.

Email: stuart.newson@bto.org

Huma Pearce is a freelance ecologist with a specialist interest in mammals. She has worked on various small- and medium-sized mammal monitoring projects in the UK and abroad, training students and citizen science volunteers in mammal survey techniques.

She developed an interest in bats in 2005, which sparked an interest in sound analysis for species identification and behavioural studies. Huma has since contributed to sound libraries (e.g. BTO Acoustic Pipeline) and has become an advocate for the application of bioacoustics for monitoring bats and other mammals. As such, Huma is never far away from either a recording device, a new site to explore, or the analysis of megabytes of data from previously deployed sessions.

Email: huma@mostlybats.org

All correspondence relating to this publication should in the first instance be sent to Neil Middleton, contact details as follows:

Email: neil.middleton@batability.co.uk
Tel: 0044 (0) 7877 570590

Preamble

Overarching spirit of intent

The purpose of this book, and the accompanying Sound Library, is to highlight both audible and ultrasonic (above the hearing range of humans) sounds produced by land-based terrestrial mammals (excluding bats) within Britain and Ireland. The examples given are by no means considered to be the full repertoire for any of the species discussed, but they provide a good starting point for identifying species from their vocalisations and characteristic sounds. We are all, collectively, still on a steep learning curve about many of the subjects covered, and hope that this book will inspire more interest and research into the study of wildlife acoustics.

We have tried, as far as possible, to be accurate and thorough when recording subjects ourselves and/or accepting recordings from others. The most important thing is to enable you to move closer to becoming more knowledgeable and spending less time trying to work out what it is you are looking at or listening to. If this book achieves this for you, even just some of the time, then that is a major step in the right direction.

Glossary

A Glossary is contained within Appendix I.

The first time that a word, term or abbreviation described within the Glossary is used within the main text of the book, it appears underlined, so as to alert the reader.

About the Sound Library

To accompany the text, we have provided a series of downloadable sound files in .wav format. The files are grouped into folders which tie in with the chapters where they are described. To download this Sound Library, go to the following link:

https://pelagicpublishing.com/pages/terrestrial-mammals-call-library

In the book, when reference is made to a file in the Sound Library, the symbol 🔊 is inserted in the text. Where the 🔊 symbol is attached to a figure (such as a spectrogram), the figure number doubles as the Sound Library file number. It is also worth noting that when you open a sound file, the file itself may be larger than the portion that is shown in the figure. The sound file thus has the potential to show you more when viewed as a spectrogram on your computer, or allow you to listen to more than is illustrated in the text.

Each track has an abbreviation (see table) included in its title to alert you to the type (audible or ultrasonic) of recording that you are about to download or listen to.

AU	Audible recording. Usually representing what you would hear with your ears if you were in the field.
FS	Full spectrum, real-time recording. Usually ultrasonic, meaning that you won't normally hear anything unless the playback speed is slowed down to bring the sound data into the audible range (i.e. you convert the call to time expansion).
TE	Time expanded recording. Slowed down by a factor of 10, meaning that what you hear is 10 times slower in terms of time, and 10 times lower in terms of frequency.

Clicking on the track will open it with your default audio software (e.g. Windows Media Player), which is usually only useful if the file is labelled 'AU' or 'TE'. Alternatively, and for all file formats, in order to open a file within a software program, we recommend that you first open the sound analysis software of your choice, and then using the software browse facility, open the track from within its folder.

Otter – Sandra Graham

CHAPTER 1

Introduction & Context

1.1 Setting the scene

We think it is fair to say that land-based, terrestrial mammal identification resources (e.g. books and web-based material) have focused primarily on the visual aspects of how we, as humans, interact with the animals and their field signs. We are, after all, predominantly visual interpreters in this respect, as we are in many aspects of our lives. How often do we hear someone (including ourselves, of course) say something like *'I'd really love to see an otter'*? The phrase *'I'd really love to hear an otter'* would be quite unusual, would it not? For some, however, hearing and recording is just as important as seeing. It all contributes to an overall appreciation and understanding of the subjects that we are passionate about or interested in surveying, studying and/or researching. Whichever of these categories you regard yourself as being in (and you are allowed more than one!), we hope this book will add an acoustic perspective to your visual horizon. We believe that sound has the potential to open up so many more possibilities. And, for those who are already exploring mammal sounds, it will give you additional reference material, to support you even further in your goals.

With all of this in mind, this book is equally valuable to hobbyists, researchers, conservationists and ecological consultants. It broadens the opportunities for engagement with wildlife from beyond the visual, to include acoustic encounters. This means that much more information can be collected, leading to better-informed naturalists, who are then more able to make better judgements.

We should also bear in mind that acoustic study, for at least some mammals, is far less intrusive than other means (e.g. live trapping), and, as such, potentially reduces our impact upon these animals as they carry out their daily lives, within their natural environment.

As far as we are aware, a comprehensive account and accompanying sound library of the vocalisations and characteristic behavioural sounds produced by all land-based terrestrial mammals that occur in Britain and Ireland has not been done before. There have been more generalist books, such as the *Collins Field Guide to Wildlife Sounds* (Sample, 2006), and smaller-scale projects that focus on target species or groups (e.g. Newson *et al.*, 2020), but nothing that has sought to collate all terrestrial mammals together in one publication. Indeed, we have struggled to find anything quite like it from anywhere in the world. Why would that be the case? Well, at least in part, it *is* difficult and there are, most definitely, challenges. For example, there is not much to refer to in terms of existing accessible knowledge, and, furthermore, 'sound' is a very difficult thing to describe in words: one person's squeak is another's cheep! Then there is the additional equipment that may be required, dependent upon the species of interest, as well as associated software packages when catering for ultrasonic sound in particular. And, of course, you can spend lots of time in the field without hearing anything, and when you do hear it (heard not seen), how confident can you be in making an accurate identification?

Developing this latter point further: once you have recorded or heard something that you have not seen, how do you actually verify what it was? It is perhaps due to this challenge of verification that few people pay much attention to sounds produced by these mammals. It is so much easier to be certain of an identification when you have seen it.

We definitely 'get it'. We understand the frustration involved in not knowing what it was. We understand the 'wishing' that there was somewhere or someone you could refer it to. Someone who would offer some explanation, or support, or guidance on the matter. We understand what it is like when your network does not include one of a very small number of people who would have a broader appreciation of mammal acoustics. In fact, we understand this so well, and having been in all of these acoustic dark holes ourselves, we decided to write a book about it – this book. In doing so, we have learned so much more ourselves, which we now want to share with you. We have learned from each other, from the process, from the experiences, from the academic world, and from the many correspondents who have assisted us along the way, without whom what follows would be lacking in so many areas.

The subject of mammal acoustics may seem a bit too complicated for some, but it is worth considering what a greater understanding of acoustics has achieved for the study of birds, orthoptera (bush-crickets, grasshoppers etc.), cetaceans and bats. There must be a benefit in empowering people interested in mammals, to be better equipped with knowledge that they can take out into the field. Even better if such information could be made available to anyone, in an accessible format, covering a range of species, and all in one place. This book is our contribution to the challenges faced by many. It is our attempt to move things forward, and to highlight what we think acoustics can offer. We are guessing by this point, if you are reading this, that we are preaching to the converted, or at the very least, the mildly interested! So, at this early stage we would like to say, thank you. Thank you for already being involved at whatever level you are. Thank you for being intrigued, or at the very least inquisitive enough to get this far.

1.2 Core objective

An important point to get across as you start to read this book is that it focuses on the use of acoustic information for identification purposes, as opposed to us attempting to describe the social functions of sounds produced. In some instances, however, it is known, or we have a very strong steer towards, what certain sounds relate to in terms of behaviour. When we feel that there is good evidence backing up such interpretation we will say so. However, there will be many occasions where we are confident with an identification, but far less so about the meaning or intent behind the sound.

For many of the species covered, especially the smaller mammals, there is still so much more research required, in order to not only study the purpose of much of the acoustic information being delivered or exchanged, but indeed to pull together a broader catalogue of each species' repertoire. Our own experiences, from the world of bats in particular (which of course are also mammals), tell us that to begin to understand far more fully what is actually going on, it will take many people, across many colonies and a wide, international, geographic spread, before we can begin to start drawing generic diagnostic conclusions about such matters.

There are of course, undoubtedly, some good academic studies, for at least some of the mammal groups, or for certain species, which have been invaluable for our own understanding. These studies contribute hugely, as we bring things together here in a more accessible format, and we will make reference to any such studies as and when appropriate to do so. Nevertheless, if you are from an academic background, please keep reminding yourself as you use this book that we are trying to make the subject accessible to anyone and everyone (which most certainly still includes you) who is interested in learning more about what is being covered. To do this in an academic style would put at risk the core objectives we are seeking to achieve.

So, how would we describe our core objectives? Simply put:

- *To help anyone who wishes to understand more about the sound identification of land-based terrestrial mammals.*
- *To encourage greater interest so that we can all benefit from each other's experience, accelerate the pace of learning, and better our field skills, survey methods and interpretation of data.*
- *To ultimately improve our understanding of the distribution and abundance of land-based terrestrial mammals in Britain and Ireland, so that we are able to make better-informed decisions.*

There are not enough people engaged with this approach currently, and without publications and resources that inform the widest of audiences, it is unlikely that there would be any significant increase in momentum. We want to put a foot on that accelerator, for anyone who wishes to start their own journey. So, whether you have a general interest in the subject, or are focusing on a particular species, or are carrying out research, conservation activities or consultancy-related work, what we would ask is that you talk about this aspect of what all mammals do. Spread the word within your networks, however small, and together, many years from now, we will all have a greater understanding of what is going on, for the benefit of the species concerned.

You may think that all of this is just wishful thinking, but it was not that long ago when, within the bat world, there were similar challenges related to the social acoustics of bats. Most bat workers were aware of bat social calls, and there were a small number of calls that were broadly appreciated and useful to a percentage of those gathering echolocation data. Within the last 10 to 20 years considerably more knowledge has been acquired, and a good percentage of what is now known within the mainstream of bat work comes from information published in books such as this. Information enabling the wider community to become more informed, leading to interest, self-discovery, discussion and further learning for all concerned. We know that what we are aiming for is reasonably contrived, because we have been part of a similar story elsewhere.

Bringing matters closer to home, in respect of land-based mammals, a publication produced by ourselves for *British Wildlife* (Newson *et al.*, 2020), was the fullest account of small land-based mammal acoustics produced up until that point. The reaction from that article alone was substantial, and was seen by some (e.g. BBC, *The Guardian*) as being an important stepping stone in the right direction. So, the journey has already, most definitely, begun.

As you get more and more into your own journey, and you start gathering, listening and looking at the sounds produced, you will likely record mammal sounds that have not been documented anywhere before, including between these covers. We are certain about this. There is no one who is anywhere close to the complete truth for almost anything covered (especially ourselves). We hope, however, that you will benefit from extra field skills that were perhaps not at your disposal until now, in that, as well as what your eyes see, and occasionally your nose smells, you have this opportunity to add what you heard into the mix. Furthermore, we will introduce you to a variety of acoustic recording equipment that is available, including the use of remote recording devices, which can gather data on your behalf over many days, or even weeks or months, when you are not there to listen first-hand. For many of the smaller species, the amount of acoustic data and information that these can provide (e.g. during nightly or seasonal activity) is considerably more useful than a fleeting glimpse of a small rodent disappearing into vegetation. The same creature that would otherwise potentially end up being undocumented, or lumped into a group, because it was all too fast for any human to be able to deal with, and/or demonstrate to anyone else any level of confidence in its identification.

1.3 Much more than '*meets the eye*'

For those interested in studying mammals, at whatever level, identification and interpretation of behaviour are important areas of study, and by and large in seeking to achieve this, the bulk of our interface with such species is based on visual experiences. Unlike birds and bats, where so much has already been written about acoustic interpretation, the same level of knowledge has not been made as accessible to those interested in land-based terrestrial mammals. In the British Isles, for many of our larger mammals this may not always necessarily be a major issue, or at least as it first appears. When you see a badger, or a fox, or a roe deer, you know what it is. As well as this, their species-specific signs (e.g. footprints, faeces, places of shelter) can (sometimes!) be quite distinctive. But so much activity occurs out of sight, either within dense vegetation or at night. Then, when we do hear something, can we honestly say that we are as comfortable with our acoustic identification for these larger mammals, as we are with visual clues? And what about the smaller mammals which are much harder to see visually? Many of these species are all around us, but go unnoticed, unless we happen to find a dead one somewhere, or have been doing live trapping. Furthermore, unlike their larger cousins, acoustically they present another challenge, in that despite them being reasonably vocal, we rarely hear them. This is because most of the sound they produce is much quieter (i.e. not travelling far through the air) and/or ultrasonic. And, even if we do happen to hear something – a squeak – it is usually so fast and high-pitched to the human ear that it would be impossible to describe the sound in any meaningful way or identify the species with any accuracy.

Now, you may not be that convinced that there are many of these smaller mammals about. To help put things into perspective, we have provided examples of population estimates for a range of mammals to which most readers will be able to easily relate (see Table 1.1). Think how often you see a deer species, for example. For many it could be fairly regularly (e.g. roe deer). Now think about how often you see a smaller mammal (brown rat or wood mouse for example). Next, compare the relative population sizes within Table 1.1. Hopefully this helps to emphasise the point we are making. There are

Table 1.1 Examples of the relative abundance to one another of land-based terrestrial mammals within Britain (Mathews *et al.*, 2018)

Group	Species	Estimated population size
Small terrestrial mammals	Brown rat	7,070,000
	Wood mouse	39,600,000
	Field vole	59,900,000
	Common shrew	21,100,000
Large terrestrial mammals	Badger	562,000
	Fox	357,000
	Roe deer	265,000

considerably more of the smaller species around you than meets the eye, and in order for you to explore their world more fully, acoustic studies may prove to be useful. And, as it happens, we do have ways of making them more accessible, acoustically. This is achieved through the use of ultrasonic recording devices (i.e. bat detectors), and when you begin to explore their acoustic world in more detail, you will see that there is far more going on than meets not only our eyes, but also our ears; more than one would ever, otherwise, envisage.

1.4 Scope of content

From the title of this book, it should be fairly clear what it is all about. But having said that, and having thought very carefully about the title, a few explanations regarding the scope of the material may be useful, as different people may be expecting different things. As such, let us now dissect the title, and offer a more detailed explanation.

Sound identification

'Sound identification' is fairly clear on first impressions, but it will be interpreted differently by individual readers. What we are covering here is both audible sound and ultrasonic sound. Some people will only be interested in one of these areas, whilst others will be interested in both, especially when there are some species which create sound that falls into both categories.

This book covers both of these acoustic experiences. The usually audible, sonic sounds made by larger mammals, and the mostly inaudible, ultrasonic sounds made by the smaller species. We will take each species group, and according to how you are most likely to engage with the sound heard or recorded, we will give guidance which will help improve your field skills.

Audible versus ultrasonic

The terms 'audible' and 'ultrasonic' themselves are not that easily definable in nature, as they are merely a human perspective on what is happening naturally.

If something is described as 'audible', that normally means audible to the human ear. Do bear in mind, however, that as far as we are aware, every sound deliberately created by any animal is audible to another animal of the same species and/or, in some cases, to other species.

Conversely, something described as 'ultrasonic' is not audible to the human ear. However, what is ultrasonic to a human may be audible to another animal. At an individual human level, we could even go so far as to say that what is ultrasonic to one person may be audible to another. As a generalisation, an older person (hearing often deteriorates with age) is encountering (or rather is not) much less higher frequency sound than someone in their teenage years.

There is, however, from a human perspective, a scientific definition for the term 'ultrasound', describing it as a sound beyond the normal upper limits of human hearing, medically established as being 20,000 hertz (Hz), or 20 kilohertz (kHz), in a healthy young adult.

At the opposite end of the spectrum we have extremely low sound, which is deemed to be below the normal hearing range of humans. This is called infrasound, which from an audible tone perspective is regarded as beginning to gradually come in at frequencies lower than c.20 Hz (note: not kHz), although humans *can* pick up (sense) sound well below this level (Persinger, 2013).

There are numerous ways in which people can engage with mammals acoustically, and these engagements are broadly determined by whether it was intentional at the time on the part of the naturalist, or it was more of an unanticipated encounter. Table 1.2 provides a more detailed explanation.

Table 1.2 Examples of acoustic engagements with mammals

Potential for acoustic engagement		Notes
Surveyor present	**Mammal heard by ear and seen**	Visual and acoustic clues are sufficient for an experienced/amateur naturalist to accurately identify common or well-known species. It may not be necessary to record sound for accurate species identification.
	Animal heard by ear but not seen	Unless the sound is known and distinctive, determining species may be difficult in the absence of visual clues and without the aid of recording for later referral.
	Animal/mammal sound is recorded	Surveyor is at the scene, with equipment to hand, and has been able to capture sound recordings. Mammal vocalisations can be recorded using specialist acoustic equipment (audible or ultrasonic), or simply by using voice memo or video features on a mobile phone (audible sound only) to record sound for later referral and/or using acoustic software to help with species identification. Animal may be seen during acoustic recording and visual cues will assist with species identification.
Surveyor absent	**Intentional monitoring:** **Remote recording of target species/group**	Surveyor is not present, but a recording device has been deployed to gather data remotely over a defined period of time. Sound recordings can be analysed using acoustic software to help identify or narrow down the species recorded.
	Unintentional monitoring: **Mammal vocalisations recorded incidentally as 'by-catch' whilst carrying out other unrelated work**	Recording equipment in place for one reason inadvertently records the vocalisations of (non-target) mammal species. Sound data can be analysed audibly or by using acoustic software to identify the species and/or better understand the behavioural context of sounds made by species. Examples: – Bat detector survey, recording a shrew. – Security systems (CCTV)/trail cameras recording video footage incidentally also recording sound, whilst triggered events are taking place. – Colleagues, friends and family send you wildlife sound recordings because they know you might be interested or able to help with species identification.

Terrestrial mammals

What do we mean by terrestrial mammals? Well, maybe not quite what you would expect. For the purposes of this book, we are narrowing things down from the norm, whereby bats are not included within these pages. As it happens, acoustic identification of our bat species is already well catered for elsewhere (e.g. Barataud, 2015; Russ, 2021; Middleton *et al.*, 2022). Bats, which are correctly defined and broadly speaking regarded as terrestrial mammals, could also be considered as being semi-aerial, in a similar way that seals can be referred to as being semi-aquatic.

To be clear, bats and seals are not included here, nor are cetaceans (marine mammals). This means we are left with those mammal species that are purely land-based, with their feet firmly on the ground, at least a lot of the time. Hence why, up until now, you may have noticed that we have used the term 'land-based terrestrial mammals'. Having now explained the scope of the species covered more fully, we will no longer always include 'land-based' in the pages that follow.

Having decided upon the scope, this allowed us to consider for inclusion 46 species/subspecies of terrestrial mammal which would be regarded as being resident and breeding within Britain and Ireland (see Tables 1.3 to 1.7). These species are representatives from 14 families, comprising 41 genera.

Within Tables 1.3 to 1.7 we describe the species we considered for inclusion, within the definition as described, and whether or not they are catered for. Additionally, when they are covered in this book, the final column directs you to the appropriate chapter.

Table 1.3 Even-toed ungulates (order – Artiodactyla) of Britain and Ireland

Family	Genus	Species name	Further information included	Chapter reference
Suidae	*Sus*	Wild boar *S. scrofa*	Yes	5.4.1
Cervidae	*Muntiacus*	Reeves' muntjac *M. reevesi*	Yes	5.3.5
	Cervus	Red deer *C. elaphus*	Yes	5.3.2
		Sika *C. nippon*	Yes	5.3.3
	Dama	Fallow deer *D. dama*	Yes	5.3.4
	Capreolus	Roe deer *C. capreolus*	Yes	5.3.1
	Hydropotes	Chinese water deer *H. inermis*	Yes	5.3.6

Table 1.4 Carnivores (order – Carnivora) of Britain and Ireland

Family	Genus	Species name	Further information included	Chapter reference
Felidae	*Felis*	Wildcat *F. silvestris*	Yes	6.3.1
Canidae	*Vulpes*	Red fox *V. vulpes*	Yes	6.4.1
Mustelidae	*Meles*	Badger *M. meles*	Yes	6.5.1
	Lutra	Otter *L. lutra*	Yes	6.5.2
	Martes	Pine marten *M. martes*	Yes	6.5.3
	Mustela	Stoat *M. erminea*	Yes	6.5.4
		Weasel *M. nivalis*	Yes	6.5.5
		Polecat *M. putorius*	Yes	6.5.6
	Neovison	American mink *N. vison*	Yes	6.5.7

Table 1.5 Rabbits and hares (order – Lagomorpha) of Britain and Ireland

Family	Genus	Species name	Further information included	Chapter reference
Leporidae	*Oryctolagus*	Rabbit *O. cuniculus*	Yes	7.3.3
	Lepus	Brown hare *L. europaeus*	Yes	7.3.1
		Mountain hare *L. timidus*	Yes	7.3.2
		Irish hare *L. timidus hibernicus*	Yes	7.3.2

Table 1.6 Rodents (order – Rodentia) of Britain and Ireland

Family	Genus	Species name	Further information included	Chapter reference
Sciuridae	*Sciurus*	Red squirrel *S. vulgaris*	Yes	8.3.1
		Grey squirrel *S. carolinensis*	Yes	8.3.2
Castoridae	*Castor*	Beaver *C. fiber*	Yes	8.4.1
Gliridae	*Muscardinus*	Hazel dormouse *M. avellanarius*	Yes	9.6.1
	Glis	Edible dormouse *G. glis*	Yes	9.6.2
Cricetidae	*Myodes*	Bank vole *M. glareolus*	Yes	9.5.2
		Skomer vole *M. glareolus skomerensis*	No	N/A
	Microtus	Field vole *M. agrestis*	Yes	9.5.3
		Orkney vole *M. arvalis orcadensis*	Yes	9.5.4
		Guernsey vole *M. arvalis sarnius*	Yes	9.5.4
	Arvicola	Water vole *A. amphibius*	Yes	9.5.1
Muridae	*Micromys*	Harvest mouse *M. minutus*	Yes	9.4.3
	Apodemus	Wood mouse *A. sylvaticus*	Yes	9.4.1
		St Kilda field mouse *A. sylvaticus hirtensis*	No	N/A
		Yellow-necked mouse *A. flavicollis*	Yes	9.4.2
	Mus	House mouse *M. musculus*	Yes	9.4.4
	Rattus	Brown (Common) rat *R. norvegicus*	Yes	9.3.1
		Black (Ship) rat *R. rattus*	Yes	9.3.2

Table 1.7 Insectivores (orders – Erinaceomorpha and Soricomorpha) of Britain and Ireland

Family	Genus	Species name	Further information included	Chapter reference
Erinaceidae	*Erinaceus*	Hedgehog *E. europaeus*	Yes	10.3.1
Talpidae	*Talpa*	Mole *T. europaea*	Yes	10.4.1
Soricidae	*Sorex*	Common shrew *S. araneus*	Yes	10.5.1
		Pygmy shrew *S. minutus*	Yes	10.5.2
		Millet's shrew *S. coronatus*	Yes	10.5.3
	Neomys	Water shrew *N. fodiens*	Yes	10.5.4
	Crocidura	Lesser white-toothed shrew *C. suaveolens*	Yes	10.5.5
		Greater white-toothed shrew *C. russula*	Yes	10.5.6

With these species in mind for potential inclusion, we have tried to cover as many of the wider-distributed species that we can, and also some where distribution is far more concentrated. There are however a few introduced species that were not considered (e.g. red-necked wallaby), and we have not included non-native species that are not considered to be established in the wild, or species that are considered to be domesticated, for example pets and farm animals such as cows, sheep, horses and goats.

We did in fact, as part of the decision-making process regarding *'what's in and what's out'*, have a thought process we went through, and it possibly helps to share this with you here. This will also provide more background as to why we feel this book is useful in different ways, dependent upon the perspective of the reader, and why one may wish to understand mammal acoustics better in the first place. The thought process is laid out within Table 1.8, and the more we could answer 'yes' to the questions posed, the more pressure (healthy pressure!) we felt to try and cater for a species, at whatever level we could.

Table 1.8 Thought process adopted in determining scope of species covered

All species	Is it difficult to engage with/ identify the species visually, in a non-intrusive way, either directly or indirectly (e.g. field signs)?	Does the species make acoustic noise (audible or ultrasonic) that would help with identification or interpretation of behaviour?	If the species' range is restricted, is there particular conservation interest, or legislative pressure, where non-intrusive acoustic study would be beneficial?
	Yes/No	Yes/No	Yes/No
Non-native species	Is the species widely distributed?	Does the species occur in an area where wider dispersal is likely, thus increasing in range over time?	Is the population established in the wild?
	Yes/No	Yes/No	Yes/No

Britain and Ireland

Who would have thought that producing a book title could become potentially difficult from a geographic perspective? There are so many terms that could be used to describe the islands in which we live, and each of these terms are legitimate and correct from whatever perspective you adopt, or wherever you sit.

From the perspective of land-based terrestrial mammals, however, they do not recognise these human descriptions of where they happen to live. And unlike cetaceans, seals and bats, they are pretty much stuck where they are, and unable to make long sea crossings to other areas. This means that we have a reasonably 'captive' subject matter, that by and large (barring unfortunate or illegal introductions and/or escapees from private collections) remains stable.

In terms of geographic scope for this book, Britain and Ireland should be taken to also include all of the islands that collectively make up what some would describe as the British Isles, this being all of the land mass and associated islands pertaining to Britain and Ireland, including the Isles of Scilly, the Channel Islands, the Isle of Man, the Western Isles, Orkney and Shetland.

Figure 1.1 Geographic scope of this book

1.5 Structure adopted

There were many ways in which we could have structured the content of this book, and after some thought we felt there was a preferred logical approach. We feel it does no harm to explain our approach to you, and what follows may help some readers as they begin to gather thoughts around the subject matter.

You are currently reading the *Introduction & Context* chapter, not a difficult concept to get your head around at the start of a book. After this we will talk about *Survey Equipment & Field Techniques* (Chapter 2), as the logical place to start with such work is the selection of equipment, and using the equipment in the field, in order to collect sound recordings. Having obtained these data, especially those collected on recording devices (be that audible or ultrasonic), the next logical step is to think about analysis of the recordings, hence Chapter 3 (*Analysis of Acoustic Recordings*). During the analysis you may need to refer to the reference material, in order to assist you with identification and interpretation – thus the remaining chapters in this book, which occur in the order that they do, for reasons which will covered within Chapter 4 (*Overview of Species Group-specific Chapters*). Towards the end of the book there are a couple of useful appendices, Appendix I providing a glossary, and Appendix II which provides a number of case study examples.

And finally, and very importantly, we are talking about sound, and as such it would be falling far short of what is now available, in terms of resources that a book such as this can deliver, to not provide you with a sound reference library. So please, if you have not done so already, download and refer to the calls within the library. You can access and download the library via the following link:

https://pelagicpublishing.com/pages/terrestrial-mammals-call-library

1.6 Is that a mouse?

Perspective is everything, and an understanding of different perspectives is extremely useful, especially when it comes to the study of sound in the natural world.

To help explain what we mean here, the book *Is That a Bat?* (Middleton, 2020) was published, in part, to inform those carrying out acoustic analysis of bat-related data about things, other than bats, that make similar structured sounds to bats. At the time and in the lead-up to the publication of that book, not many people had considered that there were other things out there that sounded and looked like bats acoustically. One area where this was demonstrated very well related to other small terrestrial mammals, and a good number of references to these were made, elaborating on where confusion could lie.

Why does this matter here? Well, conversely, the opposite is true, and we need to take a different perspective. If other mammals, for example, are capable of making 'noise' that can confuse those in the bat world, then this also means that bats can make 'noise' that can confuse those who are interested in the 'sound' produced by other mammals. Therefore, if you *are* someone who is interested in the study of small purely land-based terrestrial mammals, there is a risk that bat-related acoustic data could creep into what you may consider to have been produced by your target species. To this end, we would recommend the book mentioned above, with a view to it being an extremely useful companion to the material produced here, as there will be times when undoubtedly you have to ask the question, *'is that really a mouse, or a shrew, or a vole?'*, etc.

1.7 Let's move on!

OK, so now that we have covered all of the above, you can safely proceed to Chapter 2, *Survey Equipment & Field Techniques.*

Parabolic reflector with digital recorder – Sandra Graham

CHAPTER 2

Survey Equipment & Field Techniques

2.1 Introduction

When gathering or interpreting acoustic data relating to the species covered by this book, we have two separate approaches. These are driven by whether the sound emitted by the mammal concerned is audible to ourselves (< 20 kHz) and being captured using recording equipment capable of gathering audible acoustic data, or whether it is ultrasonic (> 20 kHz) and so requires specialist equipment (e.g. microphones capable of picking up high-frequency sound) in order to record the original sound and for it to be converted to something we can listen to and/or be processed through software packages designed to deal with acoustic data of this type.

As a broad rule of thumb, larger mammals are more likely to be creating sounds that are within human hearing range (i.e. audible), whilst smaller mammals produce sounds that are often beyond our ability to interpret at all, because they are too high in frequency and/or occur too quickly in time. However, despite this simplistic approach to division between these two groups (large and small), there is overlap where some mammals produce calls that are audible to us, as well as producing ultrasonic calls that we would not necessarily be aware of, without the use of an ultrasonic bat detector for example.

In Table 2.1 we describe the species groups that we believe, from an identification perspective, produce predominantly audible sound, and those that are predominantly ultrasonic.

Table 2.1 Species split according to whether it is audible or ultrasonic sound that could prove most useful for identification purposes

Predominantly audible (< 20 kHz)		Predominantly ultrasonic (> 20 kHz)	
Roe deer	Weasel	Brown (Common) rat	Orkney vole
Red deer	Polecat	Black (Ship) rat	Guernsey vole
Sika	American mink	Wood mouse	Hazel dormouse
Fallow deer	Rabbit	Yellow-necked mouse	Common shrew
Reeves' muntjac	Brown hare	Harvest mouse	Millet's shrew
Chinese water deer	Mountain hare	House mouse	Pygmy shrew
Wild boar	Irish hare	Water vole	Water shrew
Wildcat	Red squirrel	Bank vole	Lesser white-toothed shrew
Red fox	Grey squirrel	Field vole	Greater white-toothed shrew
Badger	Beaver		
Otter	Hedgehog		
Pine marten	Mole		
Stoat	Edible dormouse		

Until fairly recently, it was not realised that small terrestrial mammals produced calls above our hearing range (Anderson, 1954). In fact, the development of ultrasonic microphones and bat detectors helped to establish that many species of small mammals vocalise at frequencies that are not audible to humans (Brudzynski, 2009).

A number of studies have been carried out to demonstrate that small mammals can emit both audible and ultrasonic sound while communicating and interacting with each other, both in conspecific (i.e. between animals of the same species) and heterospecific (i.e. between animals of different species) situations (Sales and Pye, 1974; Stoddart and Sales, 1985; Kapusta *et al.*, 2007; Ancillotto *et al.*, 2014). Circumstances for sound to be emitted include, but are not restricted to, the following behaviours: aggression, mating, distress, contact (adult to adult, and neonatal to adult interactions) and territoriality (Sales, 2010). Naturally, there is no point in being able to emit a sound if your conspecifics cannot hear it. In many species of small mammals, it has been shown that they can hear at both audible and ultrasonic frequencies (e.g. brown rat and harvest mouse can hear up to c.60 kHz, and house mouse up to c.100 kHz). In some species, there are two regions of hearing sensitivity, one within the audible (sonic) range and another within the ultrasonic range (Thomas and Jalili, 2004). For instance, carnivores such as mustelids, felids and canids, although vocalisations between adult conspecifics are more typically within the audible sound range, they are able to hear high-frequency sounds of up to 80 kHz, which may be an adaptation to identifying and locating prey (Heffner and Heffner, 1985; Malkemper *et al.*, 2020; Kruger *et al.*, 2021).

Having now considered the likely spectrum (audible or ultrasonic) in which we are able to listen to and/or record sound from different mammal species, the next factor to consider is the maximum distance that you or your recording equipment would need to be from the source to capture useful sound data.

With this in mind, there are two characteristics of sound that will influence the detection distance:

(1) Attenuation, which from our perspective when studying acoustics, can be described as the gradual loss in intensity of sound as it travels through air (i.e. sound disperses into the atmosphere, and dissipates over distance). The influence of attenuation relates to the frequency (or range of frequencies) at which the sound is produced, where the

Table 2.2 The broad relationship (not to scale) between frequency and amplitude as they relate to the distance that sound travels through air

Lower frequency Higher amplitude (i.e. low and loud)	Lower frequency Lower amplitude (i.e. low and quiet)	Higher frequency Higher amplitude (i.e. high and loud)	Higher frequency Lower amplitude (i.e. high and quiet)
For large mammals this could equate to a considerable distance (100 m or more in the right conditions)			For small mammals this may be less than 5 m
Recordable distance from sound source			

effect of attenuation increases with frequency, and therefore has a greater impact on vocalisations produced in the ultrasonic spectrum.

(2) Amplitude (i.e. loudness) of the sound. The louder a sound is, the further it will carry through the air.

It is useful to think about these two characteristics together, in that a combination of a quiet, high-frequency sound will not travel very far at one end of the scale, whilst a loud, low-frequency sound will travel the greatest distance. Table 2.2 describes the broad relationship between these characteristics, whilst further information on attenuation is included in Chapter 3, Section 3.4.

Added to these sound characteristics, there are external environmental considerations which will impact upon the quality of the perceived or recorded sound. Table 2.3 looks

Table 2.3 Sound characteristics and external considerations influencing the distance at which sound may be heard or recorded

Sound characteristics	Notes
Frequency/Frequency range of emitted sound from source	Higher frequencies attenuate more quickly in air than lower frequencies. Lower frequencies travel further in air than higher frequencies.
Amplitude (i.e. loudness)	The louder the sound at source, the further it will carry through air, and hence may be heard or recorded from further away than quieter sound.
External considerations	**Notes**
Physical barriers	Clear line of 'acoustic sight' is beneficial. The habitat in which the species resides will impact on your ability to detect its vocalisations. The presence of ground vegetation or tree cover will reduce the distance that sound will carry. Barriers caused by terrain (contours) will impact upon sound (e.g. the presence of higher ground between the source and the surveyor/survey equipment). Barriers created through the presence of structures and the like will impact upon the distance that sound travels.
Weather related **Humidity related** **Temperature related**	Wind direction will impact upon the distance that sound carries, either positively or negatively. Rain will reduce the distance that sound will be carried. Moisture on vegetation or tree cover may impact negatively upon the distance sound is carried. As temperature rises, the distance that sound travels declines. As humidity increases, the distance that sound travels increases.
Noise	Additional background noise may influence the ability to hear or record the sought-after sound, as it is less able or unable to be picked up clearly or at all through or beyond the interference. Examples: fast-flowing water; weather-related (wind and/or rain); traffic; other anthropogenic noise.
Microphone specification	The effective frequency range of a microphone (i.e. frequency responsiveness) will affect whether or not it is better able or able to pick up certain frequencies, and at which amplitudes. For example, one microphone type may not pick up something at a particular frequency or distance that another microphone may record.

at the sound characteristics and the additional environmental and equipment considerations in more detail.

With so many variables, it is difficult to be precise when it comes to determining the maximum or optimum distance from a sound source one can afford, to collect meaningful data. On first impressions, the best approach might be getting as close as possible without causing disturbance to the animal. This approach has two drawbacks, in that (i) it is not good to cause disturbance and/or potential distress unnecessarily, and indeed to do so, for some species, may not be legal without the appropriate licence being in place, and (ii) normally you would wish to record or listen to sounds that reflect natural behaviour within the animal's environment. Also, as well as being adversely impacted by increased distance, sound clarity can similarly be degraded at very close range, whereby the sound recording becomes overloaded, and elements of the vocalisation become indistinguishable.

Recording quality – signal to noise ratio

Every time you attempt to make recordings you should be conscious that there are a number of additional factors, some of which you may have no control over (e.g. refer to Table 2.3), that could degrade the quality of the acoustic data collected. Ultimately you are seeking to acquire, as fully as possible, your targeted sound, thus aiding analysis and interpretation. Conversely, the presence of unwanted noise should be minimised. Such 'noise' can come from a variety of sources, for example speaker noise, weather-related noise, mechanical noise, electronic noise, talking and other anthropogenic sources, as well as from other wildlife that co-inhabit the environment where the target species is found. Taking account of these may prove to be the difference between a successful venture and a wasted effort, depending upon the results being sought.

The difference in amplitude between what is being sought (i.e. sound), and what is interfering (i.e. noise) is often referred to as the 'sound to noise ratio' or 'signal to noise ratio' (SNR). In Figure 2.1 we provide an example of good SNR (i.e. the target species

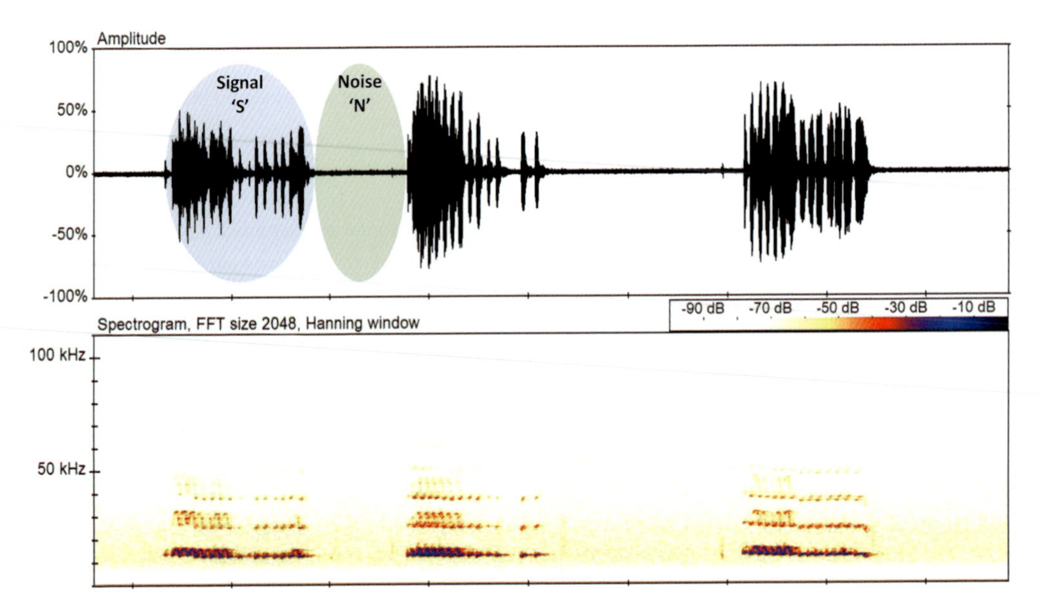

Figure 2.1 Recording of common shrew (oscillogram above/spectrogram below) showing good signal (S) to background noise (N) ratio, as demonstrated by the amplitude registrations in the oscillogram view

represents the dominant acoustic signals with almost no unwanted 'noise' recorded during or in between the discrete elements of the target vocalisations). Later on, within Chapter 3 (Section 3.3), we will explore the matter further, but for the time being we will provide an additional thought here, in that when considering 'sound' and 'noise' it is sometimes worth remembering that one person's 'sound' is another's 'noise', and of course vice versa. For example, you may be seeking to record small rodents, and your recordings are continually interrupted with insect 'noise', whilst someone else who is seeking insect 'sound' may have their recordings interrupted by 'noise' coming from mammals.

Factoring in recording distance from target species

Within each of the species-group chapters (Chapters 5 to 10) we will provide some guidance as to approximate distances from source needed in order to record the target sound. As a general rule, the larger the animal the greater the distance at which you may detect it. However, this may not always be the case, as large animals may communicate quietly in certain settings, and small mammals can, at times, be quite loud for their size. The purpose behind the sound emission in the first place may be important here, with territorial advertisement sounds and distress sounds generally being much louder than other acoustic behaviours.

For species that vocalise in the ultrasonic spectrum, typically the smaller terrestrial mammals, it is more difficult to estimate detectable distances compared to species that produce audible vocalisations. For these ultrasonic species we are reliant on specialist equipment to record at higher frequencies, with vocalisations at higher frequency being more affected by attenuation. In addition, we need to manipulate these acoustics if we wish to interpret them as audible sound (e.g. converting to, or playing in, time expansion). As a 'rough rule of thumb', Figure 2.2 (originally published in Newson and Pearce,

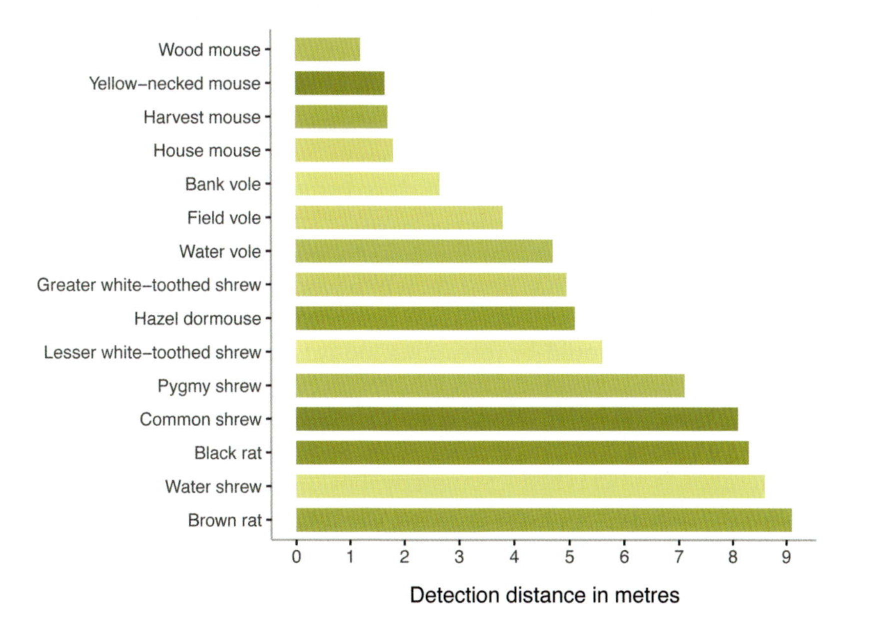

Figure 2.2 Estimated distance that sound may carry in order to be picked up and recorded by an ultrasonic microphone. The bar colours provide a visual representation of the number of calls from which the estimates have been quantified, whereby darker bars represent a greater number of call samples (taken and updated from Newson and Pearce, 2022).

2022) provides estimates of the distances at which we are likely to detect different small terrestrial mammal species. These estimates have been calculated from an extensive call library for each of the species listed, using information on the signal pressure level of the calls produced, alongside knowing the peak frequencies produced and the properties of the microphone used, as described by Agranat (2014). In carrying out such an exercise, there are quite a few assumptions made in the calculations, including what represents the average loudness in decibels (dB) of calls produced by the species, as well as needing to pre-define temperature and humidity.

There are reasons why ultrasonic noise developed in these mammals, primarily to reduce the risk of attracting predators (at least those that cannot hear at higher frequencies), as higher frequencies may not only be less likely to be heard by predators, but will attenuate more quickly over a shorter distance. Attenuation will be greater still (i.e. more effective at not being heard at a distance) when the animal is within dense vegetation or underground within a burrow.

Where do we go from here?

Considering the acoustic behaviour of these mammals, we have, as suggested above, a range of field survey equipment at our disposal, dependent upon the target sound being sought. Within Section 2.2 we will discuss this equipment (including supporting items) in more detail. Having discussed the equipment, we will then continue with Section 2.3, where we will talk about potential field survey techniques.

2.2 Survey equipment

Systems for recording small mammal ultrasonic sound

For recording most of the useful calls from smaller mammals it needs to be borne in mind that they predominantly emit their sounds at ultrasonic frequencies. In order to record ultrasound, you need equipment capable of picking up these higher frequencies.

Conveniently, bats also make sound at very high frequencies, and the emphasis placed in recent years in the development of equipment to record bats has opened up many more easily accessible recording systems for those wishing to study other ultrasonic taxa, including our small land-based terrestrial mammals.

Undoubtedly there are other ways in which recording within the ultrasonic range is possible, including the potential creation of bespoke systems, using separately sourced components, but to our mind (and that of many others) why re-invent something when the solution already exists, and has been shown to work?

Compared with even just a handful of years ago, the range of ultrasonic bat recording equipment available to the would-be researcher today is considerable, and every system has its place and purpose depending upon the requirements of the work. However, not every system will be suitable for what we are interested in here. In considering all of the systems most often used by bat workers, it must be appreciated that the bulk of the time, bat echolocation is what is being sought by that audience, which is a consistent and rapidly produced (many times a second) series of sounds, as opposed to the occasional, and random (to us) sounds produced by land-based mammals. Therefore, some of the techniques used in the aerial bat world do not transfer onto land, we are sorry to say.

So, before you go and spend your money on a bat detector, we need to provide some guidance as to what will work, and what will fall short. Table 2.4 summarises the situation, with the conclusion being that, because of the complex nature of the sounds being sought, the need to measure call parameters precisely and the presence of harmonics as a regular

Table 2.4 Comparison of different bat detector systems as they relate to recording non-bat mammal species

Bat detecting system	Description	Suitable for land-based mammals?
Tunable/Auto heterodyne	A listening output, providing a real-time, audible representation of the tone, rhythm and repetition rate of the true sound. Because the true sound has not been recorded (i.e. there is only an electronic representation) it is not possible to use this system for spectrogram analysis.	No
Frequency division	A listening output and recording format which works in real time. Divides the frequency of sound so that it becomes audible. Rhythm and repetition rate are heard. Calls can be used for spectrogram analysis, but results are not as clearly represented when compared to full spectrum or time expanded recordings. Harmonics are lacking and/or not shown accurately.	No
Zero crossing	A recording format whereby sound is recorded in real time and saved as zero crossing files. These files are very memory efficient because audible information (i.e. sound) is not retained. Recordings can be used for spectrogram analysis, but only the strongest frequency is shown at any moment in time, meaning that simultaneous noise from other sources is cloaked by the loudest sound or noise occurring at any point in time. Due to sound not being retained, amplitude cannot be measured, and files cannot be listened to during the analysis process.	No
Time expansion recording in full spectrum	A listening and recording format where original sound is slowed down (usually x10), meaning that the duration of the event is expanded and the frequency is reduced, in order to become audible. Because all of the original sound is retained (albeit slowed down), recordings are useful for analysis, as all sound is recorded, including any harmonics produced, as well as sound from other sources. A surveyor can hear calls at point of recording, thus knowing what is being recorded. Calls can be used for spectrogram analysis, giving the greatest level of quality/clarity of results, as is also the case with full spectrum systems. Detector downtime when in recording mode (i.e. during playback) means the machine is no longer functioning as a bat detector and data can be missed. Due to recordings being longer than real time (e.g. x10), memory size of saved files is much larger, with knock-on impacts upon memory storage, and time taken to load individual files on software.	Yes
Real-time recording in full spectrum	A recording format whereby all sound is recorded accurately and in real time, including any harmonics produced and/or sound from other sources occurring at the same time. Calls can be used for spectrogram analysis, giving the greatest level of quality/clarity of results, as is also the case with time expansion systems. Due to the sound being retained in the recording, it is possible with most software packages to allow calls to be listened to in a preferred format (e.g. time expansion) during analysis. It is not possible to listen to calls at point of recording unless the bat detector has a separate function for creating audible sound.	Yes

feature, only full spectrum bat detectors recording in time expansion or real time will offer the required outputs. In addition, full spectrum detectors retain amplitude (i.e. sound) within the recordings, meaning that strength of signal across the call or sequence range can be measured. The retention of sound also means that, with both of these systems, you can listen to the call during analysis on many of the mainstream software programs being used. Each of these two systems (time expansion or real time) has its strengths and weaknesses (Table 2.4), and they are usually relatively more expensive than other bat detector options (e.g. heterodyne and frequency division). However, these 'cheaper' alternatives would be a false economy for someone seeking to achieve workable results for small land-based mammals.

With time expansion detectors, the machine is set to be triggered by the presence of ultrasonic sound. In this instance the machine records for a pre-set length of time before then entering playback mode, at a multiple of time slower than the original recording. By playing the ultrasonic (i.e. high-frequency) sound back at a slower pace, in effect the sound is lowered in frequency and elongated in time, thus bringing it into audible human hearing range. This is ideal if you wish to be able to hear the sound (or the quality of the recording) when you are in the field, and people experienced with this system can differentiate many species or species groups relatively easily. The main weakness with this system, though, lies in the fact that while the detector is playing back the time expanded call for you to listen to and record, it is no longer active as a recording device. By way of example, a one-second call sequence captured and played back at a time expansion rate of 10 times will take 10 seconds to play back. Therefore, for 10 seconds of playback, out of the 11 seconds in total, the bat detector mode is no longer operational and other data (i.e. other calls) may be missed. This process also comes hand in hand with larger file memory, with corresponding challenges as described in Table 2.4.

Full spectrum real-time systems are triggered to respond to the presence of ultrasonic sound, and record calls in real time and at their originally emitted (i.e. true) frequencies. In their pure form it is not possible, in any meaningful way, for the surveyor to hear the calls in real time at the time of recording (the sound is ultrasonic). One advantage, however, of full spectrum real-time systems, compared to time expansion, is that no calls are missed during the collection process (i.e. if the call is loud enough to be picked up by the detector, it will be recorded).

Another consideration in equipment choice may relate to microphone specification. This could also be a driver as to the choice of bat detector, as microphones are normally integrated into the design of detectors. Some detectors allow the option of attaching an external microphone, thus giving you flexibility on how the microphone is positioned. Different microphone specifications can give varying levels of performance at different frequency levels. Some microphones perform better at lower frequencies, while others offer better performance at higher frequencies, and a whole range of other variations. If you are conducting serious research into any aspect of mammal-related sound you should at least be conscious of microphone sensitivities, strengths and weaknesses, and consider this in your choice of equipment, as well as in your interpretation of results.

Most of the full spectrum real-time detectors (see Figure 2.3 for an example) come with microphones suitable for ultrasonic recording, as well as built-in recording storage (.wav format), date/time-stamped files and often built-in GPS. If on the other hand you are using a time expansion system, you may need to purchase a separate digital recorder. In doing so, it is important to ensure that you are collecting good-quality recordings, that you have a recorder that is practical in the field, and that has manual control over recording input volume levels. In addition to this, it is best if it is capable of recording files in .wav format, as this gives the best-quality results and is recognisable by the software programs you are most likely to use for analysis purposes, without having the extra hassle of using separate format conversion software.

Figure 2.3 Full spectrum real-time bat detector (BatLogger M, Elekon) set up to record small mammals in an enclosure

Several time expansion and real-time bat detectors recording in full spectrum, which are also suitable for recording ultrasonic mammal vocalisations, operate as manned handheld devices, whilst others can be used for remote static recording, whereby they can be deployed at a fixed location and have the capability to record over longer and specified time periods – for example 24 hours a day, from sunset to sunrise or over specific time slots for days, weeks or months depending on the battery life. The remote static detectors can be programmed so that they are triggered by sound above specified frequencies (e.g. 16 kHz is a typical default setting), and they will record all sound at or above this level as real-time full spectrum data, which is saved to their internal memory facility (e.g. SD card). Once retrieved, the data can be downloaded and then further investigated using sound analysis software. Using this type of remote/static equipment can be very similar to how you would work with a trail camera, the only difference being that it is audio events that trigger the machine (as opposed to visual/motion), and it is purely audio that is recorded (i.e. no video footage). Examples of full spectrum bat detectors that work as manned and/or remote static ultrasonic recorders come from companies such as Wildlife Acoustics Inc., Titley Scientific, Pettersson Elektronik and Elekon AG, all of whom have considerable experience in developing bat detectors.

Systems for recording larger mammal audible sound

For recording most of the useful sounds from larger mammals it should be borne in mind that they predominantly produce sound at audible frequencies, which is very convenient for us humans, as it means that pretty much any recording system that will work within a human environment is likely, to at least some extent, to be able to cope with these sounds we are capable of hearing with our own unaided ears. This opens up a vast range of equipment choices to the would-be mammal recorder, but with some options being much better than others in terms of quality of output. But before we get into the desired kit for those serious about quality of results, let us not wholly dismiss the value that other non-specialist (and often cheaper) systems can deliver.

First of all, for the cheapest equipment available, we need look (or indeed listen) no further than our own head. On either side we have ears, and quite often when engaging with audible sound (if we were actually there, in the first place), all we may need to do is stop, not panic, and listen very carefully to what we are hearing so that we can recall it later on for identification purposes, or describe it as best we can to a more knowledgeable third party. Having said this, I am sure we can all agree that even after a short period of time our memory plays tricks on us, and it becomes difficult to be certain as to what it was you actually remember from an earlier encounter – whether from earlier today, yesterday, or previous weeks or months.

Accordingly, this then takes us to a second part of a process, whereby as well as having ears, most of us these days carry a mobile phone with us everywhere we go. No, we are not suggesting that in the moment you 'phone a friend'. What we are saying is that an audio and/or video recording function will be included in your phone. Most of us know the video function, as it is in the same area as the camera. So, what is the use of that if you cannot see the animal? Why would you want to film nothing of any visual value? Quite simply because the video function also records sound, and having recorded the sound, you can use it later to recall and compare against other references, and/or to ask an 'expert' for their opinion. The audio recorder on a phone would also be suitable for such an approach, but most of us have to really think about where to find that particular app on our phones (e.g. Android Voice Recorder or Apple Voice Memos) – for the sake of time (speed is of the essence, because the sound may stop at any moment), the camera function will be a quicker route for most phone users. Adopting this approach, you may after further investigation discover that what you have recorded is nothing particularly special, but if you or someone else can determine what it was (or even what it wasn't!), then this all contributes towards your self-development 'savings account', and makes you better informed tomorrow than you were yesterday. And that alone is very worthwhile. It's true that the quality of the recording most likely is not going to be good enough for proper sound analysis purposes, but when identification is the key objective, sounds recorded in this way can prove to be extremely valuable.

We now want to move on and discuss devices that may indirectly, as a result of their deployment or use for other purposes, record sound. These broadly fall into three areas.

Firstly, systems that are used for security purposes, such as CCTV, whereby someone reviewing such systems may come across an event while other sounds are present. As such the recorded sound may be played back and investigated further.

Next, for those of us who are bat workers, we know all too well that many of today's bat detecting systems are capable of picking up audible sound. Sometimes this is because they have a voice recording option as a feature within the machine; if so and you activate such a function, it will also record any other audible sound in the vicinity. So, for some in the bat world, bat detectors have also become very useful as on-site Dictaphones, recording voice notes and other audible sounds. Another feature of some bat detectors is that they will inadvertently record all frequencies (ultrasonic and audible) whilst recording bats, and if you choose to listen to a recording in real-time on a software package, you may very well hear what was happening audibly. This can be very useful for working out the source of confusing noise, or indeed trying to recall something audible. It can also cause potential issues, in that conversations may inadvertently get recorded without the participants being aware or remembering their machine is capable of such wizardry.

Finally, the use of trail cameras and CCTV specifically aimed at the natural world can provide a means of acquiring audible sound associated with the recorded event. For these systems, it would be the visual presence of the animal that would cause the system to become active (motion activation) and start recording, but in doing so sound may also be recorded simultaneously, and if so this can be extremely valuable – you have not only recorded the sound but can hopefully also see what made the sound in the first place, and very importantly from a behavioural perspective, you have the context in which the sound was made. Camera systems that record video typically save data as .mp4, .avi or .m4a files. In situations as discussed here (i.e. sound contained within video footage) it should be possible, if required, to create sound-only versions of such recordings in .wav format. To assist you with how to do this, we describe one viable process below, albeit other approaches may work equally well.

Converting video recordings to sound .wav recordings

Once you have recorded a wildlife sound, whether as a sound or video file on your phone, or a video file on a CCTV or trail camera system, it may be useful for identification purposes to visualise/analyse these recordings as a spectrogram. Different devices will record in different file formats, such as .m4a, .mp3, .mp4 or .avi, but software programs designed for editing and analysing audio data typically work with .wav formats and therefore it may be necessary to convert the file format of your recording.

There are several free file conversion software programs available (e.g. www.freeconvert.com; www.convertio.co) that are able to convert different file formats to .wav files. They typically have a storage limit (e.g. maximum of five files or 1 GB of data) and therefore only small numbers of recordings can be converted at any given time.

The process is fairly simple: you are directed to select the existing format of the files to convert and then the desired format of the output files. You then select and upload the chosen files and, following the process of conversion, download the data in the new format, which is typically saved to the 'Downloads' folder on your computer, unless otherwise specified.

Specialist audible acoustic recording equipment

Finally, we are now going to talk about the more specialist equipment that is available for recording audible sound, which will undoubtedly normally be expected to yield the best results. These systems broadly fall into two categories: firstly, equipment that is designed to be used with the operator present (usually handheld); and secondly, equipment that would normally be used for recording at a fixed location over longer periods of time (i.e. operational without a human present). These latter systems we will refer to as remote static recorders.

It would be difficult here to be overly focused on any particular system that would be best for mammal sound recording, as there is a huge amount of specialist kit available, which can be used in different combinations. Nevertheless, to make up such a system, you would require an appropriate high-quality microphone, usually with a windshield, and the microphone feeding into a suitable digital recording device, with the following spec being appropriate for most circumstances: 16 bit; .wav recording format; manual recording level control; 44.1 kHz (or higher) sample rate. Added to this, for some systems you may also wish to use a parabolic reflector (see Figure 2.4), which acts to focus the directionality of the microphone, as well as substantially increase volume from the source (think of this as being binoculars for your microphone).

Figure 2.4 Transparent Telinga parabolic reflector (on the right) with microphone and digital recorder, as used by one of the authors

There are many useful sources of reference available from distributors of such equipment, as well as social media groups you can participate in and ask questions to. One such group that comes immediately to mind is the long-standing (founded in 1968) and highly respected Wildlife Sound Recording Society (www.wildlife-sound.org).

Within this group there are numerous very experienced members who willingly share their experiences and thoughts on equipment choice and the like, through articles published in the society's magazines, and website resources. We would encourage you to check out the above website for yourself.

When choosing a good-quality microphone for mammal recording there are a number of considerations to factor in, which we have summarised within Table 2.5.

Table 2.5 Considerations in microphone choice for audible mammal recording

Feature description	What does it mean?
Directionality options	**Directional** – picks up sound from a particular direction, e.g. directly in front, or to the side(s). Usually, you would want it to be from directly in front of you. With this approach you are 'trading' the focus of direction of the microphone with the relative filtering out of sound coming from other directions. **Omnidirectional** – picks up sound from all around the microphone. You will record more, and possibly things of interest coming from an unexpected direction. Conversely your targeted sound quality may not be as good due to more background noise.
Shotgun microphone	Records sound from a specific direction to the detriment of noise coming from other areas. Sound is not amplified.
Parabolic dish or reflector	Not a microphone, but a dish with a microphone positioned pointing towards or at the dish's focal point. The sound is focused in and amplified from the direction in which the dish is pointed. Ideally a transparent dish is better than a solid one, as it is then possible to see through the dish towards what you are recording.
Gain	The higher the gain the further away you will record quieter sounds. Conversely, sounds emitted close to the microphone may become overloaded if the gain is set too high (refer to Chapter 3).
Microphone 'self-noise'	This relates to the sound produced by the microphone itself, even when no other sound is present. The lower the 'self-noise' the better, in order to reduce background hissing etc. being generated by the microphone itself.
Frequency responsiveness	This relates to how effective the microphone is at/across different frequencies. For audible mammal sound recording, ideally you would want it to perform effectively across the range of 20 Hz to 20 kHz.

Finally, we turn to remote static recording systems, which operate in much the same way as the remote static bat detectors whereby they are designed to be deployed at a given location for longer periods of time (e.g. from a single day to many days, covering selected time intervals). For audible sound, continuous recording (or in some cases, triggered recording) occurs over the selected time interval and all data (usually .wav files) are saved to an internal storage facility (e.g. SD card) which once retrieved can be downloaded for analysis. Some examples of remote static recorders include: Song Meter Mini, Song Meter Mini Bat with an audible microphone or the Song Meter Micro (all from Wildlife Acoustics Inc.); Chorus (Titley Scientific) and AudioMoth (Open Acoustic Devices).

Using this type of equipment is very similar to working with a trail camera, particularly if a trigger is used, with the only difference being that it is audio events that trigger the machine (as opposed to visual), and it is purely audio that is recorded (i.e. no video footage).

2.3 Survey advice and techniques

When considering field survey techniques, a key objective is to collect as high-quality recordings as possible to maximise the accuracy of species identification from sound. Within this section we will explore various factors that can contribute greatly towards attaining this goal.

Factoring in human behaviour

Given the right circumstances, pretty much any mammal out there (wild or domestic) has the potential to make noise, and probably the noisiest of all is one that is widespread, common and always present when we are around: *Homo sapiens*. So, the first consideration that should be factored in is trying to reduce or eliminate any noise that you yourself make, and of course those who are with you, as well as other noises that you can control in terms of equipment use. The sort of commonly occurring anthropogenic noise and 'controllables' we are talking about are described in Table 2.6. These should be avoided, if possible, as they not only affect recording quality but may also scare off the subject matter you are trying to record in the first place.

Table 2.6 Anthropogenic and other controllable noise to be mindful of during recording sessions

Noises created or controllables to avoid if possible	
Walking; especially through high vegetation or on gravel etc.	Mobile phone ringtones or alerts
Loose change or keys jingling in pockets	Poor choice of clothes, meaning additional unnecessary noise from rubbing
Vehicle engine and/or traffic-generated noise: – Aircraft noise – Railway noise	Poor placement of microphone, e.g. too close to (< 2 m) reflective surface, causing echoes, or too close to rustling vegetation
Eating, with associated noise coming from packaging etc.	Interference from other recording equipment
Electrical sounds produced in the vicinity of human habitation or influence	Mechanical sounds produced in the vicinity of human habitation or influence
Distant anthropogenic noise, e.g. people talking/ children playing	Speaker noise coming from device (eliminated by using headphones)
Coughing, sneezing, sniffing	Talking

Planning and preparation are other 'controllables' from a human perspective. You can choose to put a lot of thought into planning a session, with corresponding good amounts of preparation, or you can choose to just 'wing it', and with some luck it will all go OK. We would always recommend proper planning and preparation. Depending upon the survey method there will be a range of additional support equipment that could be used in order to help the session run smoothly. Some of this equipment has been mentioned previously, but for the sake of completeness we list the most useful items within Table 2.7.

Risks to survey methods and analysis during small mammal recording sessions

While recording many of the small terrestrial mammals for this publication, it was noted that they often showed inquisitive behaviour relating to new items placed within their environment, which of course included recording devices and microphones. It would

Table 2.7 Useful support items and other considerations for successful recording sessions

Item(s)/Considerations	Notes
Headphones	If you are recording directly yourself in the field, it is essential to hear the recording quality and eliminate speaker noise being recorded at the same time as microphone-produced sound.
Spare batteries	In case existing batteries fail, or session(s) lasts longer than anticipated. Rechargeable batteries are usually avoided due to unreliability.
Spare memory cards	In case existing SD cards get damaged or corrupted, or you need additional storage.
Spare leads	Spare connector leads, USB cables etc. in case existing items fail.
Calibration of equipment settings	Ensure date/time stamps on equipment are accurate and consistent, so that data can be accurately recorded and cross-referenced during analysis. If the equipment has an inbuilt GPS or the location can be set with a phone, use this functionality.
Trigger settings, gain thresholds and recording lengths	If using bat detectors or static audible recording devices, ensure that trigger settings, gain threshold and recording length settings etc. are appropriate and consistently applied across all devices.
Tripod(s)	Preferably use sturdy tripods, with means of anchoring or securing them so they do not get blown or knocked over.
Gaffer tape	To assist with securing items correctly, or carrying out repairs etc. Gaffer tape can solve a multitude of problems.
Padlocks and similar items	To protect remote/static devices from being stolen or tampered with. Always in conjunction with latent placement of devices, unless working in remote areas.
Waterproofing items	To protect equipment from weather-related events (rain, snow).
Camouflage netting	To help make items less noticeable to others, from a security point of view.
GPS device	So that you, or another team member, can find the recording equipment at a later date.
Food and drink	Always have enough (and more) to last the session and anticipate that things may go on longer than planned.
Risk assessment document	In paper format (you may lose your phone or run out of charge), with you at all times, including paper format map of site, directions to hospital, nearest police station etc.
First aid box and other related medicines	In case of an accident, or becoming unwell (e.g. migraine). Anti-allergy medication, sun cream, energy bars etc.
Mobile phone	Charged and ready should there be an emergency, or you need to communicate a change of plan to others. Always have everyone's numbers added to all phones before starting. Also useful for taking pictures, video footage, voice notes etc. if required.
Walkie-talkies	Very useful when in teams or as individuals working separately at the same site. Often more instant and effective than mobile phones.
Notebook(s) and pens	For taking notes relating to observations, equipment set-up etc.
Dictaphone (voice recorder)	For taking notes relating to observations, equipment set-up etc. Often better than a notebook, as you can continue paying attention to what you are observing.
Weather/environmental recording equipment	If required as part of survey method, but always good to have this information recorded for future reference in any case.

therefore not be unreasonable to assume that remotely placed bat detectors, used for long-term monitoring purposes, could attract curious small mammals, and as such there is the risk that equipment could be damaged, creating additional cost but also impacting upon survey deliverables. In tackling these issues, amongst others, Table 2.8 offers some potential solutions worthy of consideration.

Table 2.8 Considerations to be taken into account when recording small mammals

Subject	Problem	Potential solutions
Equipment damage	Small terrestrial mammals causing damage to equipment (e.g. wind shield, cabling or microphone).	Create a small metal gauze or plastic cage for the microphone. The gauze should not touch the microphone or the wind shield. Adopting approaches described below, under 'Inadvertently recording', may also help.
Inadvertently recording	Other small terrestrial mammals being recorded inadvertently and potentially impacting upon the sound analysis process.	Other small mammals are more likely to be recorded when in close proximity to the microphone. The greater the distance between the microphone and any potential unwanted encounter, the better. Place the entire bat detector in such a way as to prevent a small mammal making progress towards the microphone area. Place the microphone/bat detector on a smooth-surfaced pole, reducing the chances of something being able to climb up. Alternatively, position a large smooth cone, open side pointing downwards, beneath where the microphone or detector is placed.

Important note: Anything occurring immediately around or beside the microphone could impact upon how it performs or how recorded sound behaves (e.g. sound echoing off a smooth flat surface close to the microphone). Therefore, any alterations made to a microphone system, or its immediate surroundings, that could impact upon recorded sound should be tested ahead of being used on case work.

2.4 Survey constraints

To conclude this section, we would like to pull together our thoughts on survey constraints that may be applicable, depending upon which species or species group you are interested in. Table 2.9 provides an overview of some broad guidance on survey constraints worthy of consideration.

In addition to what is covered within Table 2.9, before you commence with any type of survey/recording activity the following questions should be addressed. Is what you are about to do potentially going to cause disturbance to a sensitive species, and/or at a sensitive point in time? As well as which, are there any legal implications (e.g. the requirement for a species disturbance licence) that may need to be addressed ahead of the work taking place? Accordingly, you should research your species of interest thoroughly and adopt an approach that minimises or removes any risk to disturbance, and, if the species is afforded a level of protection (e.g. from disturbance) whereby licensing is required, the necessary documentation should be applied for and granted by the appropriate licensing authority of the country concerned (e.g. Natural England; NatureScot).

Table 2.9 Guidance regarding survey constraints for consideration, relating to mammals in the British Isles

Species/ species group	Microphone: audible or ultrasonic	Temporal constraints	Spatial constraints	Behavioural constraints
Even-toed ungulates (Chapter 5)	Audible	Rutting season varies between deer species	Deploy equipment above antler and scratching post height	Large ranges
Carnivores (Chapter 6)	Audible Ultrasonic	Most vocal during breeding and/or when with young	Wildcats are rare, occur at low density and are solitary	Solitary or group-living animals? Elusive and solitary: trail cameras may be useful Deploy equipment above scratching post height
Rodents (audible) (Chapter 8)	Audible Ultrasonic	Breeding season Diurnal/nocturnal	How high to position the microphone? Is the species ground dwelling and/or arboreal?	Detection distance Frequency of vocalisations Communal or solitary species? Protect equipment from gnawing
Rodents (ultrasound) (Chapter 9)	Ultrasonic	Active all year round or hibernate? Daily activity patterns (diurnal, nocturnal, crepuscular) Breeding season Spring/daytime surveys susceptible to being overloaded by birdsong recordings	How high to position the microphone? Is the species ground dwelling and/or arboreal?	Predominance of olfactory communication (voles) Protect equipment from gnawing
Insectivores (Chapter 10)	Ultrasonic (shrews) or Audible (hedgehogs and moles)	Shrews are active throughout the year Hedgehogs are most vocal during breeding season (territorial disputes/courtship)	Hedgehogs (ground dwelling) Moles (subterranean) Shrews (arboreal and ground dwelling)	Protect equipment from gnawing

Laptop sound analysis – Sandra Graham

CHAPTER 3

Analysis of Acoustic Recordings

3.1 Introduction

In this chapter we explore what you can do with the mammal sounds that you have recorded. This could be to identify the species that produced the sound, and/or for those interested in carrying out further investigations, to look at the characteristics of or to take measurements of the sounds recorded.

What follows is intended to provide guidance for those who are using, or intend to use, software applications to interpret sound recordings, be those audible or ultrasonic at source. The capabilities of such software applications are important for looking at call structure, establishing recording quality, taking measurements of specific call characteristics and/or when you wish to listen to audible interpretations of ultrasonic recordings.

Audible sound

The process of analysing **audible** sound, at the outset anyway, is something that can be done without much applied thought. We hear a sound, which we may compare in our head to other sounds that we are familiar with, to reach a conclusion as to what made the sound. If we record the sound, we can listen to it again (e.g. for verification purposes), by simply replaying the sound using the device you recorded it on, and/or particularly for devices without a playback function, we can download the acoustic file (.wav, .mp3, .mp4) to a computer, and using software applications such as Windows Media Player or Apple Music, replay the recording. In either case, in order to hear clearly, we recommend using headphones, as there may be subtleties within the recorded sequence that go unnoticed until you have had a proper listen, in a quiet environment.

Occasionally you may encounter a very low-frequency sound which you are struggling to hear unaided with only your ears. In order to make the sound more audible you could increase the speed of the playback (e.g. x2), thus bringing it more into your audible range. This in effect is the opposite technique to time expansion (where the sound is slowed down), and you should always bear in mind when doing this that what you are now hearing is faster and higher in frequency than the true sound.

If you wish to explore the structure of the sound in detail (i.e. beyond purely a listening interface), for example to view a visual representation of the recording or take measurements (e.g. frequency levels or duration), then a specialised software package, designed to enable such work to be carried out, should be used. There are a number of software applications available, such as:

- **Anabat Insight** (Titley Scientific): www.titley-scientific.com
- **Audacity** (Audacity Open Source): www.audacityteam.org
- **BatExplorer** (Elekon AG): www.batlogger.com/en/products/batexplorer
- **BatSound** (Pettersson Elektronik AB): www.batsound.com/product/batsound
- **Kaleidoscope Lite** (Wildlife Acoustics): www.wildlifeacoustics.com/products/kaleidoscope/kaleidoscope-lite
- **Raven Lite** (Cornell Lab of Ornithology): www.ravensoundsoftware.com/software/raven-lite
- **SonoBat:** www.sonobat.com

These software packages have all been used by ourselves, and we feel they are worth considering for the subject matter of this book, albeit undoubtedly there are many more options than those we have listed. Having said this, **not all of the above are ideal for viewing and playing audible and ultrasonic sounds**, and we would always recommend that you do not purchase any software without having tested it against the hardware (e.g. bat detector or audible recording system) and the relevant data, as each software package has its own strengths and weaknesses, and as such all packages are not equally suitable across all work applications.

For carrying out detailed analysis using software applications, the remainder of this chapter is as applicable to audible sound, as it is for ultrasonic sound.

Ultrasonic sound

With **ultrasonic** sound, specialised sound analysis software is normally used, unless the recording device (i.e. a bat detector) has already converted the sound into an audible format (e.g. time expansion), during the recording process. If this is the case, then you can listen to the audible interpretation of what was encountered without using specialised software (as described above in 'Audible sound').

However, with unaltered real-time ultrasonic recordings (see Chapter 2, Section 2.2), you would normally want to investigate the characteristics of the sounds more fully. Most of the software packages available that enable you to do this also include functions that convert ultrasound into a format that allows you to listen to an audible interpretation of the recording. For recordings of terrestrial mammals that vocalise with ultrasonic sounds, it is most useful to convert ultrasonic versions to audible interpretations using the time expansion function provided by the software (x10) (see Chapter 2, Section 2.2). With this done, many of these sounds become more distinguishable, thus helping you to potentially identify what produced the sound – but if not, to at least narrow things down to a group. Therefore, with an audible version of the recording, you will have more information from which to draw your conclusions.

In addition to providing a means of playing back your audible and ultrasonic recordings, the software will take your recorded sound and present you with a visual interpretation, displayed for example as spectrograms, oscillograms and power spectrums. Having produced these visuals, you can then appreciate the structure of the sound, as well as use 'tool functions' to take measurements.

Frequency and time

To assist you with the differentiation between the audible and the ultrasonic spectrum, Figure 3.1 provides a visual interpretation of both of these, as they would be shown on a spectrogram. A spectrogram represents what sound looks like when comparing frequency against time, with frequency being measured on the y-axis, and time on the x-axis. As discussed earlier (Chapter 1, Section 1.4), from a human perspective, ultrasound is medically defined as anything higher than 20 kHz.

Figure 3.1 should also prove helpful in providing you with 'rules' as to how sounds are described within this book, and, indeed, how others often do so. As a generalisation, for small terrestrial mammals that typically vocalise in ultrasound, we usually describe frequency in kilohertz (kHz), with time often being measured in milliseconds (ms). Comparatively, the frequency of sounds produced by larger mammals, with corresponding lower, more audible and louder calls, would normally be described within the hertz (Hz) range, with time measured in seconds (sec). This is how we will proceed within the remainder of this book.

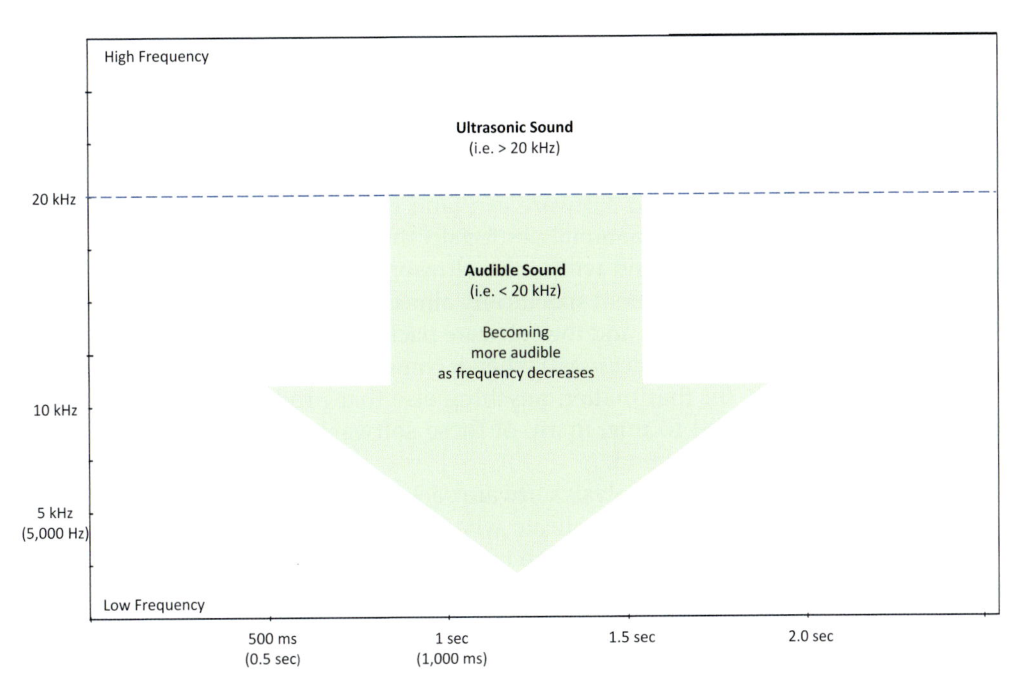

Figure 3.1 Ultrasonic and audible sound as presented on a spectrogram (y-axis: frequency, x-axis: time)

Your next question may now be *'what is the difference between kHz and Hz, and how do they relate to each other?'* Excellent question, and thanks for asking. Some helpful interpretation is contained within Figure 3.1, but to add to this Table 3.1 will help you to understand the relationship between the two (1 kHz is the equivalent of 1,000 Hz). We have also made reference to the relationship between seconds and milliseconds, as this may be your next question (1 sec = 1,000 ms). It is useful to be comfortable with these relationships, as software packages and identification keys sometimes require you to think quickly from one unit description to the other.

Table 3.1 Time/frequency unit conversion key

Time	
Seconds (sec)	**Milliseconds (ms)**
1.0	1,000
0.1	100
0.01	10
0.001	1

Frequency	
Hertz (Hz)	**Kilohertz (kHz)**
100,000	100
10,000	10
1,000	1
1	0.001

3.2 Analysis of recordings

For many of us, the most important aspect in the study of sound relates to the analysis of data recorded in the field. To carry out this analysis, as previously mentioned, we use software to look at, interpret and listen to the recordings that have been collected.

In the world of bats, sound analysis is an everyday occurrence, with bat workers using specialist equipment (i.e. bat detectors) to gather what are predominantly ultrasonic recordings. This means that sound analysis software is always required in order to work further with these sounds.

'But this is not a bat book! Why do we need to know what bat workers do?' Well, as discussed earlier, the same equipment and software can (and does) get used for collecting and studying sounds made by other mammals, especially those smaller mammals that usually communicate vocally using sound within the ultrasonic range. This means that we do not need to go and create equipment specifically aimed at non-flying, four-legged small mammals, because bat detectors and the software packages that are used to analyse bat sounds are equally suitable (in fact ideal) when it comes to the study of these ultrasonic mammals – or indeed, for that matter, anything else that produces ultrasonic sound (e.g. bush-crickets). Added to this, many of these software packages are also just as capable at handling audible sound.

Most of the suggested sound analysis software packages come with their own tutorials, and it is not the intention here to replicate what is specifically catered for elsewhere. Nevertheless, in addition to these software tutorials, we encourage you to further your knowledge on the general principles of sound analysis, as a broader appreciation will increase your confidence and understanding of the subject, and assist greatly with analysis.

There are, however, certain aspects of sound analysis as it relates to mammals that should be considered within a book such as this, and accordingly we will now continue with some thoughts and guidance.

3.3 Sound versus noise (SNR)

Within Chapter 2 (Section 2.3) we have already mentioned the importance of trying to keep unwanted noise to a minimum when carrying out recording sessions. Doing this will help greatly when it comes to the analysis of recordings, in that it is much easier to see (and hear) what was going on with your targeted sound if it is not lost (in whole, or in part) amongst simultaneous noise.

Such noise can result from countless sources of nuisance, including: speaker noise; weather-related noise; mechanical noise; electronic noise; and talking and other anthropogenic sources (see Chapter 2, Table 2.6). Being aware of and taking this into account during acoustic analysis may prove to be the difference between a successful conclusion or a wasted effort, depending upon the results being sought.

Within Chapter 2 (Figure 2.1) we also provided an example showing good sound (signal) to noise ratio (SNR). Here (Figure 3.2) we provide you with a poorer-quality recording, which although not completely unworkable, does give you an example of what you should be seeking to avoid at the outset (i.e. whilst carrying out fieldwork). With this example, notice within the spectrogram (lower window) that there is a continuous noise occurring at c.15 kHz (see black arrow), and halfway through, some louder background noise overlaps and dominates the recording. In addition, the sound produced at source is not that much louder than the background noise, as demonstrated by the weak amplitude peaks occurring within the oscillogram (upper window). Now take a quick look back at Figure 2.1, to compare Figure 3.2 with a better-quality recording.

When faced with such circumstances (i.e. poor-quality recordings) there may be a few things you could attempt in order to make things clearer, and therefore more workable, for yourself. We will explore a couple of techniques that will be useful for improving the usability of your recordings. Before we do, however, we feel that we should mention

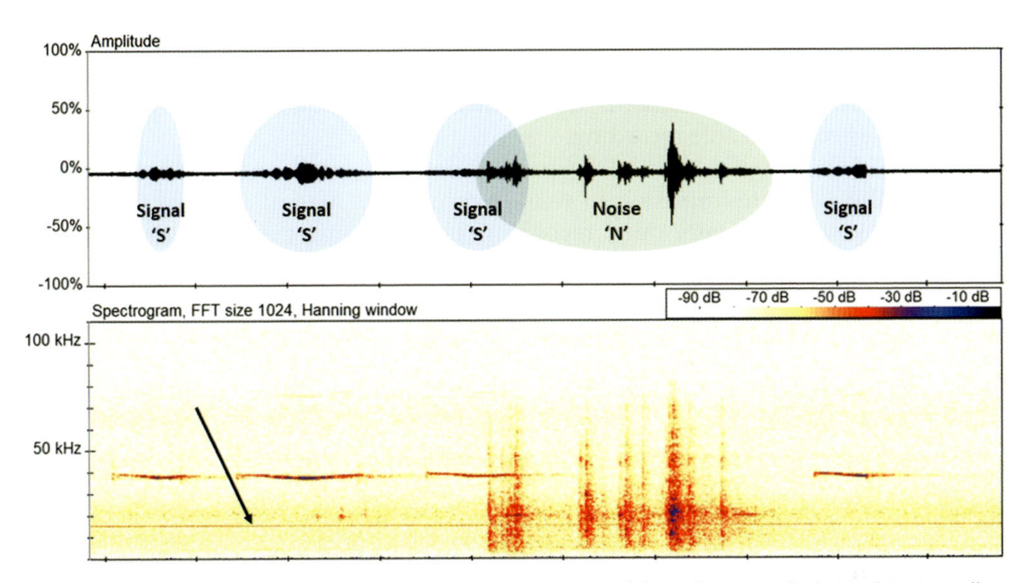

Figure 3.2 Example of a recording (stoat) showing a poor 'sound (signal) to noise' relationship. As well as the louder 'noise' shown within the green-shaded area, a continuous electronic 'noise' also occurs (see black arrow).

that irrespective of whether you have been able to improve the recording quality, you should still attempt to take measurements etc. With perseverance and experience you will often find that it is possible to use such recordings for identification purposes. Let us demonstrate this by looking again at Figure 3.2. Despite the poor quality, we can still see where our target sound is, and therefore we would be able to take measurements relating to frequency and time. As well as these measurements, we could possibly differentiate the sound from background noise whilst listening, at least at the points when it did not co-occur with the louder burst of noise.

As alluded to earlier, we will now discuss a couple of techniques you may find useful when working with recordings of poorer quality.

Adjustment of threshold settings

With most software packages you will be able to adjust the effect of 'over-loudness' through a feature that can give you a 'quieter' version of the spectrogram (i.e. as if the recording volume had been lower at the point of engagement). This may be described within the software in a number of ways, for example: threshold setting; amplitude setting; gain setting. In adjusting this setting, you may find that its impact upon unwanted noise is more dramatic, and therefore advantageous, than its impact upon the target sound. Figure 3.3 provides an example as to how Figure 3.2 would look after adjusting the threshold to make sounds appear quieter on the spectrogram. You can see that it is a bit clearer now; the sound at 15 kHz is almost completely removed and the louder background noise is substantially reduced, whilst the target sound still shows well.

There is a note of caution to be applied here. When you start messing about with such thresholds, adjustments will usually affect the whole recording, i.e. it doesn't isolate the 'fix' purely to the area of concern, and in doing so you may inadvertently lose some of the information relating to your targeted sound. In practice, when applying such techniques, you should make a number of careful adjustments, watching for any detrimental changes to your targeted sound. One example of a detrimental change is the maximum frequency of your sound appearing to be lower than is really the case, due to the amplitude setting (i.e. impression of volume) being turned down. So, please do remember that such adjustments, up or down (louder or quieter), would usually be expected to affect the whole recording.

In the previous paragraph, with the words 'up or down', we have hinted at something. Applying the opposite approach to weaker recordings may also prove useful. For example, if you have a very quiet recording, by increasing volume through use of the threshold setting, you may be able to make the target sound far more visible than originally presented.

For those of us who analyse sounds on a daily basis, we are never far away from the threshold setting tool, and it is quite normal for us to be regularly using this technique during our analysis sessions.

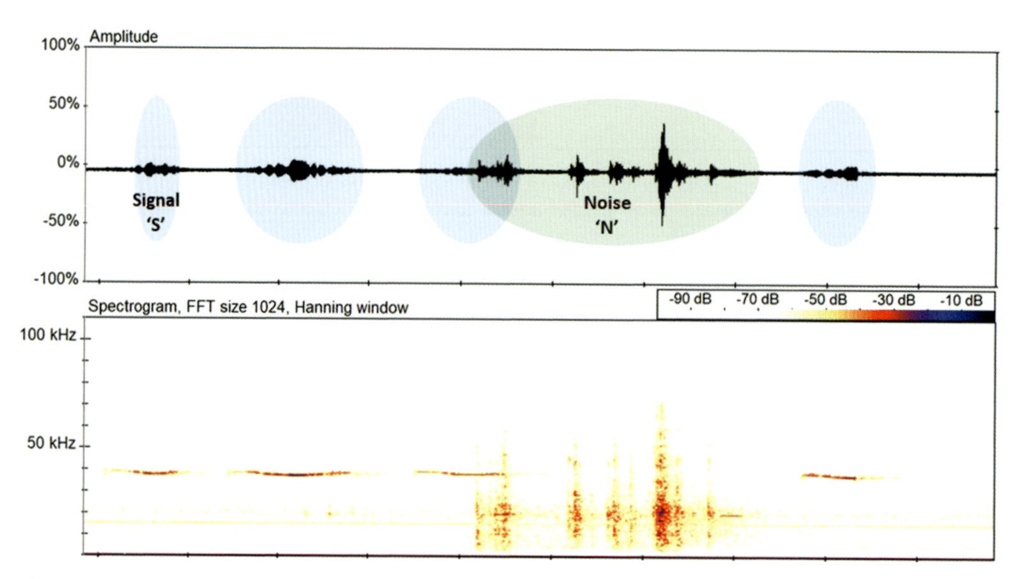

Figure 3.3 Example provided earlier within Figure 3.2, with threshold adjusted in order to improve the analysis experience

Use of editing filters

If the background noise is totally disassociated, in terms of frequency or time, from the targeted sound, with some software packages you may be able to edit the recording in order to remove (erase) the background noise (e.g. someone speaking in the background, or a sudden unexpected event, such as a car driving past).

To demonstrate this, Figure 3.4 takes our earlier example (Figure 3.2) and filters out all noise below 35 kHz, leaving us with a less cluttered recording, enhancing our ability to see and hear the sound we are interested in. Again, you need to be careful in doing this that you do not lose any information about the target sound which may be useful for making an identification, or in taking measurements.

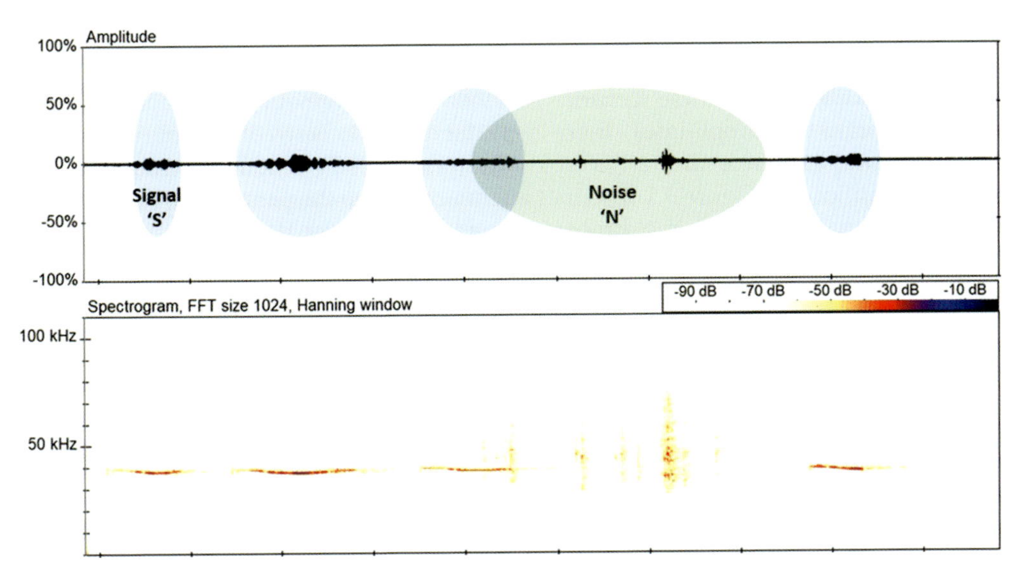

Figure 3.4 Example provided earlier within Figure 3.2, with a filter applied removing all noise below a certain frequency (in this case, 35 kHz) in order to try and improve recording quality

3.4 Attenuation and sound artefacts

The book *Is That a Bat?* (Middleton, 2020) discusses in more depth than described here the various sound artefacts that can sometimes be found in recording sessions, any of which can impact upon recording quality and the analysis thereafter. These artefacts are as follows: attenuation, clipping, aliasing, echoes and comb filtering. First of all, and separately, we will look at attenuation, after which the other artefacts will be demonstrated.

Attenuation

We mentioned in Chapter 2 (Section 2.1) that we would talk more about attenuation in this chapter. So, here goes.

A basic principle of sound physics dictates that at similar amplitude (i.e. volume or loudness), higher frequencies get lost or dissipate into the atmosphere before lower frequencies. This means that the further away you are from the source of any sound, the more likely that higher frequencies will not be recorded, and hence information may be missing (e.g. maximum frequency may no longer be accurately represented).

As a result, high-frequency sound waves that are emitted at too great a distance to reach the microphone will not be recorded in the first place (i.e. they are too distant to pick up), and hence won't be captured by your recording device, and no amount of manipulation (e.g. adjusting threshold settings) can create something that is just not there in the first place. Accordingly, for more distant sources of sound, you should always be conscious that what you are seeing/hearing is not necessarily the whole picture.

In the same way that higher frequency attenuates, weaker (low-amplitude) sound does not carry as far in the air as louder (high-amplitude) sound. This means that from a distance, there may be quieter sounds (weaker in amplitude) that have not been recorded, irrespective of the frequency (high or low) at which these sounds occurred. Simply put, the sounds were just not loud enough for your microphone to detect and therefore record. Thus, you could have a sound where the lower-frequency elements are far weaker (lower amplitude) than the higher-frequency elements. This can give the impression that the minimum frequency of the call is higher than it actually was, as the true minimum levels have not been recorded.

When you take each of these scenarios together – the combination of higher frequency and quiet sound (low amplitude) – there is greater disparity between the sound detected and the sound at source, and an even greater need to be cautious when attempting species identification. Figure 3.5 describes attenuation from the perspective of the higher frequencies that get lost in the atmosphere, ahead of lower frequencies of a similar amplitude (i.e. power or loudness).

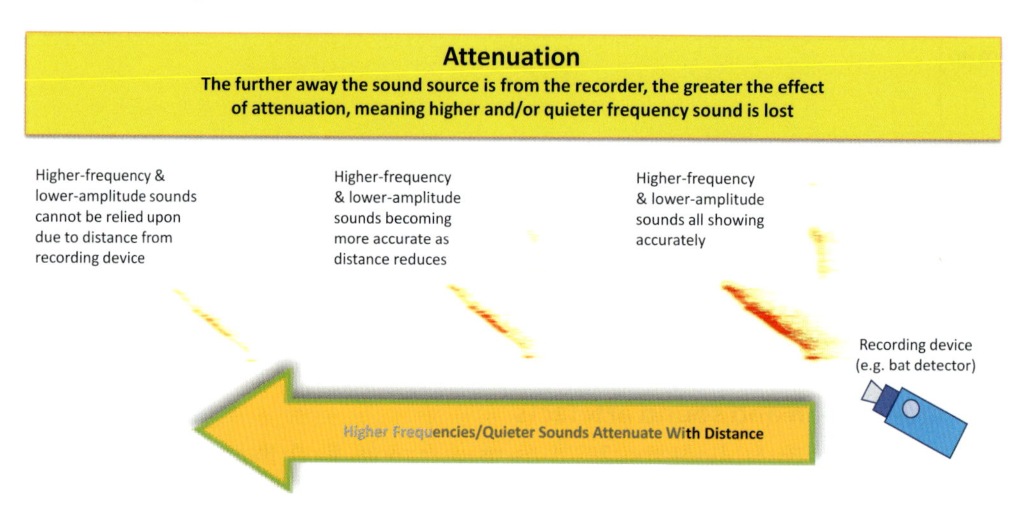

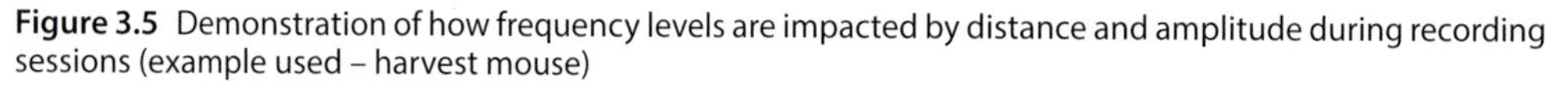

Figure 3.5 Demonstration of how frequency levels are impacted by distance and amplitude during recording sessions (example used – harvest mouse)

The potential effects of attenuation, either on its own or combined with signal to noise ratio, tend to be a fairly constant consideration whilst carrying out sound analysis. More unusual, but equally important to have an appreciation of, are the other artefacts we mentioned earlier, and we will now proceed to tackle these.

Comb filtering, echoes, clipping and aliasing

Occasionally during analysis you will come across examples that, for whatever reason, just do not look quite right. To the less experienced eye, they may look natural, and it may be difficult to grasp that what you are seeing is in fact an artefact, and does not truly represent what the sound should actually look like.

Table 3.2 provides more detail about those artefacts that you may encounter when recording terrestrial mammals. Comb filtering along with the recording of echoes, separate to the original sound, will not feature as often as they do in bats, but could still potentially occur with louder recordings, dependent upon the position of the source relative to the position of the microphone.

Clipping and aliasing, on the other hand, could easily arise more often. Clipping occurs when the source sound is too loud for your recording equipment to cope with, usually occurring as a result of the recording volume (which can be controlled by the gain setting on the recording equipment) being set too high for the circumstances. Aliasing normally occurs when the sample rate used has been set too low for the species recorded.

From hard-earned experience, we know that it is important to highlight these artefacts to you, and you should always be on the lookout for such situations. However, the good news is there are steps that you can take to prevent or reduce the risk of finding problems like these.

Table 3.2 Summary and examples of sound artefacts potentially occurring as a result of acoustic recording

Artefact	Description	Example
Comb filtering	Occurs when original sound is recorded simultaneously with the echoes of that sound (i.e. overlapping occurs). This results in what can be referred to as a constructive/deconstructive interference. It can arise because the microphone, or a bat detector, is close to a reflective surface from which echoes have bounced. To reduce the chance of problems like this, position the recording equipment away (at least 1.5 m) from any flat surfaces, vegetation or calm water.	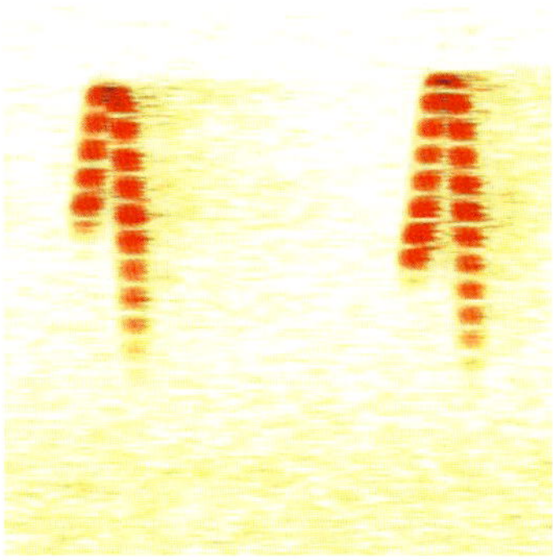
Echoes	Unlike the effect created during comb filtering, there is clear separation in time between the original sound and the returning echo. As a result, constructive and deconstructive interference does not occur, and we see the echo as a separate sound registration, appearing quickly after the original sound.	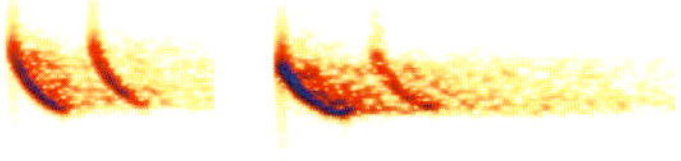
Clipping	Occurs when the sound being recorded is too loud for the recording device to cope with. This results in a recording input volume which overloads the system. It can be prevented by lowering the recording input volume or reducing the gain setting to a level where clipping no longer occurs. As a consequence of clipping, some of the original sound quality can be lost, as well as distortion and artefacts occurring. Clipping can cause even-numbered harmonics to feature less strongly than odd-numbered ones, as well as creating spurious higher-frequency harmonics beyond the accurate range of the source's sound.	Oscillogram view  Spectrogram view 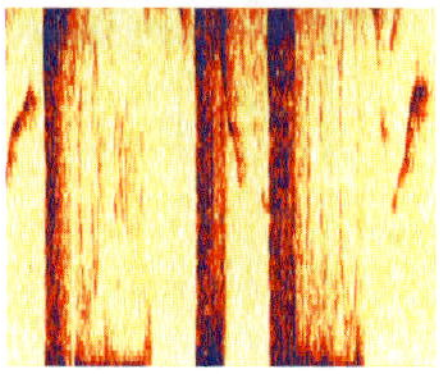
Aliasing	Occurs when the sample rate used while recording a sound is too low to cope with what's being recorded. In effect this causes an artefact that can make the spectrogram appear odd, or quite different to what you would normally expect to see. When aliasing does occur, it happens at exactly the same point in time on a spectrogram as the actual sound was recorded, and gives a visual, at least in part, of an upside-down (mirrored) version of the real call. As a general rule, the sampling rate used should be more than twice the highest frequency of the original sound.	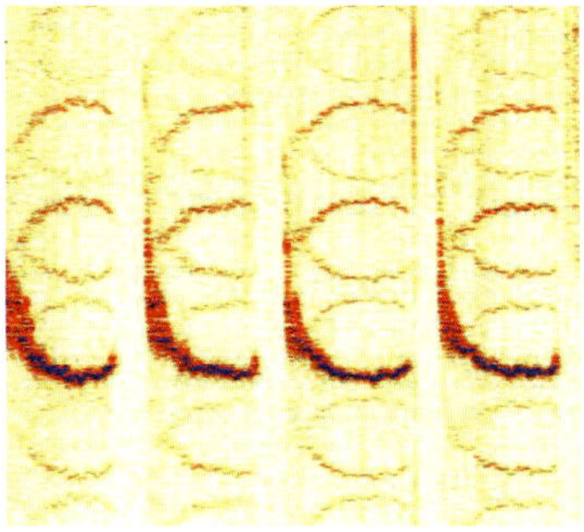

3.5 Harmonics

Harmonics are a naturally occurring feature of all sound. They are the acoustic layers that give sound its wider texture, and are always present, but because they occur simultaneously you would rarely be aware, audibly, of their presence. When exploring sound in more detail (e.g. on a spectrogram), they can at times be far more evident, and hence you need to be aware of what they look like.

For some mammal species, harmonics may be evident. They are first represented by the base frequency (i.e. the fundamental frequency or first harmonic) of the sound produced. The remaining harmonics (i.e. the second harmonic upwards) are a straight multiple of the fundamental frequencies, thus meaning, for example, that a fundamental frequency of 25 kHz will produce a second harmonic at 50 kHz, a third at 75 kHz, and so on (Briggs and King, 1998). Figure 3.6 provides a visual explanation.

It should be noted (see Chapter 2, Table 2.4) that for those using bat detecting systems other than full spectrum or time expansion (e.g. frequency division or zero crossing) to record mammals that vocalise using ultrasonic sounds, harmonics (other than usually the fundamental) are not accurately shown, if indeed they are shown at all. The apparent harmonics which can sometimes be seen in frequency division recordings are not truly placed, and are the result of 'an artefact' of the recording format (Russ, 2012) as opposed to a true representation of what was actually produced. When gathering sounds from terrestrial mammals as described in this book, we would not normally recommend the use of frequency division or zero crossing systems, but we are mentioning this here, just in case!

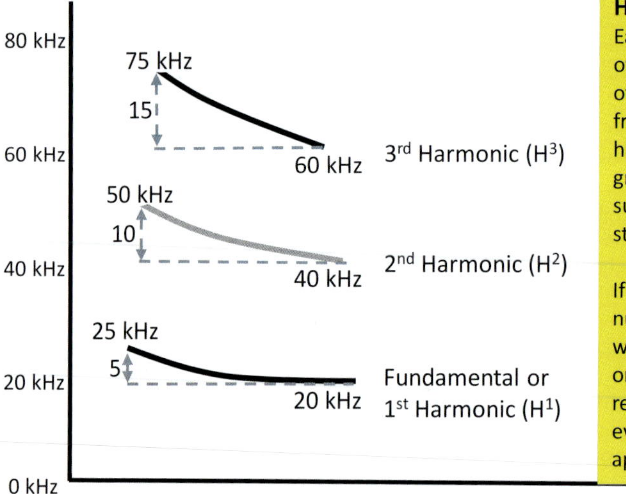

Harmonics

Each harmonic is a multiple of H^1. Due to the start frequency of H^1 being higher than the end frequency, subsequent harmonics for higher parts of the call increase at a greater rate, resulting in each subsequent harmonic becoming steeper in appearance.

If clipping has occurred, the even-numbered harmonics may appear weaker than the odd-numbered ones, or if the harmonics are faintly represented in the first place, the even-numbered ones may not be apparent.

Figure 3.6 Example showing the positioning of naturally occurring harmonics

3.6 Sniffing around

One aspect of recording mammals that can occur quite regularly in some situations relates to the spectrograms created as a result of sniffing. It is unlikely that these recordings would ever be useful for identification purposes, but they can confuse the analyst, as they can appear to be far more interesting than what they eventually turn out to be.

The following figure (Figure 3.7) provides an example of the sort of things to look out for during analysis in this respect, and often just listening to the recording in real time will give you the confidence that it nothing more than just a sniff.

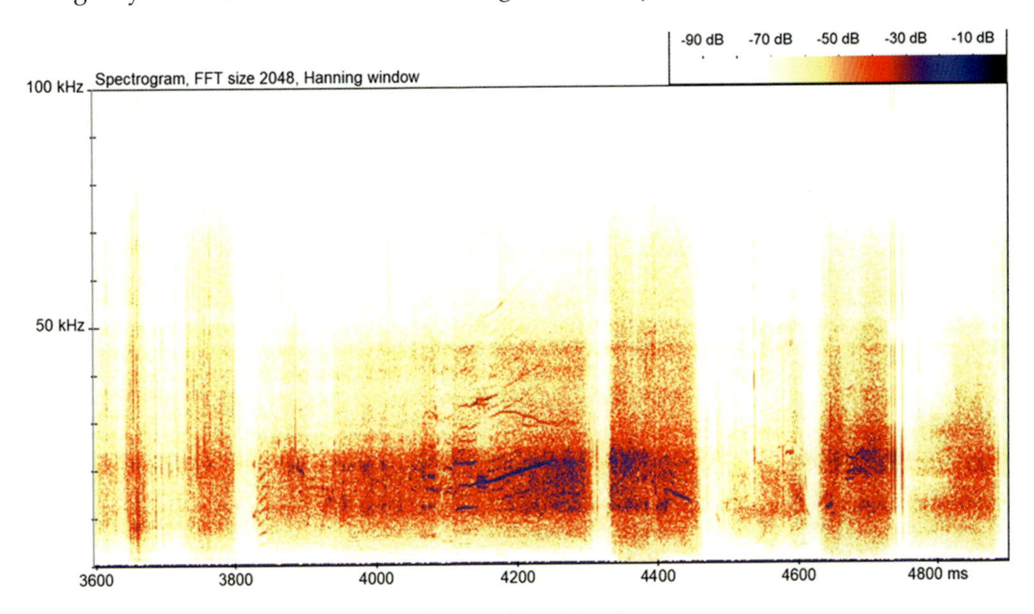

Figure 3.7 American mink sniffing around (frame width: c.1.3 sec)

3.7 Call measurements

A number of key call measurements are commonly used in the analysis of sound across the range of mammal species occurring in Britain and Ireland, with the characteristics most frequently used being shown in Table 3.3. Within the table each 'measurement' is described and linked with the software 'measurement tool' used to view the call to accurately collect the information sought.

The 'measurement tool' column within the table cross-refers to three separate figures (Figures 3.8, 3.9 and 3.11) to enable the reader to see how these measurements relate to a spectrogram, oscillogram and power spectrum. Abbreviations are then described, where combinations of these can be useful when describing sound. For example, the frequency of maximum energy (FmaxE) of the fundamental harmonic (H^1) of the third component (C^3) in a recorded sequence could be abbreviated as $FmaxEH^1C^3$.

In addition to taking measurements, it is also important to listen to the actual sound of the call (e.g. for ultrasonic calls, if slowed down using time expansion). Also, if a recording contains sounds that are bordering on ultrasonic (up to 20 kHz), we recommend that the recording is played in real time, which can be helpful to avoid misidentifying high-frequency bird calls as small mammals.

The following set of figures (Figures 3.8, 3.9 and 3.11) are referred to within Table 3.3, and are presented to show the measurements that can be taken by using the three main analysis tools available with most software. Figure 3.8 shows the spectrogram of a brown rat distress call. In the spectrogram you are able to see the number of components (nC) (i.e. individual calls) and the overall structure of the call sequence (stG), as these relate to time (i.e. milliseconds) on the x-axis and frequency (i.e. kilohertz) on the y-axis. The

Table 3.3 Key measurements associated with the analysis of mammal calls

Measurement	Description	Measurement tool	Abbreviation
General call structure	What is the overall impression of the call?	Spectrogram **Figure 3.8**	stG
Length of sequence	The total length (duration) in time (sec or ms) of an entire sequence (S) comprising two or more components (calls).	Oscillogram **Figure 3.9** Spectrogram **Figure 3.8**	durS
Number of components	The total number of individual components (calls) making up a part of or the entire sequence.	Spectrogram **Figure 3.8**	nC
Length of individual components	The length (duration) in time (sec or ms) of individual components (calls) within the sequence. Components numbered (e.g. C^1, C^2, C^3).	Oscillogram **Figure 3.9** Spectrogram **Figure 3.8**	durC$^{\#}$
Minimum frequency	Minimum frequency(ies) (Hz or kHz) of the fundamental component(s) (i.e. first harmonic: H^1) and, if desired, of any subsequent harmonics occurring (e.g. H^2).	Spectrogram **Figure 3.8**	fminH^1 fmin$H^{\#}$
Maximum frequency	Maximum frequency(ies) (Hz or kHz) of the fundamental component(s) (i.e. first harmonic: H^1) and, if desired, of any subsequent harmonics occurring (e.g. H^2).	Spectrogram **Figure 3.8**	fmaxH^1 fmax$H^{\#}$
Frequency of maximum energy	The frequency (Hz or kHz) within the call component or sequence where most energy is concentrated.	Power spectrum **Figure 3.11**	FmaxE
Inter-component interval	The distance in time (sec or ms) from the start of one component to the start of the next component (call).	Oscillogram **Figure 3.9** Spectrogram **Figure 3.8**	ici

spectrogram will also show the presence of any additional harmonics, over and above the first harmonic (also called the fundamental). This tool can be used to establish minimum frequency (fmin) and maximum frequency (fmax) of any of the components being looked at for analysis purposes. Finally, amplitude is represented by intensity of colour.

In Figure 3.9 we have introduced the oscillogram tool (the upper window within that figure), which is used to establish a more accurate reflection of time, as well as how the amplitude (i.e. power) of the sound is distributed through time within the area being displayed by the spectrogram. The parameters used to measure the length or duration of the overall sequence (durS) are shown. Component length (durC) is demonstrated through showing the corresponding parameters of the second component (C^2). It can be seen on the oscillogram where the amplitude begins to increase and then fall away again.

As well as measuring sequence and individual component (i.e. call) length, the oscillogram should be used to measure the inter-component interval (ici) – which in effect, for the example shown, is the distance in time (ms) from the start of one component to the start of the following component. All of the time-related parameters could also be

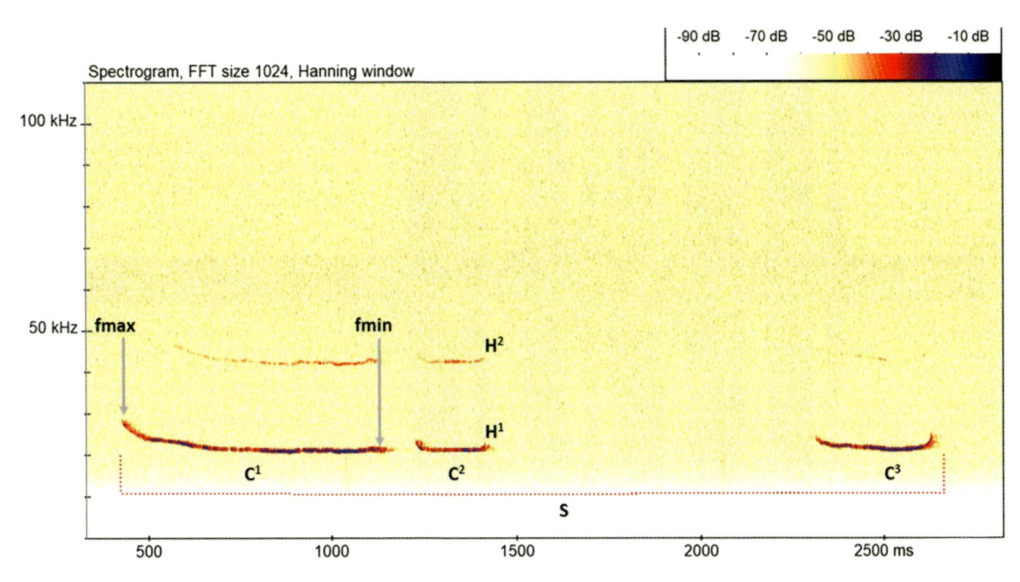

Figure 3.8 Spectrogram of brown rat distress call showing location of maximum frequency (fmax), minimum frequency (fmin), first harmonic or fundamental (H^1), second harmonic (H^2) and the call sequence (S) which comprises three components (C^1, C^2 and C^3)

measured within the spectrogram tool, but spectrograms, for various reasons, may not always allow the technician to see this measurement to the same degree of accuracy, and here it is usually considered best practice to use the oscillogram instead.

The oscillogram has further uses. In some instances, the shape of the oscillogram can give an indication as to how amplitude is distributed throughout the call, and, for some species or in some scenarios, particular shapes may prove to be of assistance in

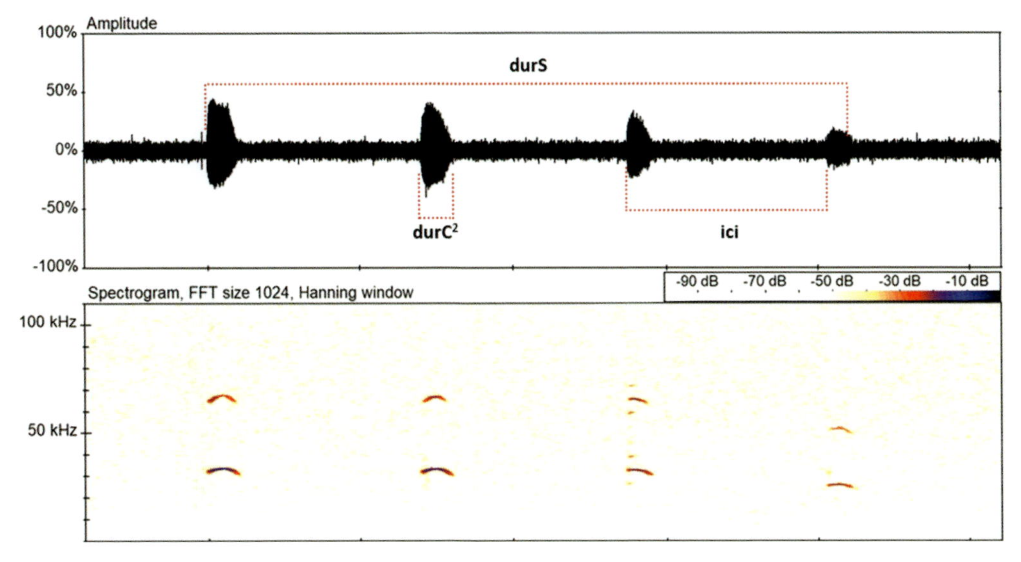

Figure 3.9 Oscillogram positioned directly above a spectrogram, showing examples of locations for component length (durC), sequence length (durS) and inter-component interval (ici) (call shown relates to a harvest mouse)

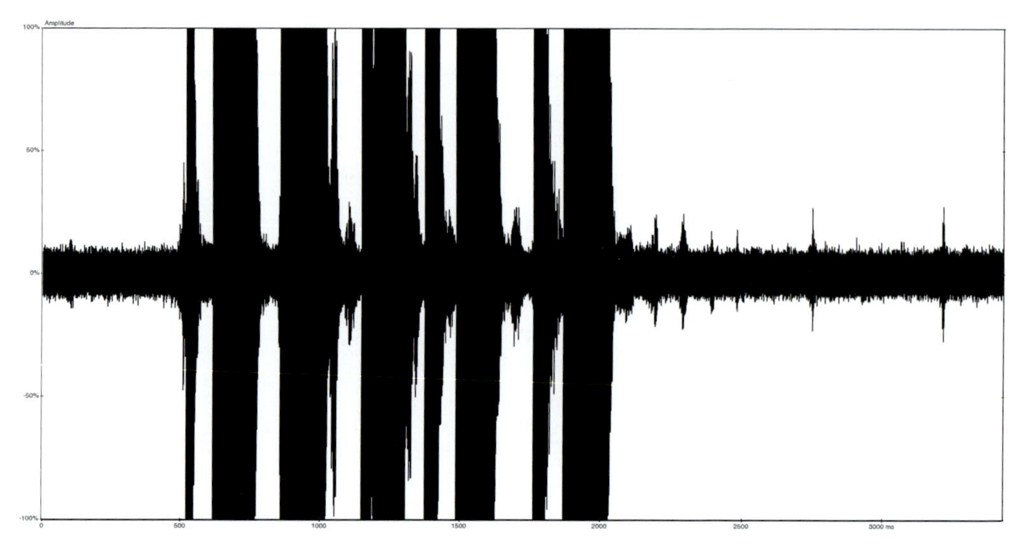

Figure 3.10 Oscillogram showing an example of clipping, whereby the sound has overloaded the recording device at time of recording

identification. To date, however, there is far more information published on this approach for bat echolocation in certain species (e.g. Barataud, 2015) compared to other mammals.

The oscillogram is also a useful tool for confirming the quality of the recording (see Section 3.3) to ensure that a good level of the call's original information has been retained. An effect known as clipping (as discussed in Table 3.2) may occur as a result of the sound recorded being too loud for the recording device to cope with. This effect can be checked using the oscillogram tool, whereby if clipping has occurred, the amplitude shapes would reach beyond the parameters of the window in the software, thus giving a straight-edged appearance (see Figure 3.10 and Table 3.2). This, when it occurs, can

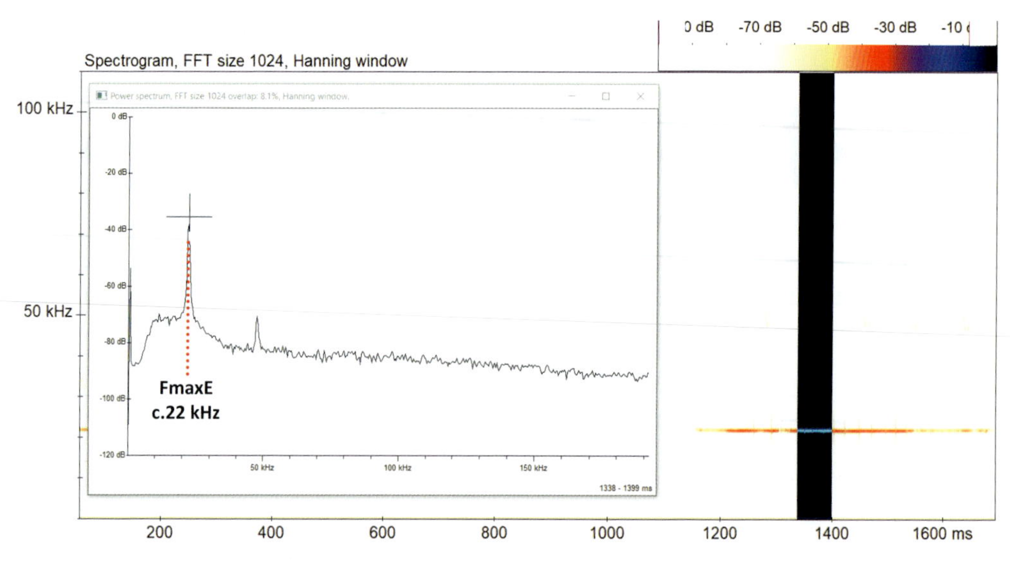

Figure 3.11 Power spectrum window within spectrogram, showing a brown rat distress call, with the location of the frequency of maximum energy (FmaxE) reading shown. The portion of the call being measured for this purpose is shaded in black.

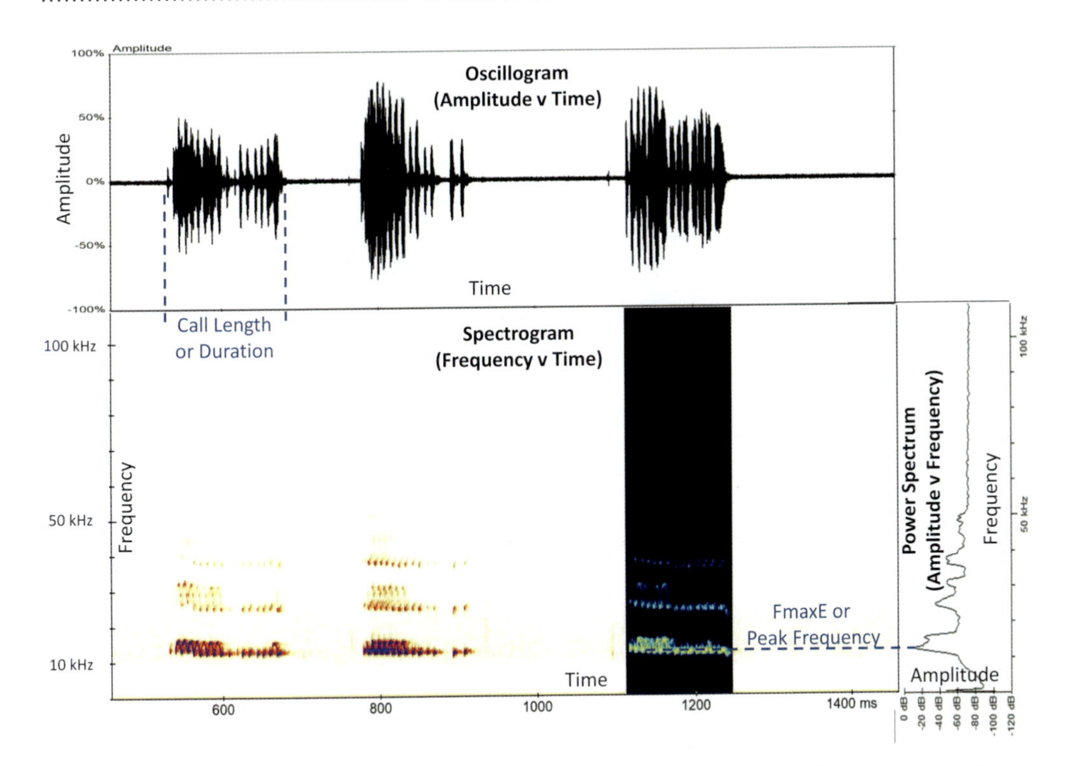

Figure 3.12 Analysis tools (as they relate to each other) with some regularly used call measurements shown, using common shrew as an example

result in distortion, as well as harmonics not being represented accurately (Middleton, 2020). An example of how this manifests itself is shown in Figure 3.10.

In Figure 3.11 the power spectrum tool is introduced. This measures the frequency (kHz), within the call component or sequence, where the mammal has exerted the most energy from an amplitude perspective. This measurement is usually referred to as the frequency of maximum energy (FmaxE) or peak frequency. The power spectrum window has been positioned to the left of the component being measured within the spectrogram, and in the example shown (brown rat) the FmaxE reading is c.22 kHz.

Finally, within Figure 3.12 we take each of the tools described (Figures 3.8, 3.9 and 3.11) and collate them in such a manner as to show how they relate to each other. A spectrogram compares frequency against time, an oscillogram compares amplitude against time, and a power spectrum takes the final perspective of comparing amplitude against frequency.

3.7 Description of calls

Having measured and recorded the various parameters of interest, it is important to be able to describe what a call, or a component, or a series of components (i.e. a sequence) looks like. The description of mammal call shapes as displayed by spectrograms has yet to be standardised, but Table 3.4 gives examples of call shapes that may be encountered during analysis, and along with these examples a written description is suggested. In the context of this book, for sake of ease and consistency, we will use these descriptions, and the reader can cross-refer any such descriptions within the species profiles in Chapters 5 to 10 with those described here within Table 3.4.

Table 3.4 Component shapes with descriptions, as viewed in spectrograms, encountered during call analysis

Shape		Shape	
Descending FM	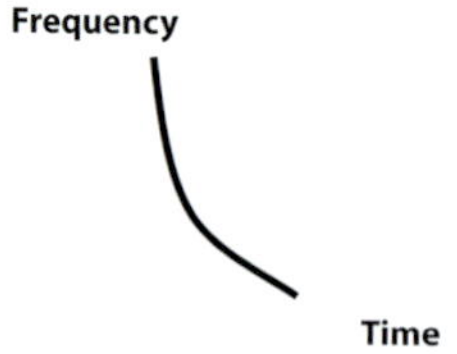 Frequency Time	**Upward hooked**	
Ascending FM		**Downward hooked 'walking stick'**	
Steep broadband FM		**Hockey stick**	
Steep narrowband FM		**U**	
Shallow narrowband FM		**V**	
CF		**W**	
Ascending QCF		**Arched**	
Descending QCF		**Modulated Oscillated**	
Single component		**Sequence (> one component)**	
Single syllable (one sound occurs)		**Multi-syllable (> one sound occurs within a component)**	

Sequence:	**Component:**	**Syllable:**
A group of closely associated components, making up an entire sequence of recorded events.	A distinct sound (i.e. individual call) separated in time from other components (calls).	Each differing aspect, in terms of structure and sound, within a single component (call).

FM: Frequency modulated (i.e. frequency changes over time)
CF: Constant frequency (i.e. frequency remains stable over time)
QCF: Quasi-constant frequency (i.e. nearly/almost constant frequency)
Broadband: Call travels through a wide range of frequencies (> 5 kHz)
Narrowband: Call travels through a short (limited) frequency range (< 5 kHz)

Following on from the descriptions provided in Table 3.4, we have found it quite useful when describing calls to others, to relate them to human speech. To this end we have found it appropriate to describe the entire sequence (e.g. a series of individual but associated components) as a sentence, with each distinct component being thought of as a word, and each word comprising one, or two, or more syllables. To demonstrate this, take another look at Figure 3.9. In this example we could therefore describe the call sequence as a sentence comprising four similar words, each word having two syllables. In describing mammal vocalisations in this manner, we are not suggesting that these animals talk in the same way as humans; we merely take this approach in order to make a difficult concept for some easier to grasp.

3.8 Application of filters and automated classifiers

It would be falling short, when covering sound analysis, to not discuss the application of filters and automated classifiers. In some areas of sound research and surveying, the use of such systems has become an important consideration on the part of the naturalist – the study of bats and birds being prime examples. This is less the case for purely terrestrial mammals, with most systems developed to date not being publicly or easily accessible, and often focusing on a single species. A notable exception is the BTO Acoustic Pipeline (http://bto.org/pipeline), a system developed by one of the authors (Stuart Newson, BTO) which although not designed to focus purely on terrestrial mammals, does cater for them, in that many of our small terrestrial mammals which produce ultrasonic sound have been factored into the outputs generated.

What is a filter?

Simply put, a filter is an application that is applied to a folder of recordings, in order to separate those files into categories/potentially species, according to a defined set of rules or criteria. Many of the specialised software packages allow the user to build filters, so that they can separate calls into different categories. Such filters are normally used in combination with manual auditing of recordings, to help simplify the analysis process, especially when the amount of data to be investigated is substantial. For example, a filter might be built to automatically scan a folder of recordings to look for those featuring sounds that fall within a particular frequency range and duration.

What is an automated classifier?

Having already discussed filters, perhaps the easiest way to now describe automated classifiers is to view them as extremely advanced filtering systems, operating in a multi-dimensional manner. The BTO Acoustic Pipeline, for example, extracts 150 measurements from every detected sound event in a recording (biological or non-biological), which are compared in an analysis with the same measurements taken from recordings where the source of the sound (again biological or non-biological) is known. This allows for quite complex measurements and analysis to be carried out, that would not be possible to carry out manually.

Importance of understanding the output from a classifier

Different classifiers use different analytical approaches to generate and then describe their outputs in different ways, which can be confusing. When using such systems, it is important to take some time to understand the outputs, and how to use them. This is particularly important in relation to any measure of error that accompanies a species identification.

Understanding this will allow you to appreciate the strengths, as well as the limitations, of the process, and importantly, how to interpret and use the outputs appropriately.

Most classifiers that have been built to support the sound identification of a range of species groups use a reference library of known species recordings that have been used to train algorithms. These algorithms can then be automatically applied to unknown recordings, and typically return an output that tells you, in some way, how similar the unknown recordings are to the species in the reference library. The analytical approach used will influence the results, but perhaps most important is how well the reference library that has been used to train the algorithms covers the range of variation in calls that could be produced by each species, as well as other confusion sounds that may be recorded in the field.

As an extreme example, if we were to build a classifier with just the calls of wood mouse, and we then gave it independent recordings of wood mouse to identify, the classifier is likely to assign these to wood mouse with high confidence. This might be encouraging, but if we now gave the same classifier recordings of hazel dormouse, it is also likely to assign these to wood mouse with high confidence, because wood mouse is the only species it has to choose from. As a slightly different real-world example, if we were to only train a classifier with bat recordings, and gave it non-bat mammal recordings, it may assign these recordings to bat species with high confidence. In other words, high confidence/low error does not necessarily mean that the species identification is correct.

In practice, it is extremely difficult to try and build a truly representative sound library for every species and all other sounds that could be recorded in the field. So, whilst the problems may not be as extreme as the wood mouse example above, there are likely to be some gaps, for example for particular call types or particular situations that are not covered fully, or at all, by the reference library. One consequence of this is that a 90% probability (or accuracy rate – however defined) for one species may not be the same as a 90% probability for another species, when it comes to the true accuracy/error in identification. Some classifiers are now working to address some of these problems by incorporating steps to try and produce more objective and standardised measures of error in identification, based on an independent assessment of error (Barre *et al.*, 2019).

It is important to understand that the performance of a regional classifier for a given species may also vary geographically, depending on the suite of potential species present in the area that you are working in. As an example, the performance of a classifier for wood mouse that has been built for a suite of species present in Scotland is likely to perform better than a classifier that includes wood mouse for southern Britain, where yellow-necked mouse also occurs and is a potential confusion species. The overall performance for a given dataset and species will also be influenced by how common the species is in the dataset that is being processed, relative to potential confusion species.

We should also appreciate that the performance of any classifier will be influenced by the quality of the recordings that you provide for processing. Some of the common problems regarding quality, and ways to address these, have already been looked at in Chapter 2 (Sections 2.1 and 2.3) and earlier in this chapter (Section 3.4). The better the quality of the recordings, the better the classifier will perform, and also the better your chance of being able to manually assign unknown recordings to species. This is something that you can influence, and it will depend on the sound recording equipment used, the recorder settings, and how/where the recorder or microphone (if separate on a lead) is positioned. If recording at higher frequencies, it's particularly important to position the microphone away from any flat surfaces, vegetation or water, as discussed above (Section 3.4).

Lastly, the repertoire of sounds made by all of these mammals, as well as anything else that could be recorded, is vast – with many overlaps and similar sounds coming from entirely different sources, both biological and anthropogenic. We have already seen numerous examples of sounds that look and sound very similar to the echolocation or social calls produced by bats (Middleton, 2020). And, as we mentioned in an earlier chapter, if something else can look and sound like a bat, then a bat can look and sound like another mammal. Indeed, there are many biological and non-biological sounds that can masquerade as the vocalisations produced by mammals. As such, caution is encouraged – but with this, we would also like to encourage a culture of critical thinking, i.e. considering the evidence and reasoning that supports the identification of one species over another. We think that such an approach has the opportunity to improve standards and push the bounds of our knowledge on species identification further, and at a faster rate than otherwise. We are after all, all of us, on a steep learning curve (in some respects a vertical cliff face) when it comes to this subject, especially as it relates to the species covered here.

Manual verification versus automated processes

As we have explored above, there are strengths and weaknesses to manual identification compared with automated approaches, but it is not necessary to choose one over the other.

If anything, such automated tools are likely to become more important for the large-scale recording of purely land-based small mammals than for bats, because unlike echolocating bats, mammals do not call repeatedly as they move around their environment, and so they are unlikely to produce the same number of recordings. Unless you are out recording the mammal species directly in the field, finding every recording that contained mammal calls – for example after deploying static bat detectors – would be like finding a needle in haystack. As an example, from a total 4,366,304 triggered recordings collected by static bat detectors deployed by the BTO as part of the Natural England-funded LandSpAES project, there were 431,887 bat recordings, but only 1,051 (0.02%) small mammal recordings (Newson and Pearce, 2022). Manually checking over 4.3 million recordings to look for the 0.02% of recordings that contain small mammal calls would be pretty tedious, if not impossible, due to time and associated costs. However, the task becomes far more appealing and workable if you only need to manually look through a couple of thousand recordings that may contain small mammal calls. Keeping with the needle in a haystack analogy, you now have a powerful magnet to find the needle, but it might pick up a few other things on the way, and as such a manual check of recordings would be required.

For all of the reasons above, if you plan to use filters or classifiers, it is important that an appropriate level of manual auditing/checking of the results is carried out. However, the approach that is taken to auditing will depend on your objectives. For example, if you are interested in producing an inventory of species recorded during a particular period of time, we would suggest starting with the recordings with the highest probability/lowest error. Once you have found the species, you no longer need to check the remaining recordings. However, if you are interested in producing a measure of activity in some way, you may want to manually check all the recordings of a species, or a random sample of recordings (e.g. some hundreds of recordings) for common species. You may also want to check a random sample of other recordings, to demonstrate that the classifier is not missing identifications, or to quantify how many are likely to have been missed in the whole dataset.

Badger – Sandra Graham

CHAPTER 4

Overview of the Species Group-specific Chapters

4.1 Overview of the species profiles

Before we start discussing species groups in more detail (Chapters 5 to 10), we feel it is beneficial to give you an overview of our approach to these chapters.

Table 4.1 provides a summary of the species we are going to cover, and this can be read in conjunction with the information contained within Tables 1.3 to 1.7 (Chapter 1). The chapters are ordered as they are, broadly taking a larger to smaller (in size) species approach, in conjunction with keeping groups that are audible (often larger species) separate to species where ultrasonic sound is more likely to be useful for identification purposes. For example, in Chapter 5 we have the audible deer species, and in Chapter 10 we end up looking at the ultrasonic shrews. Of course, this approach is not perfect, as there are some species that appear where you would not expect them in terms of size (hedgehog, for example), but there comes a point where we had to follow at least some conventional rules when separating species into groups, which should be self-evident from the chapter titles.

Table 4.1 Overview of species groups covered within Chapters 5 to 10

Chapter number	Species covered (also see Tables 1.3 to 1.7, Chapter 1)	
Chapter 5	**Even-toed Ungulates**	Deer species and wild boar
Chapter 6	**Carnivores** (including mustelids)	Wildcat, fox, badger, otter, stoat, as well as other mustelids
Chapter 7	**Lagomorphs**	Rabbit and hare species
Chapter 8	**Rodents** (large)	Squirrel species and beaver
Chapter 9	**Rodents** (small)	Rat, mouse, vole and dormouse species
Chapter 10	**Insectivores**	Hedgehog, mole and shrew species

Chapter overview

Generically, at the start of each chapter we will provide an overview of the group of species catered for within that chapter.

We will then move on to provide some further useful comparisons, as they relate to sound identification. This will include figures providing the acoustic spectrum for each species, along with information as to potential recordable distance relating to the calls most likely encountered under typical field conditions.

Having catered for the information that we are providing consistently across all the chapters, we will also discuss any other useful aspects of sound behaviour for that specific group. There is no set format for this additional information, as each group of species may offer areas to focus upon which do not necessarily transfer over or have any benefit when considering other groups.

Species profiles

Further into each chapter, we start to tackle each species individually, kicking off with a short description of the species involved, using a tabular format describing distribution, habitat preferences, as well as daily, seasonal and mating activity patterns. Any other useful notes on the species ecology and behaviour will also be included. Bear in mind, however, that we are not seeking to replicate information in this book that is more fully available and better resourced elsewhere. As such, these introductory tables are to be regarded as purely a brief overview for the species concerned.

We will then proceed to discuss acoustic identification as it relates to the species concerned. As well as notes on the matter we will also show examples of spectrograms, and where useful oscillograms and/or power spectrums. These figures will be presented using BatSound v4.4.0 (Pettersson Elektronik AB) software (see Figures 4.1 to 4.3 for examples).

Within each figure description we will include details as to the length of time, in milliseconds (ms) or seconds (sec), to which the example relates. Also, the frequency scales for spectrograms will be visible within the figure, and we will usually use the following scales: 0 to 11 kHz, or 0 to 22 kHz if the sound is audible; 0 to 110 kHz if the sound is ultrasonic. For any figures where we use a different frequency scale, we will highlight this in the figure description.

Bear in mind, of course, that once you have downloaded calls from the Sound Library to your own computer, you can analyse these yourself using your own preferred analysis software, provided it recognises .wav format. In doing so, you can then adjust file lengths, frequency scales, threshold settings etc., as well as often being able to look at a wider series of calls within the file.

All of the sound-associated figures created are done so using full spectrum .wav recordings, and for any figure accompanied with a speaker symbol (e.g. Figure 6.4 🔊), the figure number directly corresponds to its track number within the downloadable Sound Library. Within some figures you may also see a QR code alert **(QR)**, and when

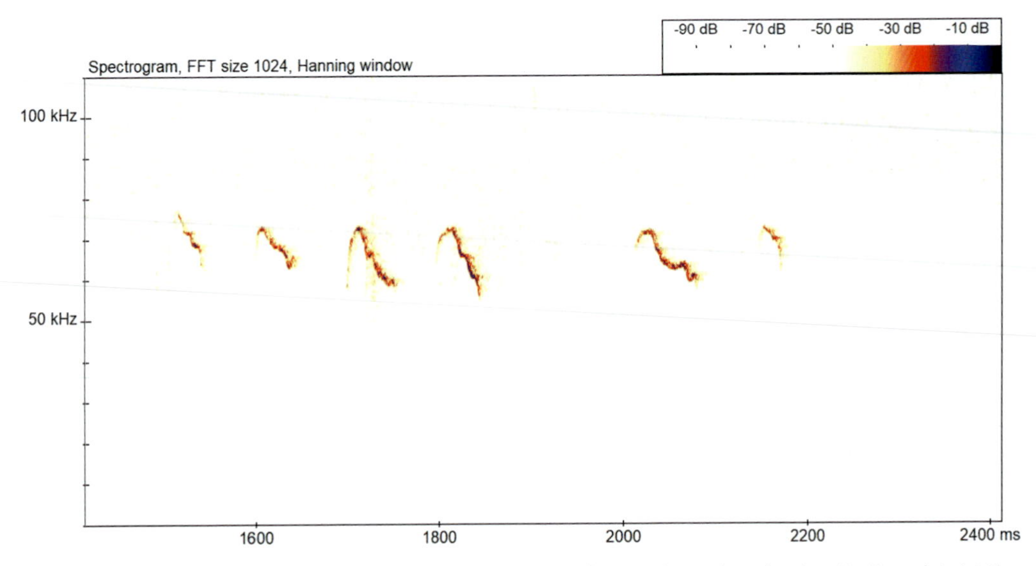

Figure 4.1 Example of ultrasonic sound spectrogram (wood mouse) produced using BatSound (v4.4.0) software (file length: 1,000 ms).

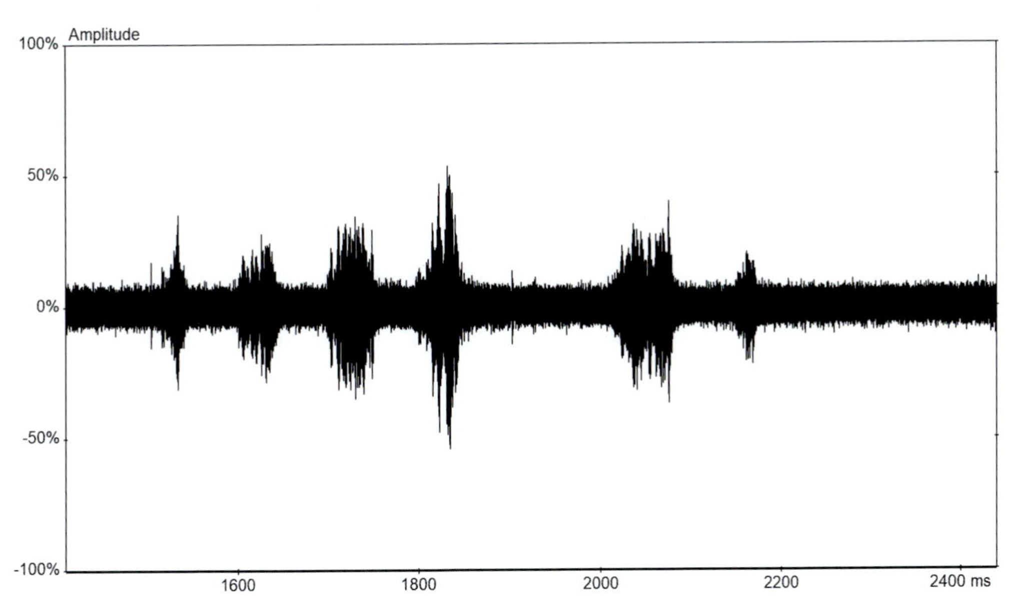

Figure 4.2 Example of ultrasonic sound oscillogram (wood mouse) produced using BatSound (v4.4.0) software (file length: 1,000 ms).

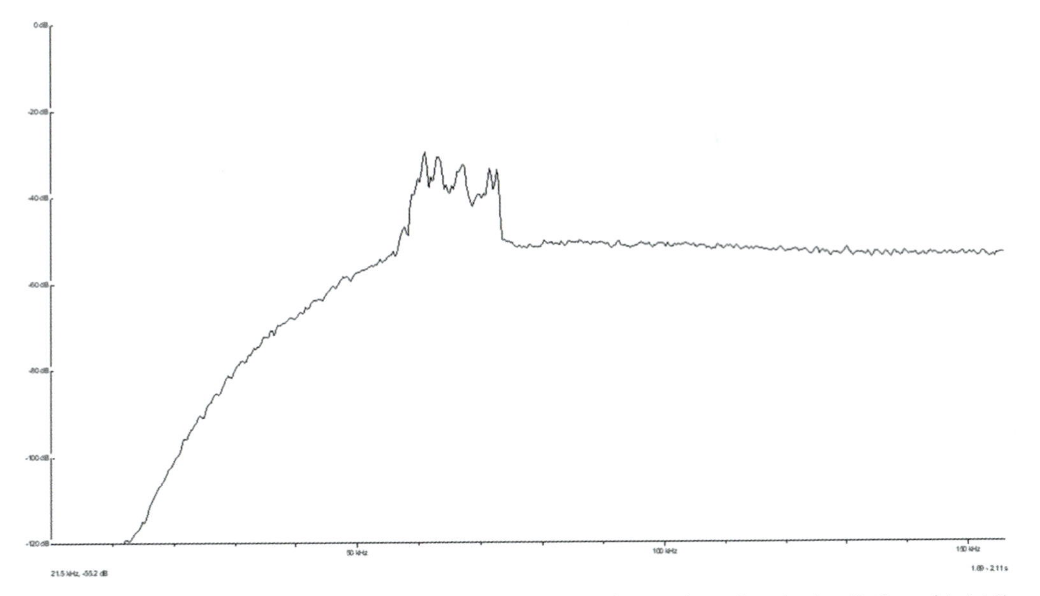

Figure 4.3 Example of ultrasonic sound power spectrum (wood mouse) produced using BatSound (v4.4.0) software (file length: 1,000 ms).

this appears you will find the QR code relating to that figure within a table at the end of the relevant section.

When the call in its naturally produced, recorded form is audible, then these QR code examples will sound as you would expect them to be heard in the field. When such examples are naturally ultrasonic, you would normally encounter these using a bat detector, and as such, these recordings will be provided in a slowed-down format (time expanded, x10), so that you will be able to hear an audible interpretation of the sounds. These QR examples have not been selected in any consistent or structured manner and should be regarded as an extra resource designed to intrigue the listener, and encourage a reader to download additional calls from our more extensive Sound Library.

Potential sources of confusion

Towards the end of each species section, we provide examples of other mammal species which produce similar sounds to those discussed. These examples may relate to other species within the same chapter, or others within different chapters. We will also highlight other species or groups, not covered within this book, that could cause confusion.

Potential application of bioacoustics for monitoring purposes

Finally, to conclude each chapter we discuss the potential application of bioacoustics for monitoring purposes as it relates to the species covered therein.

4.2 Managing expectations

For all of the species covered, a range of calls associated with different behaviours should be expected, with examples of such behaviours, when associated sound could be relevant, described within Table 4.2. Further, we would expect degrees of variation between sexes and individuals for any specific species, as well as differences occurring due to developmental stage, age or condition. Finally, it would not be surprising to find wider geographical variation (i.e. regional accents), or indeed even, perhaps, variation between different colonies at a more localised scale.

Table 4.2 Behaviours where associated sound may be produced by mammals

Examples of behaviours associated with sound		
Mate attraction/selection	Threatening	Aggressive encounters
Territorial defence	Distress	Individual contact
Copulation	Food patch defence	Group cohesion
Dominance hierarchy	Predator alert	Mother/infant interaction

Can a book such as this cover everything? In short, at this stage anyway, the answer has to be 'no'. Our 'collective' knowledge just isn't there yet, and for what is known, the constraints of time and space are also limiting factors. It would, however, be fair to say that the selection of examples we are documenting within this book provide a reasonable selection of the range of variation that we have encountered, albeit by no means would we wish to give the impression that we have covered all vocalisations for every species.

We have found, for example, that many sounds may only be produced rarely, or in extreme circumstances. There are also a good number of sounds that are probably, on their own, not as helpful when it comes to identification at species level. Remember, this book is about identification using sound, as opposed to interpretation of meaning from a behavioural context.

We have tried, therefore, to provide you with examples that are more likely to be useful when a visual identification is not possible. As such, calls produced between mothers and infants in very close proximity to each other are not included to any great extent, and likewise for calls produced by animals under severe stress or distress (e.g. whilst being preyed upon or hunted).

As described earlier (see Preface and Chapter 1), we are still very much on a steep learning curve when it comes to this subject, especially for the smaller ultrasonic emitters of sound, and this book should be seen as a solid stepping stone towards a wider and fuller appreciation of the subject, as opposed to being the final answer to every sound produced by the species covered.

Roe deer (buck) – Sandra Graham

CHAPTER 5

Even-toed Ungulates – Deer & Wild Boar

5.1 Introduction to this group

This chapter covers the even-toed ungulates (Artiodactyla), within which we consider six species of deer (Cervidae), and a single species of pig (Suidae) (see Table 5.1).

Being larger mammals, the presence of deer and wild boar can often be relatively easy to establish through visual sightings, the use of camera traps and/or finding field-sign evidence (e.g. evidence of foraging, faeces and footprints). Having said this, there are occasions when knowing some of the sounds made by the species in this group can help confirm presence, especially those which are more secretive, make use of dense habitat and/or are more likely to be active during the hours of darkness. Greater familiarity with the repertoire of sounds produced by these animals can also be useful for providing further information on behavioural and population status.

In the following sections, we will describe the vocalisations produced by each of the seven even-toed ungulate species and highlight any confusion species and how best to differentiate these. Finally, we will review some examples of where bioacoustics may be useful for the monitoring and conservation of even-toed ungulates.

5.2 Overview and comparisons of acoustic behaviour

Broadly speaking, the species in this group will be encountered acoustically through the production of vocalisations that are made audibly (see Table 5.2) and consist of sounds that can carry over a relatively long distance (see Table 5.3).

The deer species can at times be quite vocal, with both sexes producing a range of conspicuous calls, from dog-like barks to high-pitched whistles. The most familiar calls are likely to be loud mating-related rutting sounds (although not produced by solitary species such as roe deer and Reeves' muntjac), notably 'barks', 'roars' or 'whistles', produced by males during the mating season. These calls function in territorial defence, mate attraction/selection and male–male competition. Each of the gregarious deer species (e.g. red, Sika and fallow deer) has a distinctive rutting sound that once learned becomes relatively easy to identify by ear. There is also some separation between deer species distributions (see Table 5.1) and the period of their respective breeding seasons (see Table 5.4), so it is useful to be aware of where and when to expect to encounter specific deer species sounds, when considering identifications.

In addition to rutting sounds, deer produce a variety of social contact calls, mainly between dominants and subordinates and in mother–young interactions (bleats, whistles, grunts/snorts); predator deterrent and alarm calls (barking); territorial calls (barks); and distress calls (squeals), which may be heard throughout the year. These are generally produced irregularly and are more likely to be experienced in chance encounters. Nevertheless, where we have examples, these will be covered within this chapter, albeit less extensively than rutting sounds.

Wild boar tends to be quieter than deer, although in certain circumstances their presence can be established through sound. Like deer, wild boar also rut, vocalising loud grunts and growls as well as audibly champing their jaws, with the height of their seasonal rutting activity usually ranging from November to January. They also have a repertoire of growls, grunts, squeals and snorts that are vocalised singly or in combination

Table 5.1 Even-toed ungulates occurring in Britain and Ireland, including status, distribution and relative abundance. The relevant chapter sections that describe species-specific vocalisations are also listed.

Order	Family	Species common name	Genus	Species scientific name	Status	Distribution range	Abundance within distribution range	Sub-chapter reference
Artiodactyla Even-toed ungulates	Cervidae (Deer)	Roe deer	*Capreolus*	*C. capreolus*	Native	Widely distributed throughout Britain; absent from Ireland	Common	5.3.1
		Red deer	*Cervus*	*C. elaphus*		Widely distributed throughout Britain and Ireland	Common	5.3.2
		Sika deer		*C. nippon*	Non-native	Widely distributed throughout Britain and Ireland	Uncommon	5.3.3
		Fallow deer	*Dama*	*D. dama*		Widely distributed throughout Britain and Ireland	Uncommon	5.3.4
		Reeves' muntjac	*Muntiacus*	*M. reevesi*		Throughout England and Wales with isolated pockets in Scotland; localised in eastern Northern Ireland	Common	5.3.5
		Chinese water deer	*Hydropotes*	*H. inermis*		Bedfordshire, Cambridgeshire, Suffolk and Norfolk	Uncommon	5.3.6
	Suidae (Pigs)	Wild boar	*Sus*	*S. scrofa*	Native Reintroduced	South-west England and South-east England	Localised	5.4.1

(throughout the year), to communicate alarm, aggression, approachability, foraging and social interactions.

Table 5.2 Acoustic spectrum within which you would usually expect to encounter most useful sound for identification purposes within this group

Audible range		Ultrasonic range							
< 10 kHz	10–20 kHz	20–30 kHz	30–40 kHz	40–50 kHz	50–60 kHz	60–70 kHz	70–80 kHz	90–100 kHz	> 100 kHz

Table 5.3 Detectable distance for useful acoustic encounters with even-toed ungulates

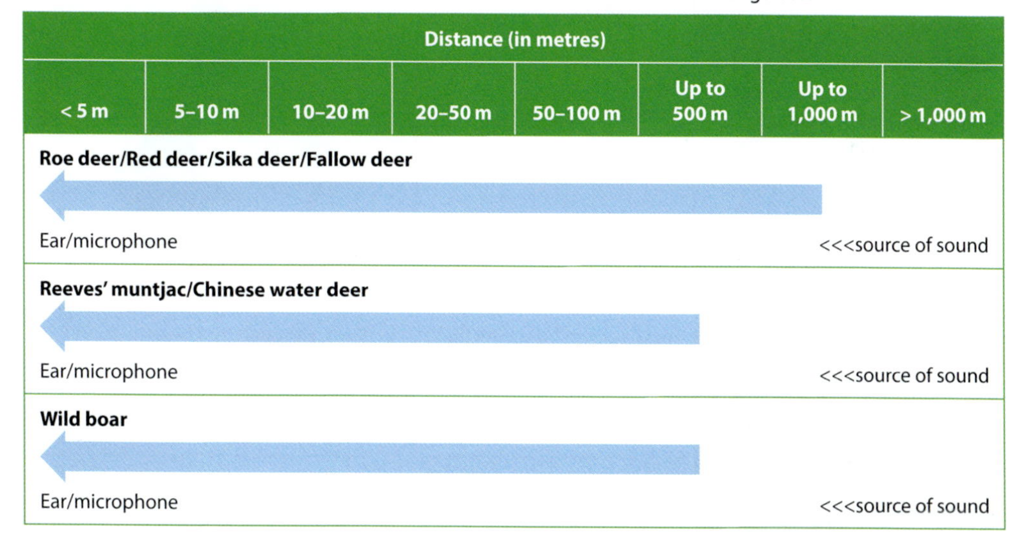

Table 5.4 Typical rutting season (shaded in blue) for ungulate species occurring within Britain and Ireland

Species \ Month	Jan	Feb	Mar	April	May	June	July	Aug	Sept	Oct	Nov	Dec
Roe deer												
Red deer												
Sika deer												
Fallow deer												
Reeves' muntjac												
Chinese water deer												
Wild boar												

5.3 Deer vocalisations

In the following sections, a brief summary of the key characteristics of each of the six deer species will be described, together with details on the species-specific vocalisations and, where known, the behavioural function of these sounds.

5.3.1 Roe deer *Capreolus capreolus*

Length/Weight	**Habitat preferences**	**Distribution**
A small- to medium-sized deer species. Both sexes are relatively similar in size, with a body length of 85 to 135 cm. Weight 16 to 35 kg.	Coniferous woodland Broadleaved woodland Habitat associated with woodland edge (e.g. fields)	Occurs throughout the British Isles, with the exception of Ireland and many offshore islands.
Diet		
Herbivorous, feeding across a vast range of food sources, varying according to season and locality.		

Daily activity pattern	Mostly active from dusk through to dawn, although may be seen during daylight hours in some locations.
Seasonal activity pattern	Active throughout the year, with males (solitary) and females (with or without young) being territorial outside of the winter period.
Mating activity pattern	Mating takes place during July to September, with males pursuing females over a period of a number of days within their territory. Young (usually twins, or less commonly one or three) are born during the period May to June.
Other notes	Most active from dusk through to dawn, and may be more easily encountered visually at dusk or dawn, especially in more open habitat adjacent to woodland. Often heard not seen during hours of darkness. Adult males grow antlers, which are shed annually (October to December).

References
Hewison and Staines, 2008; Mathews *et al.*, 2018; Crawley *et al.*, 2020.

Acoustic behaviour including spectrogram examples

Roe deer emit short, repeated barks when disturbed, either occasionally or irregularly, averaging 14 barks per minute (Hewison and Staines, 2008). Such barks, lasting typically c.0.5 sec, have a frequency range of 200 to 4,000 Hz, and at least to some extent, can be distinguishable between sexes (females are higher-pitched), and where studied intensively, potentially to an individual level (Reby *et al.*, 1998a, 1998b). The FmaxE of the fundamental tends to occur well below 1,000 Hz, with associated harmonics layered above (Figure 5.1). These barks may be singular suspicious behaviour alerts, whilst the deer is being inquisitive or suspecting danger, or a serious of barks in quick succession (Figure 5.2), the first of which, within each series, may be slightly longer (Figure 5.3). Roe deer do not have distinctive rut-related calls; however, barking by territorial males occurs relatively more often during the rutting season (Rossi *et al.*, 2002). The calls (Figures 5.1 to 5.4) could be confused with fox barks or Reeves' muntjac (the latter of which is known to make single dog-like barks that can be repeated a few times or persistently, hundreds of times) (Chapman, 2008).

Contact calls between mother and young are described as squeals, which may also occur when young are alarmed or hungry (Figures 5.5 and 5.6). We have found these calls to be c.0.2 sec in duration, repeated constantly at a rate of one call for every 1–2 seconds. Females may also produce a search squeal when trying to locate youngsters.

Figures 5.1 to 5.6 provide specific examples of spectrograms relating to this species, and where the 🔊 symbol appears, a recording of what is shown within the spectrogram is available to download from the species library. In addition, QR codes are provided for selected examples, and these can be accessed immediately for listening purposes from most mobile devices.

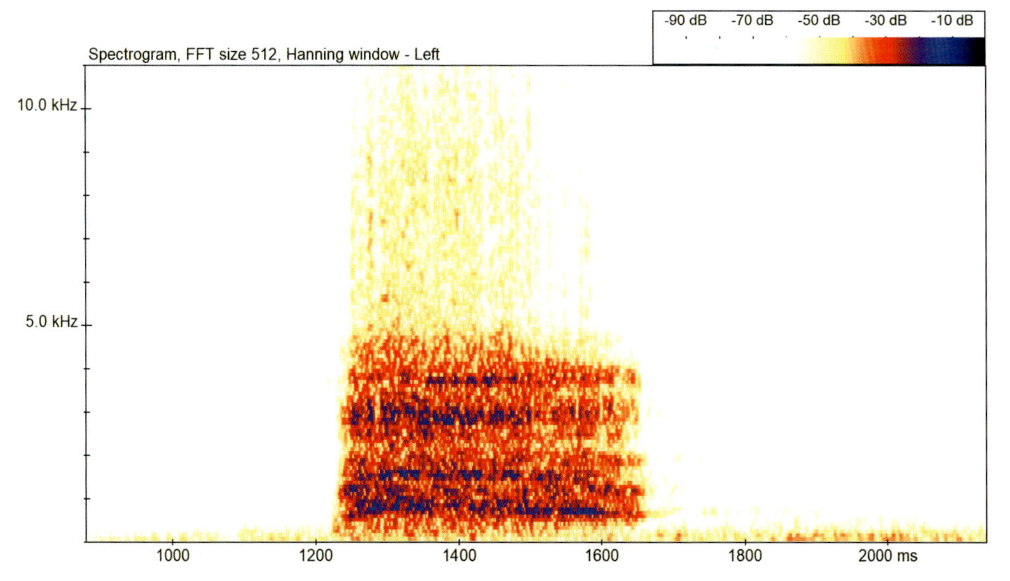

Figure 5.1 🔊 Roe deer – single bark, showing FmaxE of fundamental well below 1 kHz (courtesy of Clyde Muirshiel Regional Park) (frame width: c.1.0 sec)

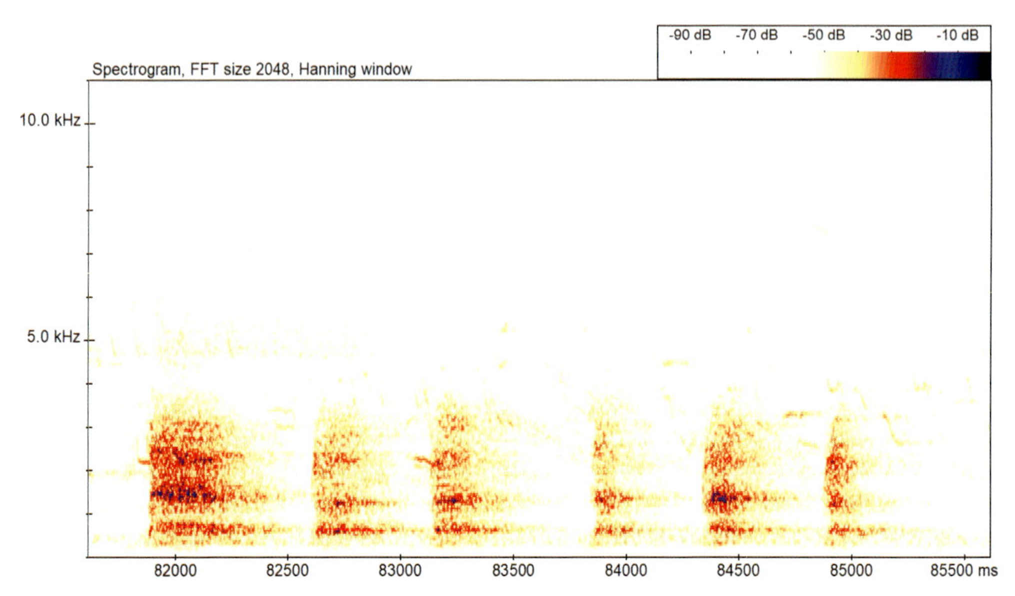

Figure 5.2 🔊 **(QR)** Roe deer – a series of six shortening barks produced by an animal retreating into cover (courtesy of M. Ferguson) (frame width: c.4 sec)

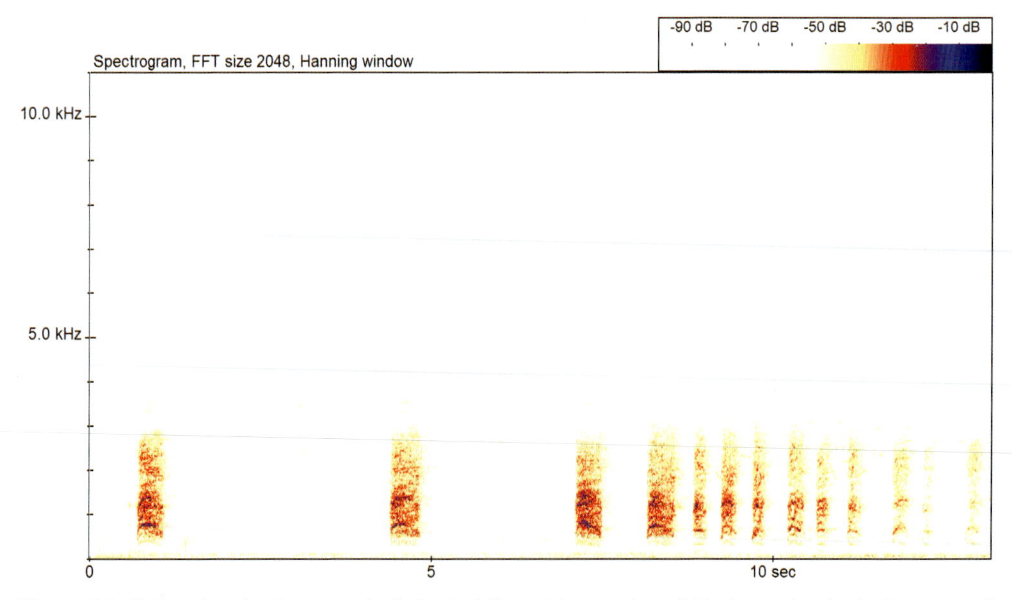

Figure 5.3 🔊 Roe deer buck – two single barks followed by a series of 11 shortening barks (courtesy of S. Elliott) (frame width: c.10 sec)

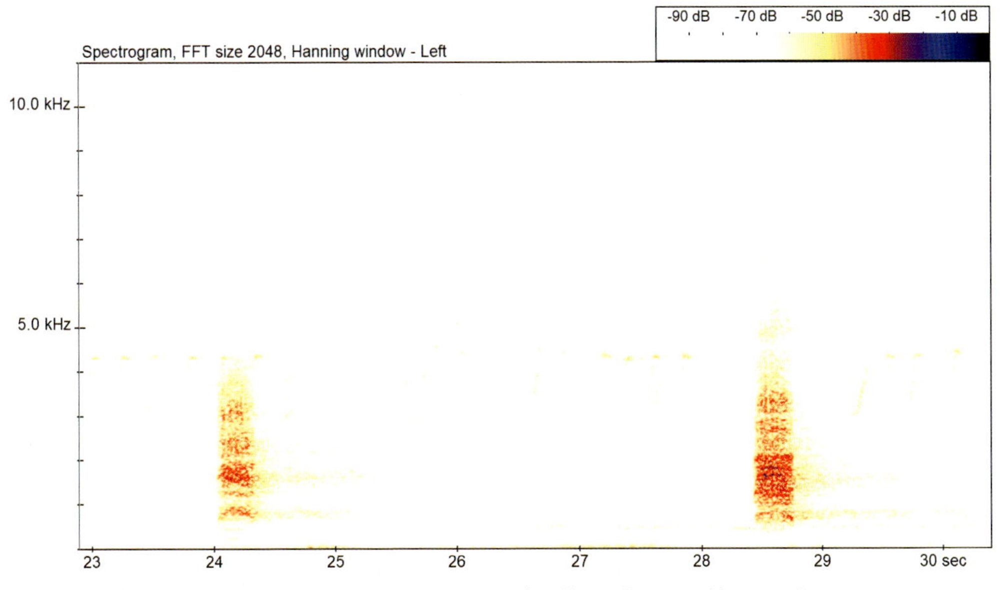

Figure 5.4 🔊 **(QR)** Roe deer doe – two barks (courtesy of S. Elliott) (frame width: c.7 sec)

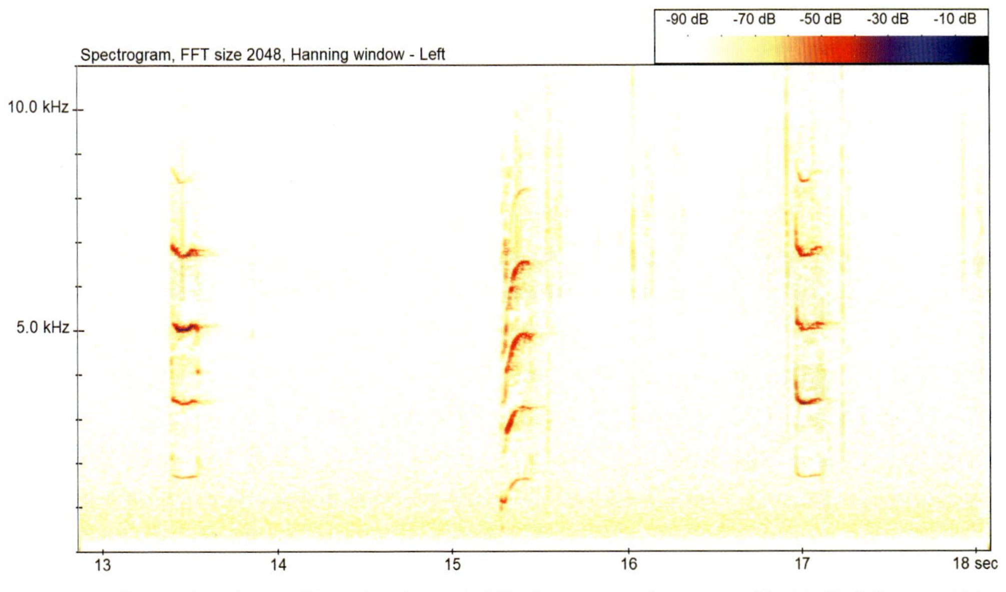

Figure 5.5 🔊 Roe deer fawn calling, showing variability in structure (courtesy of M. Findlay) (frame width: c.5 sec)

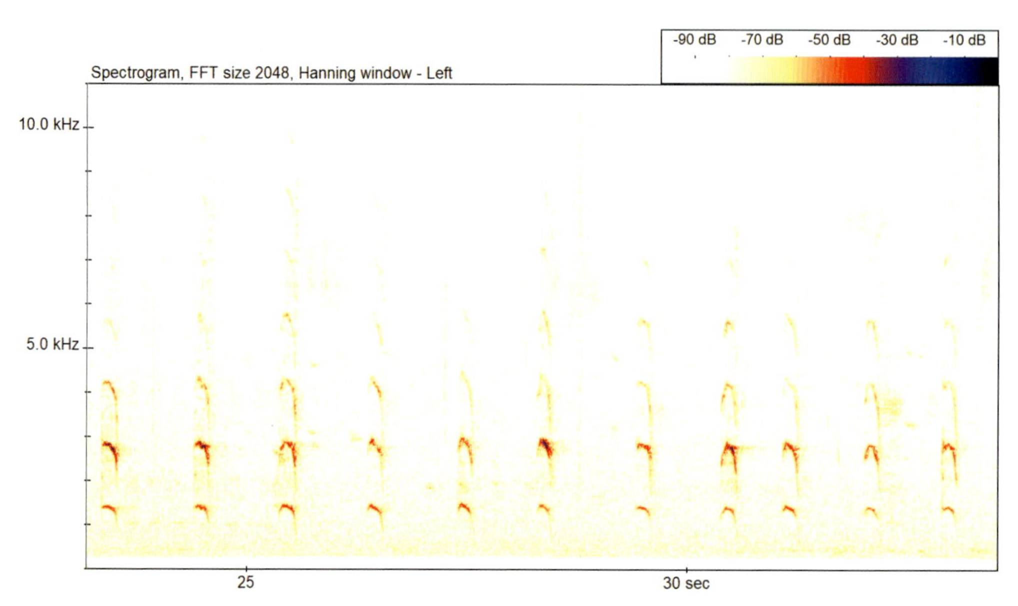

Figure 5.6 🔊 **(QR)** Roe deer fawn calling, showing a similar structure throughout (courtesy of D. Mellor) (frame width: c.10 sec)

QR codes relating to presented figures

Potential confusion between roe deer and other species

Table 5.5 Summary of potential confusion between roe deer and other species

	Potential confusion species
Confusion group	In the first instance, always consider habitat and distribution.
Other deer species	Roe deer barks are very similar to Reeves' muntjac, but roe deer barks are more intensive at the start of bark and tail off towards the end of the call.
Species in other groups	Barks produced by red fox (refer to Chapter 6, Section 6.4.1).

5.3.2 Red deer *Cervus elaphus*

Length/Weight	Habitat preferences	Distribution
Our largest deer species. Males have a body length up to 200 cm and weigh up to 250 kg. Females are smaller with a body length up to 180 cm and weight up to 150 kg. **Diet** Herbivorous feeding across a vast range of food sources, varying according to season and locality.	Coniferous woodland Broadleaved woodland Grassland Moorland Scrubland	Occurs throughout Britain and Ireland, including the Scottish Western Isles.

Daily activity pattern	Typically active during the day and overnight. When present within woodland, may move into more open areas during darkness.
Seasonal activity pattern	Active throughout the year, with male and female (with or without young) groups segregated, other than during the rut.
Mating activity pattern	Mating (rut) takes place during September to November, with males rounding up females into harems which are defended from other males. Young (usually one) are born during the period May to July.
Other notes	Due to large size and herding behaviour, can often be relatively easily seen, and at a great distance when present in more open areas of habitat (e.g. moorland). Adult males grow antlers, which are shed annually (March to April).

References
Staines *et al.*, 2008; Lysaght and Marnell, 2016; Mathews *et al.*, 2018; Crawley *et al.*, 2020.

Acoustic behaviour including spectrogram examples

Males make distinctive roaring sounds during rutting behaviour (Figures 5.7 and 5.8), occasionally in conjunction with grunting and moaning (Figures 5.9 and 5.10). Once learned, it is difficult to confuse the rutting calls of this species with anything else. These rutting calls typically drop in frequency (as easily seen in Figures 5.7 and 5.8), and are loud and repeated, with the roaring rate variable between individuals. To an extent these calls are distinctive to an individual level (Reby *et al.*, 2001). Quieter grunts can be produced by either sex when approaching each other.

Similar to other deer species, alarm calls (barks) may be emitted by either sex when disturbed (Figures 5.11 and 5.12) and these are the most likely calls where some confusion with other deer species may occur.

Females will make a 'low mooing sound' (Staines *et al.*, 2008) when searching for their young, and young may make 'bleating' contact calls (Macdonald and Barrett, 1995). Other contact sounds have also been noted (e.g. Figures 5.13 and 5.14) and likely function to assist with herd cohesion.

Figures 5.7 to 5.14 provide specific examples of spectrograms relating to this species, and where the 🔊 symbol appears, a recording of what is shown within the spectrogram is available to download from the species library. In addition, QR codes are provided for selected examples, and these can be accessed immediately for listening purposes from most mobile devices.

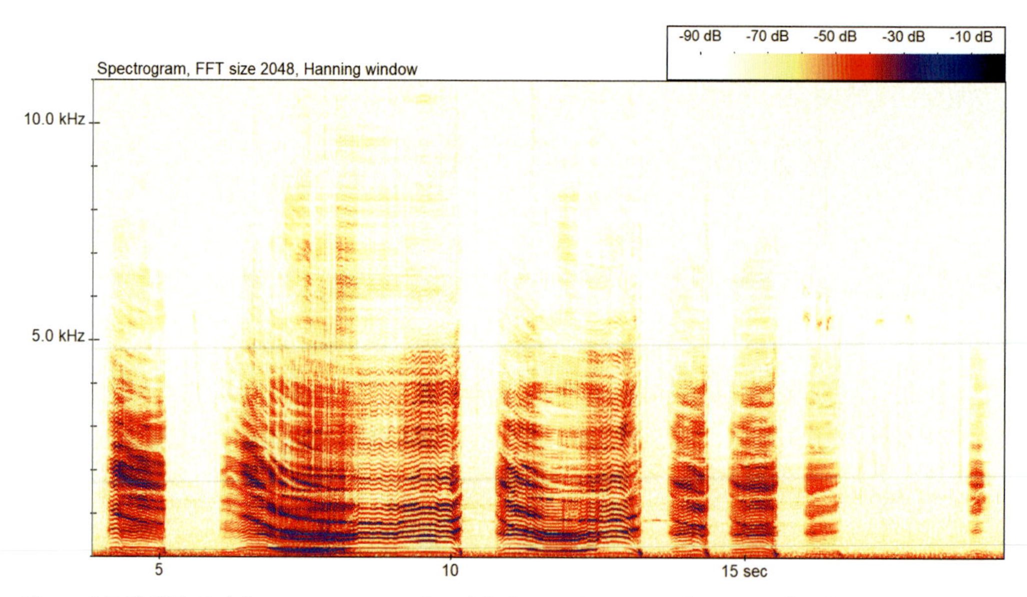

Figure 5.7 🔊 (QR) Red deer – stag roar produced during rutting season (courtesy of S. Elliott) (frame width: c.15 sec)

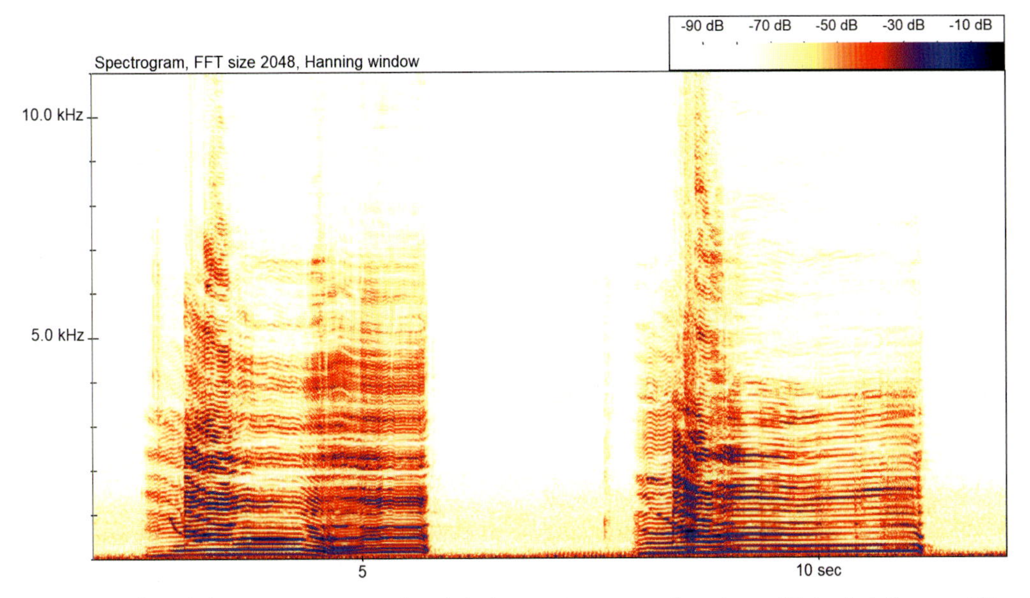

Figure 5.8 🔊 Red deer – stag roar produced during rutting season (courtesy of S. Senior) (frame width: c.10 sec)

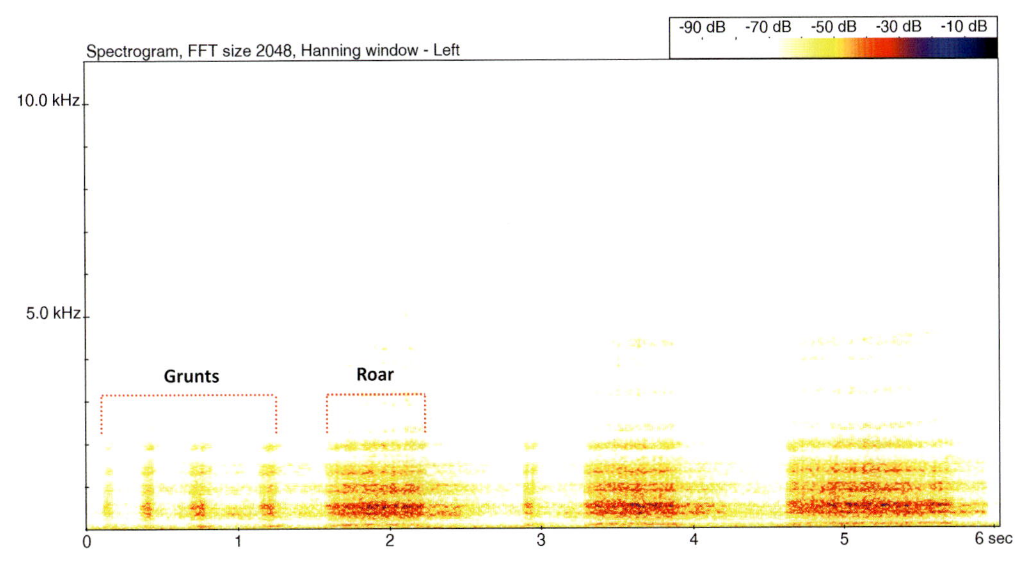

Figure 5.9 🔊 **(QR)** Red deer – grunts and roars (courtesy of P. Mill) (frame width: c.6 sec)

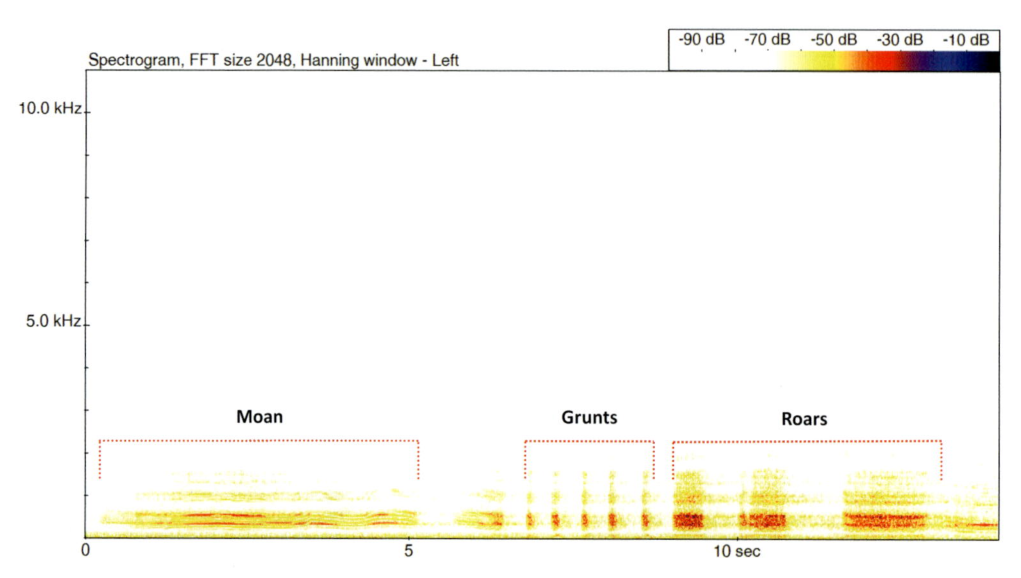

Figure 5.10 🔊 Red deer – moaning, grunts and roars (courtesy of P. Mill) (frame width: c.14 sec)

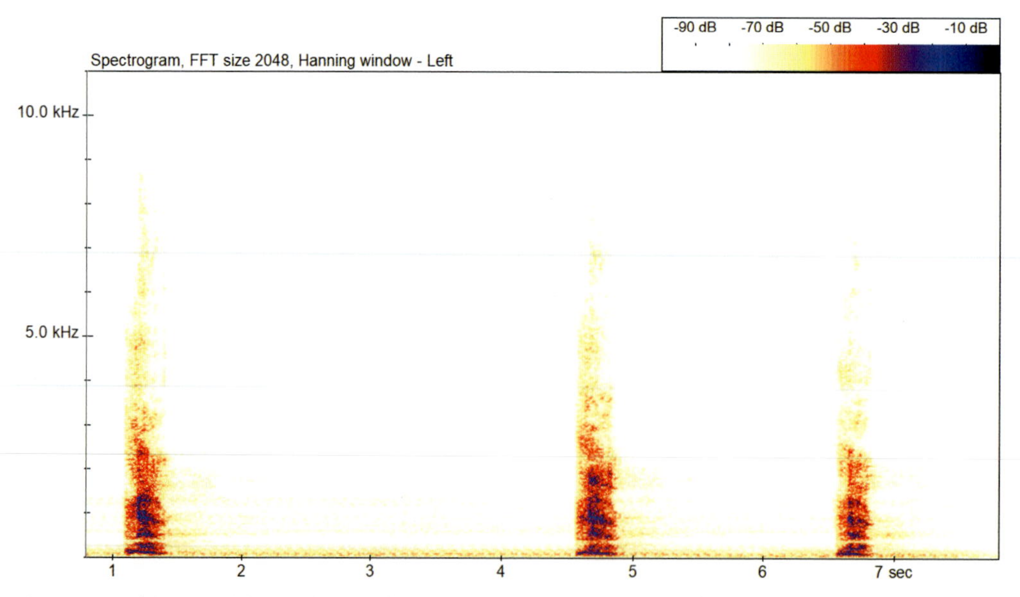

Figure 5.11 🔊 **(QR)** Red deer – alarm barks produced by female (courtsey of S. Elliott) (frame width: c.8 sec)

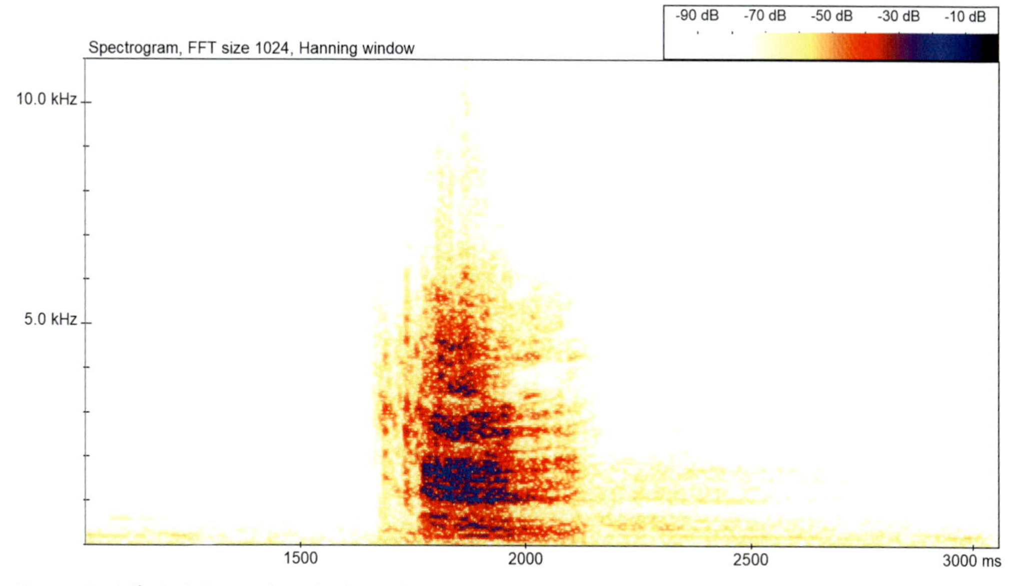

Figure 5.12 🔊 Red deer – alarm bark produced by male (courtesy of S. Elliott) (frame width: c.2.3 sec)

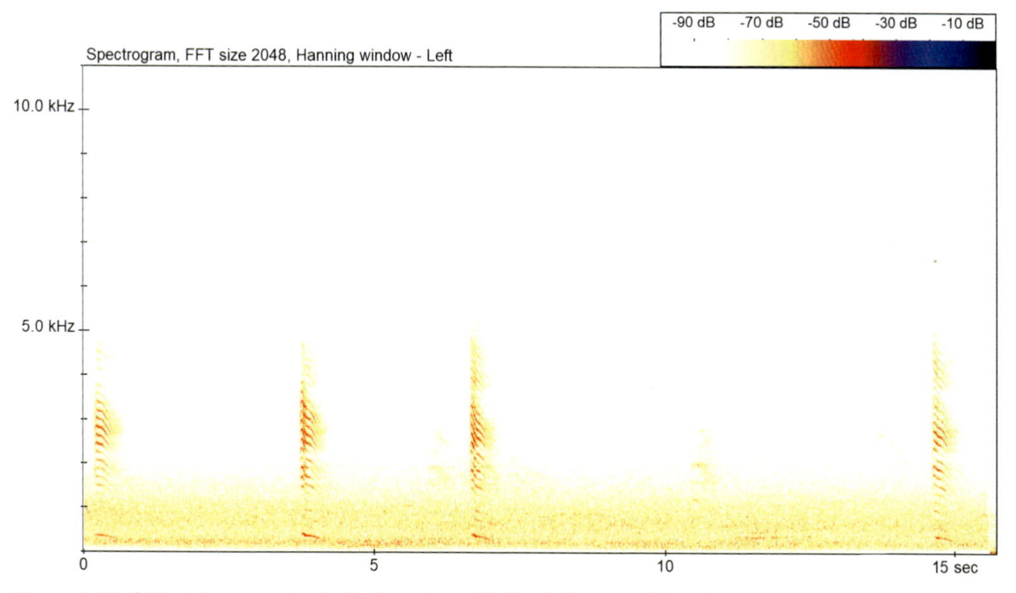

Figure 5.13 🔊 Red deer – contact calls (courtesy of A. Fure) (frame width: c.15 sec)

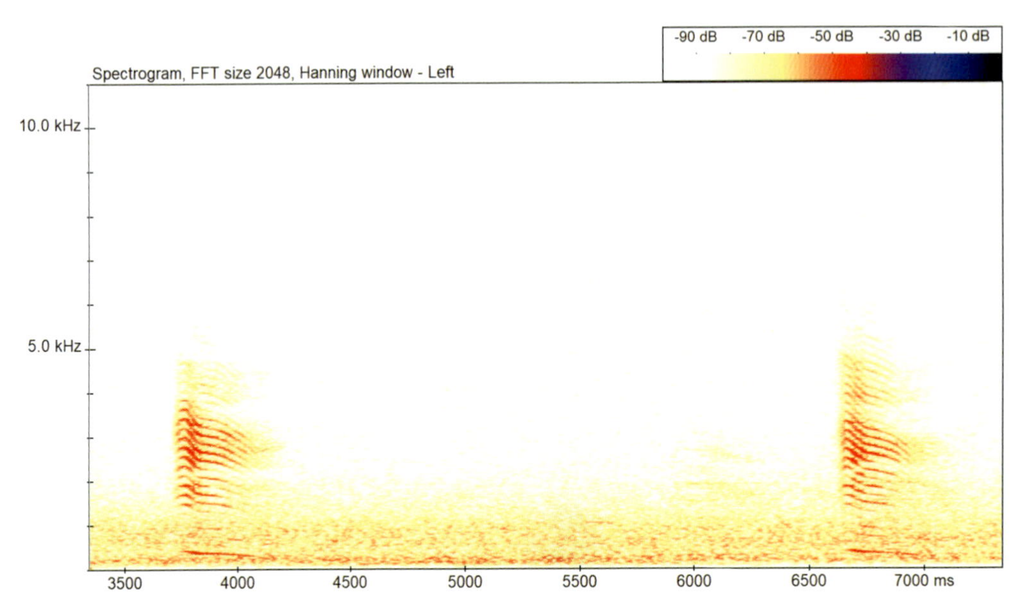

Figure 5.14 🔊 Red deer – single contact call (courtesy of A. Fure) (frame width: c.4 sec)

QR codes relating to presented figures

Figure 5.7: Red deer roar	**Figure 5.9:** Red deer groan	**Figure 5.11:** Red deer bark

Potential confusion between red deer and other species

Table 5.6 Summary of potential confusion between red deer and other species

	Potential confusion species
Confusion group	In the first instance, always consider habitat and distribution.
Other deer species	Short 'bark' alarm calls are most likely to be confused with the barks/belch calls of fallow deer.
Species in other groups	Barks produced by red fox (refer to Chapter 6, Section 6.4.1).

5.3.3 Sika deer *Cervus nippon*

Length/Weight	Habitat preferences	Distribution
A medium-sized deer species. The sexes are similar in size (males slightly bigger), with a body length up to c.145 cm. Males weigh more (up to 64 kg) than females (up to 44 kg).	Coniferous woodland Broadleaved woodland Scrubland	Occurs throughout large parts of mainland Scotland, with distribution more localised and patchy in England, Wales and Ireland.
Diet		
Herbivorous feeding across a vast range of food sources, varying according to season and locality.		

Daily activity pattern	Typically active during the day and overnight.
Seasonal activity pattern	Active throughout the year, with males and females (with or without young) forming segregated groups over winter, but more solitary at other times, with exception of the rutting season.
Mating activity pattern	Mating (rut) takes place during September to November, with males adopting different strategies. Within the defence of a rutting territory, they may form and defend a harem against other males, or they may patrol an area in search of individual females, and pursue females over several days, within their territory. Young (usually one) are born during the period May to July.
Other notes	Can be active throughout the 24-hour period, but more likely to be secretive in areas where human disturbance is a factor. As such may be more easily encountered visually at dusk or dawn. Adult males grow antlers, which are shed annually (March to May).

References
Putman, 2008; Lysaght and Marnell, 2016; Mathews *et al.*, 2018; Crawley *et al.*, 2020.

Acoustic behaviour including spectrogram examples

During the breeding season, males produce 'high-pitched whistles' (Macdonald and Barrett, 1995) that are most often emitted in a series of three (Figure 5.15), followed by a pause, before being repeated. However, a series of more whistles can occur (e.g. four or five) (Figures 5.16 and 5.17). A low-frequency moan may also be given in conjunction with such behaviour, as heard when listening to the full sound file relating to Figure 5.18. These rutting whistles may become longer and more scream-like later in the season, with Figure 5.19 providing an excellent example.

Females will make grunting noises, and contact calls (bleats) are also made between females and their young.

Distress calls, made by both males and females, have been described as a brief high-pitched squeal, which may be repeated a number of times, and may be followed by a bark before departing. Barking in isolation may also occur. In addition, further calls have been described between adults interacting with each other (Putman, 2008).

Figures 5.15 to 5.19 provide specific examples of spectrograms relating to this species, and where the 🔊 symbol appears, a recording of what is shown within the spectrogram is available to download from the species library. In addition, QR codes are provided for selected examples, and these can be accessed immediately for listening purposes from most mobile devices.

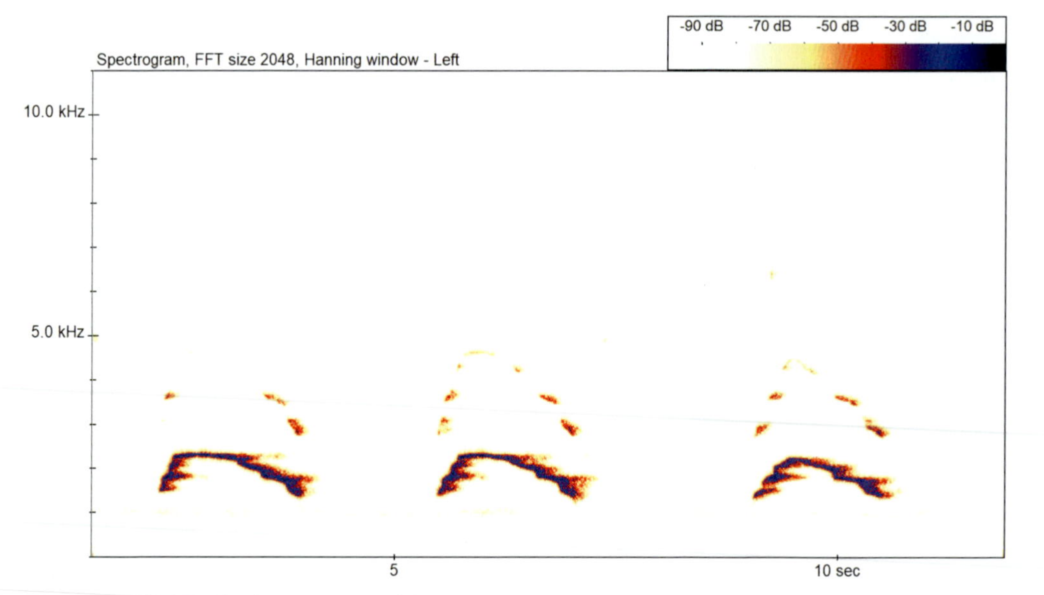

Figure 5.15 🔊 (QR) Sika deer – a series of three whistles produced by a stag during rutting season (courtesy of N. Hull) (frame width: c.10 sec)

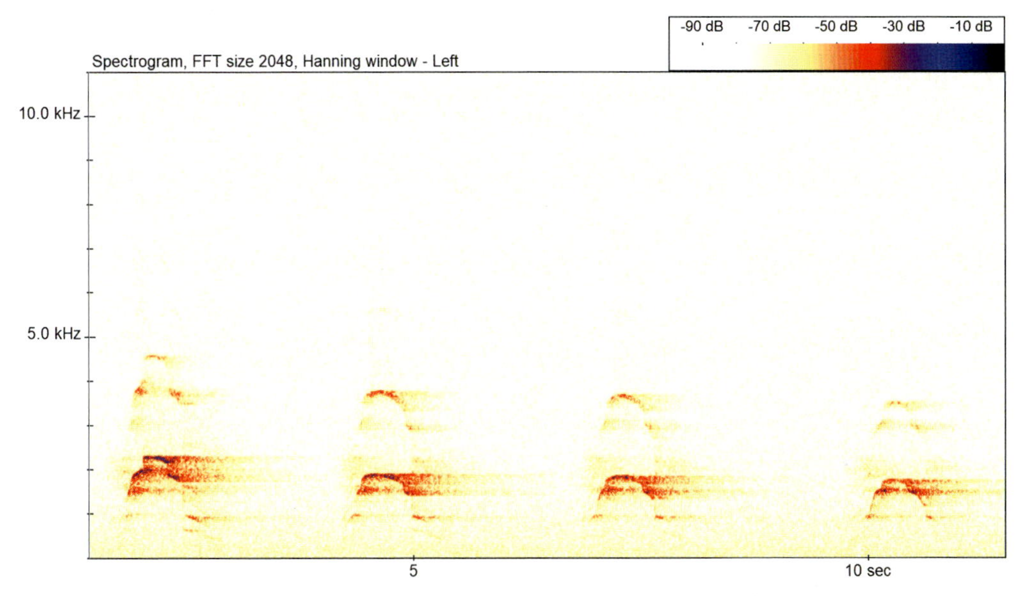

Figure 5.16 🔊 Sika deer – a series of four whistles produced by a stag during rutting season (courtesy of A. Perez Grandi) (frame width: c.10 sec)

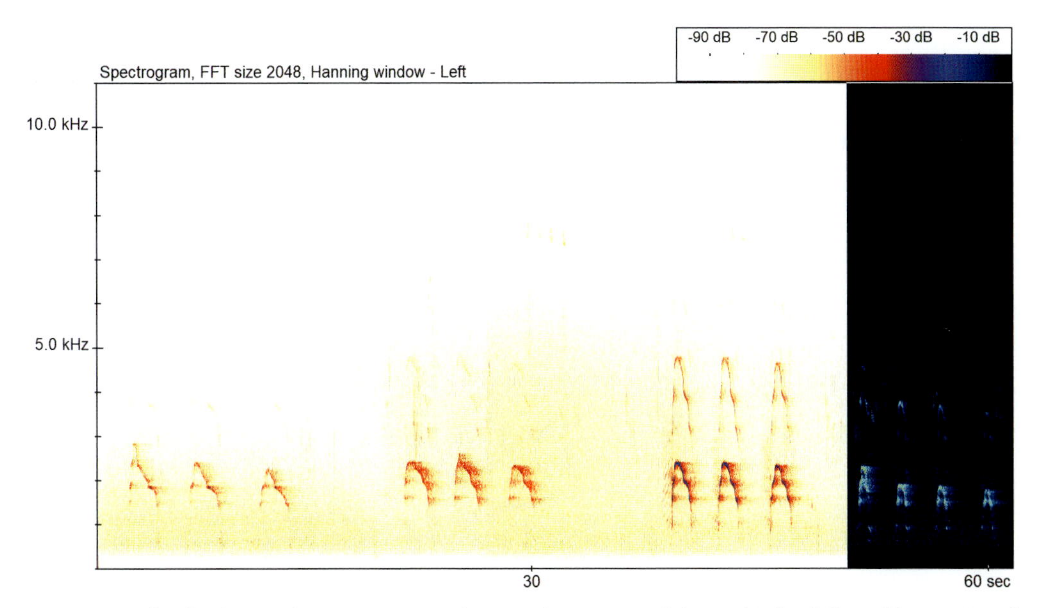

Figure 5.17 🔊 Sika deer – a longer sequence showing three series of three whistles, followed by a series of four whistles (highlighted area) (courtesy of A. Perez Grandi) (frame width: c.65 sec)

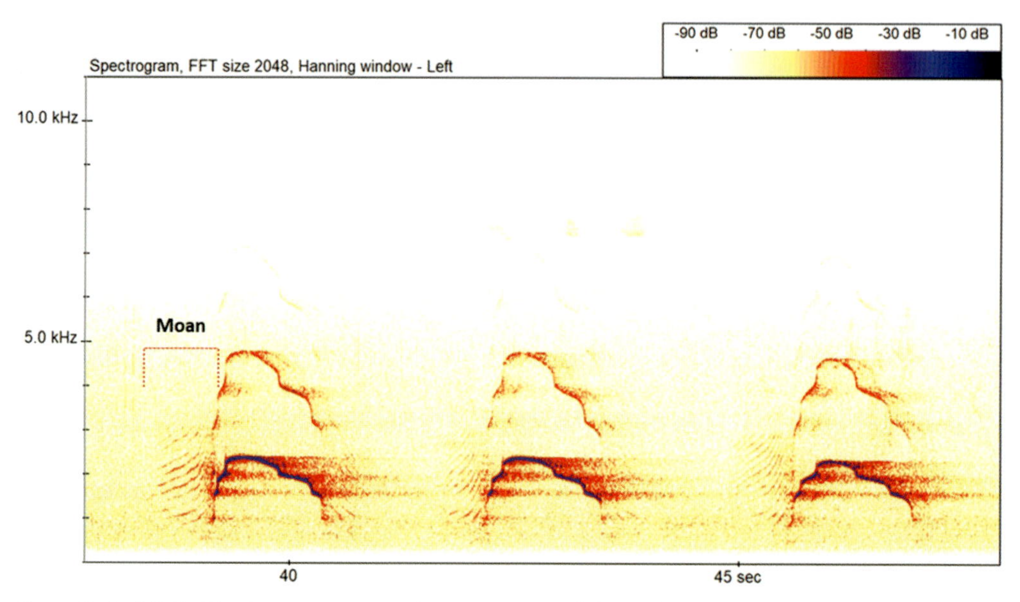

Figure 5.18 🔊 **(QR)** Sika deer – a sequence showing a series of three whistles, each of which are immediately preceded by a brief moan (courtesy of A. Perez Grandi) (frame width: c.10 sec)

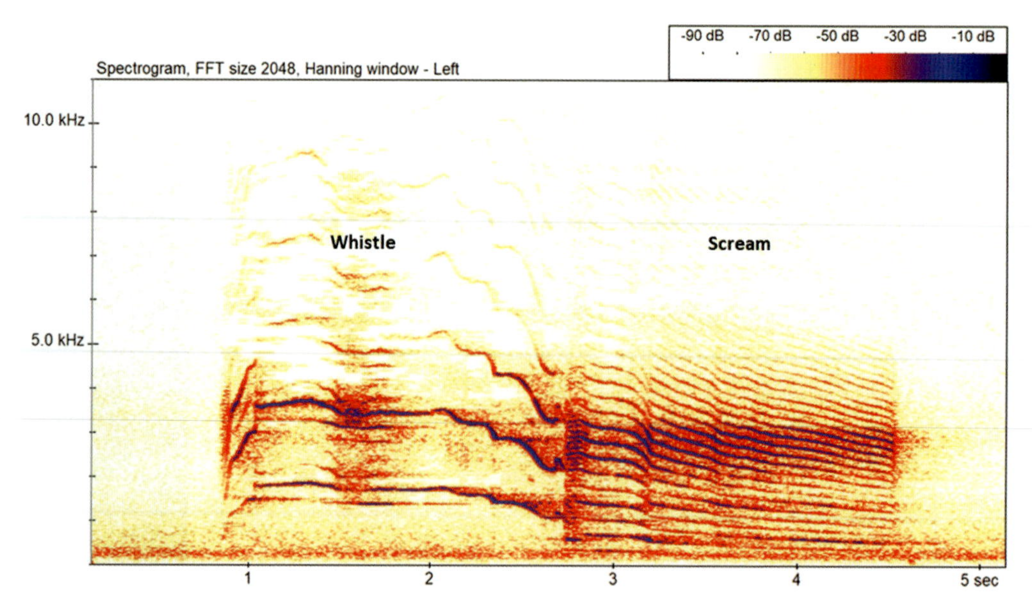

Figure 5.19 🔊 Sika deer – longer rutting call, with a whistle leading immediately on to a scream (courtesy of M. Baldwin, wildlifeonline.me.uk) (frame width: c.5 sec)

QR codes relating to presented figures

Figure 5.15: Sika deer whistle

Figure 5.18: Sika deer moan-whistle

Potential confusion between sika deer and other species

Table 5.7 Summary of potential confusion between sika deer and other species

	Potential confusion species
Confusion group	In the first instance, always consider habitat and distribution.
Other deer species	The bark calls of other deer species may cause confusion.
Species in other groups	Barks produced by red fox (refer to Chapter 6, Section 6.4.1).

5.3.4 Fallow deer *Dama dama*

<table>
<tr><th>Length/Weight</th><th>Habitat preferences</th><th>Distribution</th></tr>
<tr><td>A medium-sized deer species.
Males have a body length of 150 to 180 cm, and are larger than females.
Males weigh 55 to 70 kg.
Females weigh 35 to 50 kg.</td><td rowspan="3">Broadleaved woodland
Mixed woodland
Open coniferous woodland
Habitat associated with woodland edge (e.g. grassland and arable fields)</td><td rowspan="3">Occurs throughout Ireland, England and Wales, with patchy distribution within Scotland (e.g. Central Belt, Dumfries and Galloway, and Isle of Mull).</td></tr>
<tr><th>Diet</th></tr>
<tr><td>Herbivorous feeding across a vast range of food sources, varying according to season and locality.</td></tr>
<tr><th>Daily activity pattern</th><td colspan="2">Typically active from dusk through to dawn, unless in areas where disturbance is less likely.</td></tr>
<tr><th>Seasonal activity pattern</th><td colspan="2">Active throughout the year, with males and females (with or without young) being segregated into separate herds.</td></tr>
<tr><th>Mating activity pattern</th><td colspan="2">Males will begin to move into female territories from September, with rut taking place during October to November, with males competing for females within a defended area, or congregating at a lek in order to gather females.
Young (usually one) are born during the period June to July.</td></tr>
<tr><th>Other notes</th><td colspan="2">More active from dusk through to dawn, and may be more easily encountered visually at dusk or dawn, especially in more open habitat adjacent to woodland.
Often heard not seen during hours of darkness.
Adult males grow palmate antlers, which are shed annually (April to May). No other deer species in the area covered by this book have palmate antlers.</td></tr>
<tr><td colspan="3">References
Langbein et al., 2008; Lysaght and Marnell, 2016; Mathews et al., 2018; Crawley et al., 2020.</td></tr>
</table>

Acoustic behaviour including spectrogram examples

Males make a belch-like groaning sound during rutting activity (Figures 5.20 and 5.21). These have been described in different ways by researchers, for example 'common groans' and 'harsh groans', and, at least to some extent, have been shown to be individually distinctive (Vannoni and McElligott, 2007 and 2008).

Mauri *et al.* (1994) found that the most commonly produced vocalisations during the rutting season were, understandably, male 'roars' (i.e. the belch-like groaning) and female submissive calls. The 'roars' were described as single events, each lasting c.0.50 sec within the frequency range of 100 to 8,000 Hz. A shorter roar, at c.0.35 sec, was also described. The female submissive calls were noted as being continuous, within the frequency range of 300 to 500 Hz, each lasting between 0.12 and 0.70 sec. Contact calls between mothers and fawns were also mentioned, stating that these ranged from 1 to 6 kHz, lasting less than 0.50 sec.

Females and their young are known to produce 'meeping' or 'chirping' contact calls (Figure 5.22), which may be heard while on the move (Baldwin, 2022).

Both sexes produce an alarm bark (Figures 5.23 and 5.24), which may be emitted repeatedly (Macdonald and Barrett, 1995; Langbein *et al.*, 2008).

Figures 5.20 to 5.24 provide specific examples of spectrograms relating to this species, and where the 🔊 symbol appears, a recording of what is shown within the spectrogram is available to download from the species library. In addition, QR codes are provided for selected examples, and these can be accessed immediately for listening purposes from most mobile devices.

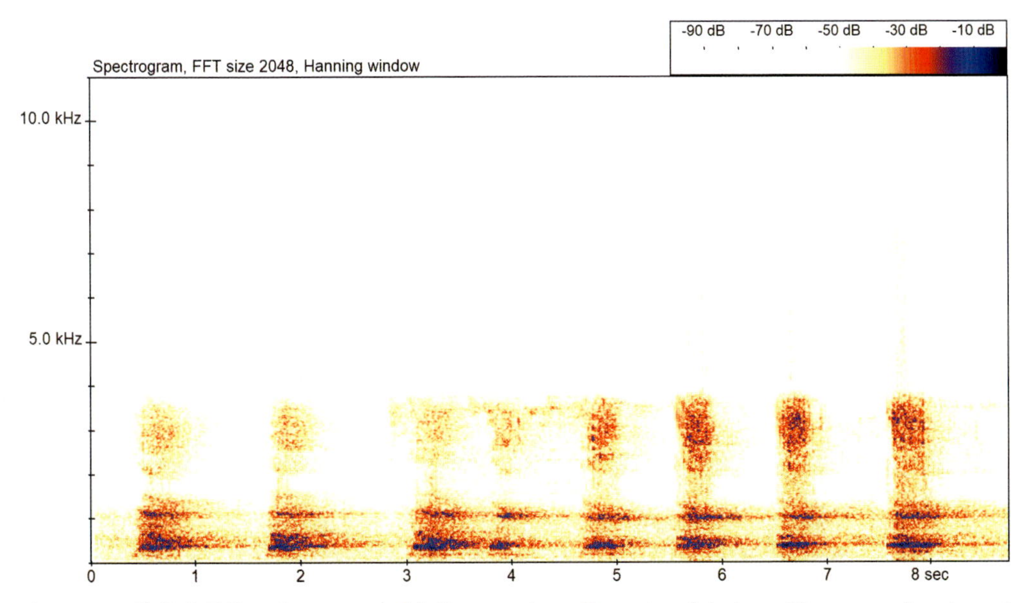

Figure 5.20 🔊 (QR) Fallow deer – stag belch-like groaning calls produced during rutting season (courtesy of M. Baldwin, wildlifeonline.me.uk) (frame width: c.9 sec)

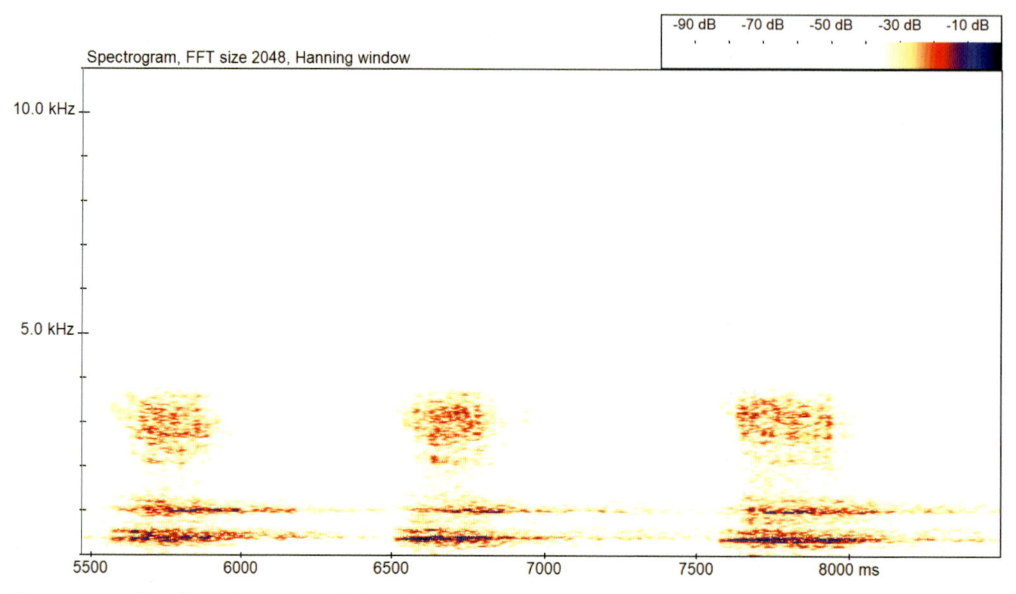

Figure 5.21 🔊 Fallow deer – stag belch-like groaning calls produced during rutting season (courtesy of M. Baldwin, wildlifeonline.me.uk) (frame width: c.3 sec)

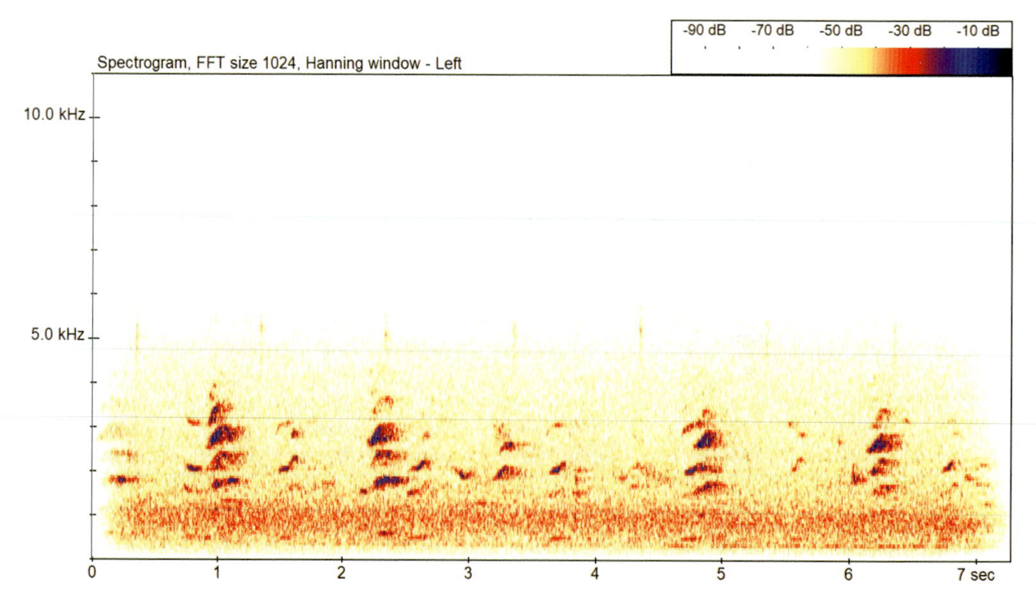

Figure 5.22 🔊 **(QR)** Fallow deer – contact calls produced by females with young (courtesy of M. Baldwin, wildlifeonline.me.uk) (frame width: c.7 sec)

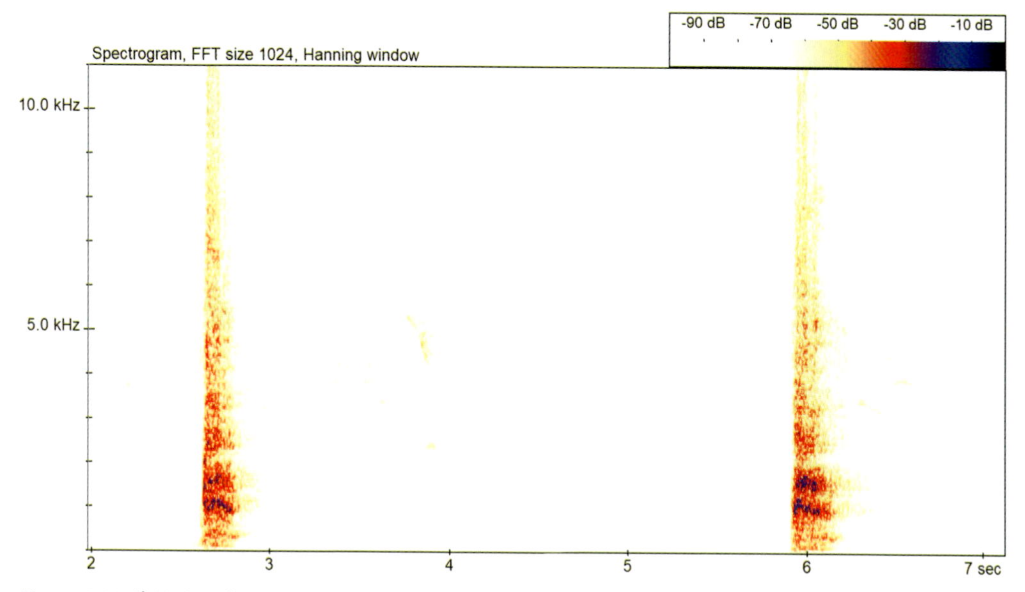

Figure 5.23 🔊 **(QR)** Fallow deer – bark (courtesy of D. Bagur) (frame width: c.5 sec)

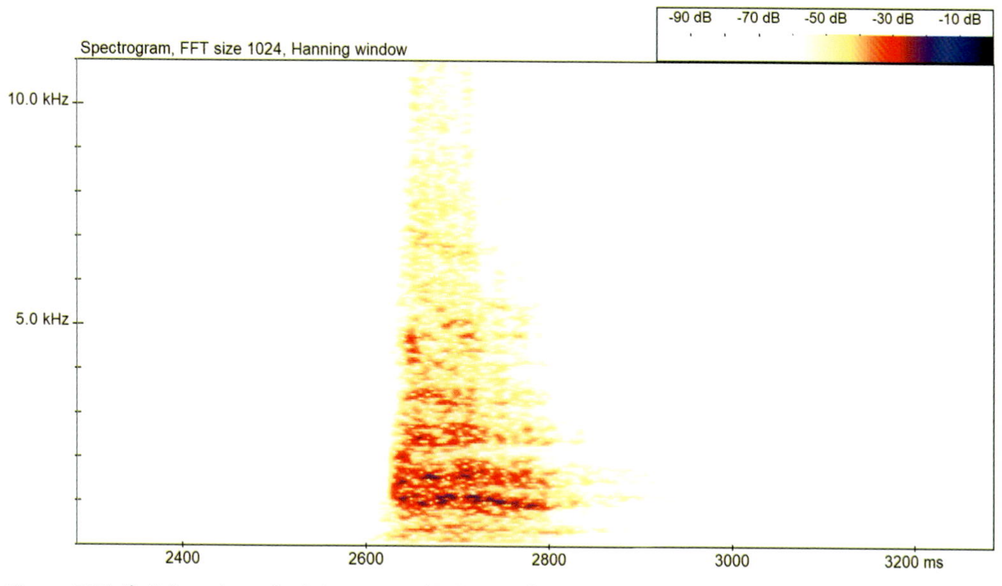

Figure 5.24 🔊 Fallow deer – bark (courtesy of D. Bagur) (frame width: c.5 sec)

QR codes relating to presented figures

Figure 5.20: Fallow deer groan	**Figure 5.22:** Fallow deer contact	**Figure 5.23:** Fallow deer bark

Potential confusion between fallow deer and other species

Table 5.8 Summary of potential confusion between fallow deer and other species

Confusion group	Potential confusion species
	In the first instance, always consider habitat and distribution.
Other deer species	The bark calls of fallow deer are most likely to be confused with short bark calls of red deer.
Species in other groups	Barks produced by red fox (refer to Chapter 6, Section 6.4.1).

5.3.5 Reeves' muntjac *Muntiacus reevesi*

Length/Weight	Habitat preferences	Distribution
Our smallest deer species, often portraying a humped-back appearance. Males and females are similar in size, with a body length of 90 to 100 cm. Weight ranges from 10 to 17 kg (males being heavier than females).	Dense broadleaved woodland Dense mixed woodland Scrubland Overgrown gardens and other dense urban habitat	Occurs throughout England and Wales, with occasional records from Scotland and Ireland.
Diet		
Herbivorous feeding across a vast range of food sources (e.g. shrubs, fruits), varying according to season and locality.		

Daily activity pattern	Typically most active during dusk and dawn, but can also be active during daytime and overnight.
Seasonal activity pattern	Active throughout the year, and usually solitary, unless female with young.
Mating activity pattern	Males are territorial, potentially mating with any female with an overlapping range. Breeding may occur at any time during the year, meaning that young (usually one) can be born at any time throughout.
Other notes	Its small size and preference for denser habitat can make it harder to encounter visually. Can be active during the day, but more so at dusk and dawn, and may be more easily encountered visually in more open habitat within or adjacent to woodland. Often heard not seen. Adult males grow antlers, which are shed annually (May to July), and their upper canine teeth grow to form tusks.

References
Chapman, 2008; Lysaght and Marnell, 2016; Mathews *et al.*, 2018; Crawley *et al.*, 2020; Mammal Society, 2022.

Acoustic behaviour including spectrogram examples

Males produce a single dog-bark-like sound (duration: c.0.5 sec, with an FmaxE of c.0.7 to 1.4 kHz) that may be repeated a few times or persistently, hundreds of times (Chapman, 2008). Barking by both sexes (Figures 5.25 and 5.26) occurs when alarmed (Yahner, 1980), as can clicking sounds, as well as squeaks and/or screams (Figure 5.28) under similar circumstances (in distress) (Macdonald and Barrett, 1995; Chapman, 2008).

Reeves' muntjac do not have distinctive rut-related calls; however, females, in season, may produce a 'scream-bark' (Figure 5.27), it is thought in order to attract males (Baldwin, 2022). During courtship, males have been heard to produce a short buzzing sound when approaching a female, and females have been heard to produce a whining noise (Yahner, 1980).

Figures 5.25 to 5.28 provide specific examples of spectrograms relating to this species, and where the 🔊 symbol appears, a recording of what is shown within the spectrogram is available to download from the species library. In addition, QR codes are provided for selected examples, and these can be accessed immediately for listening purposes from most mobile devices.

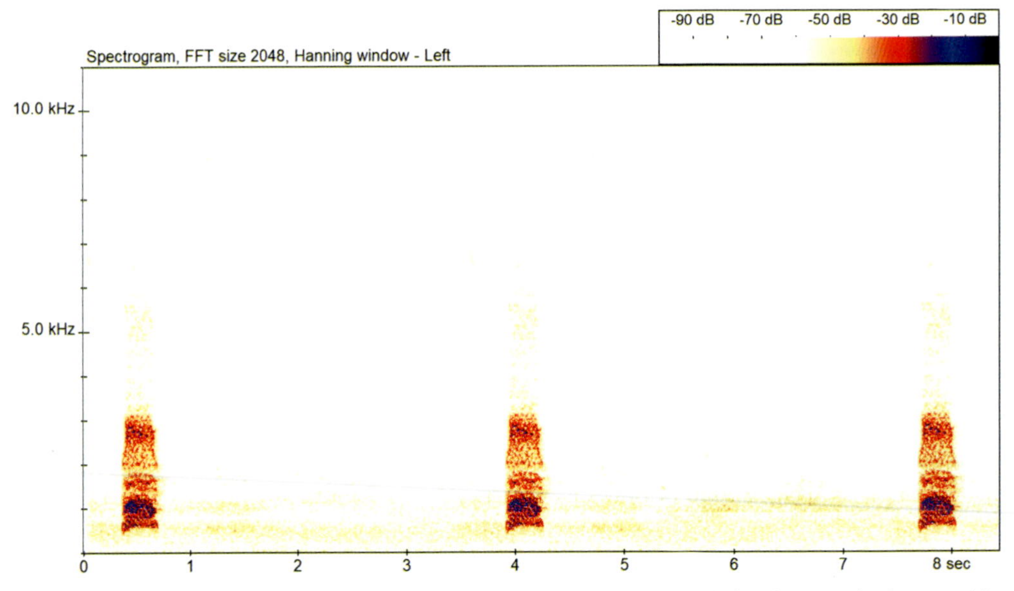

Figure 5.25 🔊 (QR) Reeves' muntjac – barks (x3) (courtesy of M. Baldwin, wildlifeonline.me.uk) (frame width: c.8 sec)

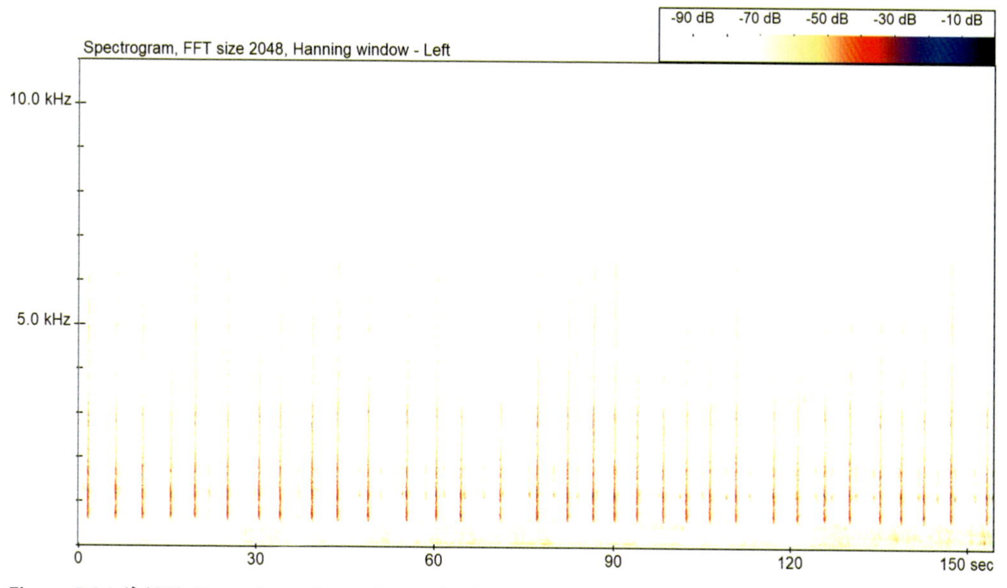

Figure 5.26 🔊 (QR) Reeves' muntjac – a longer barking sequence (courtesy of P. Mill) (frame width: c.150 sec)

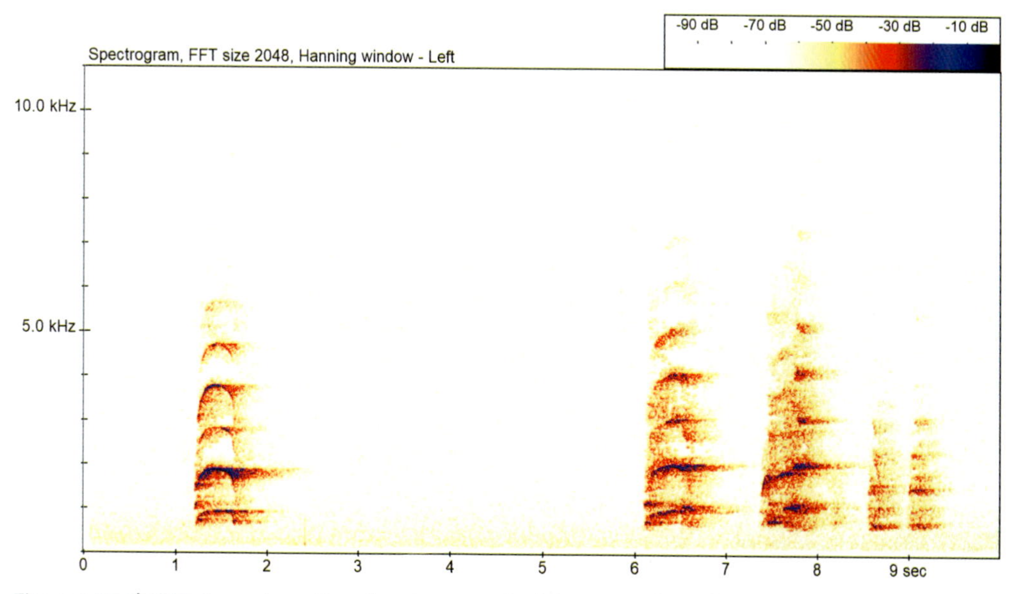

Figure 5.27 🔊 (QR) Reeves' muntjac – female scream-bark (courtesy of K. Kilfeather) (frame width: c.10 sec)

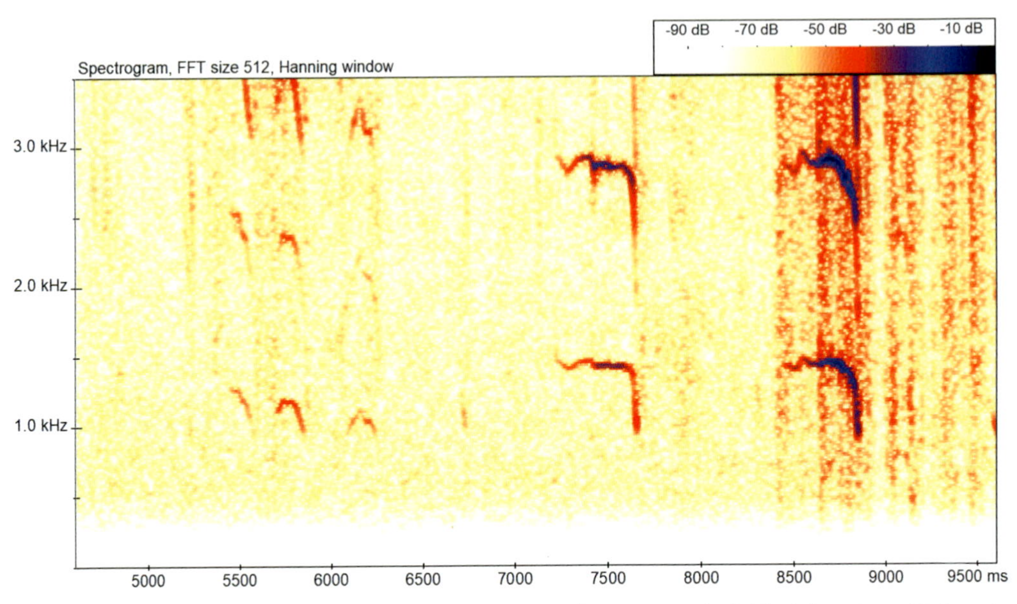

Figure 5.28 Reeves' muntjac – higher-frequency interaction between two individuals (courtesy of S. Green) (frequency scale: 0–3.5 kHz/frame width: c.5 sec)

QR codes relating to presented figures

Potential confusion between Reeves' muntjac and other species

Table 5.9 Summary of potential confusion between Reeves' muntjac and other species

Confusion group	Potential confusion species
	In the first instance, always consider habitat and distribution.
Other deer species	Reeves' muntjac deer barks can be confused with roe deer and Chinese water deer. Reeves' muntjac barks are comparatively more constant in intensity and end abruptly, when compared to roe deer. Chinese water deer barking is more 'cat- or whelp-like' rather than the more 'dog-like' muntjac.
Species in other groups	Barks produced by red fox (refer to Chapter 6, Section 6.4.1).

5.3.6 Chinese water deer *Hydropotes inermis*

Length/Weight	Habitat preferences	Distribution
A small deer species. Males and females have similar body length (80 to 105 cm). Males weigh 12 to 18.5 kg, with female weight typically ranging from 14 to 17.5 kg.	Wetlands (e.g. reedbeds, marshland) adjoining arable, fen, broadleaved woodland and unimproved grassland River banks Coastal areas	Only occurs in suitable habitat within the south-east of England, especially within Norfolk, Cambridgeshire and Bedfordshire.
Diet		
Herbivorous, foraging upon a vast range of food sources, varying according to season and locality.		

Daily activity pattern	Typically active from dusk through to dawn.
Seasonal activity pattern	Active throughout the year, and often solitary, unless female with young.
Mating activity pattern	Territorial and either solitary or in a family group comprising female with young. Rut occurs mainly during December (but can last well into new year), with males fighting over access to females. Fawns (two to five per mother) are born during May to July.
Other notes	More active at dusk and dawn, and may be more easily encountered visually at these times, especially in more open habitat associated with wetlands. Usually solitary or family group, unless in particularly good feeding areas, and when food elsewhere is less plentiful. Adult males do not have antlers, but upper canines can grow to form conspicuous tusks (up to 7 cm).

References
Cooke and Farrell, 2008; Mathews *et al.*, 2018; Crawley *et al.*, 2020; Mammal Society, 2022.

Acoustic behaviour including spectrogram examples

Males make whistling and squeaking calls during the rut (Figure 5.29 and Figure 5.31), as well as a 'mechanical sound described as clicking, whittering or chittering' (Cooke and Farrell, 2008) which can be heard during chase sequences (Figures 5.30 and 5.31). 'Snuffling' sounds have also been recorded (Figure 5.32).

When alarmed, a 'fox-like bark, often aimed at the source of the disturbance' is produced (Figures 5.33 and 5.34), and occasionally this may sound more 'whelp-like' (Figure 5.35) (Baldwin, 2022).

Females can produce squeaks and whistling sounds when associating with their fawn.

Figures 5.29 to 5.35 provide specific examples of spectrograms relating to this species, and where the 🔊 symbol appears, a recording of what is shown within the spectrogram is available to download from the species library. In addition, QR codes are provided for selected examples, and these can be accessed immediately for listening purposes from most mobile devices.

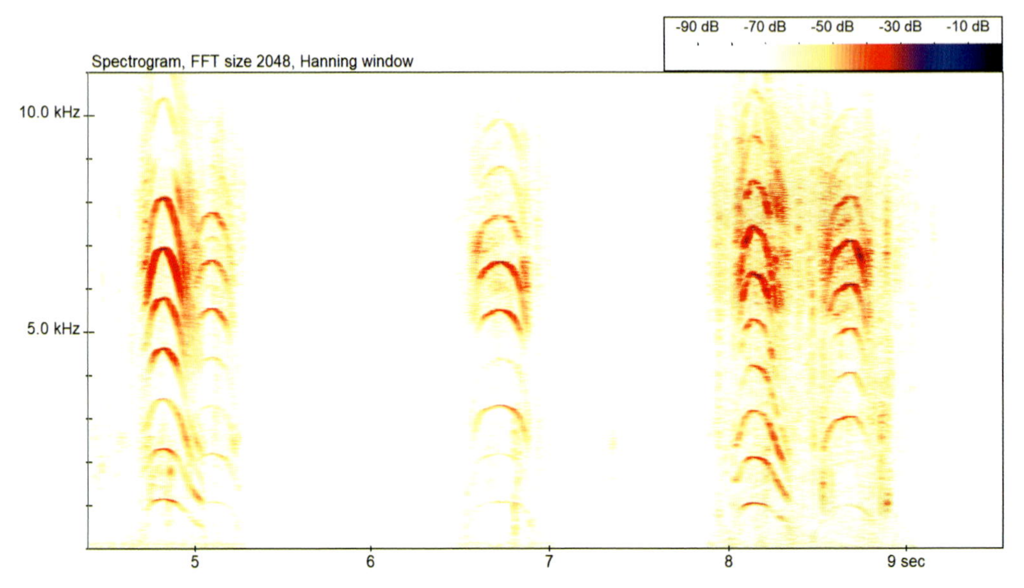

Figure 5.29 🔊 **(QR)** Chinese water deer – squeaking sounds produced during rutting activity (frame width: c.5 sec)

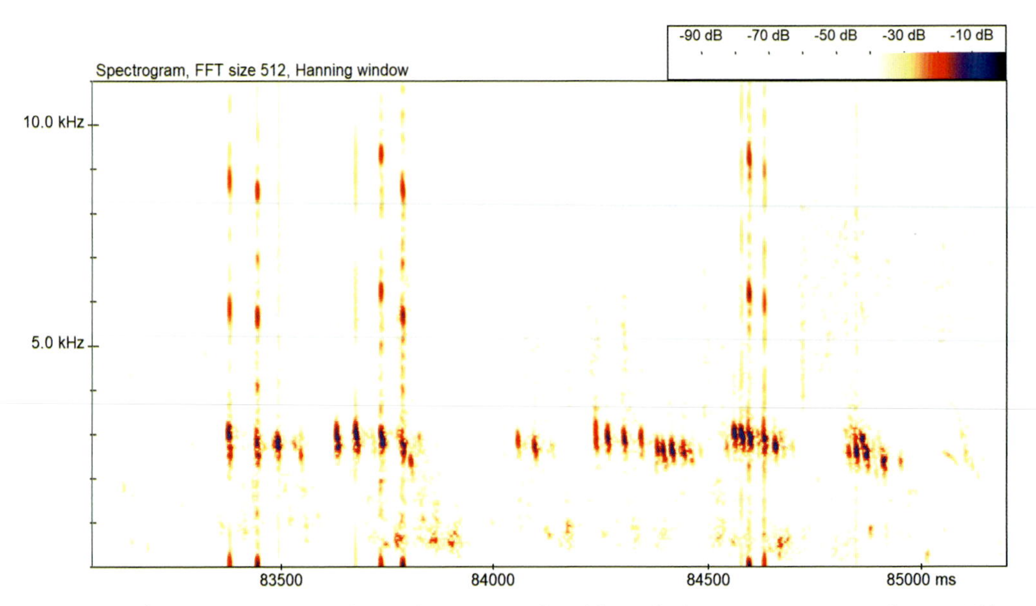

Figure 5.30 🔊 **(QR)** Chinese water deer – chittering produced by male during rutting activity (frame width: c.2 sec)

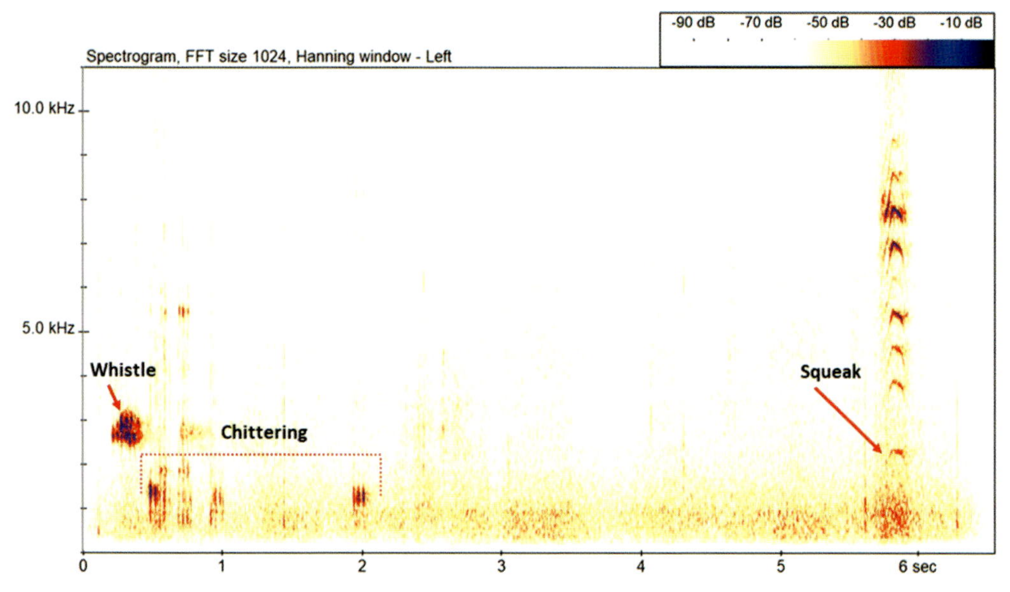

Figure 5.31 🔊 Chinese water deer – whistle and chittering, followed by a squeak, produced by male during rutting activity (courtesy of M. Baldwin) (frame width: c.7 sec)

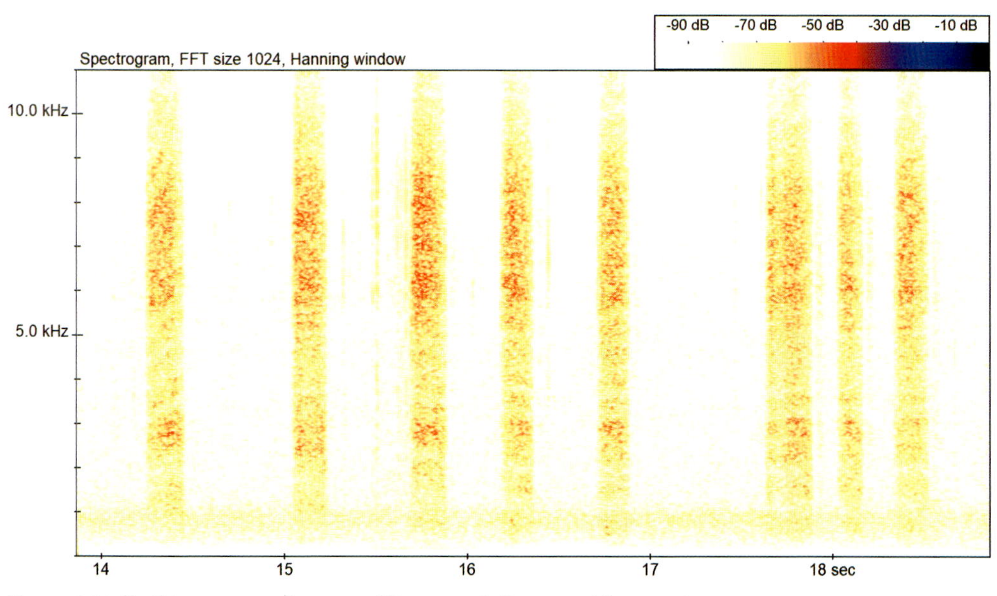

Figure 5.32 🔊 Chinese water deer – snuffling sounds (frame width: c.3 sec)

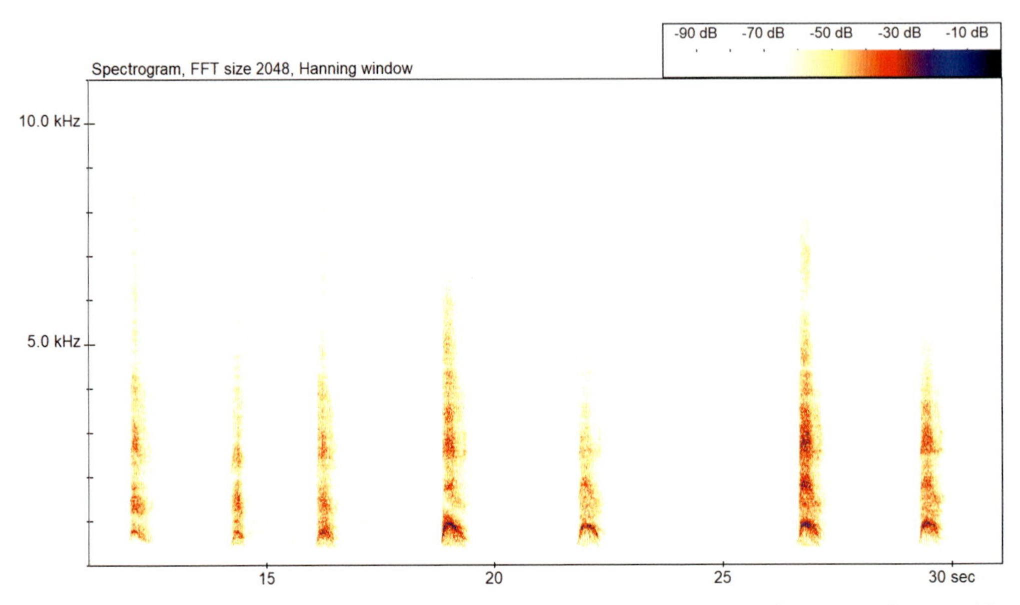

Figure 5.33 🔊 **(QR)** Chinese water deer – longer series of alarm barks (courtesy of S. Senior) (frame width: c.20 sec)

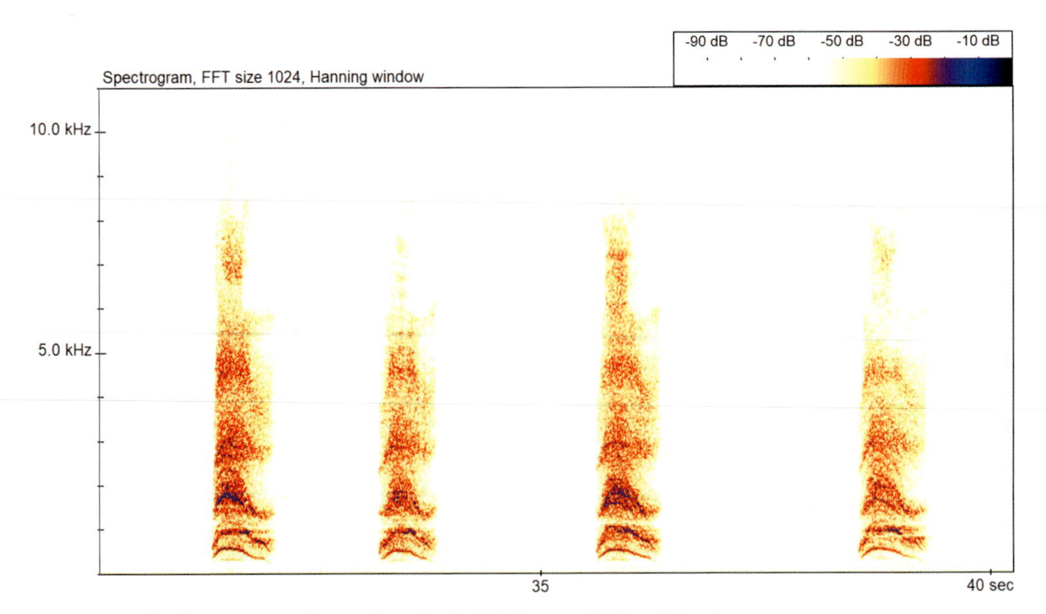

Figure 5.34 🔊 Chinese water deer – alarm barks (x4) (frame width: c.6.5 sec)

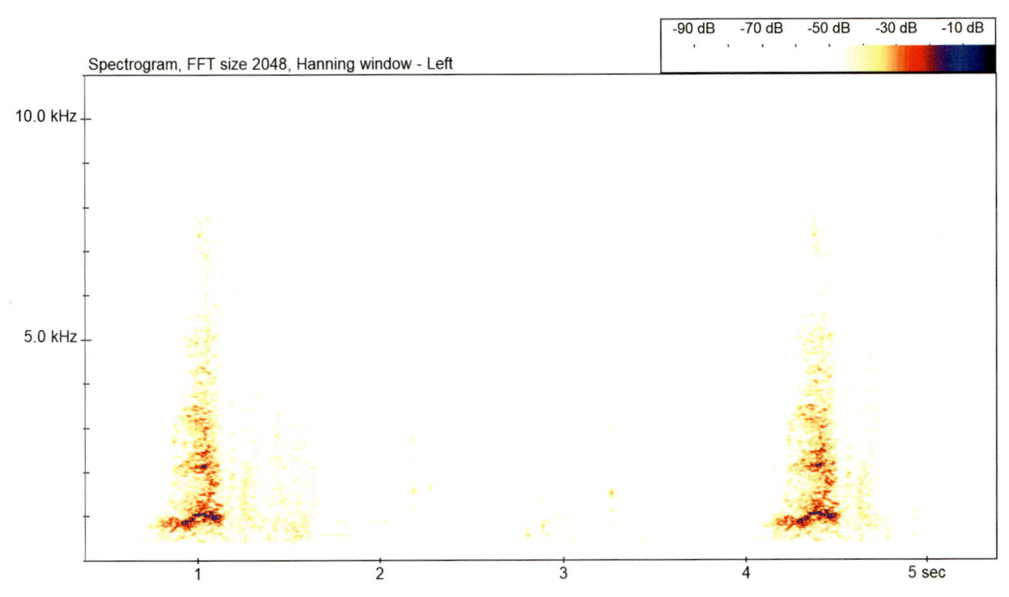

Figure 5.35 🔊 Chinese water deer – whelp-like call (x2) alarm barking variation (courtesy of S. Scott) (frame width: c.5 sec)

QR codes relating to presented figures

Potential confusion between Chinese water deer and other species

Table 5.10 Summary of potential confusion between Chinese water deer and other species

	Potential confusion species
Confusion group	In the first instance, always consider habitat and distribution.
Other deer species	Chinese water deer barks can be confused with roe deer and Reeves' muntjac. Chinese water deer barking is more 'cat- or whelp-like' rather than 'dog-like'.
Species in other groups	Barks produced by red fox (refer to Chapter 6, Section 6.4.1).

5.4 Pig vocalisations

In the following section, a brief summary of the key characteristics of the single species of pig will be described, together with details on the species-specific vocalisations and, where known, the behavioural function of these sounds.

5.4.1 Wild boar *Sus scrofa*

Length/Weight	Habitat preferences	Distribution
Body length up to 170 cm (shoulder height: up to 80 cm). Weight up to 100 kg. Males are c.10% larger and up to 30% heavier than females.	Coniferous woodland Broadleaved woodland Agricultural land	Isolated populations occurring in the far south of England (e.g. a firm footing within the Forest of Dean, as well as Sussex/Kent border area), southern Wales, north-west Scotland, and Dumfries and Galloway.
Diet		
Omnivorous, but largely focusing their feeding on vegetation, seeds and fruits.		

Daily activity pattern	Typically active from dusk through to dawn.
Seasonal activity pattern	Active throughout the year.
Mating activity pattern	Males are usually solitary outside of breeding season, which is generally regarded as the period from October through to May (height of season: November to January). During breeding (rutting) season males will travel long distances to seek out females, driving away young animals and fighting off other adult males in order to establish mating rights. Young (ranging from 3 to 10 piglets) are born usually in spring/early summer. On occasions a second litter may be produced later in the year.
Other notes	Adult males grow prominent tusks from two years old. Adult females and their young live in groups called 'sounders', whilst adult males tend to be solitary, unless seeking out females during the mating season, which takes place predominantly in winter. More easily encountered at dusk or dawn, especially in more open habitat adjacent to woodland.

References
Goulding *et al.*, 2008; Mathews *et al.*, 2018; Crawley *et al.*, 2020.

Acoustic behaviour including spectrogram examples

A research study (Garcia *et al.*, 2016) categorised the vocal repertoire of wild boar into four main call structure types: grunts (Figures 5.36 to 5.38, and 5.44), grunt-squeals (Figure 5.43), squeals and trumpets (Figures 5.39 to 5.41). This study did not attempt to conclude the behavioural context for each of the structure categories, other than to record which of the following behaviours may have applied against a given recording: alarm, alert/nervous, attacked, chased, contact, scared/threatened and submissive. Garcia *et al.* (2016) found that 'grunts' and 'squeals' were the most commonly produced call types (from a structural perspective) and that 'contact' and 'scared/threatened' were the most common contexts in which vocalisations were produced.

Taking the above information in conjunction with approaches taken for other species, calls, broadly speaking, may fall into five categories: contact calls, alarm calls, distress calls, threat calls and combative calls. These are considered in more detail below. Other instances where vocalisations are notable include feeding activity, when animals will typically emit a series of grunts and sniffs/snorts (Figures 5.37, 5.38 and 5.42).

Contact calls: Adult males are not usually vocal, whereas females will often be heard to make grunting-type sounds. Young animals (piglets) produce whining-type sounds.

Alarm calls: Warning calls can be produced when threatened, for example when humans encounter a sounder, and one of the females produces a snorting call prior to the group heading off to safety.

Distress calls: Far-carrying screeching-type sounds can be heard when an animal is in a state of heightened distress.

Threat calls: Non-breeding-related behaviour, with calls produced by an animal threatening a conspecific or a heterospecific.

Combative calls: High-pitched calls produced by males during breeding-related behaviour.

We were very fortunate to be given access to excellent sound recordings by Marc Anderson of Wild Ambience (www.wildambience.com) for the purposes of providing examples here for this species. These recordings relate to encounters in Spain. Some of the following spectrograms (not available within the Sound Library) have been created from Marc's recordings, which are thoroughly recommended and can be listened to via the following link:

https://wildambience.com/wildlife-sounds/wild-boar/

Separate to this, we have included within the Sound Library (no spectrograms shown) a couple of audible recordings 🔊 relating to foraging activity (courtesy of P. Ribeiro and P. Stancliffe).

Figures 5.36 to 5.44 provide specific examples of spectrograms relating to this species, and where the 🔊 symbol appears, a recording of what is shown within the spectrogram is available to download from the species library. In addition, QR codes are provided for selected examples, and these can be accessed immediately for listening purposes from most mobile devices.

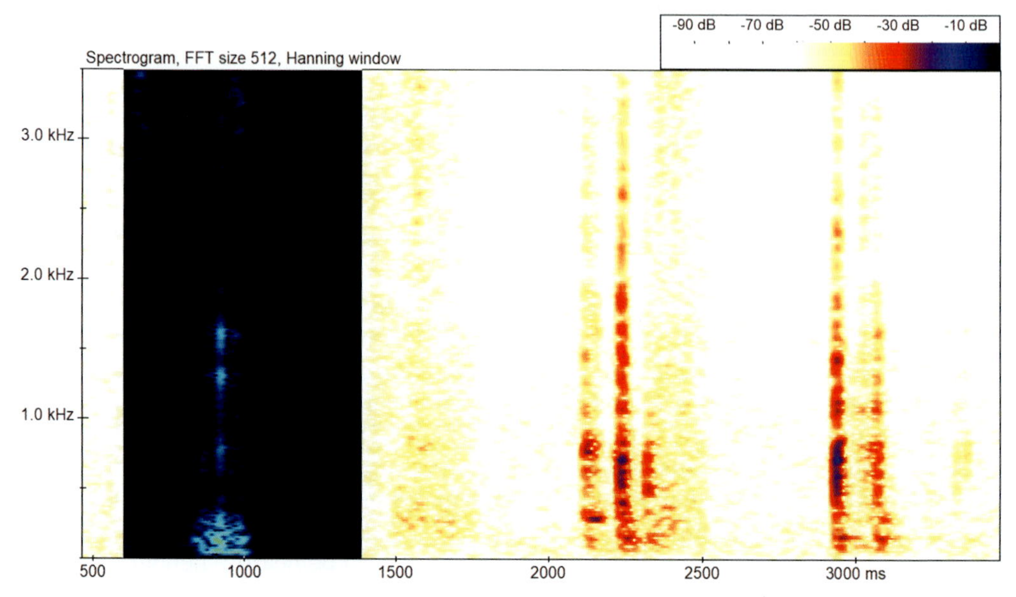

Figure 5.36 🔊 (QR) Wild boar – grunt (highlighted) (courtesy of S. Green) (frequency scale: 0 to 3.5 kHz/frame width: c.3 sec)

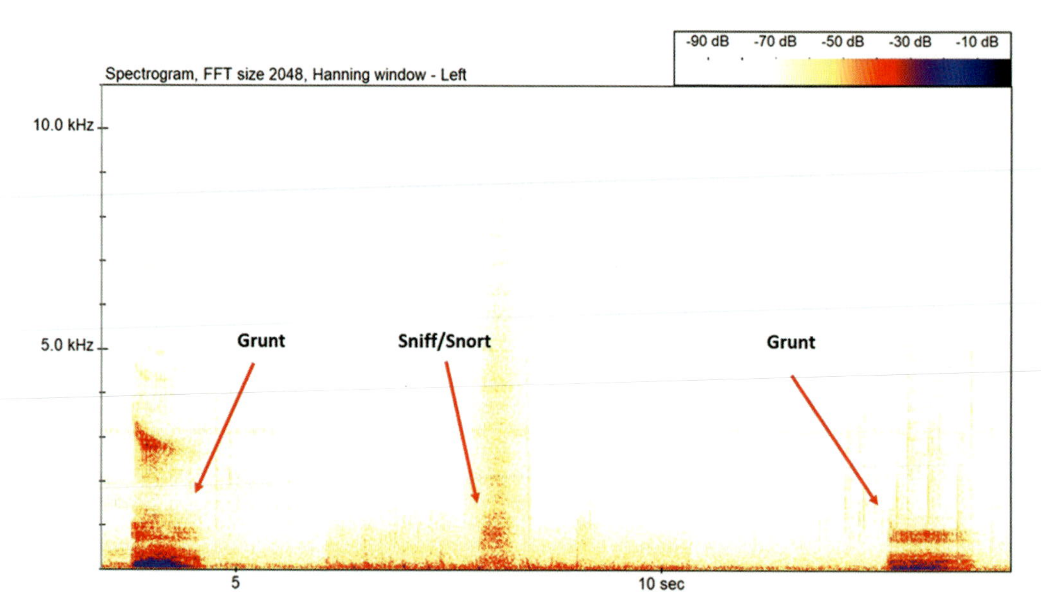

Figure 5.37 Wild boar – grunting (x2) with sniff/snort (courtesy of M. Anderson, Wild Ambience) (frame width: c.10 sec)

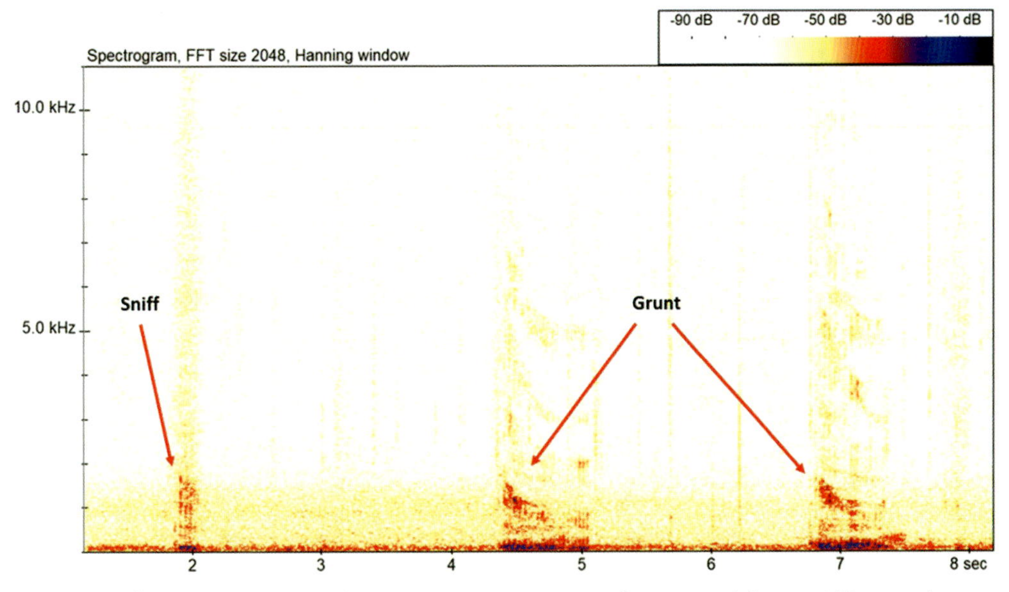

Figure 5.38 🔊 **(QR)** Wild boar – sniff with grunting (x2) (courtesy of H. Manistre) (frame width: c.8 sec)

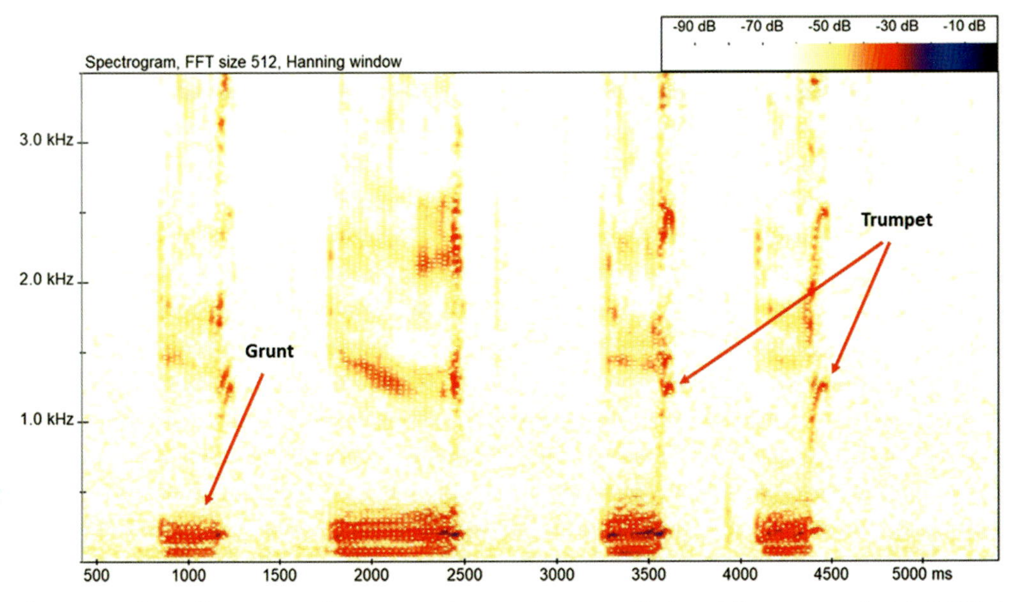

Figure 5.39 🔊 Wild boar – grunt/trumpet combination (courtesy of S. Green) (frequency scale: 0 to 3.5 kHz/ frame width: c.5 sec)

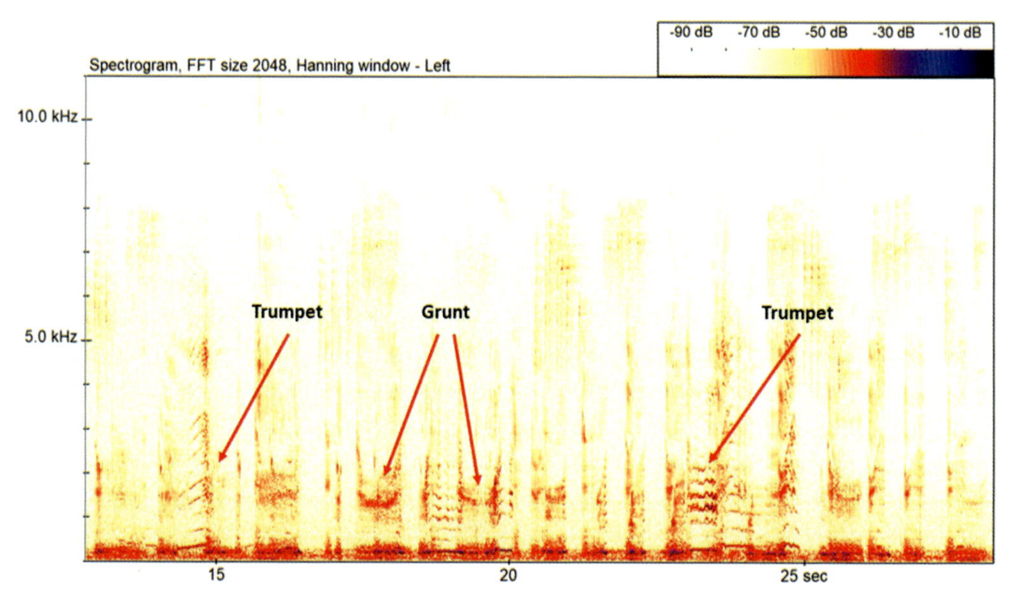

Figure 5.40 🔊 **(QR)** Wild boar – series of grunts with associated squeals/trumpeting (courtesy of H. Manistre) (frame width: c.10 sec)

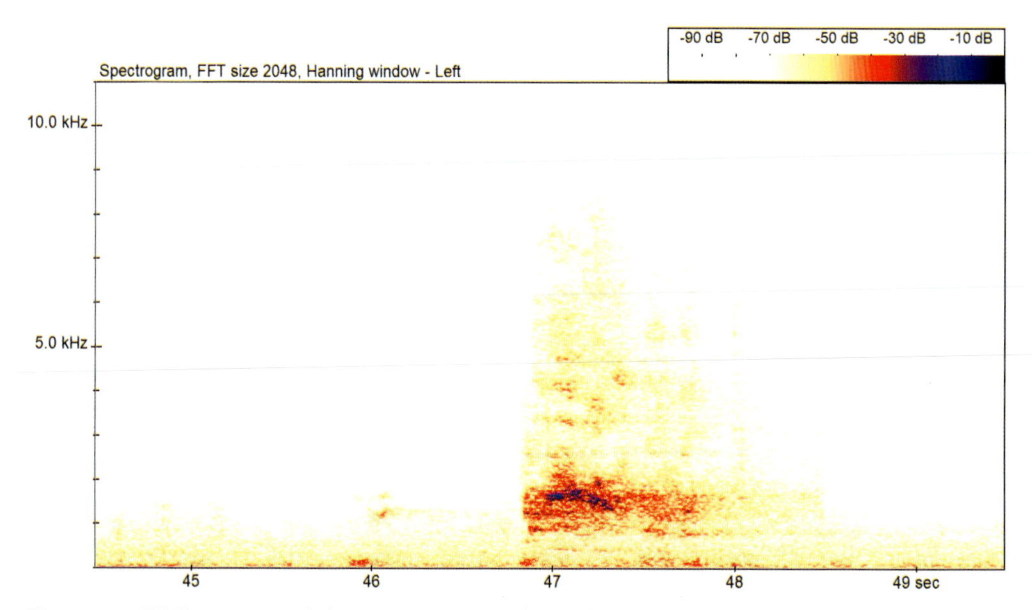

Figure 5.41 Wild boar – extended trumpet (courtesy of M. Anderson, Wild Ambience) (frame width: c.3 sec)

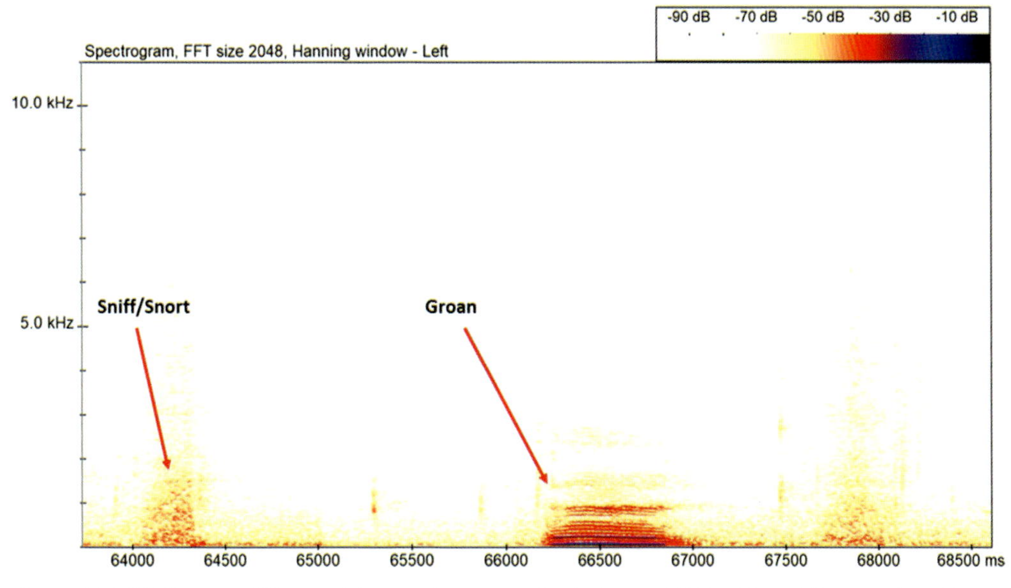

Figure 5.42 Wild boar – sniff/snort with groan (courtesy of M. Anderson, Wild Ambience) (frame width: c.5 sec)

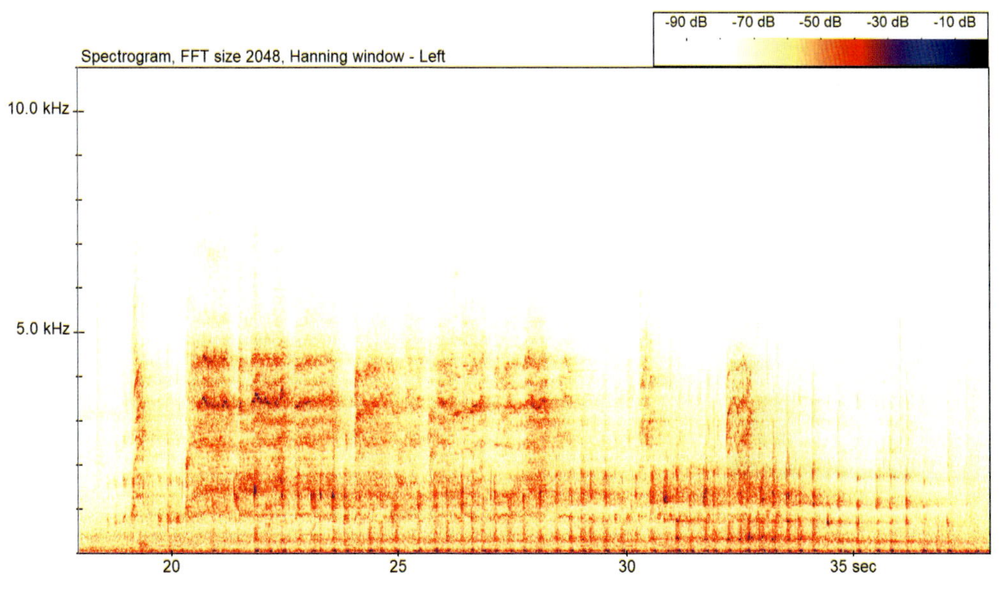

Figure 5.43 Wild boar – series of squeals and short grunts (courtesy of M. Anderson, Wild Ambience) (frame width: c.20 sec)

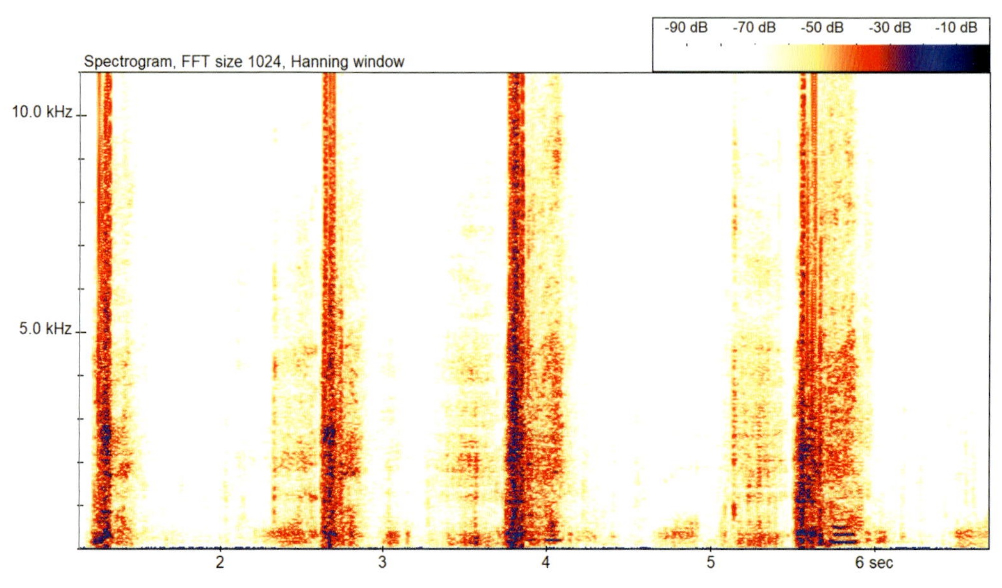

Figure 5.44 Wild boar – aggressive episode of grunts (courtesy of P. Ribeiro) (frame width: c.6 sec)

QR codes relating to presented figures

Potential confusion between wild boar and other species

Table 5.11 Summary of potential confusion between wild boar and other species

	Potential confusion species
Confusion group	In the first instance, always consider habitat and distribution.
Other pig species	The only wild pig in Britain. Most calls produced by this species are unlikely to be confused with other species, other than domestic pigs *Sus scrofa domestica*.
Species in other groups	Unlikely to be confused with other species.

5.5 Potential application of bioacoustics for monitoring even-toed ungulates

Given that this group of species are wide-ranging, supporting relatively large territories, and that outside of the breeding season they do not vocalise on a regular or consistent basis, significant amounts of resources would likely need to be deployed over long periods of time for acoustic surveys on their own to be an effective method for monitoring their populations.

In most instances, the application of acoustic surveys is likely to be more effective when used in combination with outputs from other survey methods (trail cameras, transect field sign or spotlighting surveys), and is likely to be most appropriate for behavioural research or distribution studies.

Studies have been carried out to evaluate the validity of using sound as a census survey method. Acoustic studies of roaring counts of red deer during the rut were not found to be an accurate method of estimating deer abundance when compared to other methods – for example, spotlighting surveys (Ciucci *et al.*, 2009; Douhard *et al.*, 2013; Tullo *et al.*, 2015). Comparatively, the frequency of bark alarm calls emitted by roe deer, in response to disturbance from surveyors walking transects, did provide a successful method of estimating absolute population size for this species (Reby *et al.*, 1998b).

Bioacoustic surveys may serve as a tool for collecting information on species distribution. This could be via recordings collected incidentally as part of other audible acoustic monitoring surveys such as nocturnal bird migration (Nocmig) or specifically for the monitoring of species in otherwise visually inaccessible habitats. For example, for some quieter and less visually obvious species, passive recording may give clues to presence over greater distances than would normally be detectable through trail cameras or field sign surveys. However, as with observer-based monitoring programmes, passive acoustic surveys of this kind would likely only allow effective monitoring of a sample of the whole population. At a larger population scale, acoustic surveys would generate a huge amount of data, the analysis of which is unlikely to be cost-effective until species recognition algorithms have been developed to extract and accurately classify species.

At a more localised level, acoustic surveys may be appropriate for studying the dynamics of individual herds or sounders. Roars or barks emitted by deer during the rut have been found to have inter- and intra-species phenotypic originality, i.e. the characteristics of analogous calls are inherited and are significantly different between unrelated individuals of the same deer species, as well as different deer species (Reby and McComb, 2003). Bioacoustics may therefore provide a non-invasive method for monitoring the lineage of a herd, and/or call analysis may offer a method for identifying instances of hybridisation such as red–sika hybrids (Long *et al.*, 1998).

Lastly, an example of bioacoustics working as an effective tool for deer species conservation can be seen in the use of rut call characterisation which confirmed the phenotypic originality of Mesola red deer *Cervus elaphus italicus*, the only native red deer population in peninsular Italy. Thus there is potential for individual call recognition to provide a useful tool for monitoring this endangered population (Libera *et al.*, 2015).

Red fox – Sandra Graham

CHAPTER 6

Carnivores (Including Mustelids)

6.1 Introduction to this group

This chapter covers the carnivores (Carnivora), within which we consider a single species of cat (Felidae), a single species of fox (Canidae), and seven species of mustelid (Mustelidae) (see Table 6.1).

The carnivores are predominantly nocturnal and elusive in character and, despite being some of the larger mammals found in Britain and Ireland, establishing their presence through purely visual sightings often remains difficult, albeit the use of camera traps and/or finding field-sign evidence (e.g. places of shelter, food remains, fur, faeces and/or footprints) can in many respects provide reliable results. There are of course exceptions, for example certain localities where sightings of the widely distributed red fox and badger would certainly not be uncommon, though almost all such engagements occur during the hours of darkness and as such work against our visual ability. Comparatively, encounters with all other species – for various reasons, most notably their solitary nature – are much less likely to occur, even if they are relatively well established in an area.

Taking all this on board, there are undoubtedly occasions when knowing at least some of the sounds made by the species in this group can help confirm presence. Being social/group animals, the red fox and badger are the most vocal species and their audible calls may be heard from far beyond the range of eyesight or any night vision equipment being deployed. By comparison, the solitary wildcat is considerably less vocal and has a restricted range, meaning encountering this species either visually or acoustically is likely to be extremely difficult. All the remaining species are audibly vocal at times, whilst some (e.g. stoat and weasel) also produce calls in the ultrasonic range, which can only be detected using specialist equipment.

Despite the potential difficulties in recording some of our carnivores, a greater familiarity with the repertoire of sounds produced by these animals can be useful for acquiring further information on behavioural status, in addition to determining species presence and population density (discussed later within this chapter – Section 6.6).

6.2 Overview and comparisons of acoustic behaviour

Vocalisations produced by the carnivores encompass a wide range of frequencies. Broadly speaking, all of the species in this group could be encountered acoustically through the production of audible sounds, whilst some species may also produce useful ultrasonic vocalisations (Table 6.2). The variability in call frequency between carnivore species, in addition to overall amplitude of the vocal response, will influence the distance over which they can be detected acoustically (Table 6.3).

The likelihood of detecting carnivores acoustically is also strongly influenced by the sociality of the species, their periods of activity (diurnal/nocturnal/crepuscular), and inter- and intraspecific differences in species population densities. For example, in areas of the country where red foxes are controlled, territories will be larger, group sizes will be smaller and the number of litters of cubs will be reduced, such that the likelihood of encountering territorial or contact calls will be lower than for suburban populations that occur at higher densities and where opportunities for fox–fox interactions are more likely (Sample, 2006). Suburban foxes can also be active during the day, further increasing the

Table 6.1 Carnivores occurring in Britain and Ireland, including status, distribution and relative abundance. The relevant chapter sections that describe species-specific vocalisations are also listed.

Order	Family	Species common name	Genus	Species scientific name	Status	Distribution range	Abundance within distribution range	Sub-chapter reference
Carnivora Carnivores	Felidae (Cat)	Wildcat	*Felis*	*F. silvestris*	Native	Restricted to northern and western Scotland, excluding the islands	Rare	6.3.1
	Canidae (Fox)	Red fox	*Vulpes*	*V. vulpes*		Widely distributed throughout Britain and Ireland, including some offshore islands	Common	6.4.1
	Mustelidae (Mustelids)	Badger	*Meles*	*M. meles*		Widely distributed throughout Britain and Ireland	Common	6.5.1
		Otter	*Lutra*	*L. lutra*		Widely distributed throughout Britain and Ireland	Common	6.5.2
		Pine marten	*Martes*	*M. martes*		Mainly within Scotland and Ireland, with populations more restricted within northern England and Wales	Scarce, but locally common	6.5.3
		Stoat	*Mustela*	*M. erminea*		Throughout Britain and Ireland	Common	6.5.4
		Weasel		*M. nivalis*		Throughout Britain (absent in Ireland)	Common	6.5.5
		Polecat		*M. putorius*		Wales; more thinly distributed throughout parts of England and isolated populations in Scotland	Rare	6.5.6
		American mink	*Neovison*	*N. vison*	Non-native	Scattered throughout Britain and Ireland, including some Scottish islands	Common	6.5.7

chances that they may be heard, especially where cubs are present. Similarly, interactions and associated vocalisations will be far less frequent for solitary species compared to those that live in social groups (e.g. badger and fox), and for solitary species, such as wildcat, which occur only at low densities, the likelihood of an acoustic encounter will be even further reduced.

The carnivores produce a diverse range of sounds that provide a variety of communication functions. These include: screams, growls, groans and snorts (reproductive calls); purring, hissing, whines, chirps, chittering and whistles (contact calls, parent–young sibling interactions); and growls, huffs and barks (territorial defence, alarm calls). Although research on the function of ultrasonic calls of stoats and weasels appears not to have been carried out, it is plausible that high-frequency calls serve as a predatory and/or predator-avoidance strategy.

It is important to note that, whilst the species in this group can be extremely vocal, olfaction is typically the main method of communication and therefore acoustic encounters are generally infrequent. Since intraspecific interactions and associated acoustic encounters are most common during the mating season (including the period of rearing young), it is useful to consider the mating periods for this group, as this can alert you to when certain vocalisations are more likely to be encountered. To this end, Table 6.4 provides an overview.

Each species will be discussed in more detail within its species-specific section, as presented later in this chapter.

Table 6.2 Acoustic spectrum within which you would usually expect to encounter most useful sound for identification purposes within this group

Audible range		Ultrasonic range							
< 10 kHz	10–20 kHz	20–30 kHz	30–40 kHz	40–50 kHz	50–60 kHz	60–70 kHz	70–80 kHz	90–100 kHz	> 100 kHz
↑ (Blue)	↑ (Red)	↑ (Red)	↑ (Green)	↑ (Green)	↑ (Green)	↑ (Green)			

Key
Coloured arrows relate to the following species:
Blue: all carnivore species / **Red:** stoat, weasel, pine marten, American mink / **Green:** stoat, weasel

Table 6.3 Detectable distance for useful acoustic encounters with carnivores

Distance (in metres)							
< 5 m	5–10 m	10–20 m	20–50 m	50–100 m	Up to 500 m	Up to 1,000 m	> 1,000 m
Red fox							
Ear/microphone							<<<source of sound
Badger							
Ear/microphone							<<<source of sound
Wildcat/Otter/Pine marten							
Ear/microphone							<<<source of sound
Weasel/Stoat/Polecat/American mink							
Ear/microphone							<<<source of sound

Table 6.4 Typical mating season (shaded in blue) for carnivore species occurring within Britain and Ireland

Species / Month	Jan	Feb	Mar	April	May	June	July	Aug	Sept	Oct	Nov	Dec
Wildcat	■	■	■									■
Red fox	■	■										■
Badger		■	■	■	■							
Otter	■	■	■	■	■	■	■	■	■	■	■	■
Pine marten						■	■	■				
Stoat				■	■	■	■					
Weasel				■	■	■	■					
Polecat			■	■	■							
American mink			■	■								

6.3 Cat vocalisations

In the following section, a brief summary of the key characteristics of the single species of cat will be described, together with details on the species-specific vocalisations and, where known, the behavioural function of these sounds.

6.3.1 Wildcat *Felis silvestris*

<table>
<tr><th>Length/Weight</th><th>Habitat preferences</th><th>Distribution</th></tr>
<tr><td>Males are larger than females.
Body length (excluding tail) up to 65 cm.
Weight up to c.5.3 kg.</td><td rowspan="3">Coniferous woodland – young plantations
Broadleaved woodland
Unimproved grassland
Moorland
Upland
Scrubland</td><td rowspan="3">Restricted to areas within northern and western Scotland (excluding the islands).</td></tr>
<tr><th>Diet</th></tr>
<tr><td>Carnivorous, feeding across a range of mammal (predominantly rabbits and voles) and bird prey.</td></tr>
</table>

Daily activity pattern	Typically most active from dusk through to dawn.
Seasonal activity pattern	Active throughout the year, and generally solitary and territorial unless mating or when females are nurturing young.
Mating activity pattern	Mating takes place during late winter and into spring. Male territories tend to overlap with a number of female territories. Young (two to six) are born usually during the period April to May, but may occur as late as September.
Other notes	Our only wildcat species is rarely encountered since they occur at very low population density (population estimated at 200 individuals). This species is critically endangered and heavily threatened due to hybridisation with feral/domestic cats. Daytime activity is influenced by season, being highest in autumn and lowest in spring. Potential identification using trail or thermal cameras, but differentiation from domestic/feral cats can prove difficult. Dens found in tree cavities, rock crevices, boulders, disused rabbit burrows, fox dens and badger setts, and mainly used during colder months.

References
Macdonald and Barrett, 1995; Kitchener and Daniels, 2008; Mathews *et al.*, 2018; Crawley *et al.*, 2020.

Acoustic behaviour including spectrogram examples

Calls produced by wildcat are broadly similar to the vocalisations of domestic and feral cats, and as such it should always be borne in mind that anything 'cat-like' recorded may not necessarily relate to a wildcat, particularly given their rarity.

A study by Nicholas Nicastro (2004) of the comparison of 'meow' calls produced by domestic cats and African wildcat Felis *silvestris lybica*, which is closely related to the British wildcat, found that domestic cat meows were shorter in mean duration, showed higher mean formant frequencies and higher mean fundamental frequencies. Similarly, several studies on the vocalisations of domestic and feral cats (e.g. Yeon *et al.*, 2011; Owens *et al.*, 2017; Schötz *et al.*, 2017) have demonstrated that domestic cat calls are typically higher in frequency compared to feral cats, and that domestic cats vocalise more often and for longer periods, which is thought to be associated with their greater sociality, albeit cat–human as well as cat–cat interactions. Although specific studies have not been carried out on our wildcats, based on these findings and from analysing the recordings collected for this book, the same is likely to be true when comparing wildcats with domestic cats (i.e. wildcats likely call at lower frequencies and vocalise less often than domestic cats). Whether or not it is possible to differentiate between wildcats and feral cats however remains unknown, but they are likely to be hard to separate.

Vocalisations by wildcats are rarely encountered in field conditions, although 'screams' have been heard during the mating season (Kitchener and Daniels, 2008). Wildcat calls are poorly documented and consequently the recordings collected for this book were obtained from a captive breeding family group (two adults with five dependent young). Vocalisations that we encountered included a variety of meows with peak frequencies of between 2 and 2.75 kHz: short (Figure 6.1), long (Figure 6.2) and irritable (Figure 6.3). These are at times accompanied by lower-frequency (FmaxE of c.1–1.5 kHz) groans, snarls (Figure 6.4) and/or whines (FmaxE of c.2 kHz) (Figure 6.5). In the examples provided, growls and snarls relate to play-aggression, whilst whines are more often related to submissive behaviour or when young are attempting to solicit food from parents or siblings.

In addition, low-frequency hisses (Figure 6.6) and growls (Figures 6.7 and 6.8) (FmaxE of c.1 Hz) may occur when threatened or showing aggressive behaviour, and separate to these occasions much lower-frequency, barely audible to some (< 1 kHz) grumbles (Figure 6.9) have been recorded.

Figures 6.1 to 6.9 provide specific examples of spectrograms relating to this species, all of which relate to a captive family group of two adults and five kittens. Where the 🔊 symbol appears, a recording of what is shown within the spectrogram is available to download from the species library. In addition, QR codes are provided for selected examples, and these can be accessed immediately for listening purposes from most mobile devices.

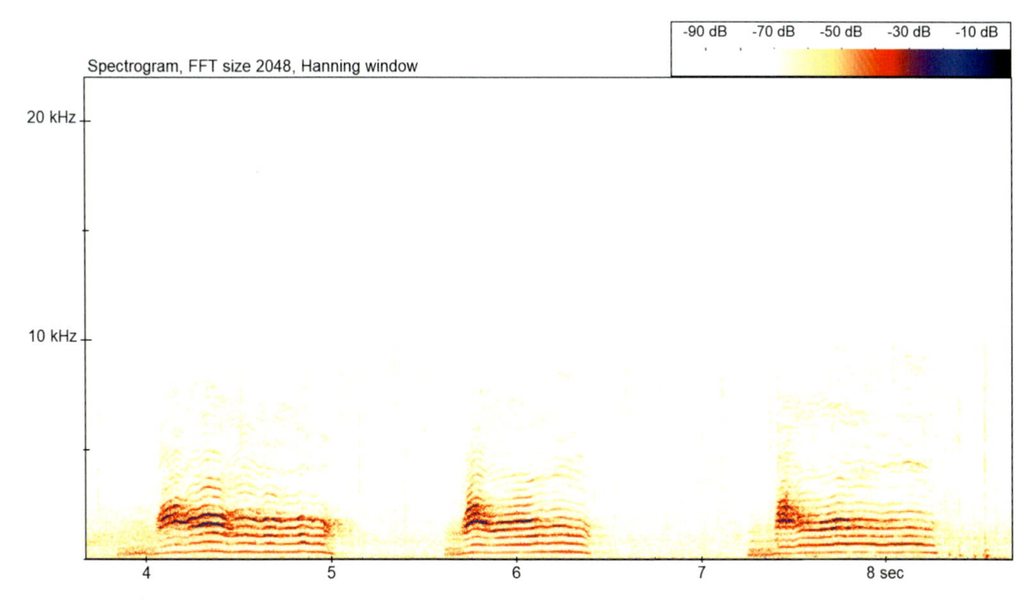

Figure 6.1 🔊 Wildcat – short meows (x3) (frame width: c.5 sec)

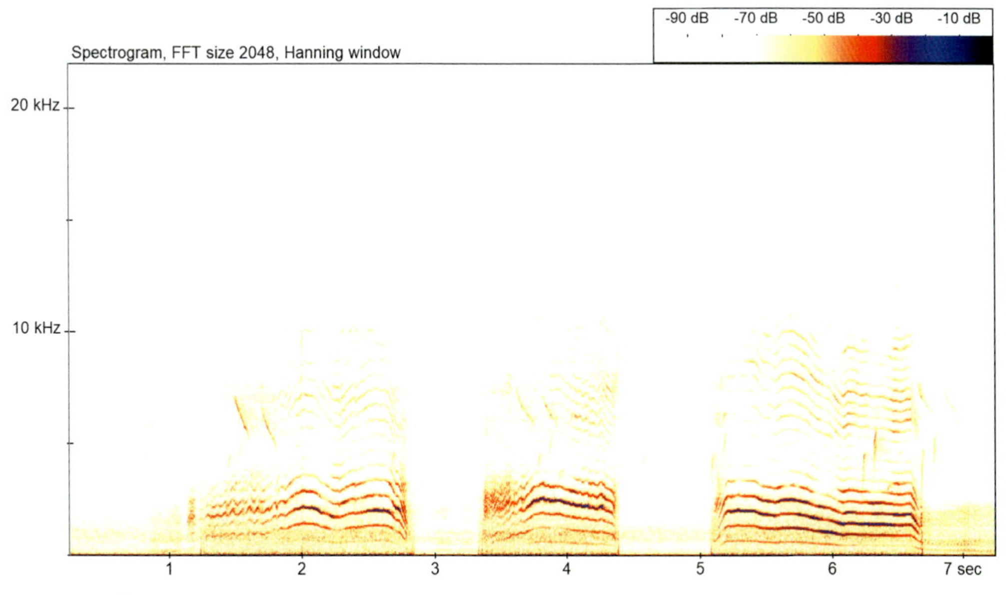

Figure 6.2 🔊 **(QR)** Wildcat – long meows (x3) (frame width: c.7 sec)

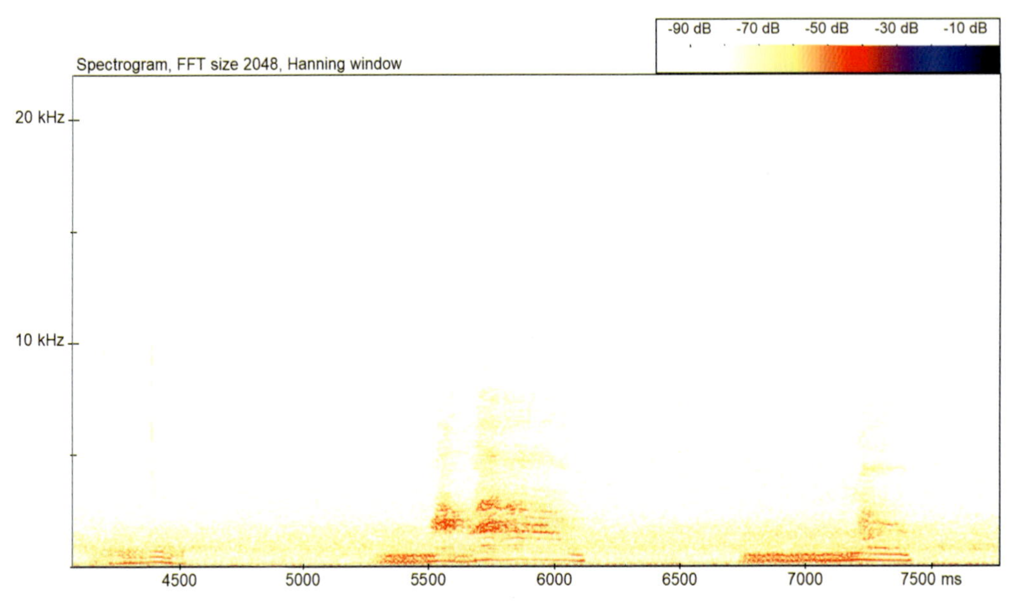

Figure 6.3 🔊 Wildcat – irritable meow (frame width: c.3.5 sec)

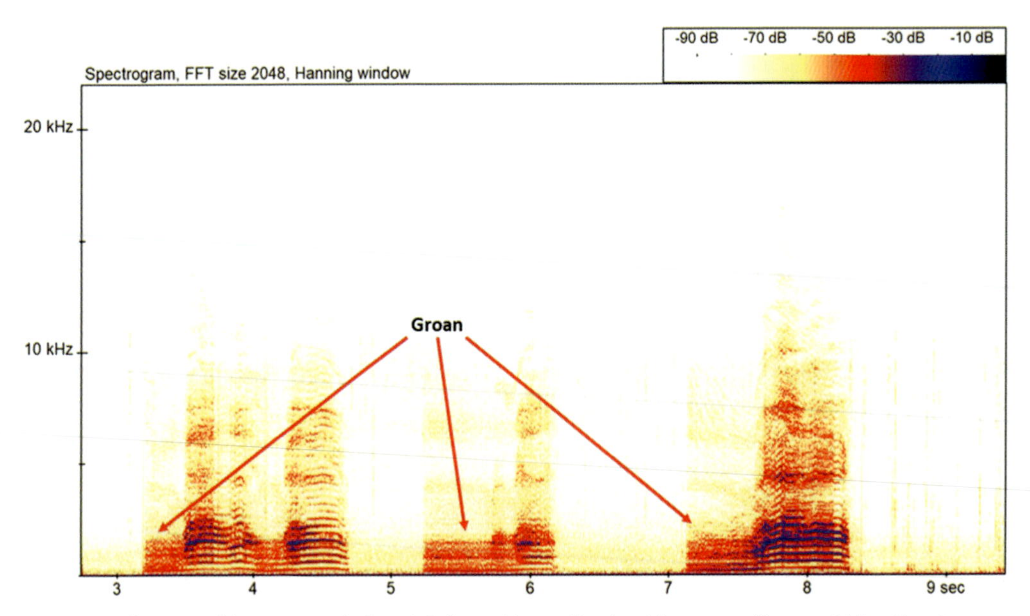

Figure 6.4 🔊 **(QR)** Wildcat – groan (x3) each followed immediately with a meow (frame width: c.7 sec)

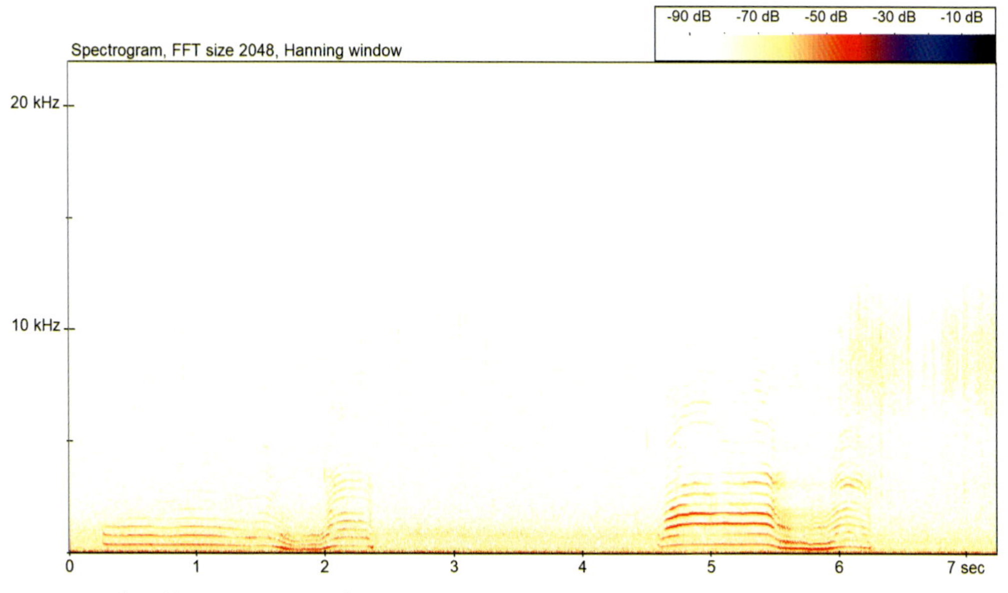

Figure 6.5 Wildcat – whines (x2) (frame width: c.7 sec)

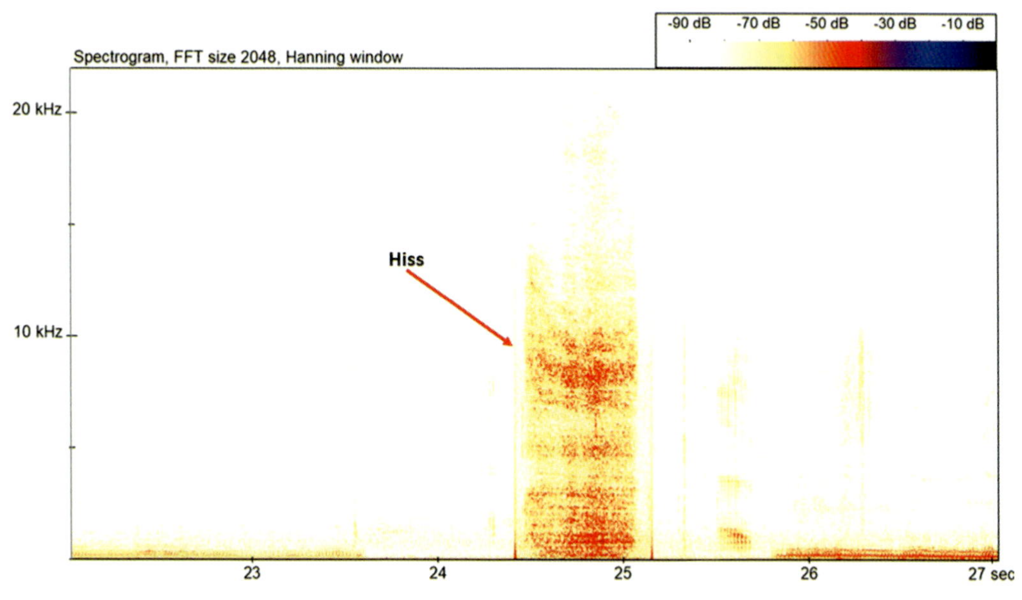

Figure 6.6 Wildcat – hiss (frame width: c.5 sec)

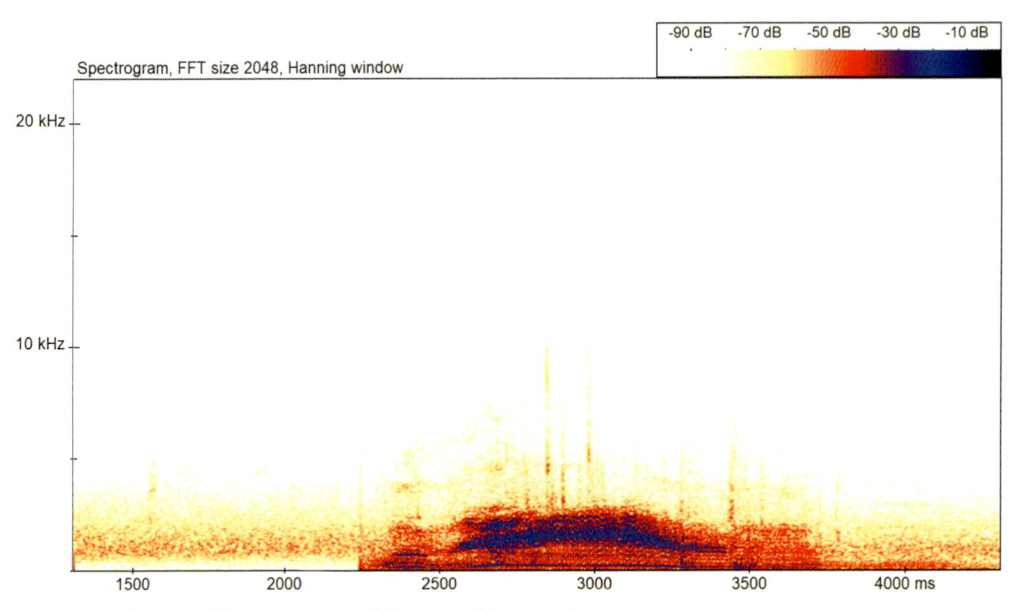

Figure 6.7 🔊 **(QR)** Wildcat – short growl (frame width: c.3 sec)

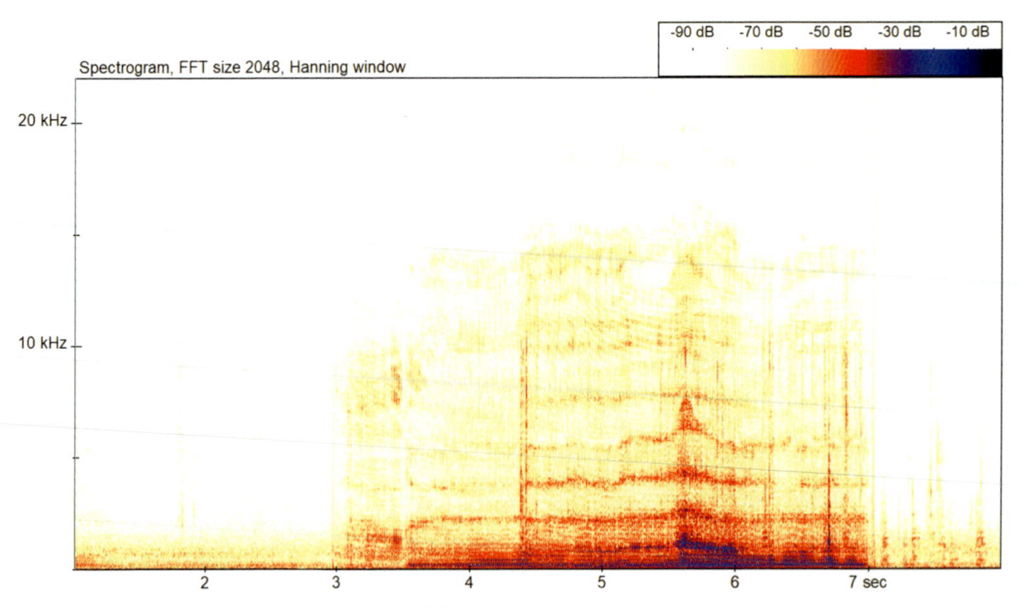

Figure 6.8 🔊 Wildcat – long growl (frame width: c.7 sec)

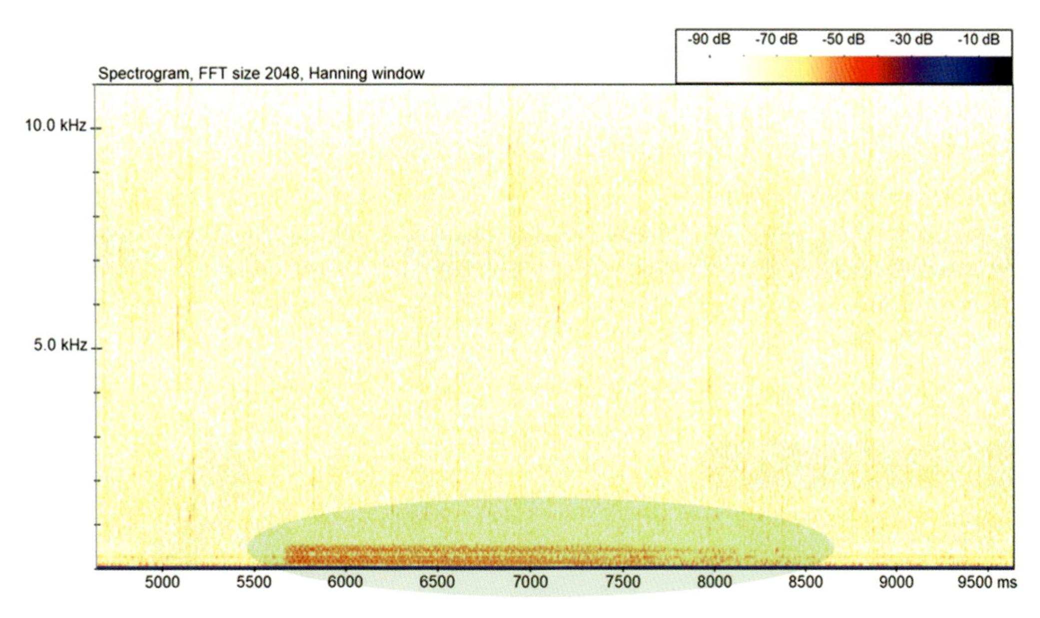

Figure 6.9 🔊 Wildcat – grumble (within shaded area) (frame width: c.5 sec)

QR codes relating to presented figures

Potential confusion between wildcat and other species

Table 6.5 Summary of potential confusion between wildcat and other species

	Potential confusion species
Confusion group	In the first instance, always consider habitat and distribution.
Other cat species	Domestic cat and feral cat.
Species in other groups	Unlikely to be confused with any other mammal species.

6.4 Fox vocalisations

In the following section, a brief summary of the key characteristics of the single species of fox will be described, together with details on the species-specific vocalisations and, where known, the behavioural function of these sounds.

6.4.1 Red fox *Vulpes vulpes*

<table>
<tr><th>Length/Weight</th><th>Habitat preferences</th><th>Distribution</th></tr>
<tr><td>Males are larger than females.
Body length (excluding tail) up to 78 cm.
Weight 5.5–7.5 kg, occasionally up to 9 kg.</td><td rowspan="3">Generalist utilising a range of habitats where prey and shelter is available, including urban environments.</td><td rowspan="3">Occurs throughout Britain and Ireland and on some offshore islands (e.g. Isle of Skye).</td></tr>
<tr><th>Diet</th></tr>
<tr><td>Carnivorous, feeding upon a variety of prey including small mammals (particularly voles), rabbits and birds.
Opportunistic foraging of eggs, insects (e.g. beetles), earthworms, fruits and berries, and carrion. In urban areas, frequently consume food supplied deliberately or accidentally by humans, including wild bird food.</td></tr>
<tr><th>Daily activity pattern</th><td colspan="2">Typically most active from dusk through to dawn.</td></tr>
<tr><th>Seasonal activity pattern</th><td colspan="2">Active throughout the year, living in small territorial family groups that consist of a mated pair and their offspring, plus other adults in some habitats (c.6–8 individuals).</td></tr>
<tr><th>Mating activity pattern</th><td colspan="2">Mating usually takes place during December to February.
Young (usually four or five) are usually born during March to May. Both adult female and male tend to raise the young, along with other adult members of the family group.</td></tr>
<tr><th>Other notes</th><td colspan="2">Can be heard during hours of darkness, especially during the mating period.
Dens will be used during the breeding season and in colder winter months, otherwise they may be found resting above ground within dense cover. Dens can be excavated into banks, adapted rabbit burrows, badger setts, naturally occurring holes and human-made features (e.g. under structures).</td></tr>
<tr><td colspan="3">References
Macdonald and Barrett, 1995; Baker and Harris 2008; Mathews et al., 2018; Crawley et al., 2020.</td></tr>
</table>

Acoustic behaviour including spectrogram examples

Red foxes have a wide range of vocalisations, with one publication (Newton-Fisher *et al.*, 1993) suggesting 12 adult and 8 cub call types, with each call type showing variation. Broadly speaking, cub calls were categorised as murmurs, warbles, whines, ratchet calls, wow-barks and growls. Table 6.6 provides a summary of the adult calls described by Newton-Fisher *et al.*, 1993, and for anyone interested specifically in this species we would strongly suggest reviewing the paper quoted. The descriptions we will use in this section will broadly attempt to follow those given in that publication.

Triple barking (Figure 6.10) is a characteristic contact call sequence that enables group members to keep in touch even over larger distances. Lower-frequency and more complex wow-wow barks (Figure 6.13), shrieks (Figure 6.14) and whines (Figure 6.15) may also be encountered during communication within a group.

The most productive period for encountering vocalisations is normally November through to March, when both peak dispersal (resulting in territorial disputes) and mating occurs. Agonistic calls known as gekkering or ratchet calls (Figure 6.17), screams (Figure 6.16) and yell-whines are commonly heard during this period, together with vocalisations described as barks and yell-barks (Figures 6.10–6.12). Gekkering/ratchet calls are used in aggressive encounters associated with both mating attempts and territorial disputes, whereas screams are not restricted to the mating season, and although considered agonistic, the true meaning of this call remains unclear. Barks and yell-barks are typically used for territoriality.

Contact calls and other vocalisations between mother and young would also be evident in such settings, including coughs and soft grunts and growls (Figure 6.18).

Call descriptions for this species can be problematic; one person's description of a particular sound may not necessarily accord with the same sound description given in another source. The words we each choose when describing sound are subjective and are not always transferable to another person's interpretation. As such we have to admit that we have struggled to assign calls confidently to descriptions provided by various sources, and we would suggest caution when trying to describe certain calls for this species (e.g. separating 'bark' from 'yell-bark'). We would suggest that future work is carried out in order to consolidate these descriptions.

Figures 6.10 to 6.18 provide specific examples of spectrograms relating to this species, and where the 🔊 symbol appears, a recording of what is shown within the spectrogram is available to download from the species library. In addition, QR codes are provided for selected examples, and these can be accessed immediately for listening purposes from most mobile devices.

Table 6.6 A summary of adult red fox vocalisations as described by Newton-Fisher *et al.*, 1993

Call description	Structure and considered context of vocalisation
Type 1 – Barks	A warning call. A single-component call, often emitted in conjunction with other barks, or other call types. Mean duration c.830 ms, occurring with an fmin of c.0.95 kHz and an FmaxE at c.1 kHz. (Figure 6.10)
Type 2 – Yell-bark Often described elsewhere as 'alarm barks'	Agonistic. A single-component call, often emitted in conjunction with other barks, or other call types. The call drops slightly in frequency towards the end. Mean duration c.760 ms, occurring with an fmin of c.0.91 kHz and an FmaxE at c.1.5 kHz. (Figures 6.11 and 6.12)
Type 3 – Shrieks	Function not certain, perhaps some sort of contact behaviour. A single-component call with a mean duration c.675 ms, occurring with an fmin of c.0.86 kHz and an FmaxE at c.1.5 kHz. (Figure 6.14)
Type 4 – Whines	A single-component call, sometimes emitted after Type 1 or Type 2 calls, with a mean duration c.650 ms, occurring with an fmin of c.1.1 kHz and an FmaxE at c.1.1 kHz. This call type usually commences at a lower frequency, rising slightly to an almost CF portion, and then concluding with a trill at the end. (Figure 6.15)
Type 5 – Ratchet calls Often described as 'gekkering'	Agonistic. Often occur as a burst of components, producing an untidy spectrogram image with multiple frequencies occurring simultaneously. Mean duration c.170 ms, occurring with an fmin of c.1.85 kHz and an FmaxE at c.2.2 kHz. (Figure 6.17)
Type 6 – Staccato barks Described as 'long-range multiple-component barking'	Possibly a longer-range contact call. A multi-component call sequence, with up to or more than three components being normal, with frequency increasing progressively during a sequence. Mean duration 930 ms, occurring with an fmin of c.0.73 kHz and an FmaxE at c.0.8 kHz.
Type 7 – Wow-wow barks	Contact signature (i.e. identity) call perhaps. A multi-component call sequence, with up to or more than five components being normal. Each individual component appears to drop in frequency. Mean duration 1,240 ms, occurring with an fmin of c.0.53 kHz and an FmaxE at c.0.76 kHz. (Figure 6.13)
Type 8 – Yodel barks	A 'softer' multi-component call sequence than Type 6 and Type 7, with up to or more than four components being normal. Each component rises and then drops in frequency. Mean duration 1,140 ms, occurring with an fmin of c.0.78 kHz and an FmaxE at c.0.78 kHz.
Type 9 – Growls	Threat call. A single-component call with a mean duration of c.1,170 ms, occurring with an fmin of c.0.12 kHz and an FmaxE at c.0.2 kHz. (Figure 6.18)
Type 10 – Coughs	Alert calls emitted in close proximity to cubs. Distinctive multi-component call sequence, with up to or more than five components being normal. Mean duration c.590 ms, occurring with an fmin of c.0.36 kHz and an FmaxE at c.0.4 kHz. (Figure 6.18)
Type 11 – Screams	Agonistic/defensive/threat. A single-component call with a mean duration of c.730 ms, occurring with an fmin of c.1.45 kHz and an FmaxE at c.1.84 kHz. (Figure 6.16)
Type 12 – Yell-whines	Agonistic/submissive. A single-component call, initially rising and then falling in frequency, with a mean duration of c.860 ms, occurring with an fmin of c.0.98 kHz and an FmaxE at c.2.3 kHz.
Note: The call parameters quoted above are approximate, and represent mean values, thus meaning values either side of those given are to be expected, and all calls have been found to be variable in reality.	

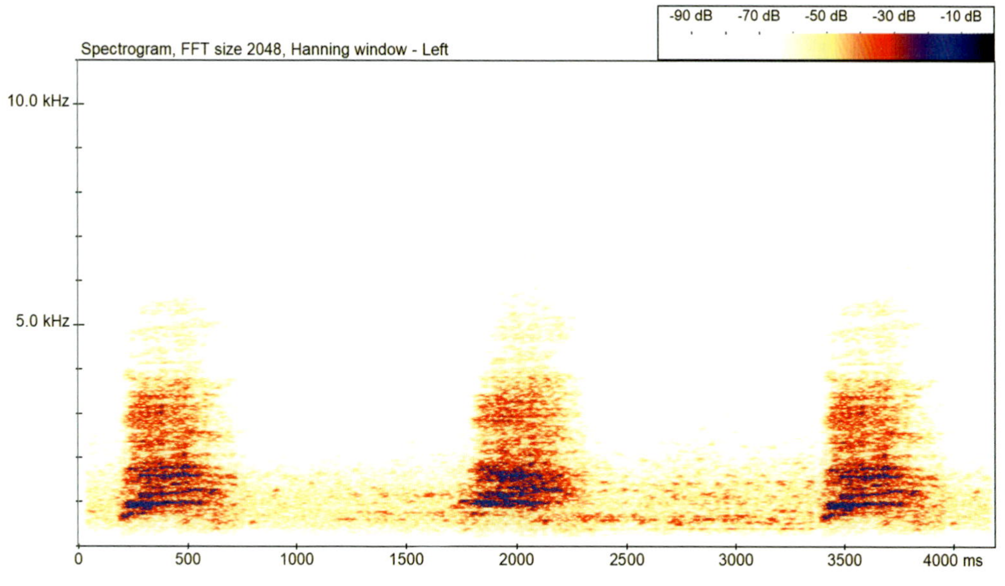

Figure 6.10 🔊 **(QR)** Red fox barks (x3) (courtesy of M. Baldwin) (frame width: c.4 sec)

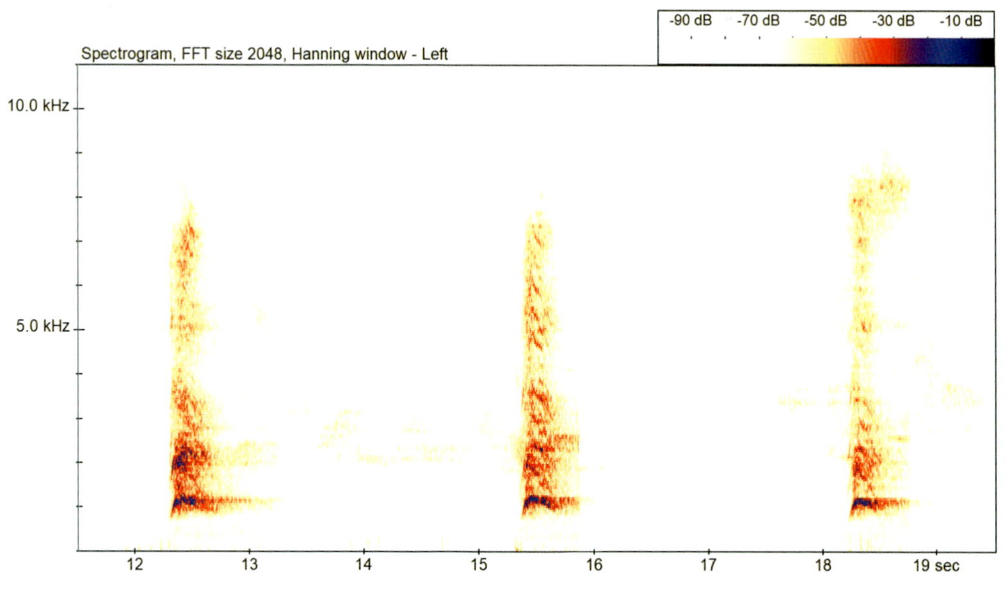

Figure 6.11 🔊 Red fox yell-barks (x3) (courtesy of D. Darrell-Lambert) (frame width: c.8 sec)

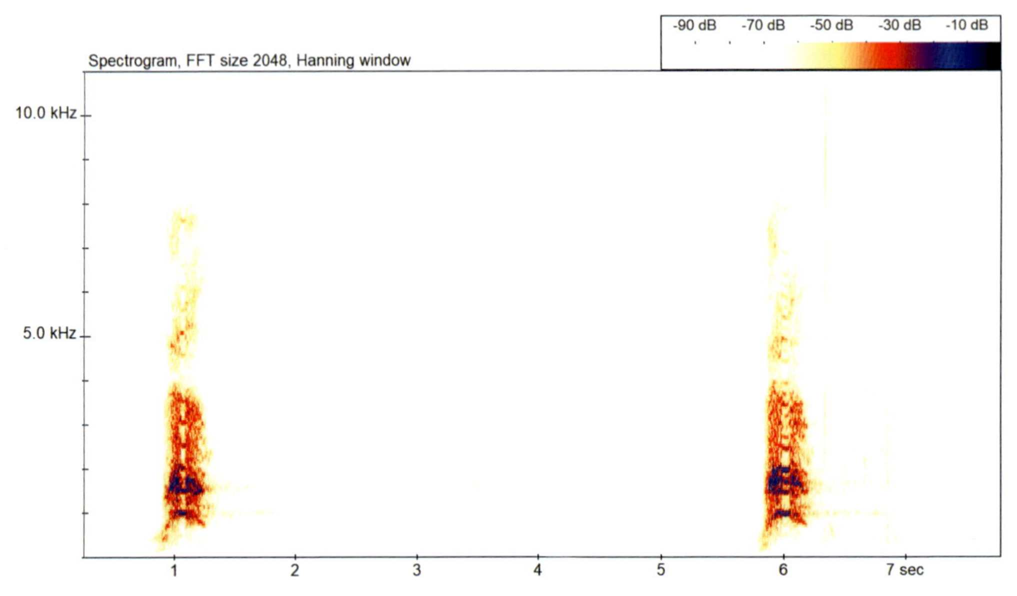

Figure 6.12 🔊 **(QR)** Red fox yell-barks (x2) (courtesy of M. Ferguson) (frame width: c.8 sec)

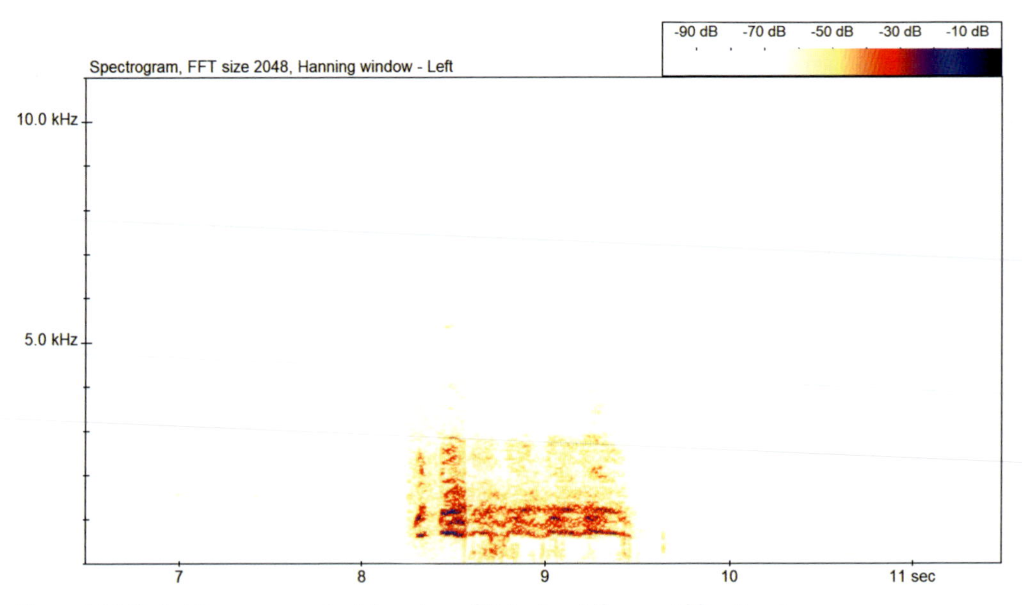

Figure 6.13 🔊 Red fox wow-wow bark (courtesy of M. Ingham) (frame width: c.5 sec)

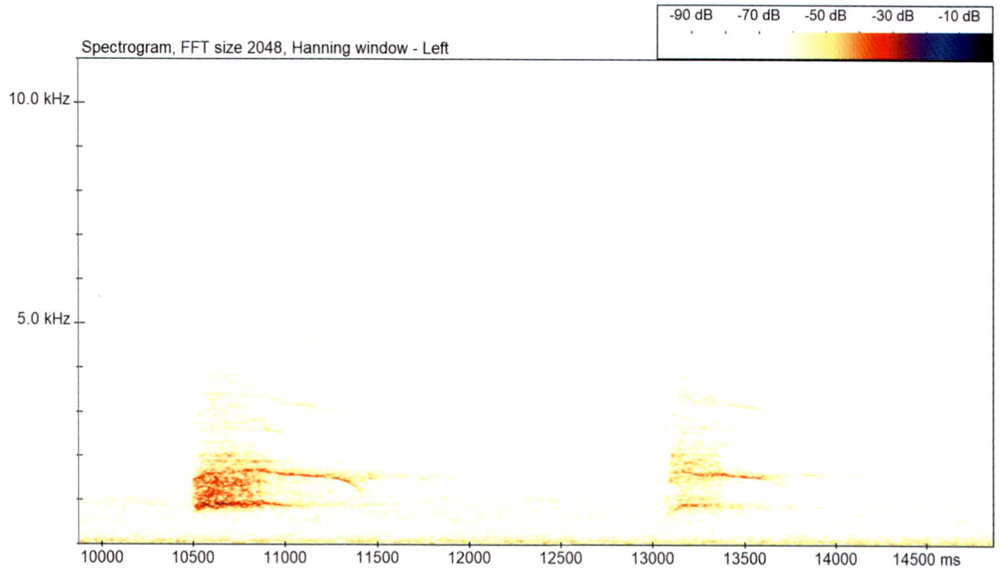

Figure 6.14 Red fox shriek (courtesy of S. Elliott) (frame width: c.5 sec)

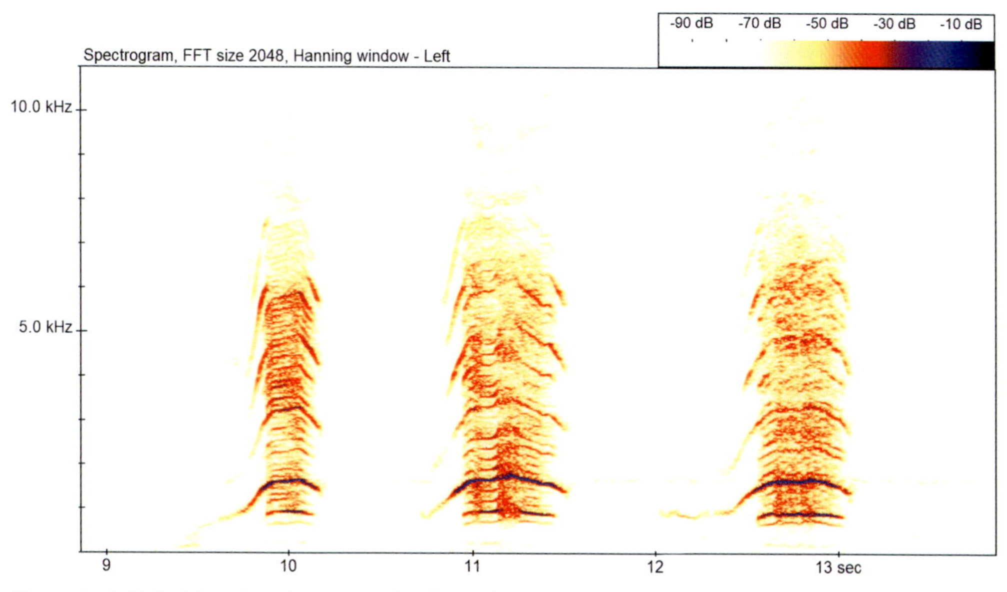

Figure 6.15 Red fox whine (courtesy of S. Elliott) (frame width: c.5 sec)

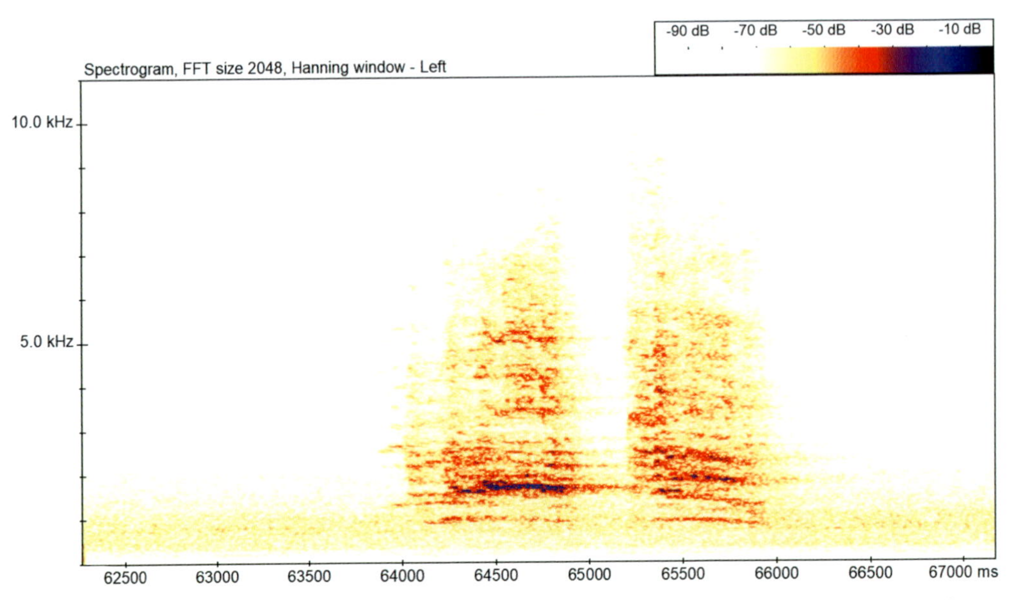

Figure 6.16 🔊 **(QR)** Red fox screaming occurring during aggressive behaviour (courtesy of S. Gillings) (frame width: c.5 sec)

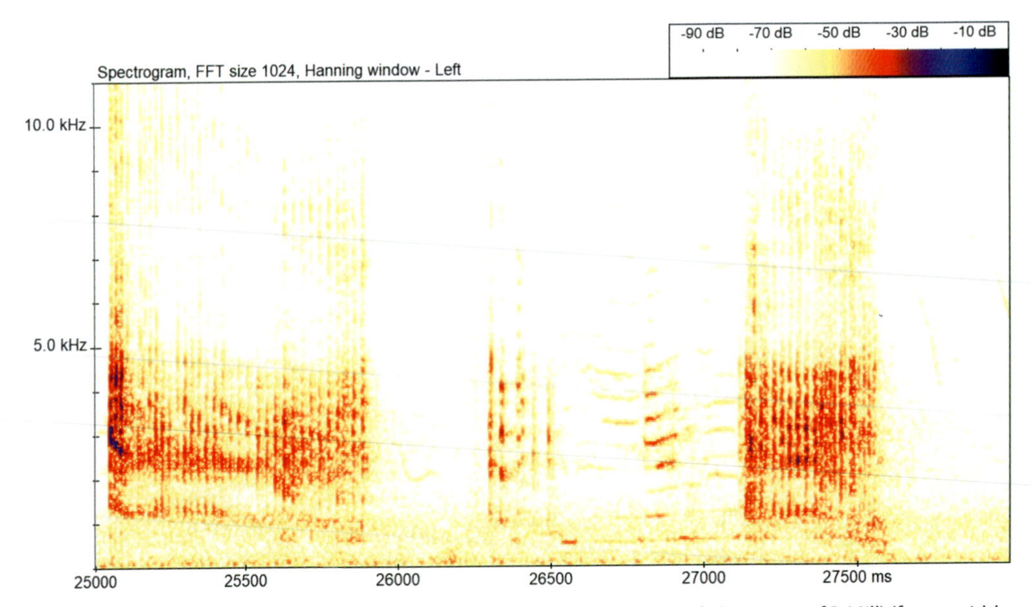

Figure 6.17 🔊 Red fox ratchet calls (gekkering) between two young animals (courtesy of P. Mill) (frame width: c.3 sec)

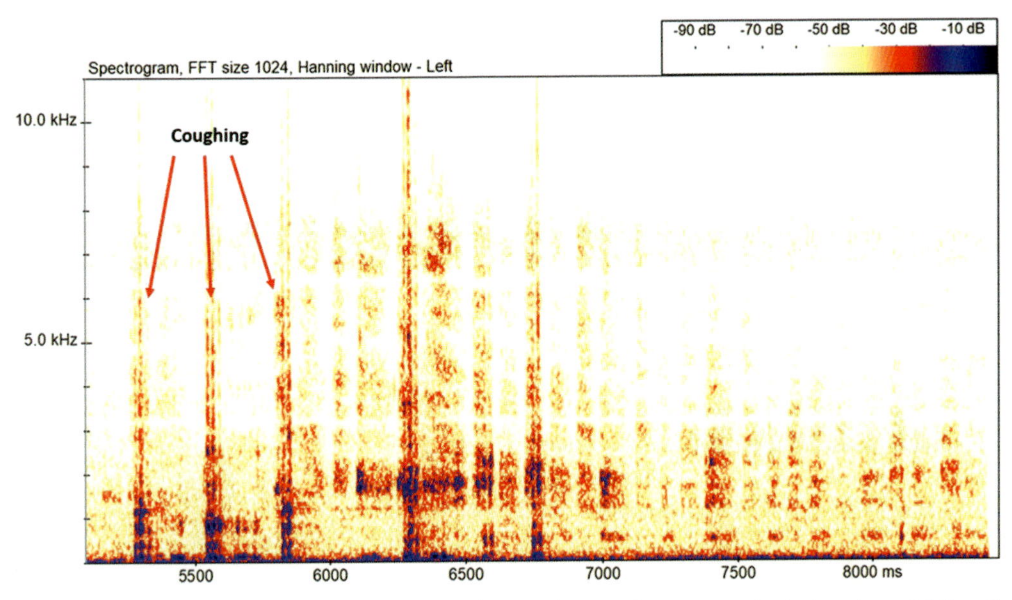

Figure 6.18 Red fox vixen coughs and various other vocalisations in presence of cubs (courtesy of C. Boxall) (frame width: c.3 sec)

QR codes relating to presented figures

Potential confusion between red fox and other species

Table 6.7 Summary of potential confusion between red fox and other species

Confusion group	Potential confusion species In the first instance, always consider habitat and distribution.
Other canine species	Domestic dog.
Species in other groups	Roe deer, Reeves' muntjac and Chinese water deer.

6.5 Mustelid vocalisations

In the following sections, a brief summary of the key characteristics of the seven species of mustelid will be described, together with details on the species-specific vocalisations and, where known, the behavioural function of these sounds.

6.5.1 Badger *Meles meles*

Length/Weight	Habitat preferences	Distribution
Males are slightly bigger than females. Body length up to 80 cm. Weight up to 16 kg.	Broadleaved woodland Sheltered vegetated habitat Parkland, gardens and urban areas Habitat associated with woodland edge (e.g. fields) Unimproved grassland Farmland	Occurs throughout Britain and Ireland, with the exception of many offshore islands.
Diet		
Although a carnivore, where birds, eggs and small mammals feature in their diet, they notably also feed on earthworms, insects and a variety of non-meat food sources (e.g. fruits, nuts, cereals). As such they are arguably better described as omnivorous, since their consumption of non-meat food is far greater than our other carnivores.		

Daily activity pattern	Typically most active from dusk through to dawn, although may be seen during daylight hours in some locations.
Seasonal activity pattern	Less active during colder winter months, when they may remain underground for long periods of time (several days). February to June appears to be the period when they are most active vocally.
Mating activity pattern	Mating usually takes place during the period February to May, although may also occur during July to September. They have delayed implantation of 2–10 months and a gestation of 7 weeks. Cubs are born mostly in early February, appearing above ground from April onwards.
Other notes	Badgers typically live in small social groups known as clans, that average six adults but can include as many as 25. They occupy setts, which can be easy to encounter, meaning that recording vocal activity in the vicinity of setts can be relatively straightforward.

References
Delahay *et al.*, 2008; Woods, 2010; Mathews *et al.*, 2018; Crawley *et al.*, 2020.

Spectrogram examples of acoustic behaviour

Badgers, being active mainly during darkness, are not regarded as being especially visual in how they interpret their environment. In fact, it is well documented that their eyesight is poor. Accordingly, this means that the information they can glean from olfactory or acoustic cues is extremely important for communication within and between clans and individual animals. It should thus come as no surprise that acoustic communication plays a more significant role when compared to some of the other carnivore species.

Badgers tend to be more vocal during the period February to June (Delahay *et al.*, 2008), this coinciding with the animals becoming more active after winter, cubs emerging for the first time from their underground setts, and mating behaviour taking place.

The book *Badger* by Timothy J. Roper (2010) refers to acoustic studies carried out by Wong *et al.*, 1999 and their vocalisations comprising 16 different call types, some of which relate only to either adults or cubs. These acoustic emissions seem to be produced in connection with one of five situations, described as being: aggression, distress, affiliation, mating and play. Table 6.8 provides more detail regarding adult vocalisations, with example call descriptions as they relate to these behaviours. The adult calls we have most commonly encountered ourselves relate to chittering (Figures 6.19 and 6.20), purring (Figures 6.21 and 6.28) and churring (Figure 6.22), as well as squabbling/fighting sequences (Figure 6.23).

Badgers do not appear to have any sort of acoustic alarm signalling, and the snorting sound they sometimes emit when disturbed is most likely an aggressive signal directed towards the intruder, as opposed to signaling any nearby badgers to potential danger (Roper, 2010).

In addition to the adult sounds described within Table 6.8, vocalisations produced only by cubs have been described as cub chittering (Figure 6.28), distress wails (Figures 6.26 and 6.27) and squeaks; affiliation chirps, clucks and coos; and playful squeaks, chirps and clucks.

Some of the calls produced by badgers sound very similar to each other, churrs and purrs for example, and although distinctively badger, they may be difficult to associate directly with behaviour in the absence of context or visual cues.

This is an area of study where one person's description of a particular sound may not necessarily accord with the same sound description given in another source. The words we each choose when describing sound are subjective and are not always transferable to another person's interpretation. As such we have on occasions struggled to assign calls confidently to descriptions provided by various sources, and we would suggest caution when trying to describe certain calls for this species.

Figures 6.19 to 6.28 provide specific examples of spectrograms relating to this species, and where the 🔊 symbol appears, a recording of what is shown within the spectrogram is available to download from the species library. In addition, QR codes are provided for selected examples, and these can be accessed immediately for listening purposes from most mobile devices.

Table 6.8 Adult badger vocalisation adapted and summarised from Christian, 1993; Wong *et al.*, 1999; Roper, 2010 and Charlton *et al.*, 2020

Behaviour	Description	Notes
Aggression	Bark Hiss Growl Snarl Kecker	**Barks** are produced in a series of two to four emissions. **Hisses** are usually produced in isolation, and are described as being similar to a 'sharp, sibilant, cat-like' hiss. **Growls** are often lengthy and lower in frequency (c.32 Hz) than snarls, and can be heard in association with snarls or hisses. Barks, hisses and growls are considered to be lower-intensity threat signals, which usually result in the recipient withdrawing from the situation. **Snarls** (Figure 6.24) are often lengthy (> 1 sec), more complex and higher in frequency (61 Hz) than growls. Snarling is usually indicative of a higher level of threat, and can be accompanied with upright posture from the animal making the sound, these being indicative of a physical attack about to commence. **Keckering** (Figure 6.25) is a quick series of staccato sounds, varying in frequency, and once heard is quite distinguishable. It is thought to be emitted only by adults. Keckering is usually noticeable during physical disputes, and can become more intense as the fighting continues.
Distress	Bark Snort Yelp Screaming	These sounds are connected to startling, distress or pain. **Snorts** are similar to barks, and likewise they are noisy and last longer, but they are emitted through the nose – they may also be associated with threat behaviour. A snort is usually emitted only once. They are often associated with the badger being startled by a potential danger and can be produced in conjunction with upright posturing. **Yelps** are shorter sounds of higher frequency, produced in a series and usually associated with distress or pain (dur. Mean 0.14 sec). **Screaming** is described as almost human-like, and is thought to be produced by animals in extreme distress.
Affiliation	Grunt Purr Purr-click	These are considered to relate to contact and/or 'reassurance' behaviours. They tend to not carry far, and as such you or your recording equipment would need to be close (a matter of a few metres away) in order to pick them up. **Grunts** are short, low-frequency, 'blunt', noisy sounds. **Purrs** (Figure 6.21) (a softer version of the similar-sounding churr) occasionally terminate with a click, and when doing so, are described as a 'purr-clicks'.
Mating	Churr Chitter	**Churrs** (Figure 6.22) are produced by adult males (individually distinctive), exclusively during the mating season in order to attract a female. They are described as lower-frequency, insistent purring, of an 'oily, bubbling' nature. **Chittering** (Figures 6.19 and 6.20) is produced by females during mating when a male is harassing the female, and she wants him to back off. It has also been noted as being produced by the female during copulation, when the male is biting her, which again could be indicative of a 'back off'-type signal.
Play	Bark Yelp Chitter Growl	These sounds are all associated with play behaviour, each individually also perhaps conveying a message within such contexts, similarly to how they have been described elsewhere within this table. During play, combinations of these sounds would be expected, with the various sounds all running quickly into each other, making it more difficult to distinguish acoustically the various call types as described.

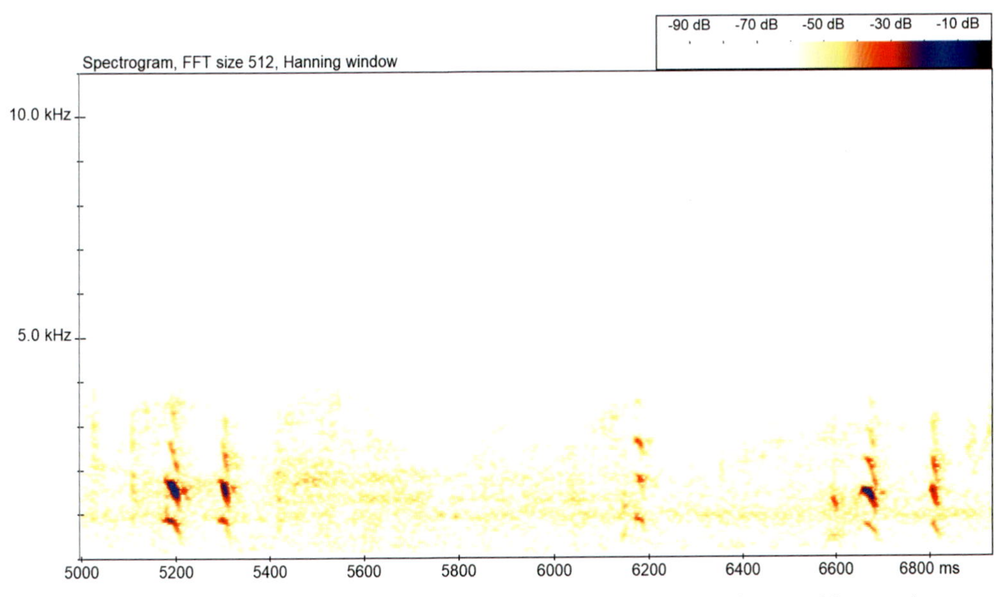

Figure 6.19 🔊 Badger – short sequences of chittering (courtesy of M. Ingham) (frame width: c.2 sec)

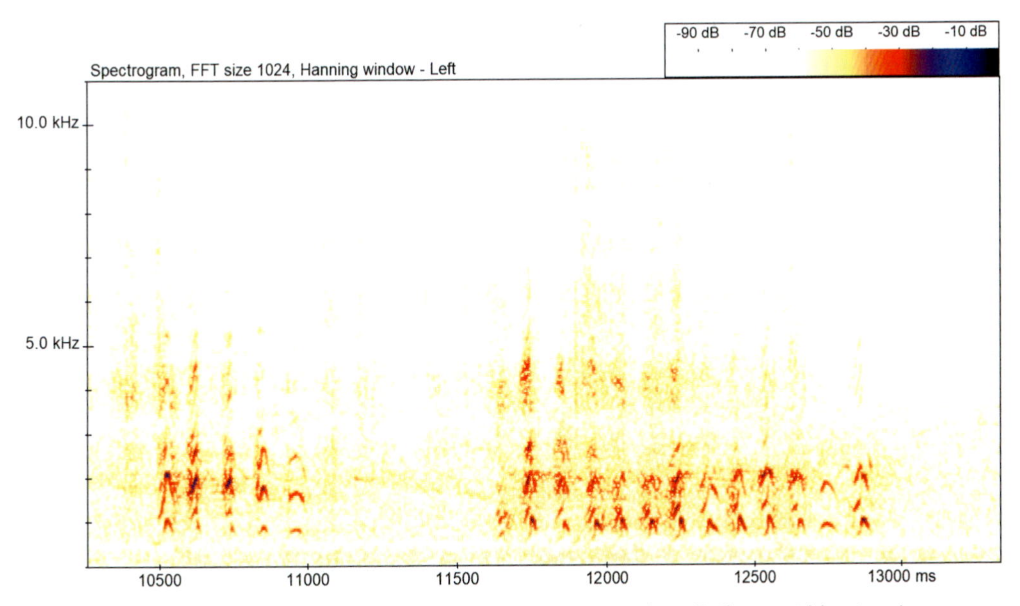

Figure 6.20 🔊 **(QR)** Badger – longer chittering sequence (courtesy of P. Mill) (frame width: c.3 sec)

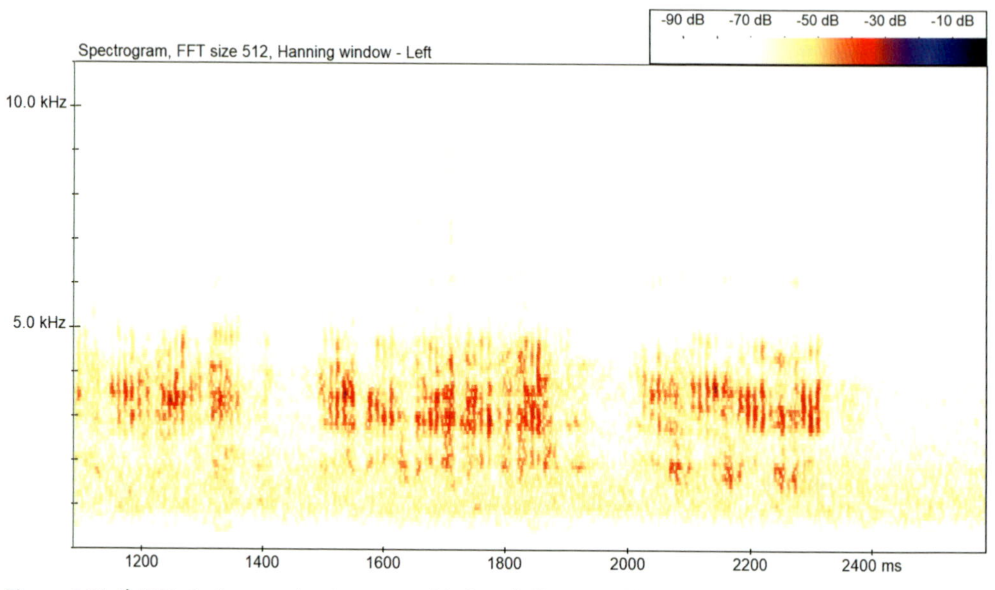

Figure 6.21 🔊 **(QR)** Badger purring (courtesy of A. Brand) (frame width: c.1.5 sec)

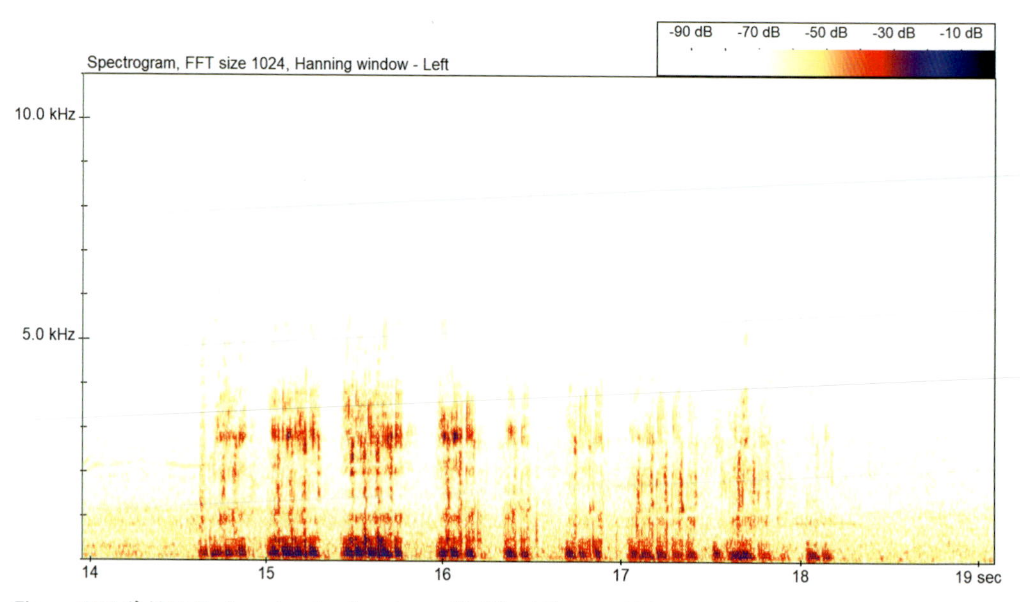

Figure 6.22 🔊 **(QR)** Badger churring (courtesy of S. Elliott) (frame width: c.5 sec)

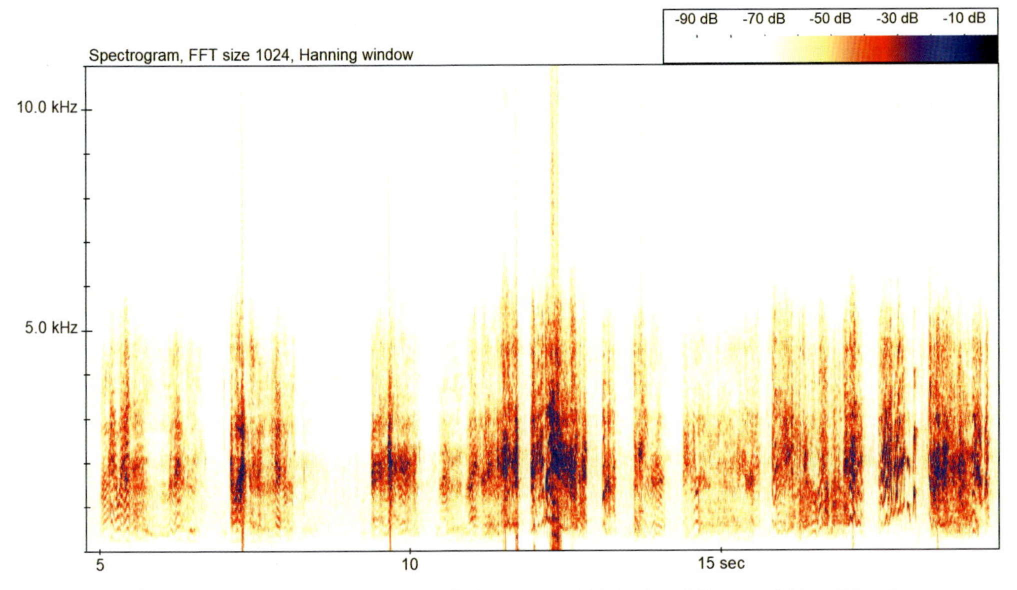

Figure 6.23 Badgers squabbling – various calls (courtesy of M. Ingham) (frame width: c.15 sec)

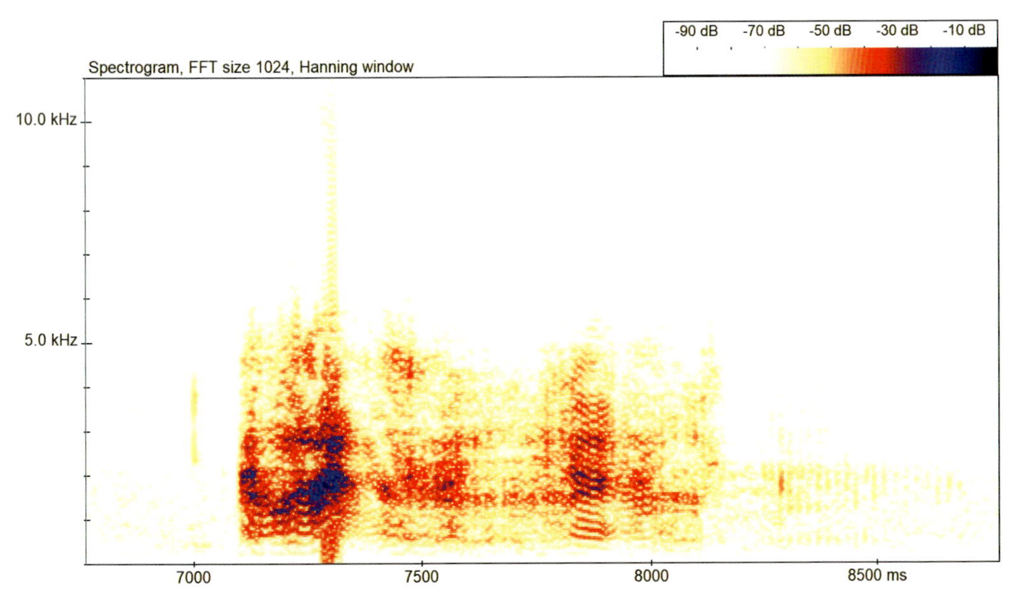

Figure 6.24 Badger snarl (courtesy of M. Ingham) (frame width: c.2 sec)

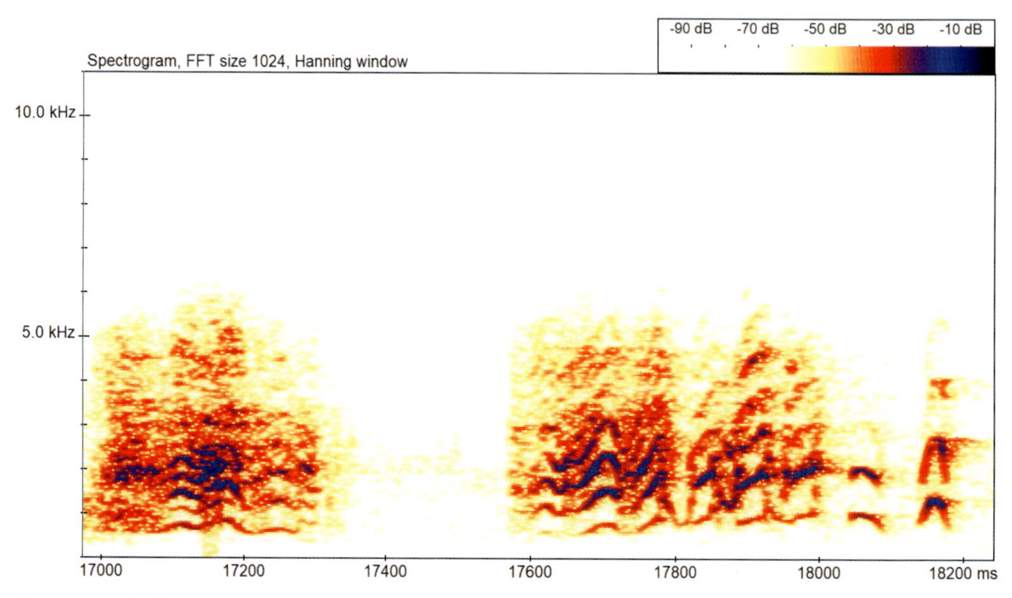

Figure 6.25 Badger keckering sequences (courtesy of M. Ingham) (frame width: c.1.25 sec)

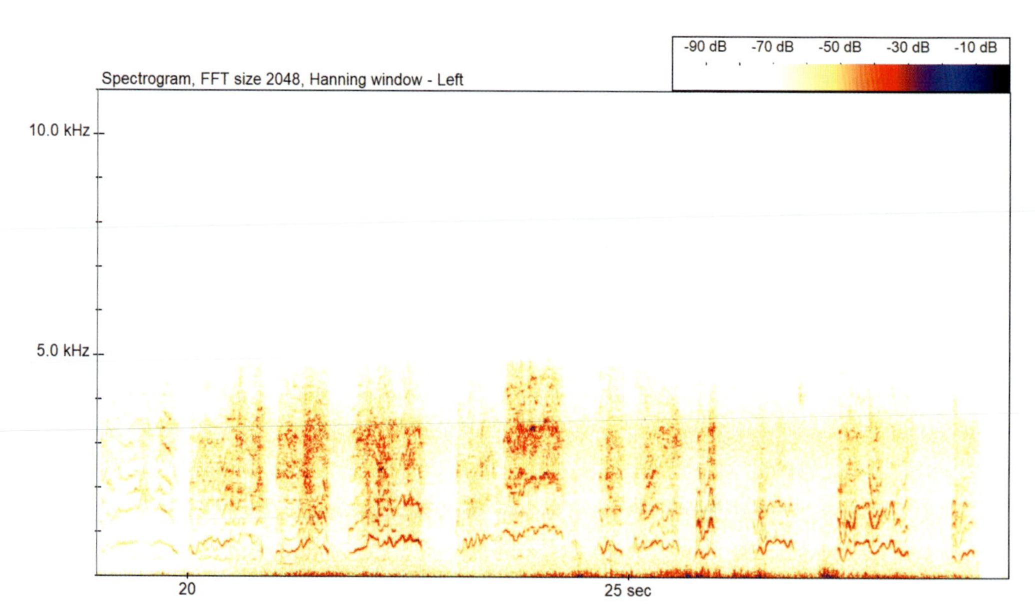

Figure 6.26 Badger keckering/wailing (courtesy of A. Middleton and L. Spence) (frame width: c.10 sec)

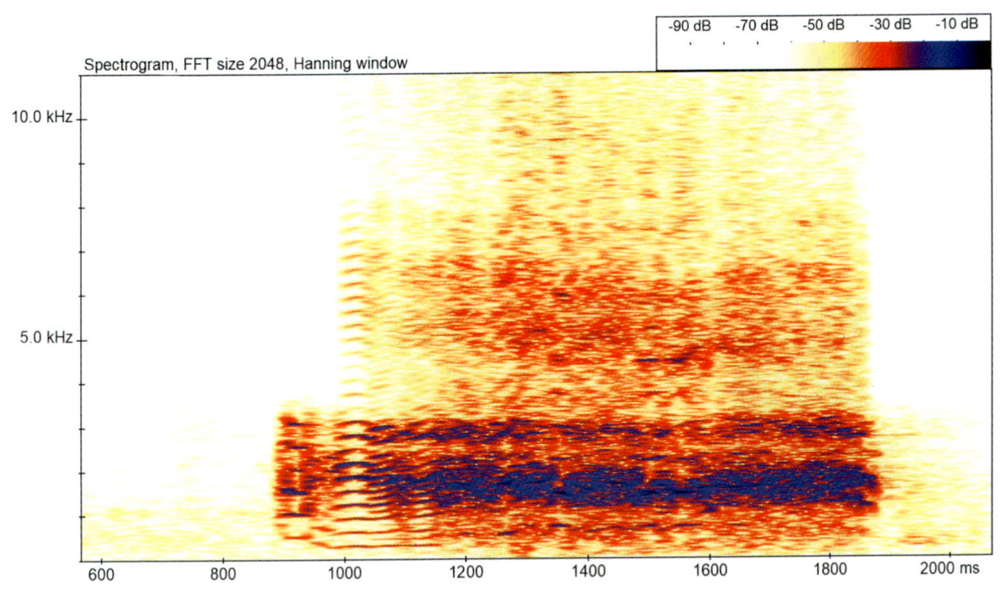

Figure 6.27 🔊 Badger wailing (courtesy of C. Wren, trogtrogblog.blogspot.com) (frame width: c.1.5 sec)

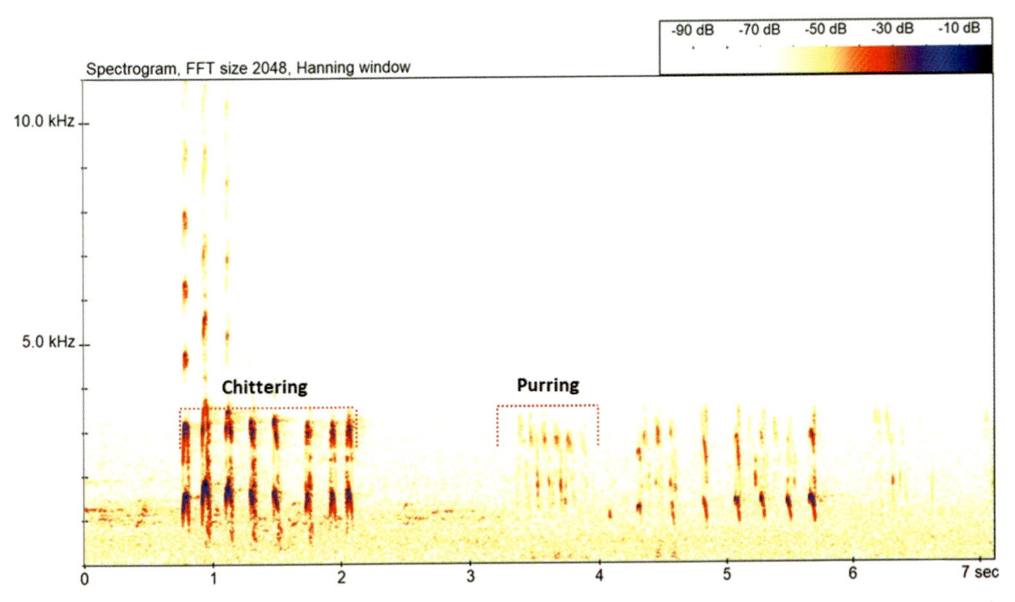

Figure 6.28 🔊 Badger cub chittering, with adult purring (courtesy of C. Wren, trogtrogblog.blogspot.com) (frame width: c.7 sec)

QR codes relating to presented figures

Figure 6.20: Badger chittering	**Figure 6.21:** Badger purring	**Figure 6.22:** Badger churring

Potential confusion between badger and other species

Table 6.9 Summary of potential confusion between badger and other species

Confusion group	Potential confusion species In the first instance, always consider habitat and distribution.
Other badger species	The only species of badger in our area.
Species in other groups	Unlikely to be confused with other mammals.

6.5.2 Otter *Lutra lutra*

Length/Weight	Habitat preferences	Distribution
Males are larger than females. Body length (excluding tail) up to c.70 cm. Weight c.10 kg.	River valleys and tributaries Water bodies Wetlands Coastal	Occurs throughout Britain and Ireland including many of the Scottish islands, notably Orkney and Shetland. Scarcer within southern Ireland.
Diet		
Carnivore feeding across a range of fish and other water-related food sources, such as eels, mussels, amphibians, water voles, waterbirds and crayfish.		
Daily activity pattern	Typically most active from dusk through to dawn, although may be seen during daylight hours in some locations (e.g. coastal areas).	
Seasonal activity pattern	Active throughout the year, with males occupying larger territories than females (with or without young).	
Mating activity pattern	Mating activity can occur at any time of the year. Males play no part in rearing cubs (usually two or three) which can be born any time of the year. Young usually become independent at c.10 months.	
Other notes	More easily encountered visually at dusk or dawn. Holts are usually within natural cavities (e.g. river banks under tree roots or boulders), and in wetland habitats they also rest above ground in 'couches'. They tend to have long linear territories associated with water habitat (e.g. river valleys and the coastline).	

References
Kruuk, 2006; Jefferies and Woodroffe, 2008; Mathews *et al.*, 2018; Crawley *et al.*, 2020.

Spectrogram examples of acoustic behaviour

Otters can be more vocal during mating activity (Woodroffe, 2007), which can take place at any time of the year (Green *et al.*, 1984). Typically, vocalisations occur at frequencies < 10 kHz (Gnoli and Prigioni, 1995).

Calls described in adults include a 'whistle' contact call (Figures 6.29 and 6.30). A similar version of this contact call is made by cubs calling when separated from their mother (Figure 6.31), and our example suggests that the calls of cubs are produced at a higher sound frequency.

'Chittering' threat calls and similar sounds emitted during mating encounters are regularly recorded (Figures 6.32 and 6.33) and are also associated with prolonged aggressive engagements, as well as playful encounters between youngsters (Figures 6.34 and 6.35). A 'hah' or huff-sounding alarm call (Figure 6.36) is also notable (Gnoli and Prigioni, 1995; Jefferies and Woodroffe, 2008).

A useful study carried out within the British Isles (Gordon, 2015) used camera trap technology in conjunction with sound analysis software to gather and describe calls made by otters in Fife (Scotland). In total, 81 sounds were extracted from video clips for which 11 distinct vocalisation categories were defined: (1) Chitter, (2) Hiss, (3) Huff, (4) Lip Smack and Lip Smack Variant, (5) Mew, (6) Scream, (7) Short Whistle and Short Whistle Variant, (8) Toot Toot and Toot Toot Variant, (9) Whicker, (10) Whoomph and (11) Snort (single, double and triple).

In order to provide some first-hand detail about otter vocalisations we approached Melanie Findlay (Findlay Ecological Services), whose camera trap video footage contributed to the previously mentioned research carried out by Shona Gordon (Gordon, 2015):

> *The contact call, a high-pitched whistle between mothers and cubs (Woodroffe, 2007) is the one that is usually heard. It's a repeated short whistle, often sounding like 'an insistent dunnock', and carries for some distance. The other common noises are very quiet. One is a very quiet 'murmur/chitter' between mum and cub when they are very close to each other, and the other is the 'huff'. These are unlikely to be heard in isolation, without the contact noise. There are also various 'mewing' noises, but again rare in isolation. Young will repeatedly call for mother if separated, with the call sounding a bit like someone air kissing, or 'lip smacking' as described in Gordon, 2015. (Melanie Findlay, Findlay Ecology Services, pers. comm., 2023)*

Figures 6.29 to 6.36 provide specific examples of spectrograms relating to this species, and where the 🔊 symbol appears, a recording of what is shown within the spectrogram is available to download from the species library. In addition, QR codes are provided for selected examples, and these can be accessed immediately for listening purposes from most mobile devices.

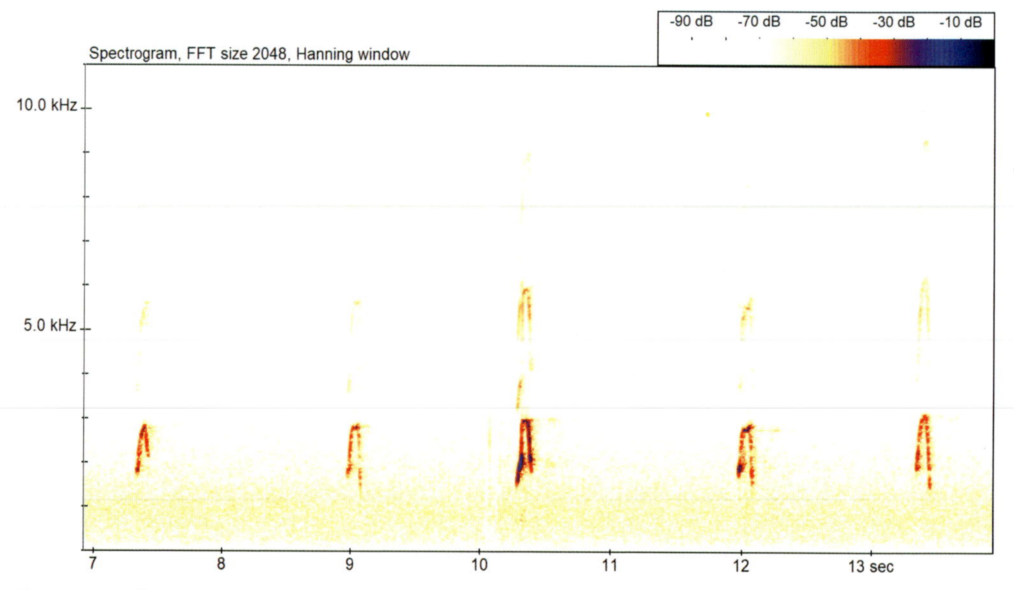

Figure 6.29 🔊 (QR) Otter whistles (male) (courtesy of C. Wren, trogtrogblog.blogspot.com) (frame width: c.7 sec)

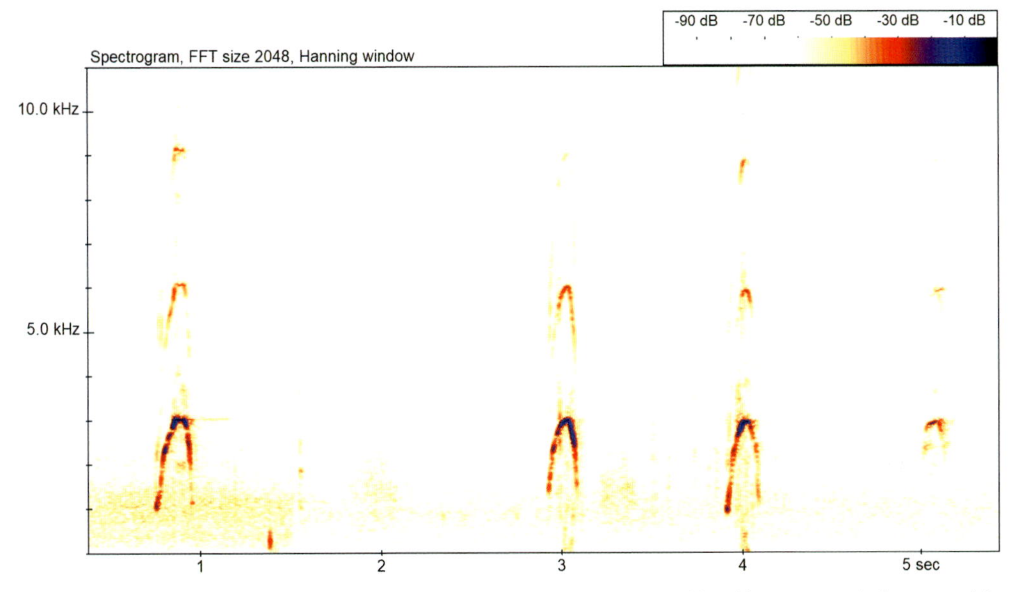

Figure 6.30 🔊 Otter whistles (female) (courtesy of C. Wren, trogtrogblog.blogspot.com) (frame width: c.5.5 sec)

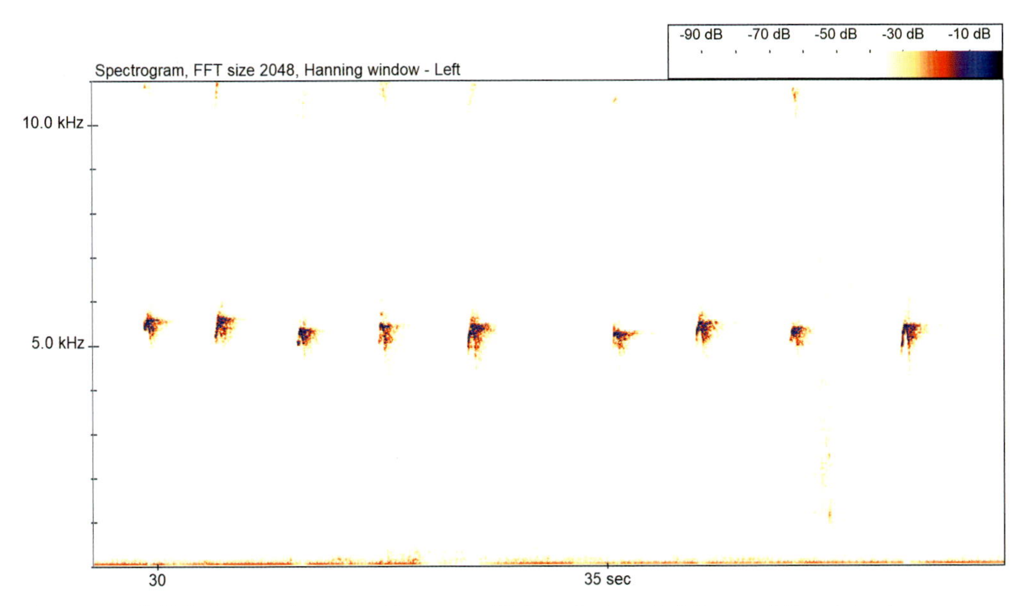

Figure 6.31 🔊 **(QR)** Otter whistles (cub) (courtesy of Echoes Ecology Ltd) (frame width: c.10 sec)

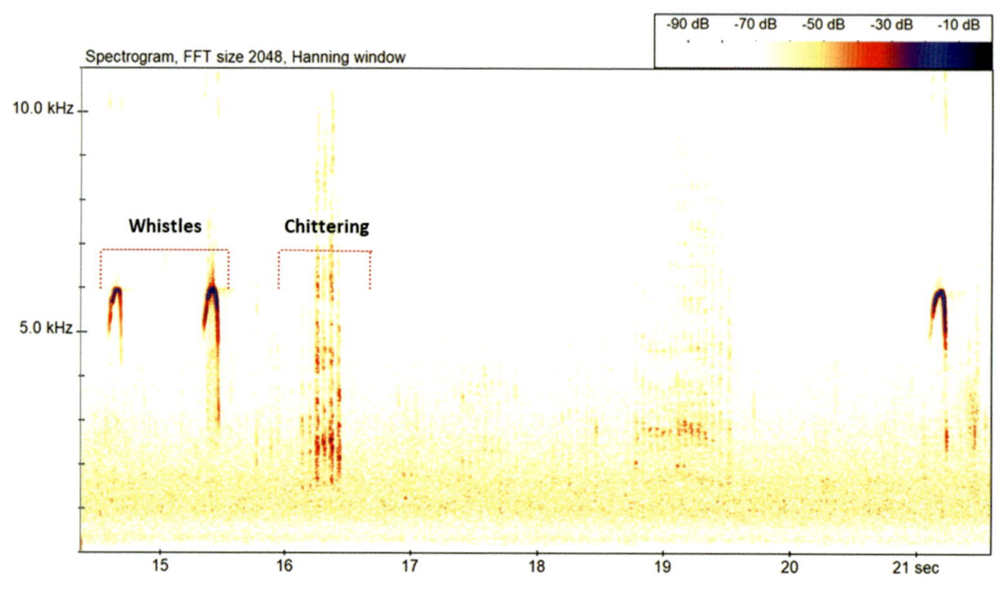

Figure 6.32 🔊 Otter whistles with short bout of chittering (courtesy of S. Elliott) (frame width: c.7 sec)

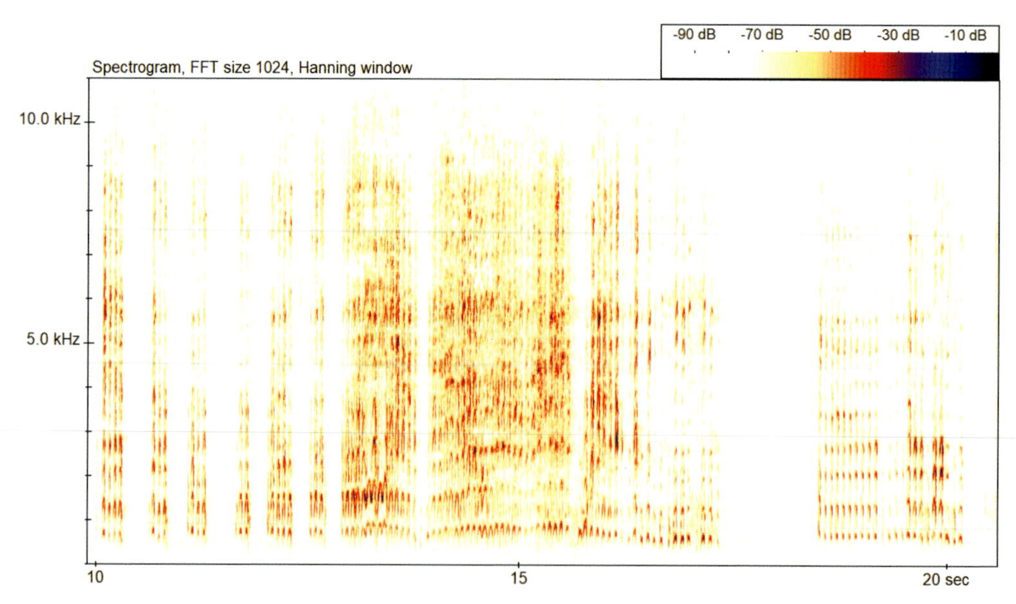

Figure 6.33 🔊 **(QR)** Otter chittering during mating activity (courtesy of M. Findlay) (frame width: c.10 sec)

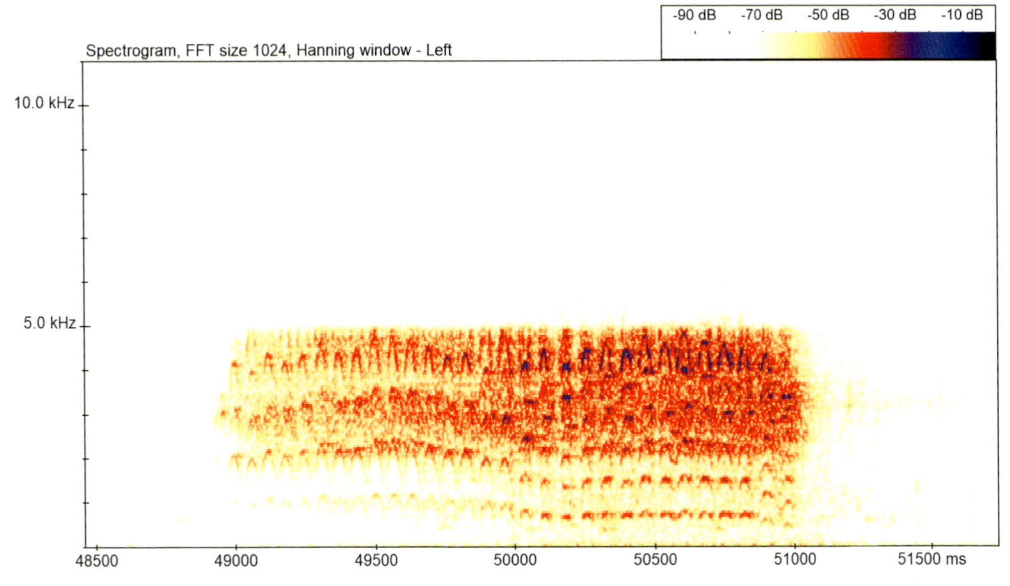

Figure 6.34 🔊 Otter female – fast chittering, whilst chastising youngster (courtesy of M. Ingham) (frame width: c.3 sec)

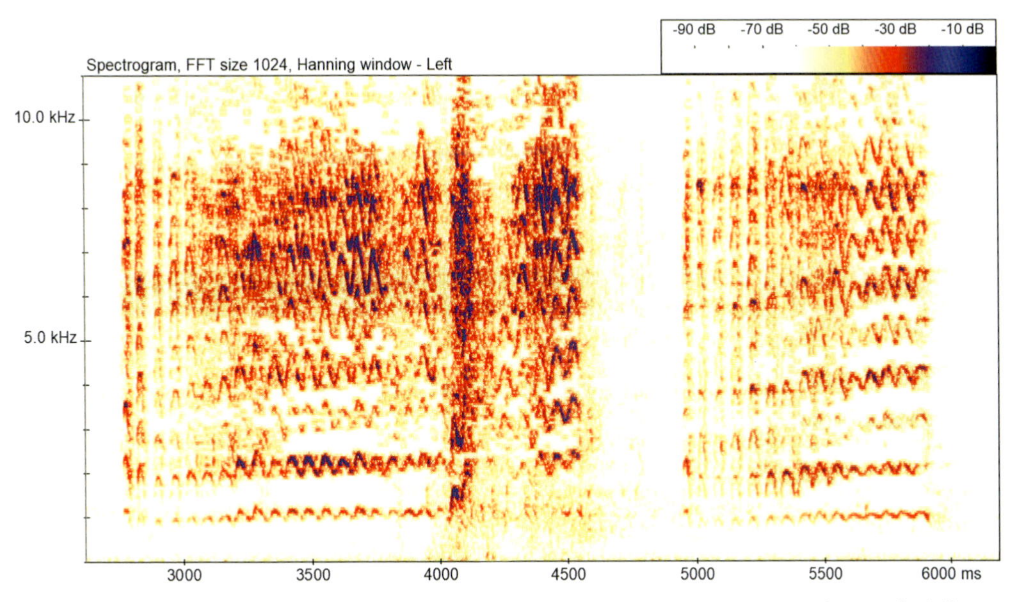

Figure 6.35 🔊 Otter – excitable, perhaps aggressive, chittering encounter (courtesy of M. Findlay) (frame width: c.4 sec)

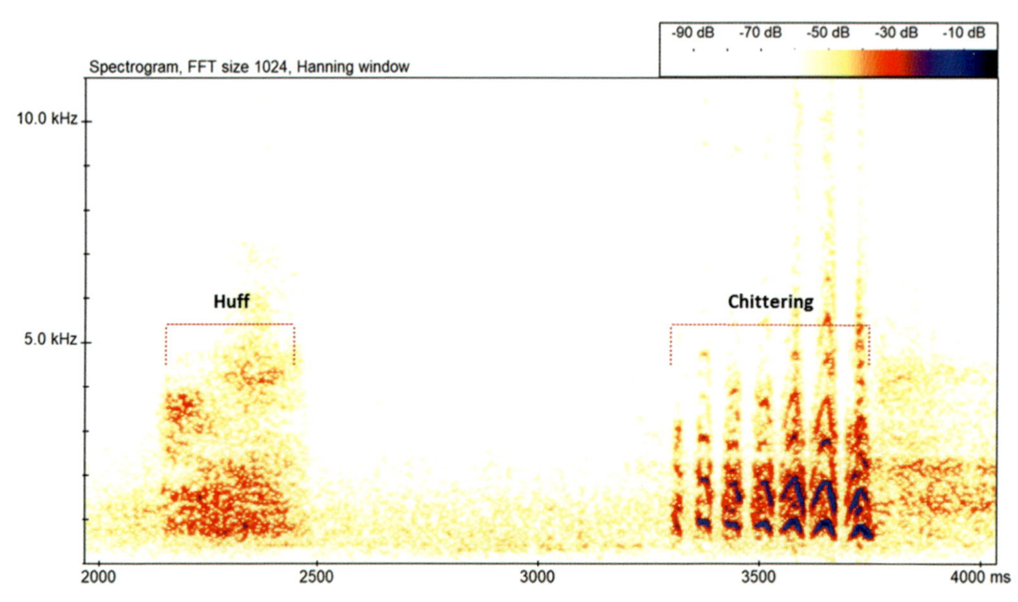

Figure 6.36 🔊 Otter hah/huff call followed by chittering (courtesy of C. Wren, trogtrogblog.blogspot.com) (frame width: c.2 sec)

QR codes relating to presented figures

Figure 6.29: Otter whistle (male)	**Figure 6.31:** Otter whistle (cub)	**Figure 6.33:** Otter chittering

Potential confusion between otter and other species

Table 6.10 Summary of potential confusion between otter and other species

Confusion group	Potential confusion species
	In the first instance, always consider habitat and distribution.
Other mustelid species	Unlikely to be confused with other mustelid species.
Species in other groups	Unlikely to be confused with other mammals. May be confused with bird species (e.g. some passerine and wader calls).

6.5.3 Pine marten *Martes martes*

Length/Weight	**Habitat preferences**	**Distribution**
Males are larger than females. Body length (excluding tail) 41 to 52 cm. Weight up to c.2.15 kg.	Coniferous woodland Broadleaved woodland Habitat associated with woodland edge	Occurs across much of Scotland, northern England and parts of Wales, as well as established in many areas across Ireland. Other pockets of population occur elsewhere. They are absent from offshore islands. 
Diet		
Predominantly carnivorous, feeding across a vast range of food sources, including small mammals, birds, eggs, carrion, amphibians and insects, they supplement their diet with honey, fungi and berries when available. Notably, in Ireland, pine martens are much more frugivorous and insectivorous than elsewhere in our Isles.		

Daily activity pattern	Typically most active from dusk through to dawn, although may be seen during daylight hours more often during summer months.
Seasonal activity pattern	Active throughout the year (although activity reduced during winter), with males and females (with or without young) being territorial, either with territories overlapping or adjacent.
Mating activity pattern	Mating usually takes place during June to August, and females may mate with several males. Births take place the following spring, utilising a rearing den with young (typically two or three, but can range from one to five) which gain independence at c.6 months.
Other notes	They typically use dens within tree cavities, old squirrel dreys, bird nests or built structures, albeit less frequently. Being solitary animals, they are usually quiet, but emit a range of known vocalisations that may be encountered dependent upon circumstances.

References
Balharry *et al.*, 2008; Birks, 2017; Mathews *et al.*, 2018; Crawley *et al.*, 2020.

Acoustic behaviour including spectrogram examples

Although not regarded as being particularly vocal, a variety of calls have been noted for this species, including a *'tok-tok-tok'* call (Macdonald and Barrett, 1995), as well as various vocalisations made during the mating season. When territories overlap, as they often do with male territories being larger than female territories, encounters between pine martens can be more vocal. Despite this, it is actually very rare to hear pine martens calling in the wild (J. Birks, pers. comm., 2023).

During the breeding season, growls (Figure 6.37), screaming/squealing (Figure 6.38), chattering (Figure 6.39) and snarling (Figure 6.40) may be heard as a result of aggressive encounters and male/female interaction. Other calls recorded and included here are a more aggressive roar during a feeding dispute (Figure 6.41), groaning sounds (Figure 6.42), as well as a wet purr emitted regularly by an isolated adult in response to finding food (Figure 6.43). Murmurs (Figure 6.44) produced by a male approaching a female during the breeding season have also been noted. Most of these close-quarter calls do not carry over a long distance, so one needs to be nearby to hear them.

Contact calls between mother and young are also to be expected, including vocalisations from young separated from mother, described as a *'sharp, rasping sound like paper being torn'* (Balharry *et al.*, 2008), begging calls (Figure 6.45) and distress calls (Figure 6.46).

We were very fortunate to be given access to the excellent YouTube documentary *The Secret Life of the Pine Marten* by Dan Bagur for the purposes of providing examples for this species. Many of the following spectrograms have been created from this video, which is thoroughly recommended and can be viewed via the following link:

www.youtube.com/watch?v=azqUn4K1M3k

Figures 6.37 to 6.46 provide specific examples of spectrograms relating to this species, and where the 🔊 symbol appears, a recording of what is shown within the spectrogram is available to download from the species library. In addition, QR codes are provided for selected examples, and these can be accessed immediately for listening purposes from most mobile devices.

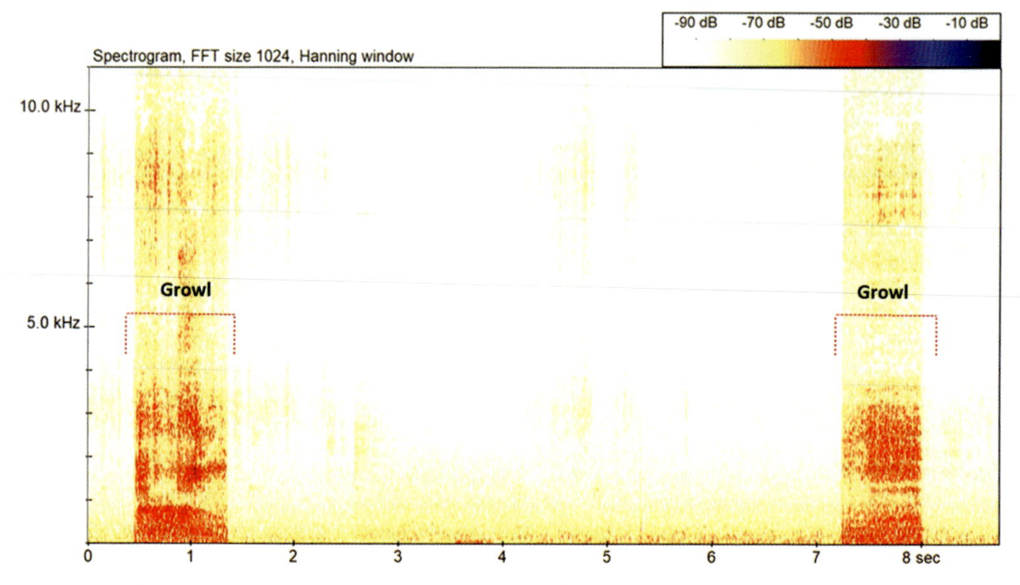

Figure 6.37 🔊 Pine marten – growls (courtesy of D. Bagur) (frame width: c.8 sec)

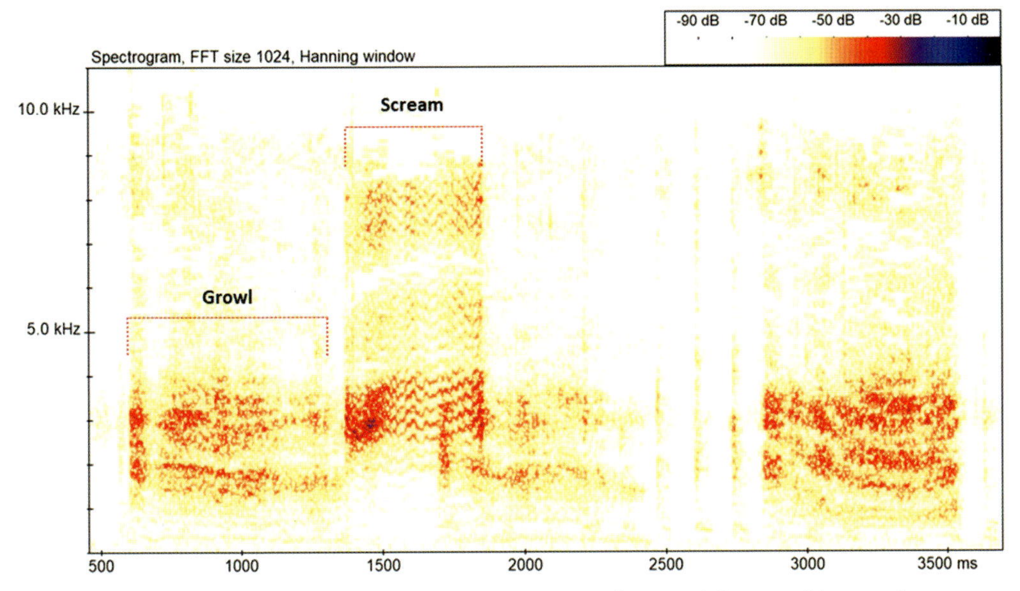

Figure 6.38 🔊 **(QR)** Pine marten – scream and growls (courtesy of D. Bagur) (frame width: c.3 sec)

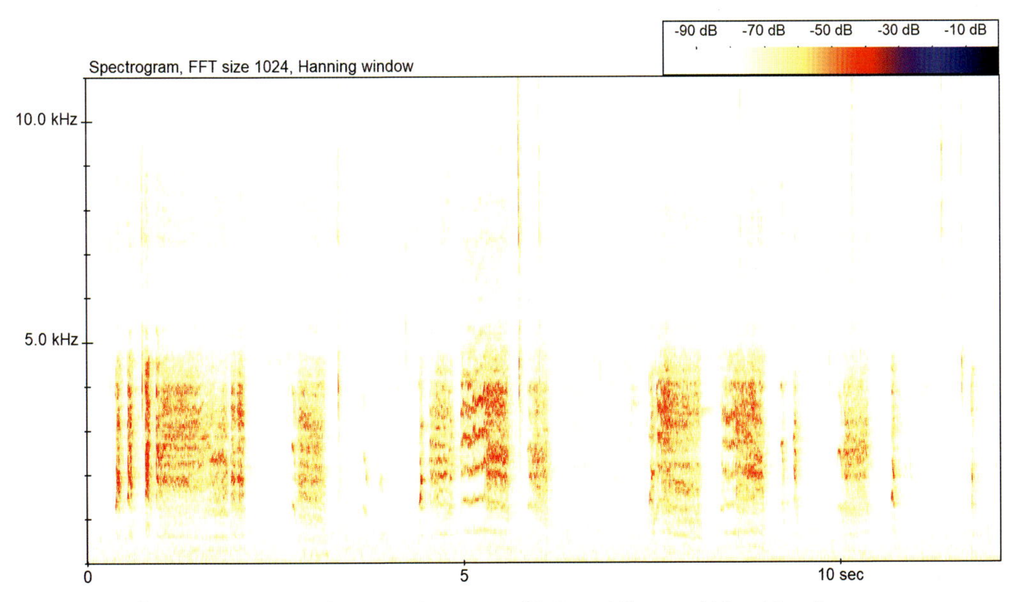

Figure 6.39 🔊 **(QR)** Pine marten chattering (courtesy of D. Bagur) (frame width: c.12 sec)

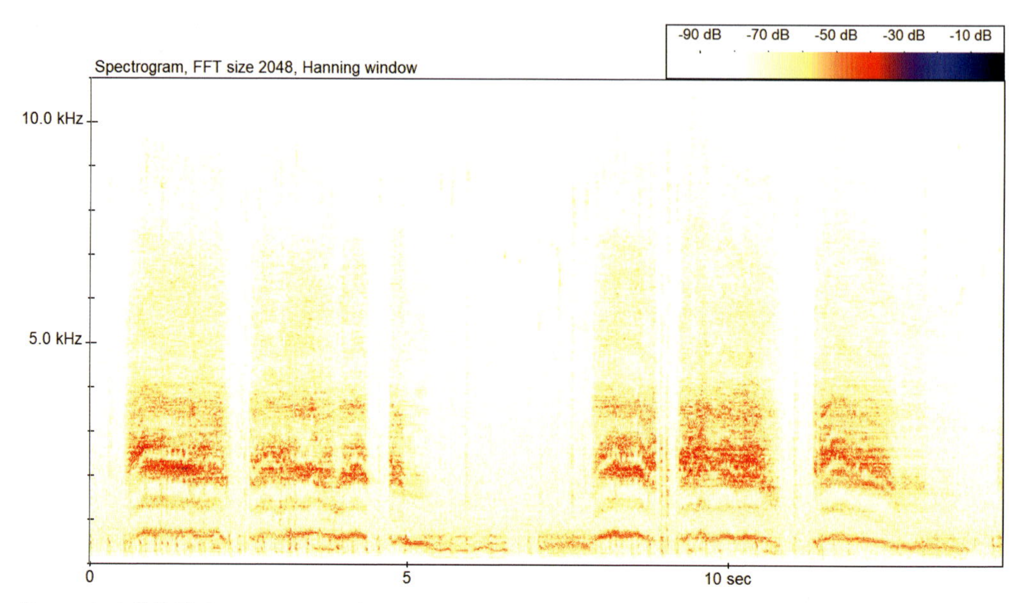

Figure 6.40 🔊 **(QR)** Pine marten snarling (courtesy of B. Jackson) (frame width: c.14 sec)

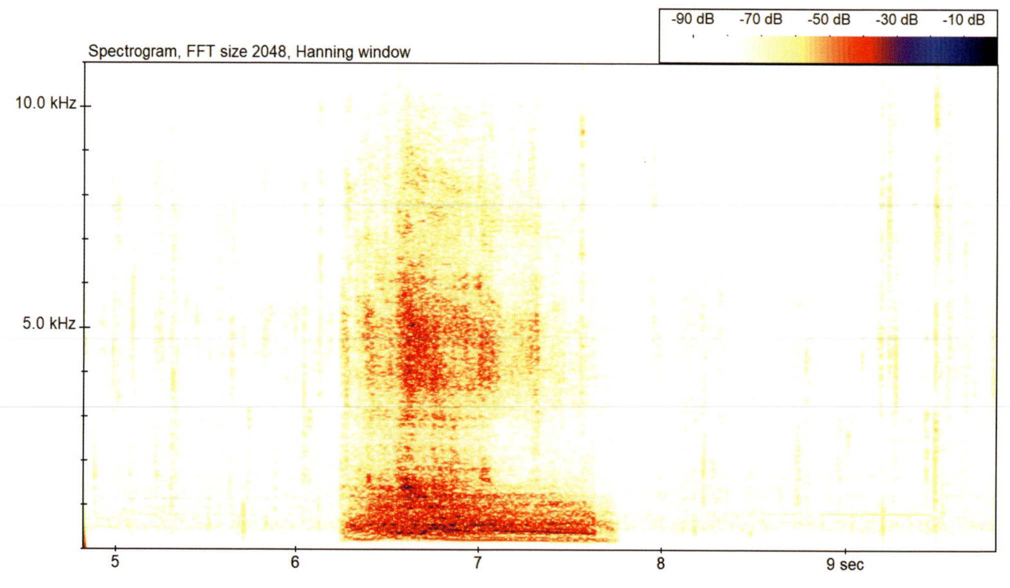

Figure 6.41 🔊 Pine marten roar (courtesy of P. Howden-Leach) (frame width: c.5 sec)

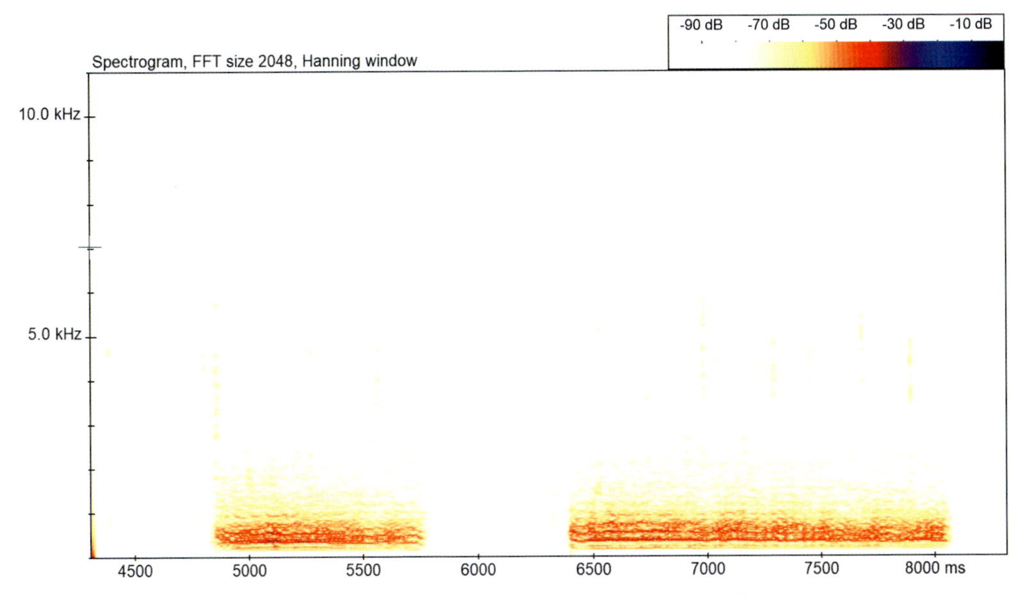

Figure 6.42 Pine marten groaning (courtesy of P. Howden-Leach) (frame width: c.4 sec)

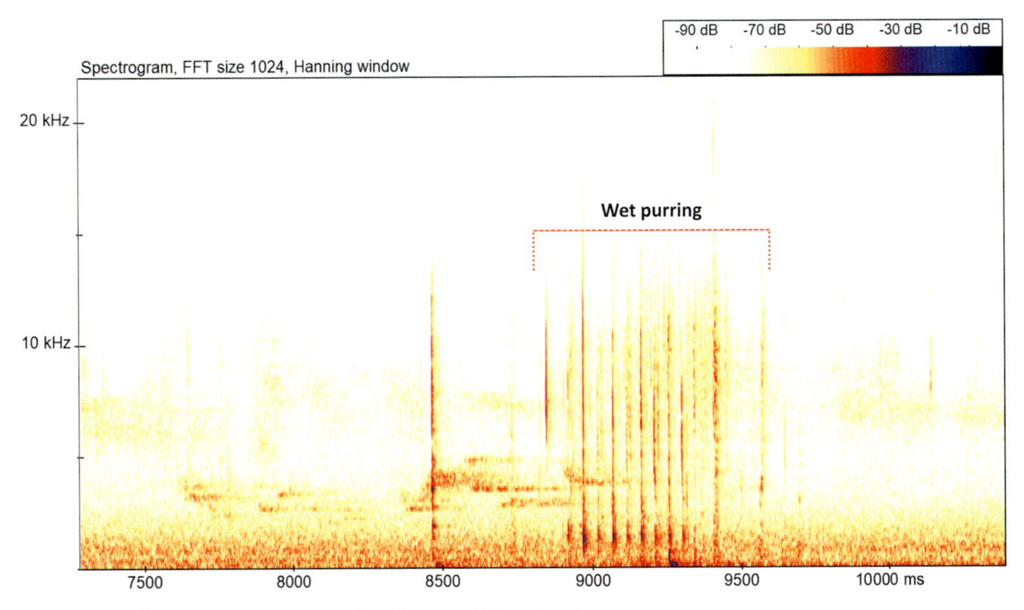

Figure 6.43 Pine marten wet purring (frame width: c.3 sec)

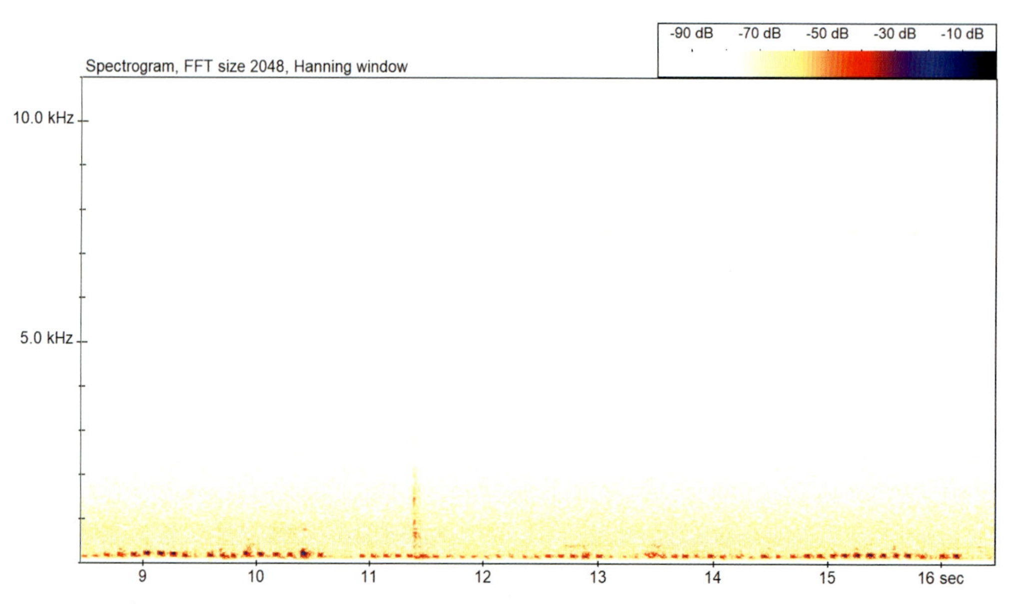

Figure 6.44 🔊 Pine marten murmurs (courtesy of D. Bagur) (frame width: c.8 sec)

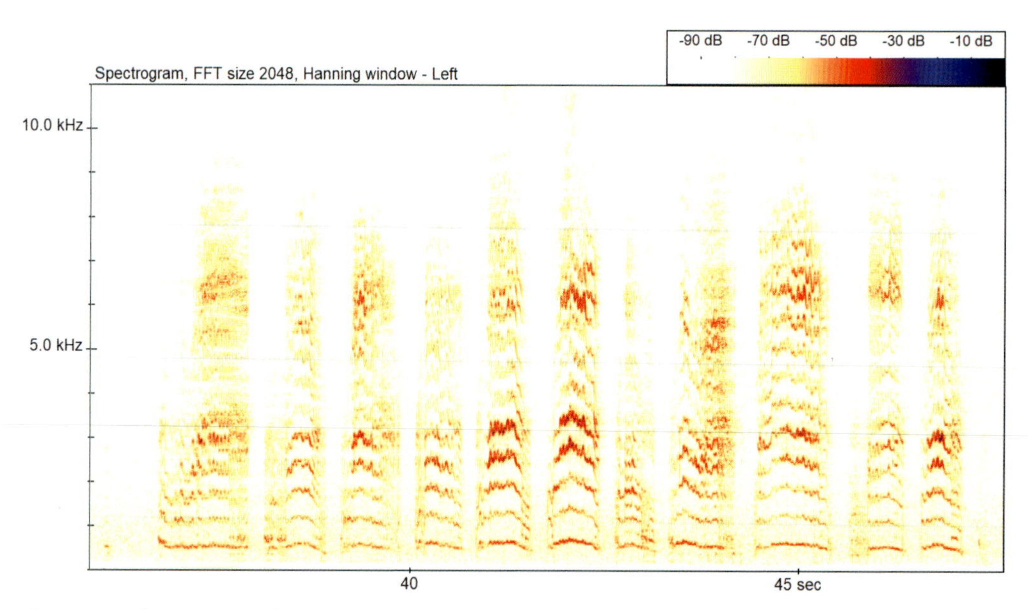

Figure 6.45 🔊 Pine marten kit begging calls (courtesy of B. Jackson) (frame width: c.11 sec)

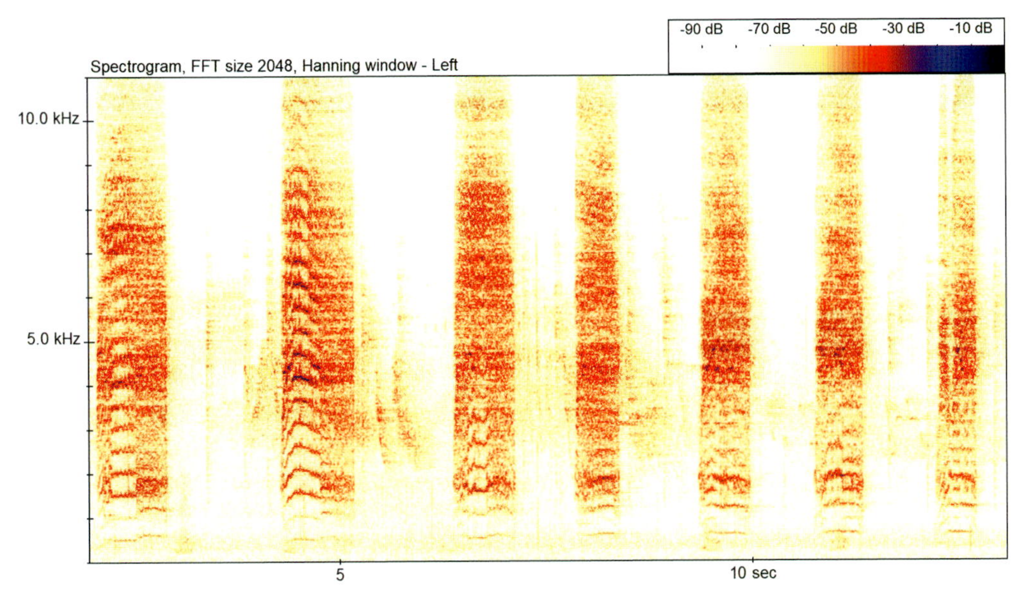

Figure 6.46 🔊 Pine marten juvenile distress calls (courtesy of D. Alder) (frame width: c.10 sec)

QR codes relating to presented figures

Figure 6.38: Pine marten screams, squeals and growls	**Figure 6.39:** Pine marten chattering	**Figure 6.40:** Pine marten various, including snarling

Potential confusion between pine marten and other species

Table 6.11 Summary of potential confusion between pine marten and other species

	Potential confusion species
Confusion group	In the first instance, always consider habitat and distribution.
Other mustelid species	All of the smaller mustelids make a range of noises that may be difficult to pin to one species or another, especially as most of us are unfamiliar with them anyway! In the absence of a sighting or field evidence to confirm species, the setting up of a trail camera is recommended.
Species in other groups	Unlikely to be confused with other mammals.

6.5.4 Stoat *Mustela erminea*

<table>
<tr><th>Length/Weight</th><th>Habitat preferences</th><th>Distribution</th></tr>
<tr><td>Larger than weasel, with a longer black-tipped tail.
Males are larger than females.
Body length (excluding tail) up to 286 mm.
Weight up to 342 g.</td><td rowspan="3">Coniferous woodland
Broadleaved woodland
Unimproved grassland
Farmland
Upland/Moorland</td><td rowspan="3">Occurs throughout Britain and Ireland, with the exception of some of the offshore islands.</td></tr>
<tr><th>Diet</th></tr>
<tr><td>Carnivorous, feeding across a vast range of food sources, including rabbit, songbirds, eggs and small mammals.</td></tr>
</table>

Daily activity pattern	Typically active during the day and night, with individuals taking breaks from activity when appropriate.
Seasonal activity pattern	Active throughout the year, with males (solitary) and females (with or without young) being territorial. Male territories are bigger and include smaller female territories within.
Mating activity pattern	Mating occurs during April to July, and most births take place the following spring. Litter size of six to nine. Young normally gain independence at c.12 weeks when the family group breaks up.
Other notes	Shelters within dens at ground level, these usually being burrows already created by other mammals, or under fallen branches or rocks. Not usually heard, but some audible vocalisations can be encountered, as well as ultrasonic emissions.

References
McDonald and Harris, 2006; McDonald and King, 2008a; Mathews *et al.*, 2018; Crawley *et al.*, 2020.

Acoustic behaviour including spectrogram examples

Stoats have been heard to make a range of audible vocalisations (trill- and chatter-type sounds) connected with mating activity, aggressive behaviour and interactions between mother and young. Figures 6.47 to 6.51 provide examples of audible vocalisations (chatters, snarls, rasping sounds and squeaks), albeit we are not in a position to determine behavioural context regarding each of these recordings.

In addition, a range of ultrasonic sequences have been recorded by the authors (Figures 6.52 to 6.58); again, however, it is uncertain what behaviours these relate to. Although no academic research appears to have been carried out specifically on the effective hearing range of stoats, it may be suggested that perhaps it could be similar to that of weasel (Heffner and Heffner, 1985) and American mink (Powell and Zielinski, 1989; Brandt *et al.*, 2013), where it has been shown that high-frequency sounds (e.g. up to 60.5 kHz for weasel, and 70 kHz for mink) are detectable. In our examples here, we appear to have recorded emissions coming from stoat beyond the anticipated possible detectable hearing range, which suggests that this species may also have the ability to hear well beyond 60.5 kHz – otherwise there would be no reason for or benefit to producing sound that cannot be heard by conspecifics. Some of these high-frequency calls exhibit similarities with their prey (e.g. brown rat and voles) and may serve as a predatory function, or in a different context, as communication with neonates. Our recordings relate to a captive family, comprising one adult female (mother) and five kits.

When considering the ultrasonic emissions, certain ultrasonic styles were noted regularly, these being as follows. Type 1 (Figure 6.52) occurs at a frequency beyond 60 kHz and is quite complex, as it rapidly changes frequency throughout, usually falling for a short period before rising again. Type 2 calls (Figure 6.54) appear overall to be similar to Type 1, but a much smoother transition, with the rise in frequency at the end being more sharply defined. Type 3 calls (Figures 6.55 and 6.56) appear at c.35/40 kHz and tend to be a series of short emissions, sometimes with the first component of longer duration than the rest.

Other ultrasonic sounds (Figures 6.57 and 6.58), similar to rat species, were also collected, these appearing to be emitted at higher frequencies than would be expected for our rat species. Finally, Figure 6.59 shows a variety of call structures within a single sequence.

Figures 6.47 to 6.51 provide specific audible examples of spectrograms relating to this species, and Figures 6.52 to 6.59 provide ultrasonic examples. Where the 🔊 symbol appears, a recording of what is shown within the spectrogram is available to download from the species library. In addition, QR codes are provided for selected examples, and these can be accessed immediately for listening purposes from most mobile devices.

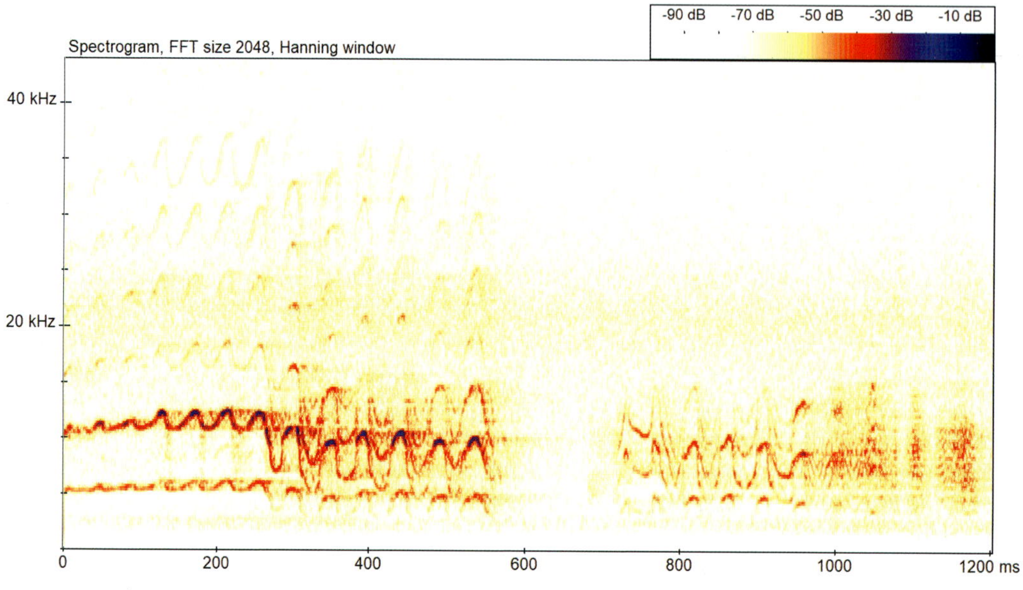

Figure 6.47 🔊 Stoat – complex audible calls (frame width: c.1.2 sec)

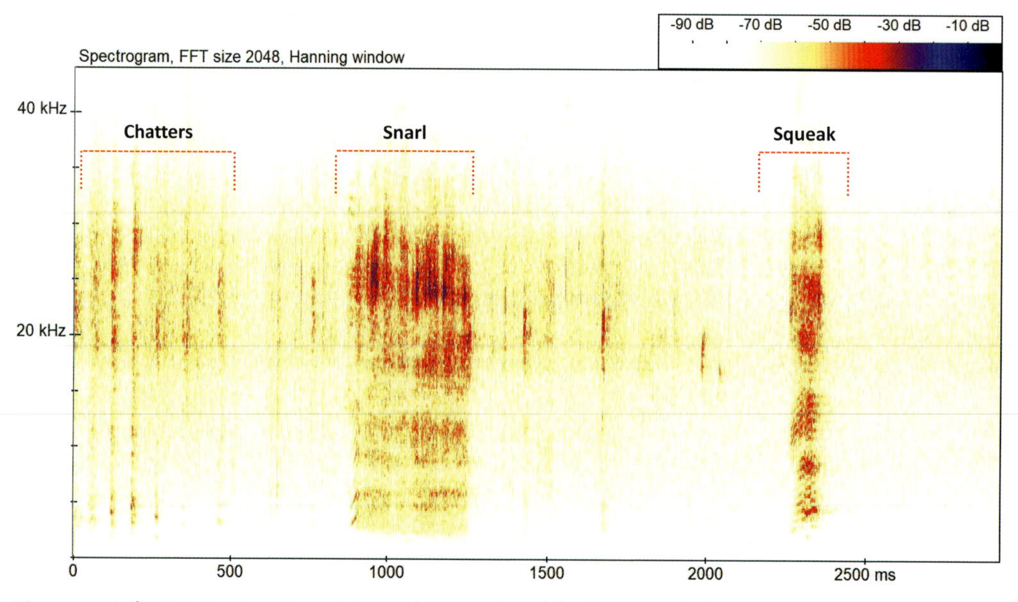

Figure 6.48 🔊 **(QR)** Stoat chatters, followed by a snarl, and finally a squeak (frame width: c.3 sec)

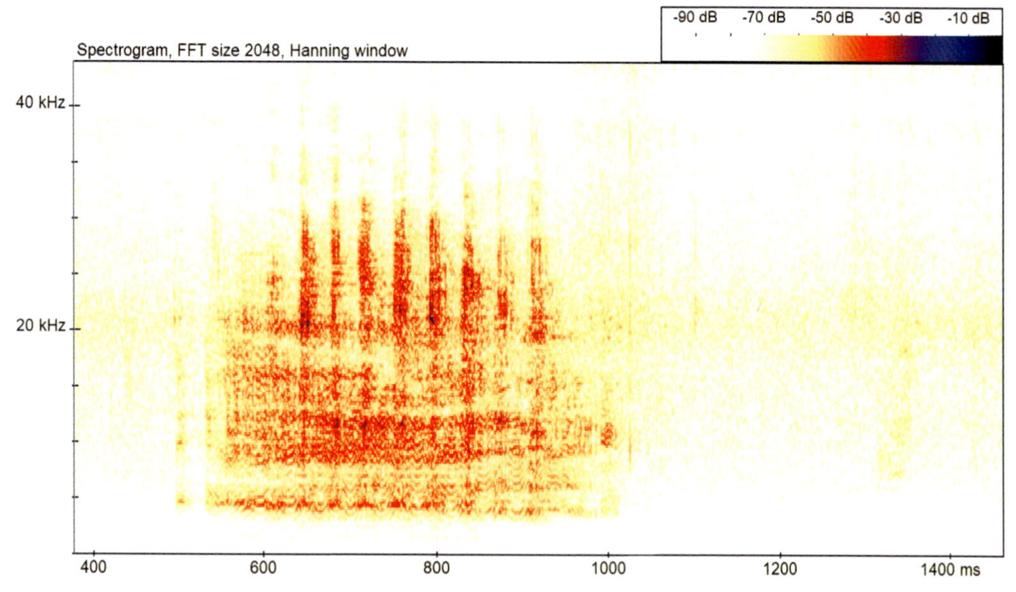

Figure 6.49 🔊 Stoat – rasping sound (frame width: c.1 sec)

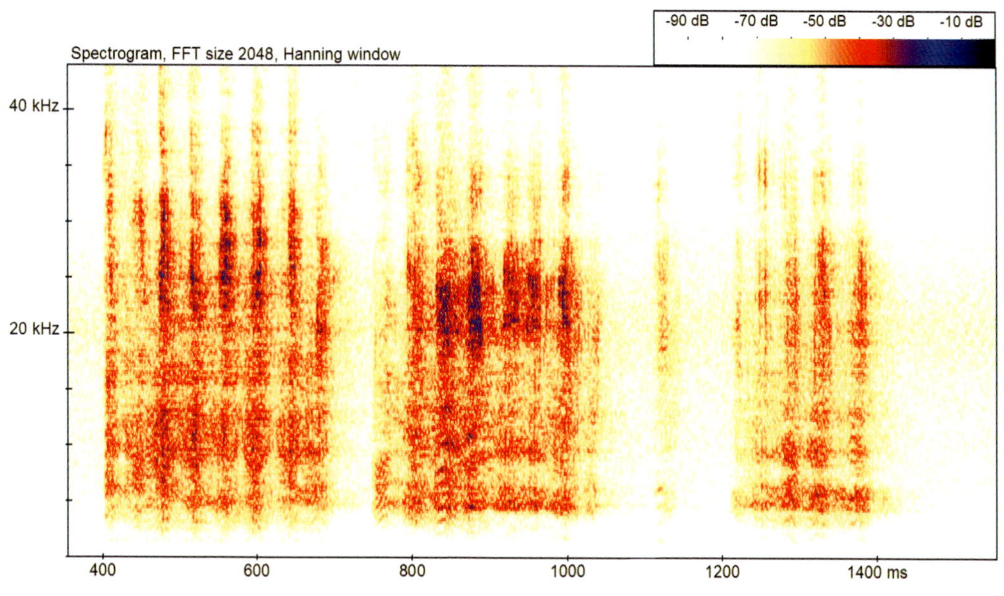

Figure 6.50 🔊 Stoat – series of rasping sounds (frame width: c.1.2 sec)

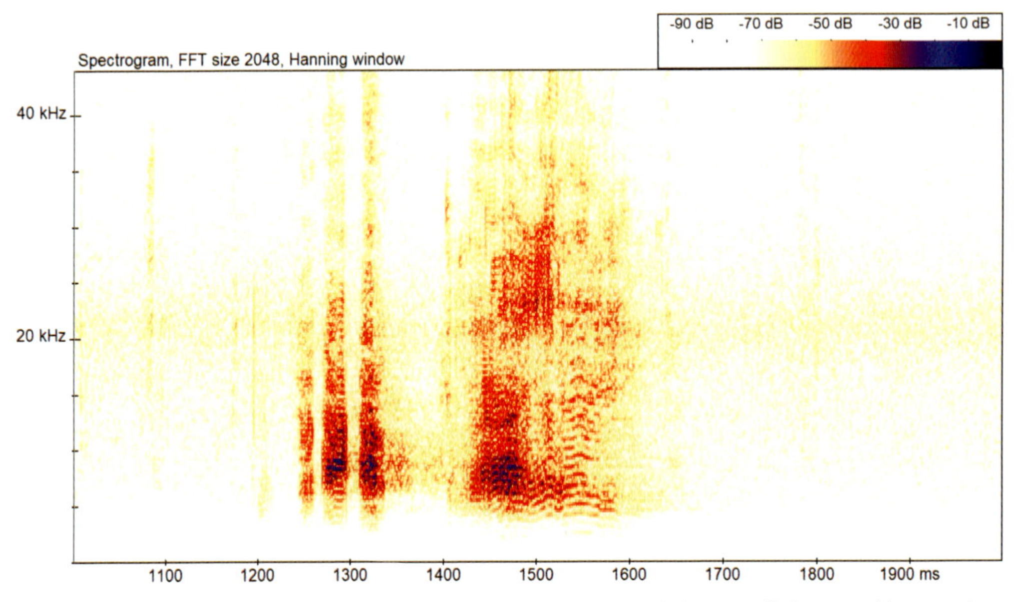

Figure 6.51 🔊 Stoat – sequence of three quick calls, followed by a single longer call (frame width: c.1 sec)

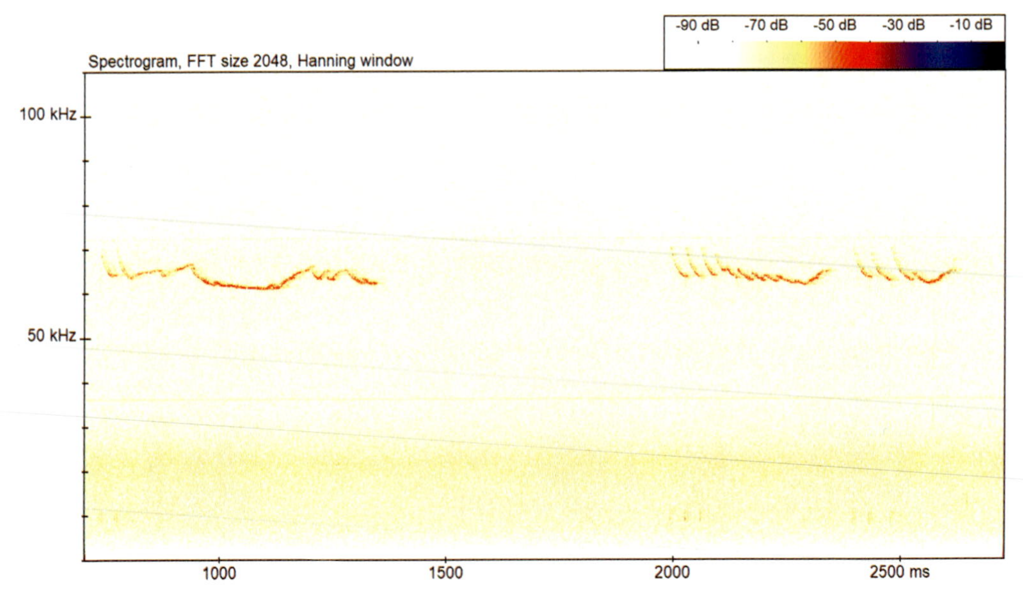

Figure 6.52 🔊 Stoat – Type 1 60 kHz ultrasonic vocalisations (frame width: c.2 sec)

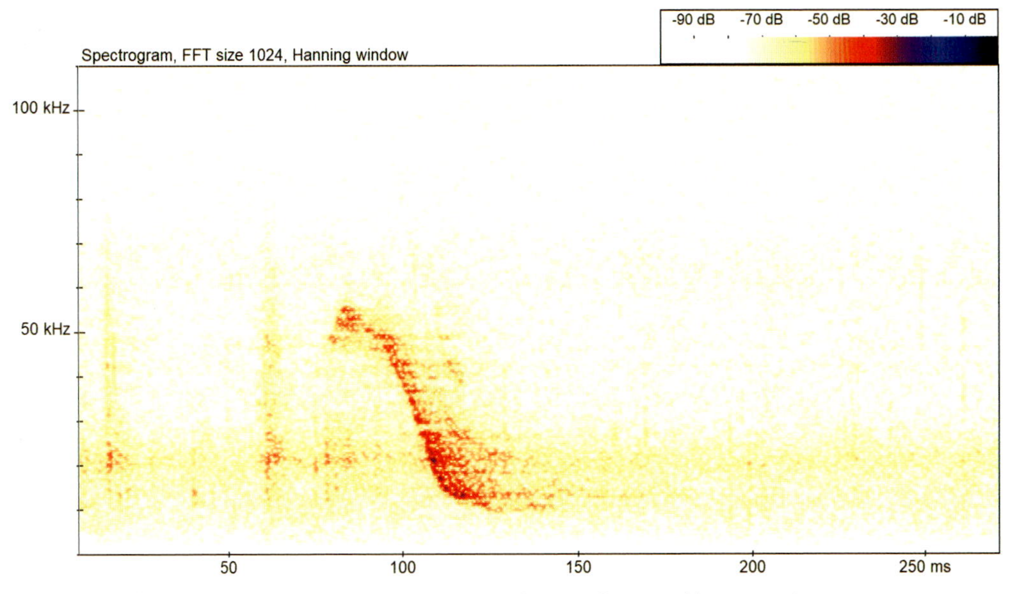

Figure 6.53 Stoat – CF steep FM QCF ultrasonic vocalisation (frame width: c.0.3 sec)

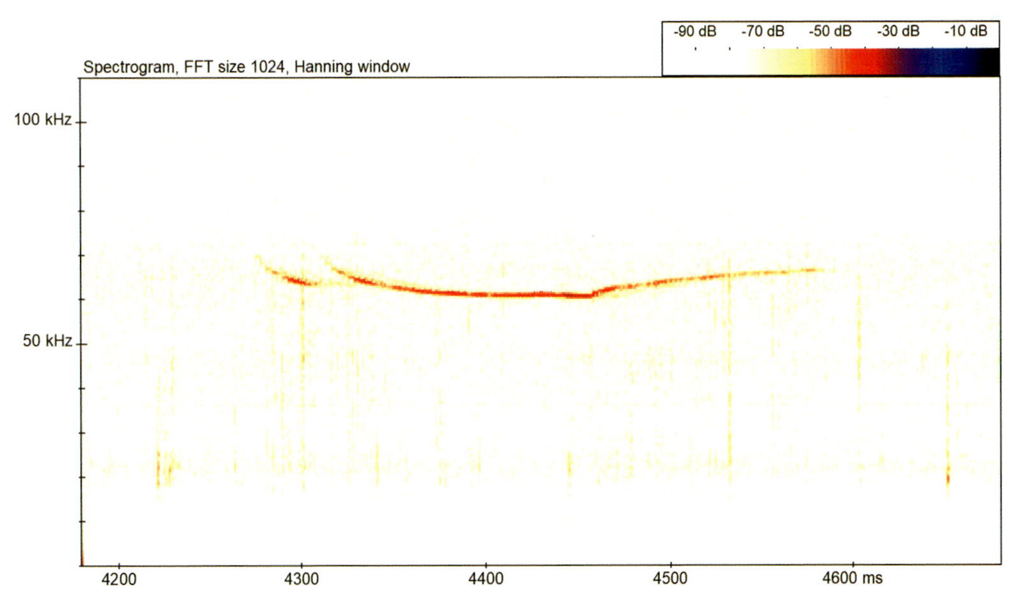

Figure 6.54 Stoat – Type 2 60 kHz falling/rising ultrasonic vocalisation (frame width: c.0.5 sec)

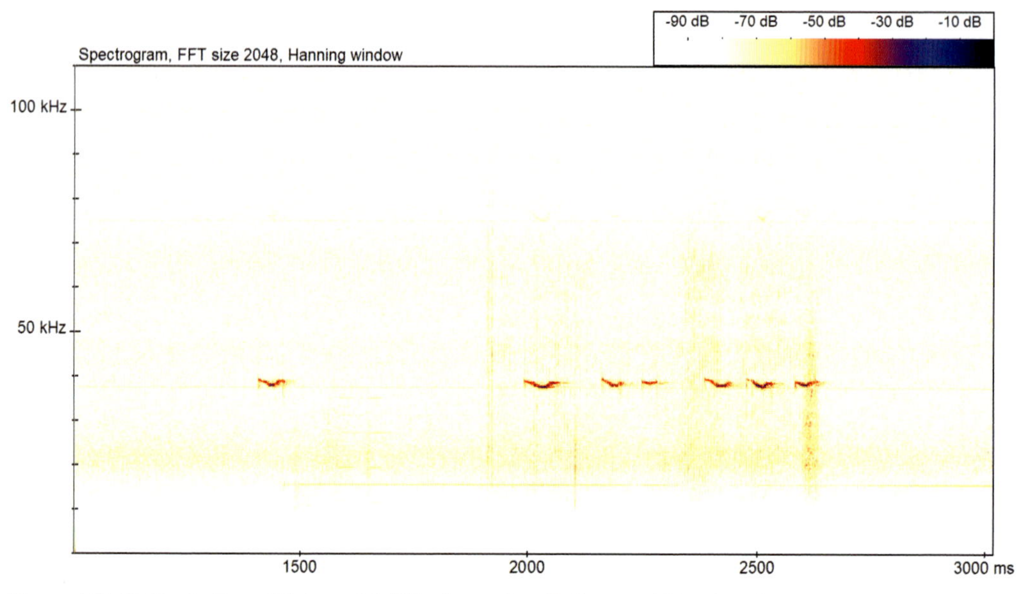

Figure 6.55 🔊 Stoat – Type 3 V-shaped 40 kHz ultrasonic calls (frame width: c.2 sec)

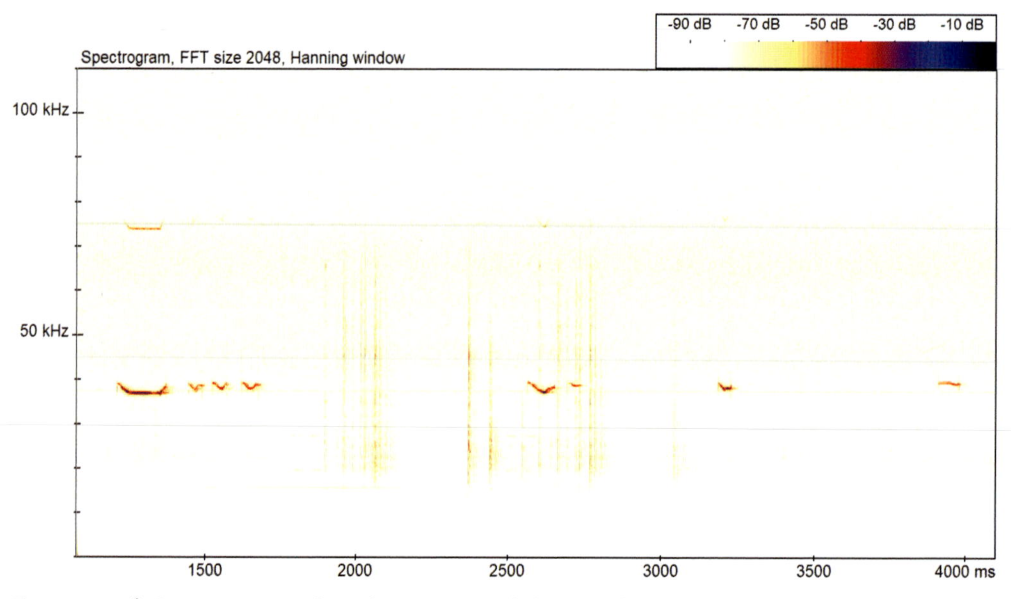

Figure 6.56 🔊 Stoat – Type 3 V-shaped variation (single long call followed by short calls) ultrasonic vocalisation (frame width: c.3 sec)

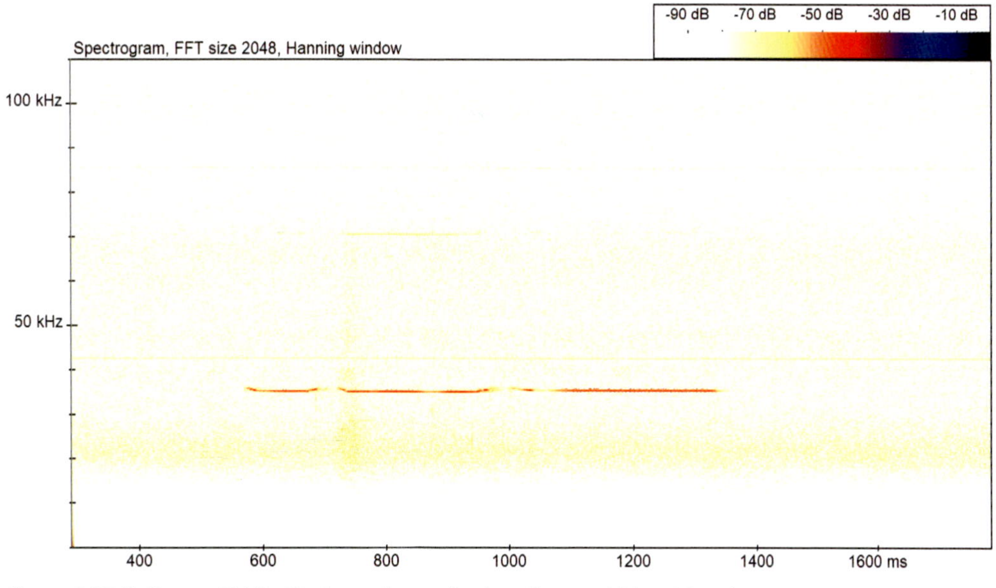

Figure 6.57 Stoat – 35 kHz CF ultrasonic vocalisations (frame width: c.1.3 sec)

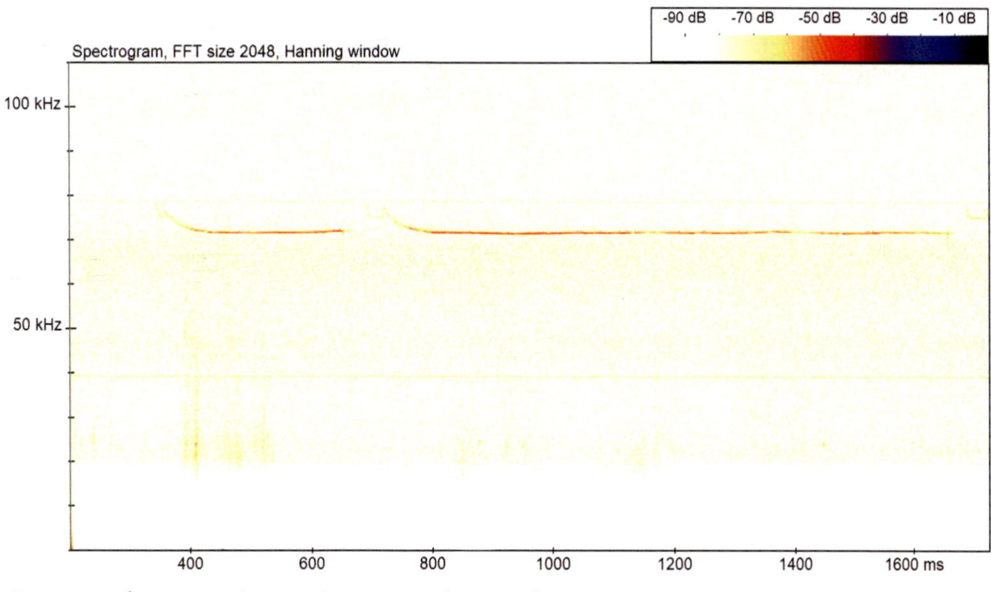

Figure 6.58 Stoat – 75 kHz CF ultrasonic vocalisations (frame width: c.1.3 sec)

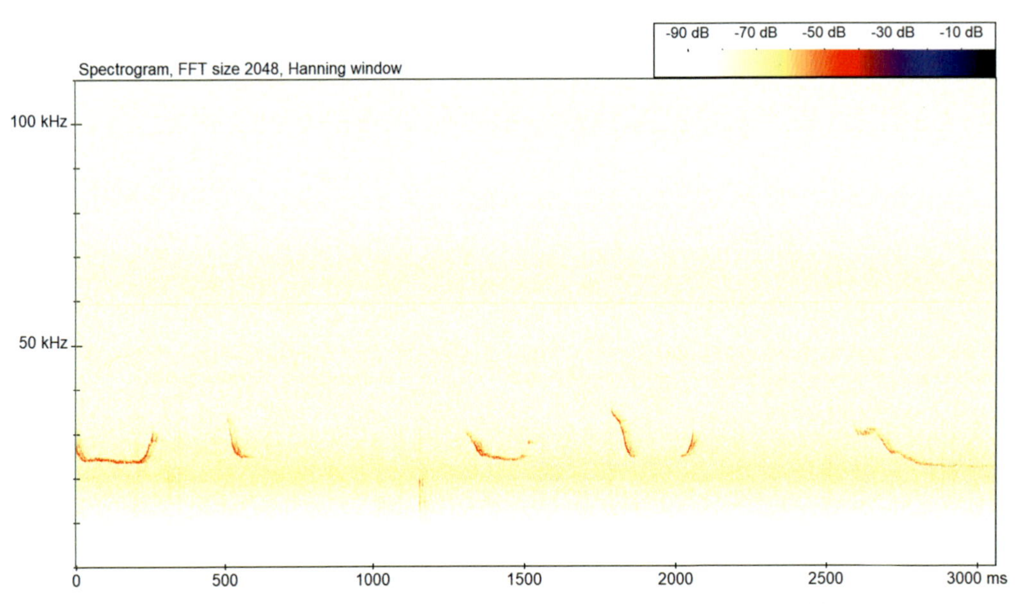

Figure 6.59 🔊 **(QR)** Stoat – variety of call structures within a single sequence (frame width: c.3 sec)

QR codes relating to presented figures

Potential confusion between stoat and other species

Table 6.12 Summary of potential confusion between stoat and other species

Confusion group	Potential confusion species In the first instance, always consider habitat and distribution.
Other mustelid species	All of the smaller mustelids make a range of noises that may be difficult to pin to one species or another, especially as most of us are unfamiliar with them anyway! In the absence of a sighting or field evidence to confirm species, the setting up of a camera trap is recommended.
Species in other groups	Rat and vole species.

6.5.5 Weasel *Mustela nivalis*

Length/Weight	Habitat preferences	Distribution
Smaller than stoat, with a short tail, lacking the black tip. Males are larger than females. Body length (excluding tail) up to 215 mm. Weight up to c.195 g.	Coniferous woodland Broadleaved woodland Variety of other habitats (e.g. farmland, parkland) where small mammal or bird-related egg prey may be abundant, and cover is provided	Occurs throughout mainland Britain, Skye, the Isle of Wight and Anglesey.
Diet		
Carnivorous, feeding most of the time on small rodents, but may also take a broader range of food sources including rabbit, songbirds, eggs, reptiles, amphibians and invertebrates.		

Daily activity pattern	Typically active during the day and night, with individuals taking breaks from activity when appropriate. Tend to forage under cover (e.g. matted grass) both to avoid predation by foxes and raptors, and because that is where they find their small rodent prey.
Seasonal activity pattern	Active throughout the year, with males (solitary) and females (with or without young) being territorial. Male territories are bigger and include smaller female territories within.
Mating activity pattern	Mating takes place during April to July, with young being born during the period April to August. Typical litter size of four to eight kits. Young normally gain independence at c.9 to 12 weeks when the family group breaks up.
Other notes	Shelters within dens at ground level, these usually being burrows created by other mammals. Not usually heard, but some audible vocalisations can occur.

References
McDonald and King, 2008b; Mathews *et al.*, 2018; Crawley *et al.*, 2020.

Acoustic behaviour including spectrogram examples

Weasels have been described as making four sounds: an alarm call (guttural hissing), short screaming barks (or shrieks) when under provocation, a defensive squeal (Figures 6.60 and 6.61), and high-pitched trills during interactions between mother and young (McDonald and King, 2008b). Fighting/squabbling between individuals may also produce a series of rapid vocalisations (Figure 6.62).

In addition, a range of ultrasonic sequences have also been recorded by the authors (Figures 6.63 to 6.66), albeit it is uncertain what behaviours these relate to – they may serve as a predatory function, or in a different context, as communication with neonates. Typical sequences encountered more often in the vicinity of weasels are low-end frequency steep FM sweeps, followed by higher FM calls, as shown in Figure 6.63, albeit the higher-frequency emissions may occur without the lower calls, as in Figure 6.64.

A useful study (Heffner and Heffner, 1985) was carried out to determine the effective hearing range of weasel, showing that they were capable of hearing at frequencies from 51 Hz to 60.5 kHz. As such any ultrasonic sound produced by conspecifics could be picked up by an individual, albeit at relatively close range due to attenuation. That study, however, did not suggest that this species communicated ultrasonically.

In addition to weasel, studies carried out on another mustelid (American mink) have shown that they have the ability to hear up to 70 kHz (Powell and Zielinski, 1989), and indeed youngsters for that species communicated within the ultrasonic range (Brandt *et al.*, 2013).

Figures 6.60 to 6.62 provide specific audible examples of spectrograms relating to this species, and Figures 6.63 to 6.66 provide ultrasonic examples. Where the 🔊 symbol appears, a recording of what is shown within the spectrogram is available to download from the species library. In addition, QR codes are provided for selected examples, and these can be accessed immediately for listening purposes from most mobile devices.

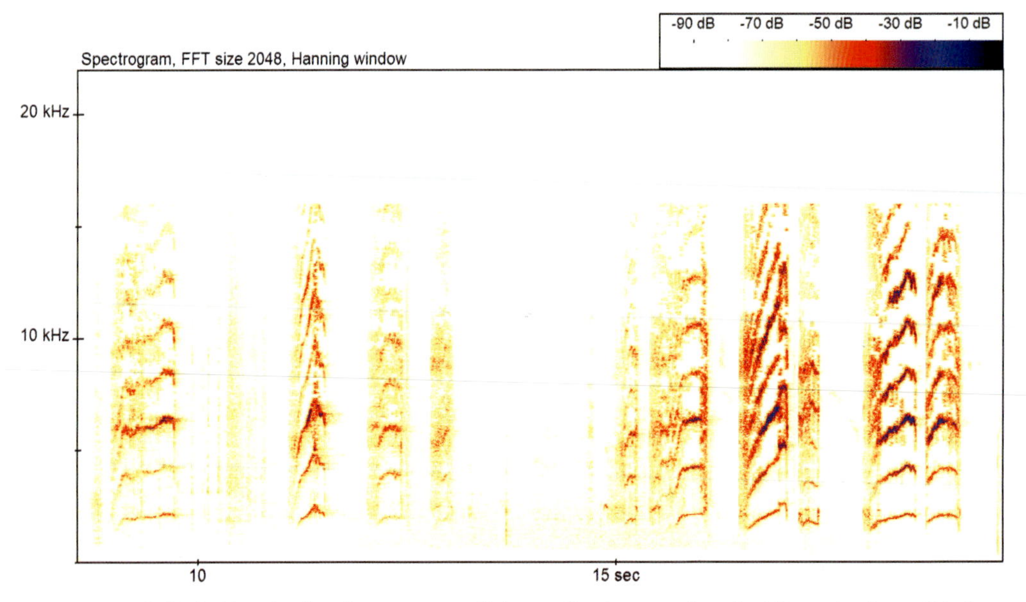

Figure 6.60 🔊 (QR) Weasel – low-frequency audible vocalisations produced whilst animal possibly in a distressing or threatening scenario (courtesy of M. Binstead, British Wildlife Centre) (frame width: c.10 sec)

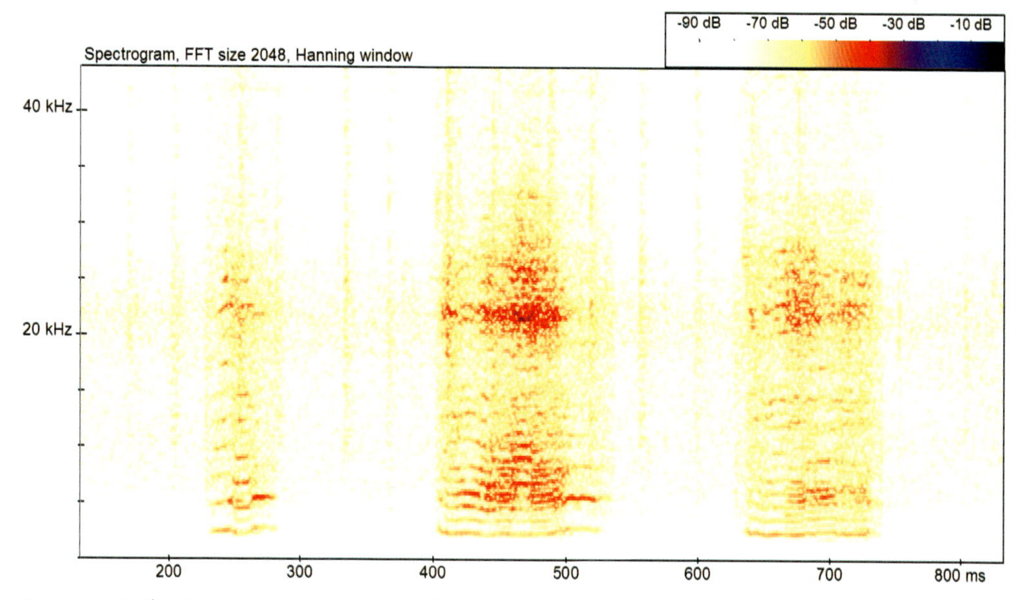

Figure 6.61 Weasel – low-frequency audible vocalisations from an animal possibly in a distressing or threatening scenario (frame width: c.0.7 sec)

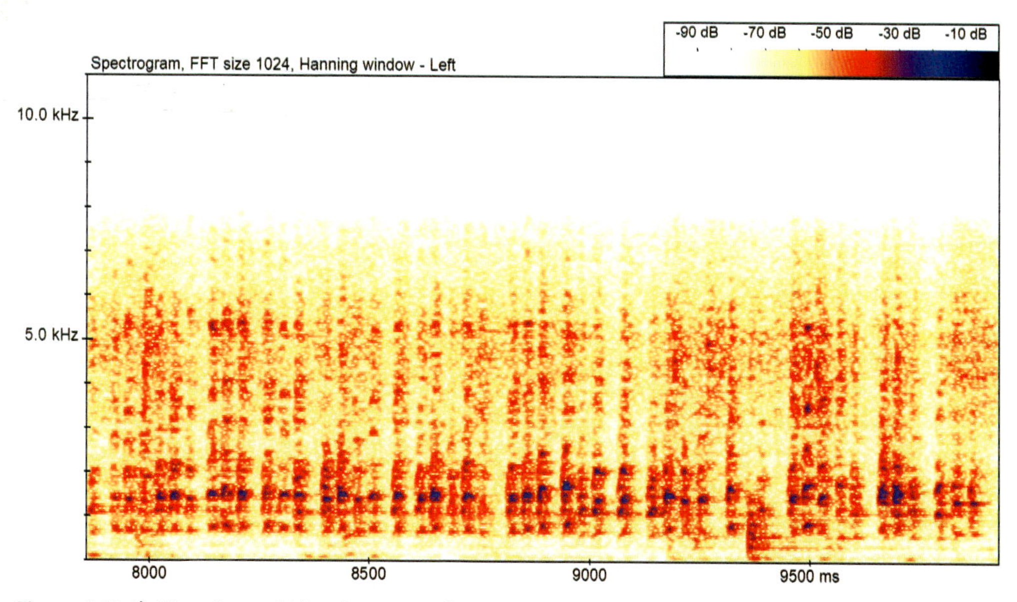

Figure 6.62 Weasels squabbling (courtesy of L. Croose, Vincent Wildlife Trust) (frame width: c.2 sec)

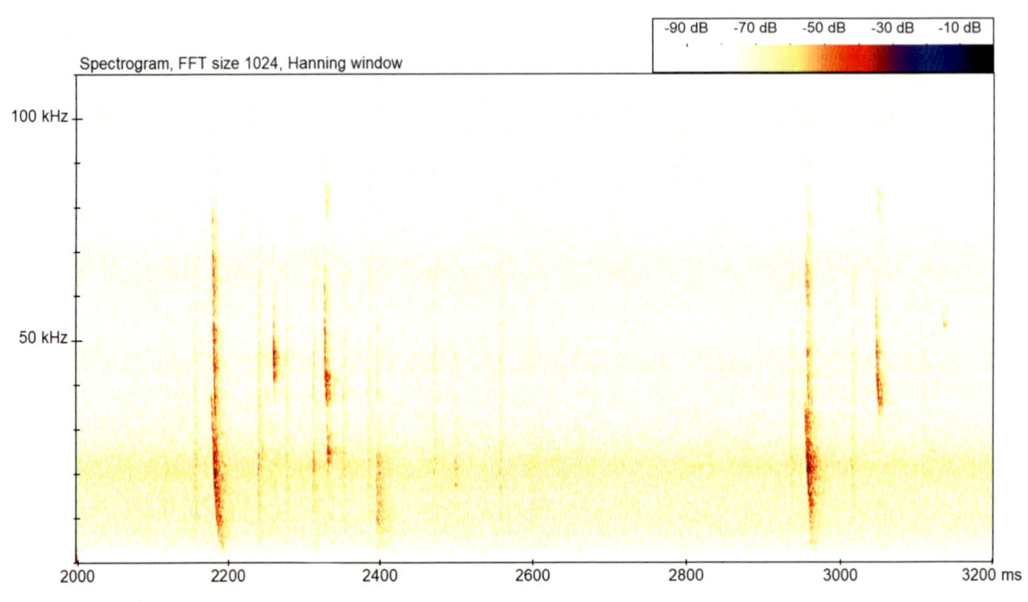

Figure 6.63 Weasel – a typical example of low-end frequency steep FM calls followed by higher-frequency FM emissions (frame width: c.1.2 sec)

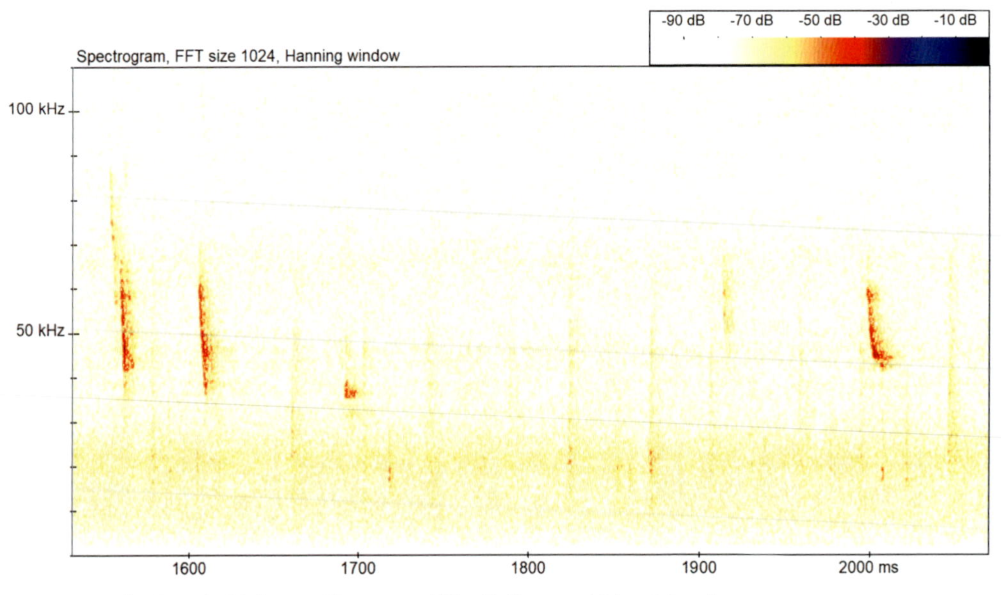

Figure 6.64 Weasel – higher-end frequency FM calls (frame width: c.0.6 sec)

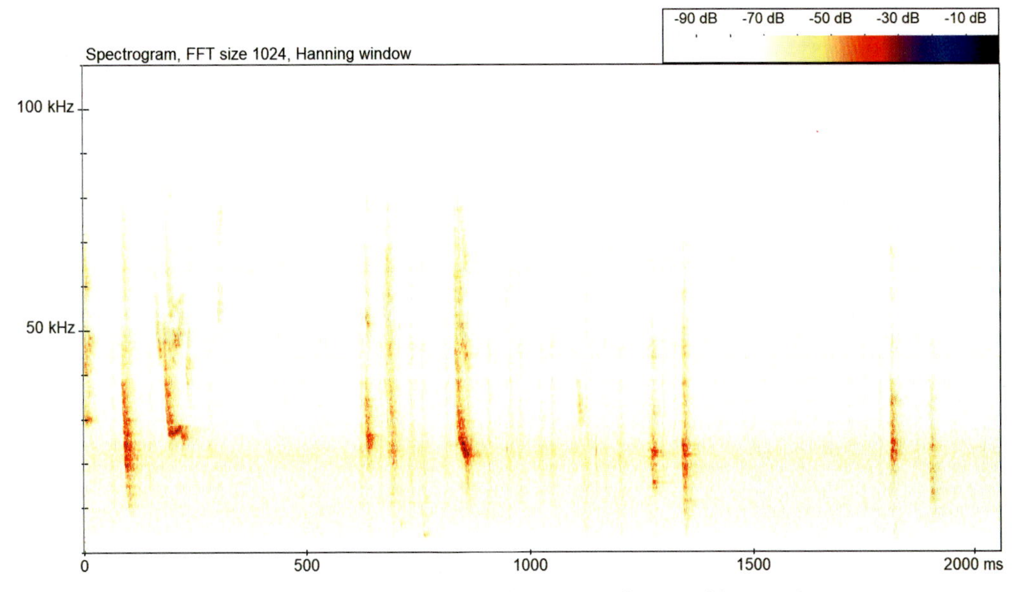

Figure 6.65 🔊 **(QR)** Weasel – a more complex series of emissions (frame width: c.2 sec)

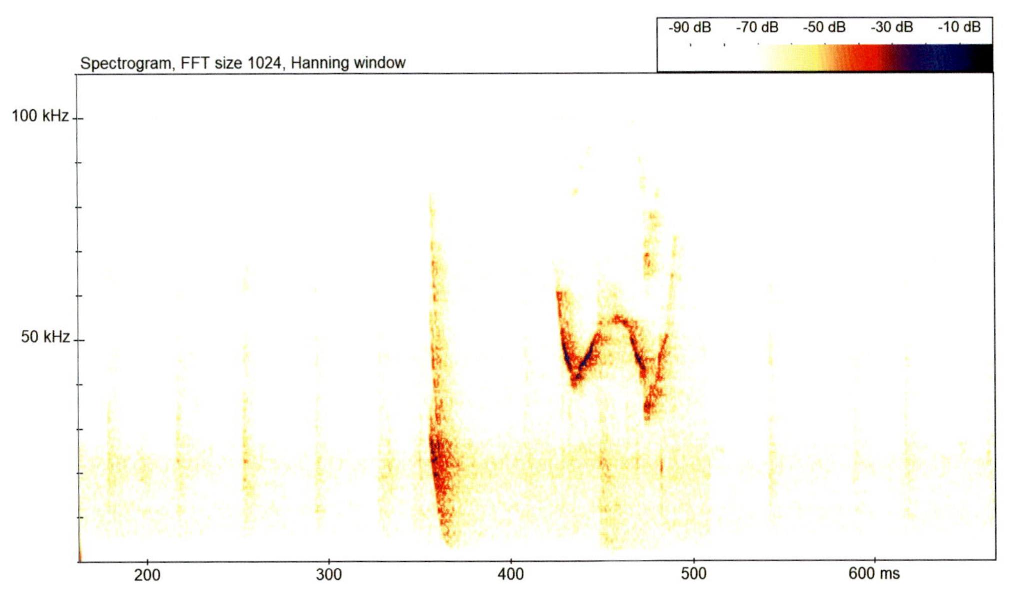

Figure 6.66 🔊 Weasel – low-end frequency steep FM call, followed by a more complex W vocalisation (frame width: c.0.6 sec)

QR codes relating to presented figures

Figure 6.60: Weasel audible	**Figure 6.65:** Weasel complex

Potential confusion between weasel and other species

Table 6.13 Summary of potential confusion between weasel and other species

Confusion group	Potential confusion species
	In the first instance, always consider habitat and distribution.
Other mustelid species	All of the smaller mustelids make a range of noises that may be difficult to pin to one species or another, especially as most of us are unfamiliar with them anyway! In the absence of a sighting or field evidence to confirm species, the setting up of a trail camera is recommended.
Species in other groups	Unlikely to be confused with other mammals.

6.5.6 Polecat *Mustela putorius*

Length/Weight	Habitat preferences	Distribution
Males are larger than females. Body length (excluding tail) up to 42 cm. Weight up to 1.9 kg.	Lowland wooded areas Farmland Habitat associated with wetlands and river banks	Polecats occur throughout most of Wales and adjoining areas in England (Midlands), as well as further to the east and south. Other more isolated pockets of distribution occur in Scotland and north-west England.
Diet		
Carnivorous, feeding on rabbits, small mammals, birds, frogs and toads.		

Daily activity pattern	Mainly nocturnal, with diurnal activity shown only by breeding females in summer.
Seasonal activity pattern	Active throughout the year, with males (solitary) appearing to occupy a larger home range than females (with or without young).
Mating activity pattern	Mating takes place mainly during March and April. Young (usually five to ten) are born during the period May to June, becoming independent at two to three months.
Other notes	They shelter in features used by other mammals (e.g. rabbit burrows), as well as within tree debris, rocks, vegetation and built structures. Not known for being particularly vocal.

References
Macdonald and Barrett, 1995; Birks and Kitchener, 2008; Mathews *et al.*, 2018; Crawley *et al.*, 2020.

Acoustic behaviour including spectrogram examples

Audible chatters, growls, hisses and screams have all been associated with this species (Macdonald and Barrett, 1995), though adults are regarded as being 'normally silent' but having a variety of calls, these being described as threat, molesting, defensive, submissive, begging, greeting and appeasement vocalisations (Birks and Kitchener, 2008). Contact calls between mother and young are also known to occur.

Our audible examples are described as murmuring (Figure 6.67), barking (Figure 6.68), honking (Figure 6.69), yelping (Figure 6.70), guttering (Figure 6.71) and squeaks (Figure 6.72). All of these were recorded when two animals were present and appeared to have been engaging in squabbling or showing a higher level of aggressive behaviour towards each other.

Ultrasonic vocalisations have also been recorded by the authors, albeit it is uncertain what behaviours these relate to. The calls shown in Figures 6.73 to 6.76 were encountered on numerous occasions, the last two examples not being dissimilar to calls produced by hazel dormouse, but typically more variable within a component, and commencing at a lower frequency then ascending.

Figures 6.67 to 6.72 provide specific audible examples of spectrograms relating to this species, and Figures 6.73 to 6.76 provide ultrasonic examples. Where the 🔊 symbol appears, a recording of what is shown within the spectrogram is available to download from the species library. In addition, QR codes are provided for selected examples, and these can be accessed immediately for listening purposes from most mobile devices.

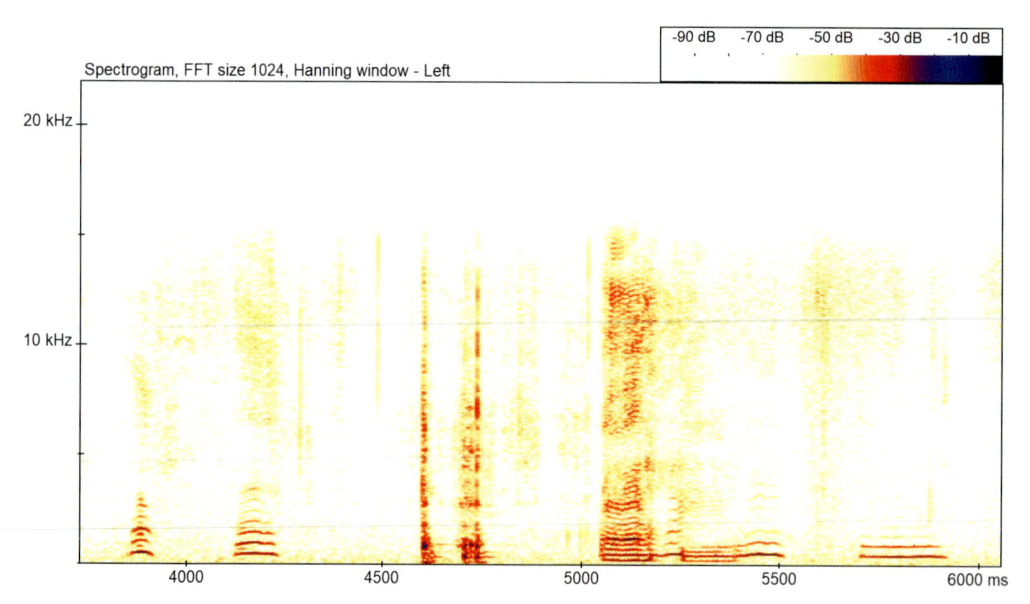

Figure 6.67 🔊 (QR) Polecat murmuring (courtesy of M. Bailey, Wildlife & Countryside Services) (frame width: c.2.5 sec)

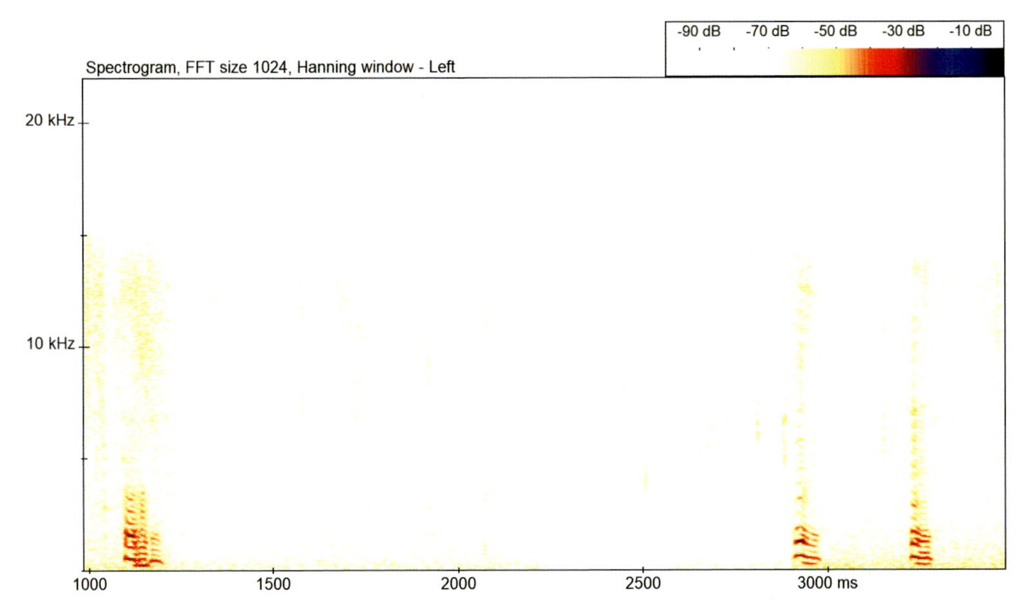

Figure 6.68 Polecat barking (courtesy of M. Bailey, Wildlife & Countryside Services) (frame width: c.2.5 sec)

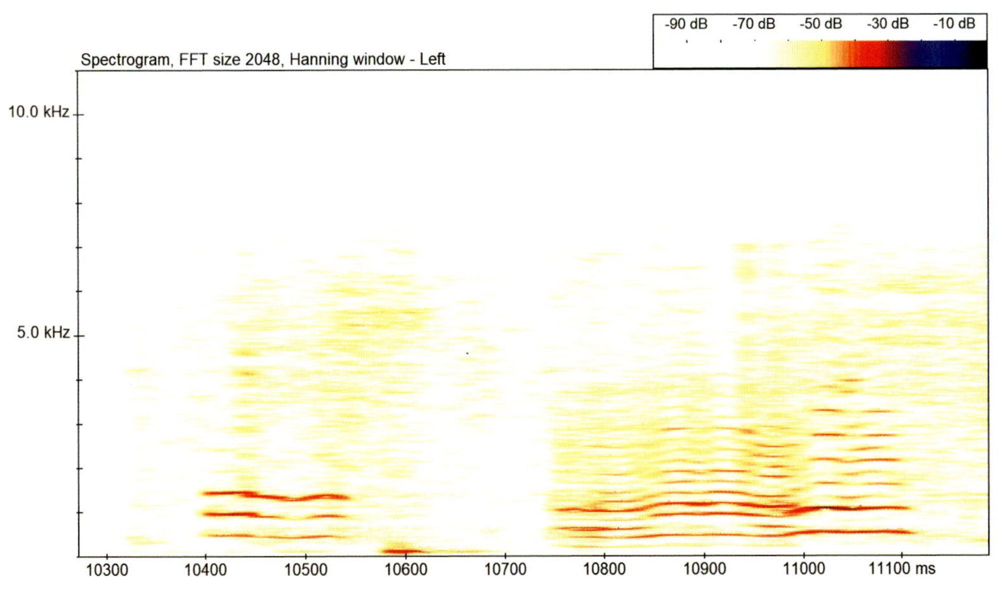

Figure 6.69 Polecat honking (courtesy of L. Croose, Vincent Wildlife Trust) (frame width: c.1 sec)

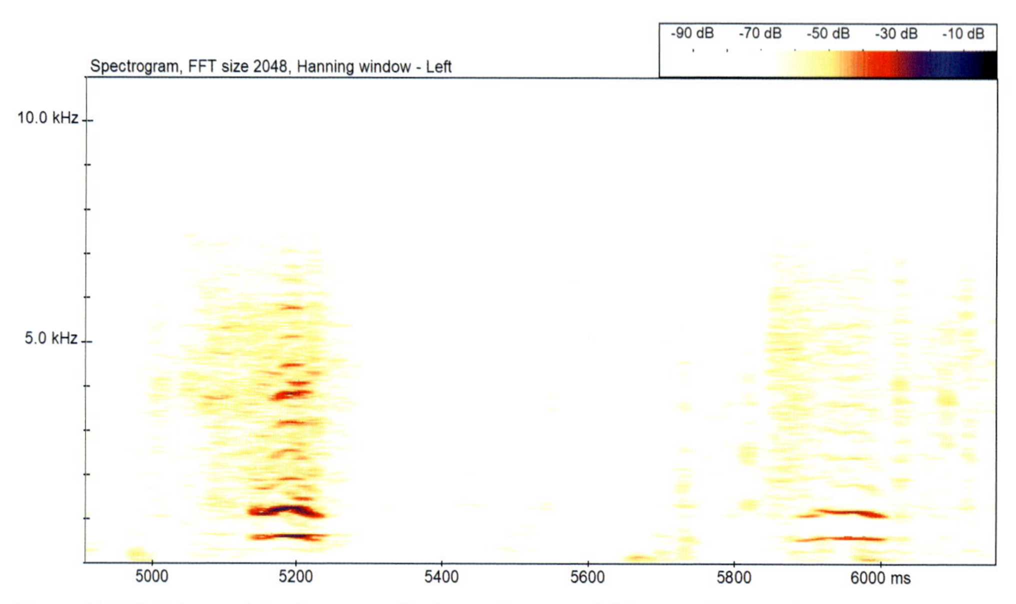

Figure 6.70 🔊 Polecat yelping (courtesy of L. Croose, Vincent Wildlife Trust) (frame width: c.1.2 sec)

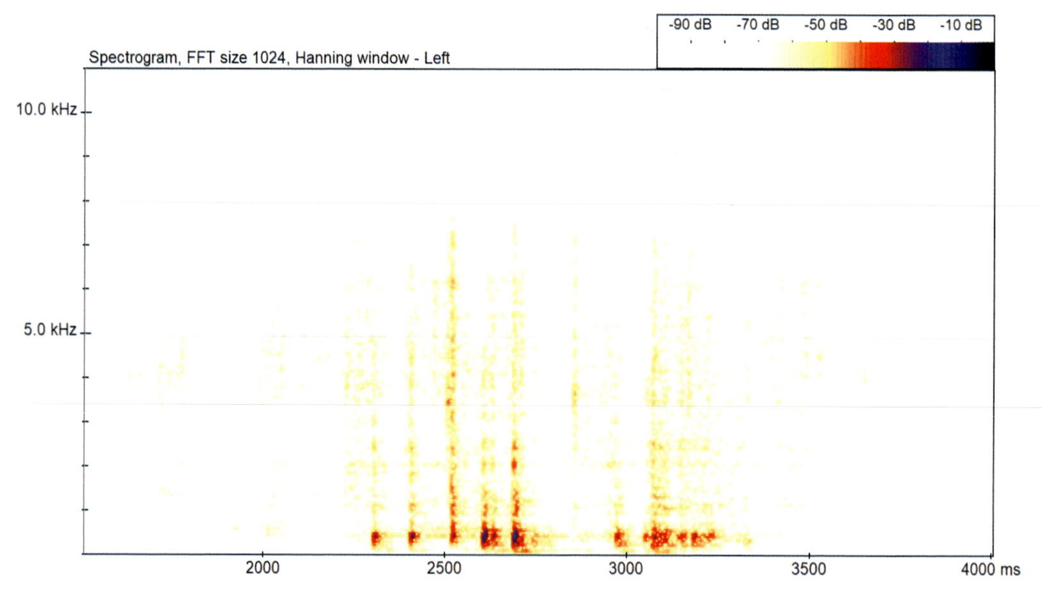

Figure 6.71 🔊 Polecat guttering (courtesy of L. Croose, Vincent Wildlife Trust) (frame width: c.2.5 sec)

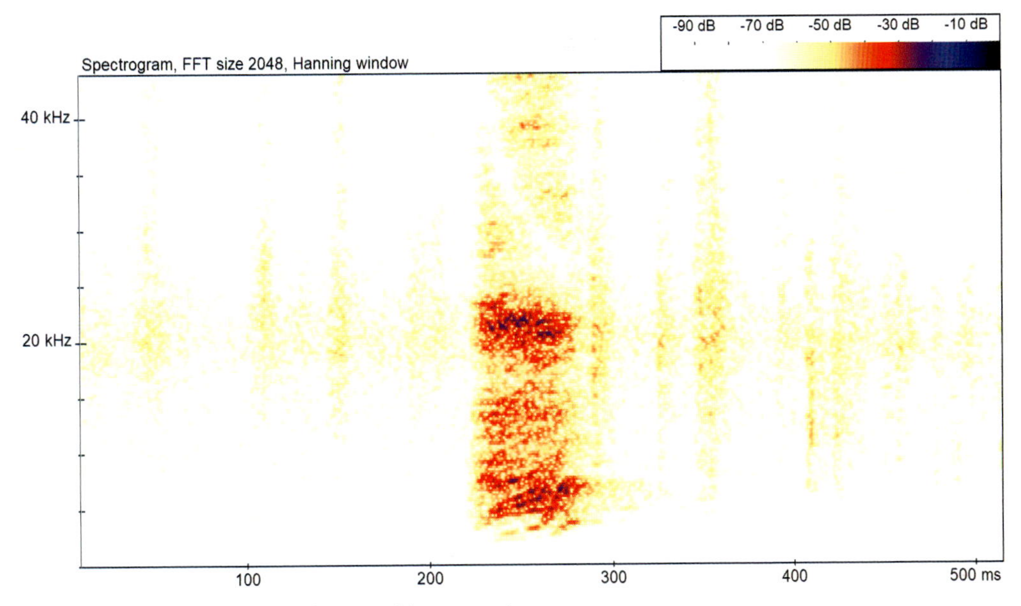

Figure 6.72 🔊 Polecat squeak (frame width: c.0.5 sec)

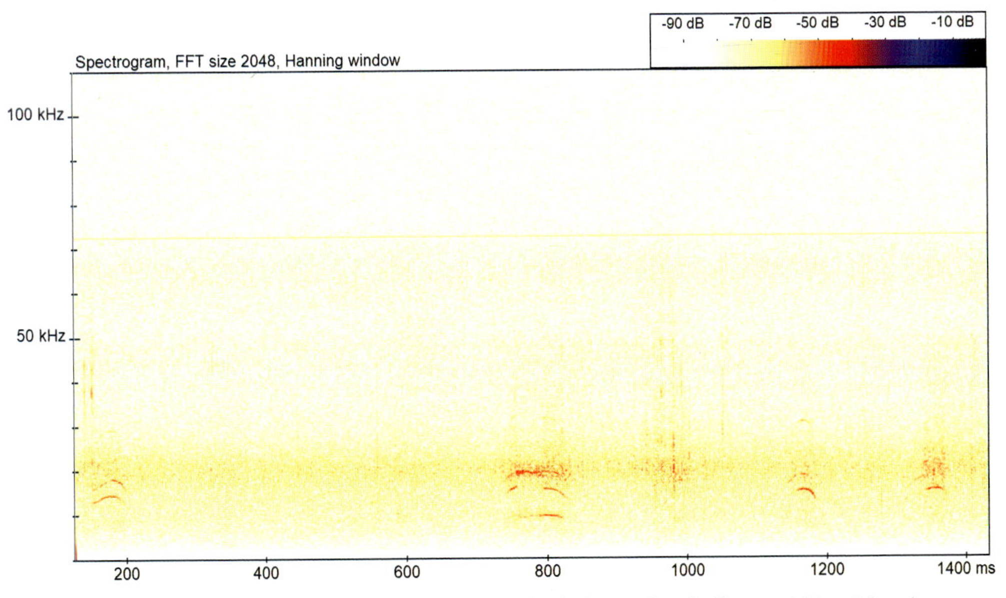

Figure 6.73 🔊 Polecat – ascending then descending (arched) ultrasonic calls (frame width: c.1.3 sec)

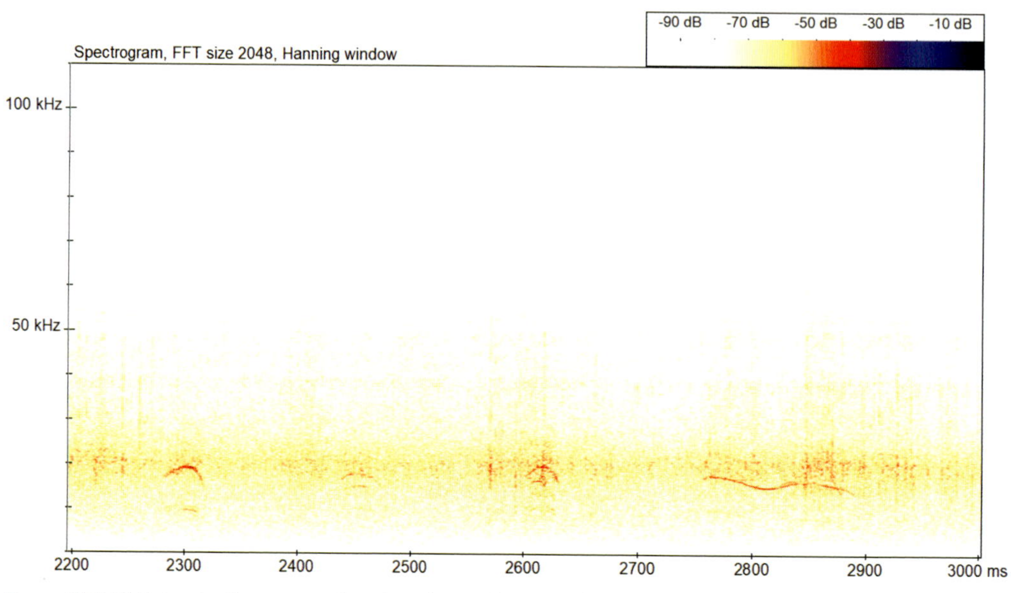

Figure 6.74 🔊 Polecat – three ascending then descending calls, followed by a more complex variation (frame width: c.0.8 sec)

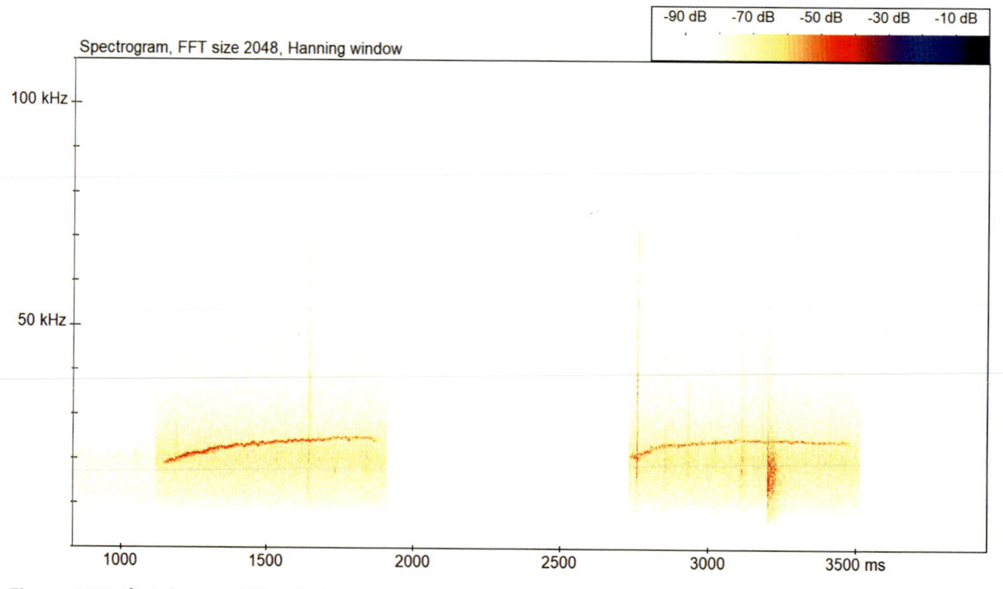

Figure 6.75 🔊 Polecat – QCF calls (x2), gentle rising FM structure (frame width: c.3 sec)

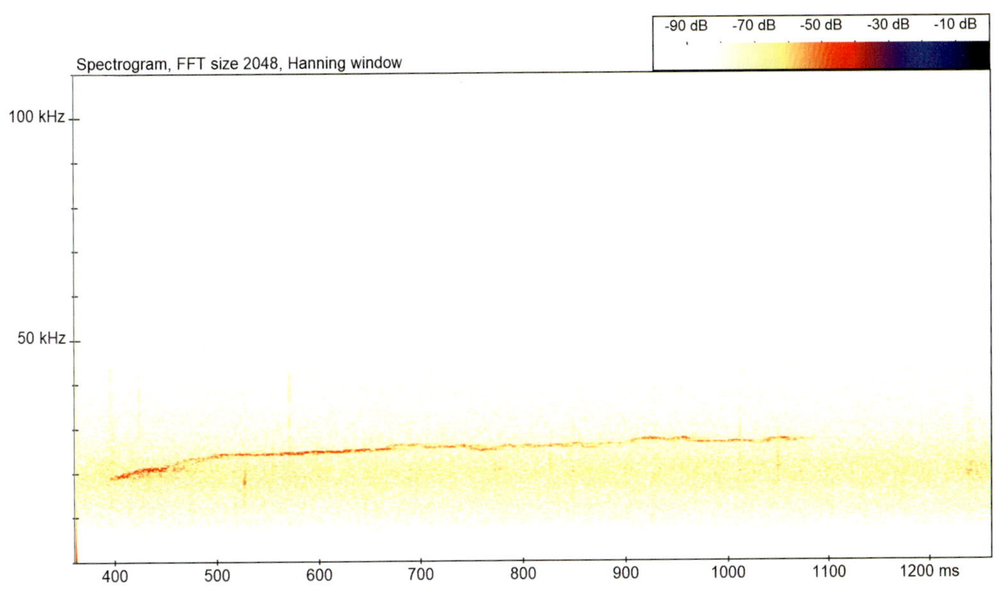

Figure 6.76 🔊 **(QR)** Polecat – QCF call, gentle rising FM structure (frame width: c.0.9 sec)

QR codes relating to presented figures

Potential confusion between polecat and other species

Table 6.14 Summary of potential confusion between polecat and other species

Confusion group	Potential confusion species In the first instance, always consider habitat and distribution.
Other mustelid species	Polecat–feral ferret hybrids are not uncommon, and the presence of such should be borne in mind. All of the smaller mustelids make a range of noises that may be difficult to pin to one species or another, especially as most of us are unfamiliar with them anyway! In the absence of a sighting or field evidence to confirm species, the setting up of a trail camera is recommended.
Species in other groups	Brown rat, black rat, hazel dormouse.

6.5.7 American mink *Neovison vison*

Length/Weight	Habitat preferences	Distribution
Males are larger than females. Body length (excluding tail) up to 45 cm. Weight up to 1.5 kg.	Rivers Wetlands Coastal	Occurs throughout much of Britain and Ireland, including some islands, with population in many areas possibly (not proven) in decline, due to expanding otter populations.
Diet		
Carnivorous, feeding across a vast range of food sources associated with aquatic habitats, including fish, waterfowl and small mammals (e.g. water vole). Also frequently preys upon terrestrial species such as rabbits.		

Daily activity pattern	Activity bouts occur both by day and night, with some evidence that mink have become more diurnal to avoid competition with recovering otter and polecat populations.
Seasonal activity pattern	Active throughout the year, with males (solitary) and females (with or without young) being territorial. Male territories overlap female home ranges.
Mating activity pattern	Mating takes place during March to April. Young (usually four to six) are born during May, becoming independent and dispersing within three to four months.
Other notes	An invasive species, breeding in the wild as a result of escapes from mink farms. Can occupy habitat similar to otter, but is much smaller in size. Not known for being particularly vocal.

References
Dunstone and Macdonald, 2008; Lysaght and Marnell, 2016; Mathews *et al.*, 2018; Crawley *et al.*, 2020; Harrington *et al.*, 2020.

Acoustic behaviour including spectrogram examples

Adults are regarded as usually silent but having a range of calls when in close association with each other, including shrieks, screams, squeaks (Figures 6.77 and 6.78), hisses and chuckling (Gilbert, 1965; Dunstone and Macdonald, 2008). Such vocalisations probably relate to threat, defensive, submissive, mating and greeting scenarios. Mink in cage traps are known for producing a *'powerfully ear-piercing defensive threat shriek'* (J. Birks, pers. comm., 2023). Contact calls between mother and young are also known to occur.

The intensive rearing of American mink in captive mink farms has prompted some specific research into their hearing ability, as well as vocalisations. Although much more study would be beneficial, compared to many other mustelids, some very useful information is available, including what follows.

Their ability to hear prey emitting frequencies up to at least 40 kHz has been documented (Powell and Zielinski, 1989). Clausen *et al.* (2008) demonstrated that kits vocalised with frequencies more strongly pronounced at 40 to 50 kHz, thus meaning that adult females should be able to hear within that range in order to react to their offspring. Indeed, this is the case as shown by Brandt *et al.* (2013), whereby the hearing frequency range for adult females was described as being 1 to 70 kHz, with peak hearing being within the range of 8 to 10 kHz. In conclusion, therefore, this species has the ability to both hear and emit sound at frequencies beyond our audible range.

Ultrasonic vocalistions have been recorded by the authors (Figures 6.79 to 6.84), albeit it is uncertain as to what behaviours these relate to.

Figures 6.77 to 6.78 provide specific audible examples of spectrograms relating to this species, and Figures 6.79 to 6.84 provide ultrasonic examples. Where the 🔊 symbol appears, a recording of what is shown within the spectrogram is available to download from the species library. In addition, QR codes are provided for selected examples, and these can be accessed immediately for listening purposes from most mobile devices.

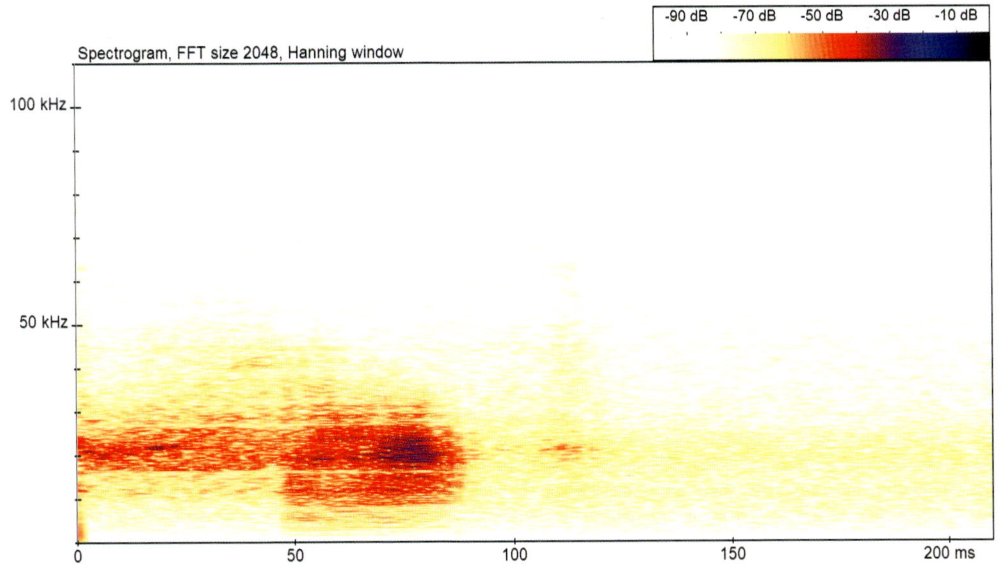

Figure 6.77 🔊 American mink – single squeak (frame width: c.200 ms)

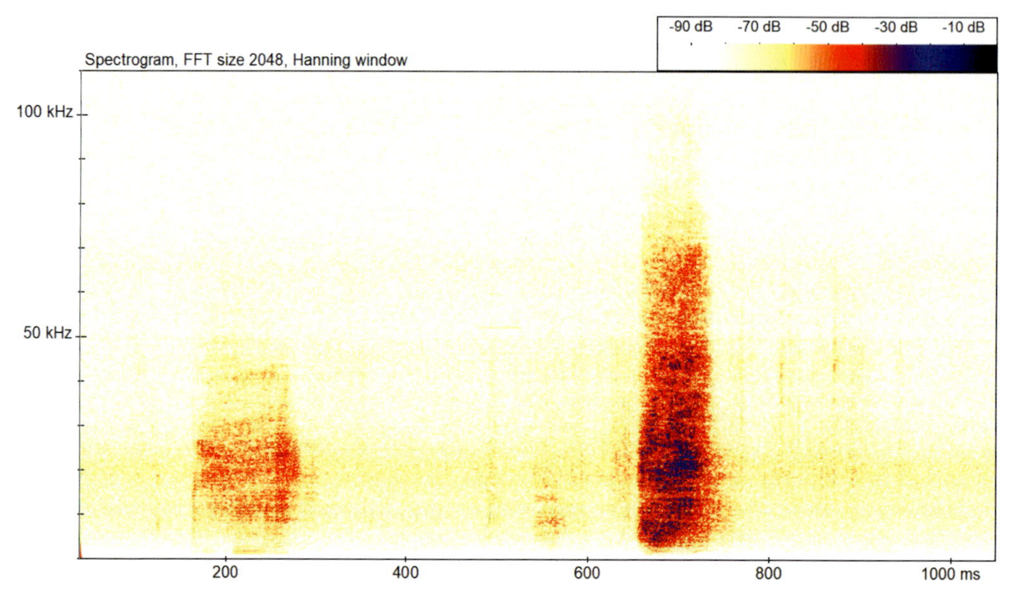

Figure 6.78 American mink – two squeaks (frame width: c.1 sec)

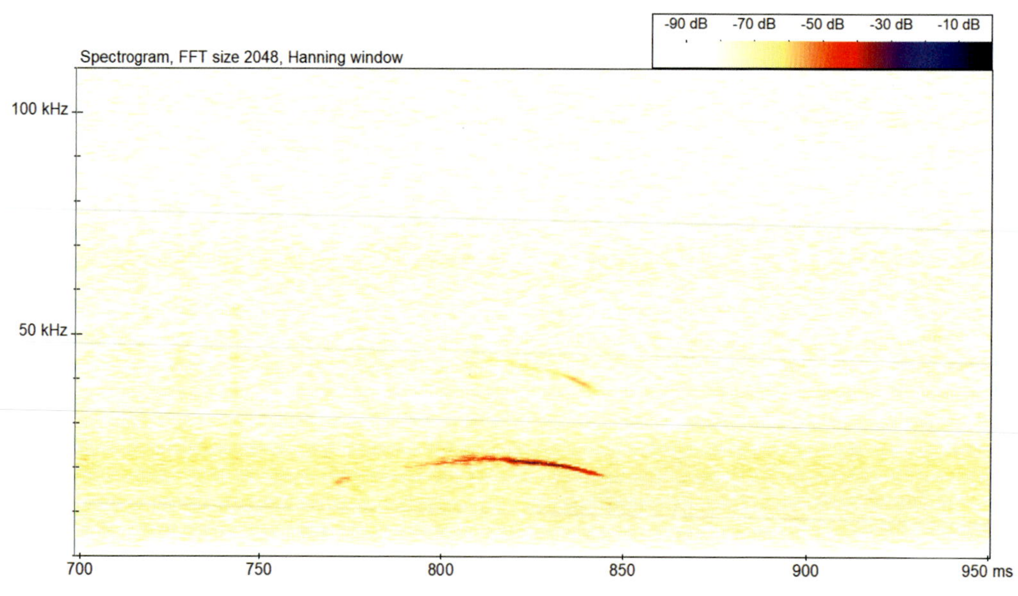

Figure 6.79 American mink – QCF, arched (frame width: c.250 ms)

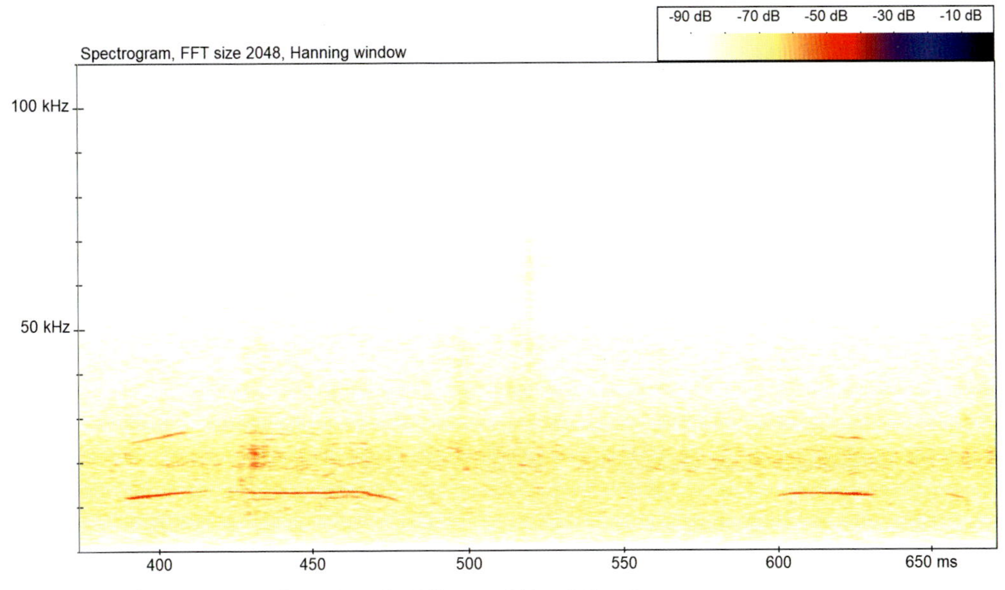

Figure 6.80 American mink – QCF, arched (frame width: c.250 ms)

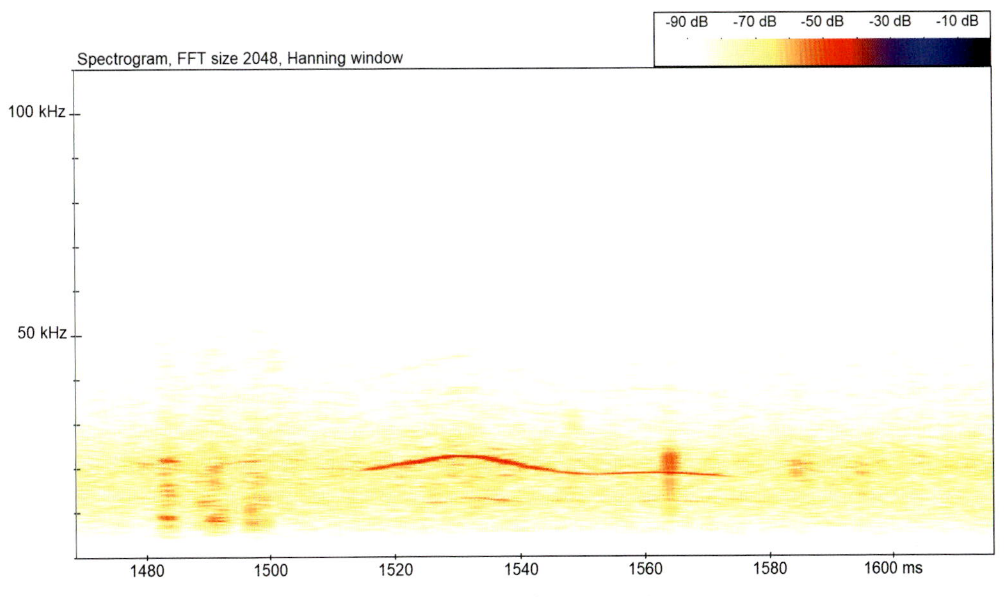

Figure 6.81 American mink – two-tone QCF (frame width: c.150 ms)

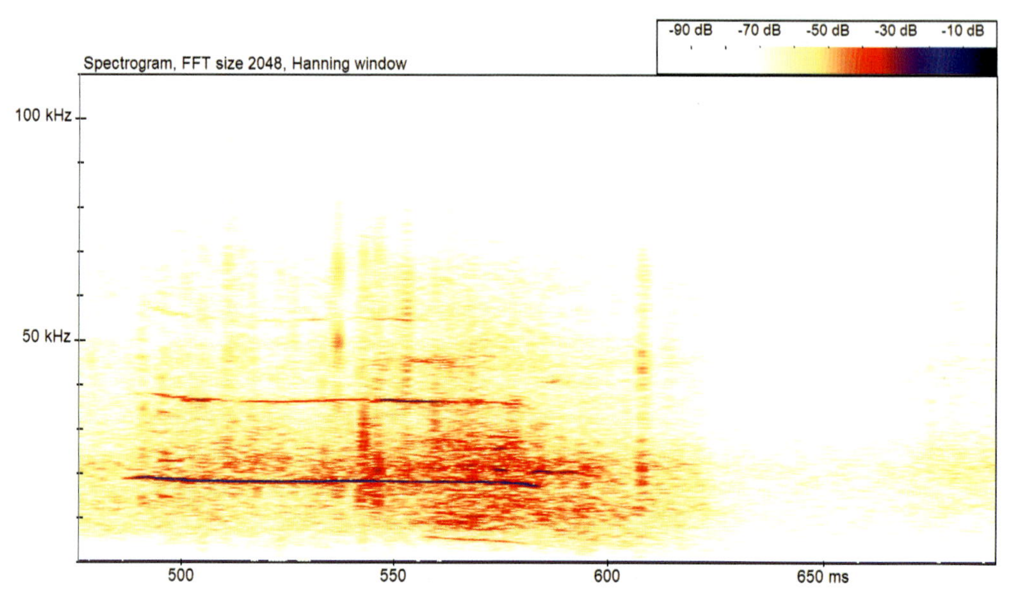

Figure 6.82 🔊 American mink – QCF whistle (frame width: c.250 ms)

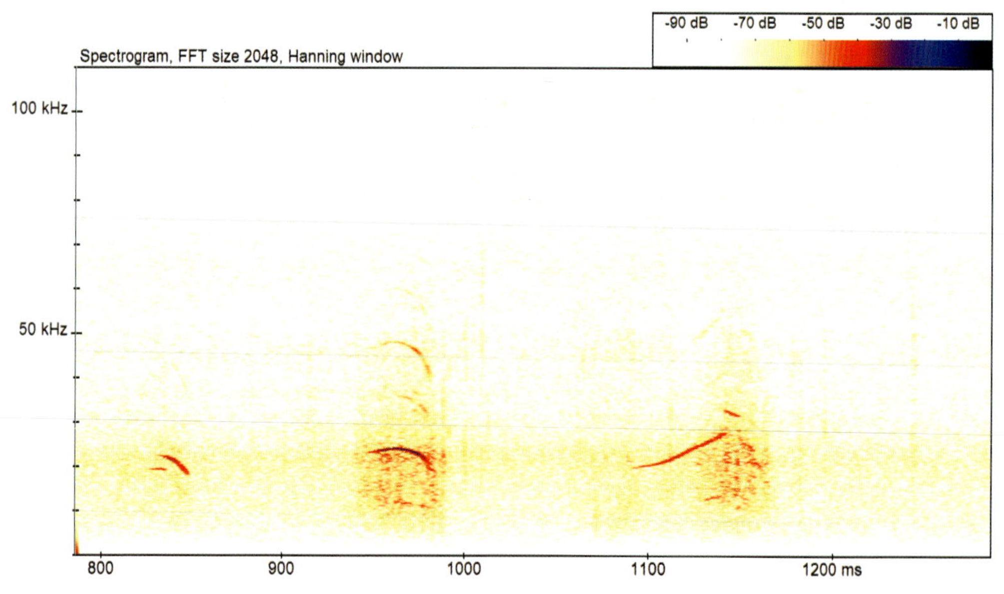

Figure 6.83 🔊 American mink – nasal whistles/sniffing (frame width: c.500 ms)

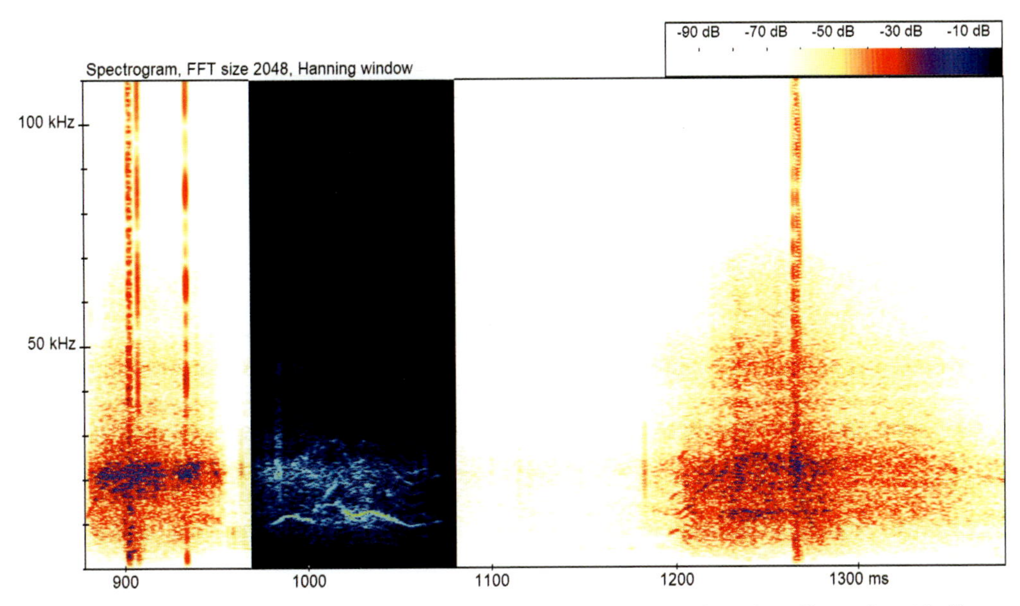

Figure 6.84 🔊 **(QR)** American mink – nasal two-tone whistle (highlighted), with sniffing either side (frame width: c.500 ms)

QR codes relating to presented figures

Figure 6.84: American mink

Potential confusion between American mink and other species

Table 6.15 Summary of potential confusion between American mink and other species

Confusion group	Potential confusion species In the first instance, always consider habitat and distribution.
Other mustelid species	All of the smaller mustelids make a range of noises that may be difficult to pin to one species or another, especially as most of us are unfamiliar with them anyway! In the absence of a sighting or field evidence to confirm species, the setting up of a trail camera is recommended.
Species in other groups	Unlikely to be confused with other mammals.

6.6 Potential application of bioacoustics for monitoring carnivores

Historically, the best monitoring data for carnivore populations in Britain and Northern Ireland has been through records collected via shooting and gamekeeping activities, on the number of each quarry species shot annually ('bag-data') on shooting estates, with national trends being produced for many species of carnivore from the 1960s (Aebischer *et al.*, 2011). In more recent years, a much broader range of survey methods have been used for monitoring carnivores, which includes recording field signs of species presence (e.g. badger setts and otter spraints), and increasingly the use of trail cameras.

Although trail cameras offer a useful tool for monitoring carnivore populations (ENETWILD *et al.*, 2020), they are not without their limitations. Several studies (Wegge *et al.*, 2004; Meek *et al.*, 2014, 2015, 2016; Henrich *et al.*, 2020) have found that the trigger mechanisms of cameras and their associated IR/flash systems generate visual and audio cues that impact upon the behaviour of some carnivores (and other animals), so that trail cameras cannot be considered to be a wholly non-intrusive survey technique.

In comparison, bioacoustics is less intrusive, but since carnivores are not making sound on a regular or consistent basis, surveys using acoustic methods would require a large effort over a longer period of time to collect useful data. Nevertheless, a single surveyor can deploy many acoustic recorders, and there are opportunities for volunteer participation/citizen science projects to obtain a large survey coverage. Because carnivores can be detected over a reasonable distance (a greater detection distance than for small terrestrial mammals and a larger detection distance than trail cameras), the survey area over which an acoustic recorder is able to collect data is greater.

Considering the low call rate of carnivores, for bioacoustics to be a cost-effective tool, it will be necessary to develop algorithms that are able to process and find likely carnivore calls within large acoustic datasets. For this it will be important for tools like the BTO Acoustic Pipeline to be developed alongside bioacoustics practice. In the same way that small terrestrial mammals are already being identified as 'by-catch' during bat surveys, it may be possible to extend identification to include the high-frequency calls of mustelids. With the increasing popularity of bird sound recording, in particular of 'Nocmig' (https://nocmig.com) or Night Flight Call (NFC) recording (as it is known in North America), there may also be opportunities for identifying low-frequency/audible carnivore calls as 'by-catch' during bird recording. With the miniaturisation of microphones, there may also be useful opportunities for deploying microphones on animals along with other sensors.

For many species, it is likely necessary to target acoustic surveys towards the appropriate seasons and locations/habitats to obtain successful recordings. For example, the European otter is largely solitary and occupies a large home range. Outside of mating or territorial disputes, vocalisations from otter are infrequent. Whilst attempting to gather data ourselves for otter from along river corridors, we applied the approaches used by both field surveys and trail camera studies for this species, whereby recorders were deployed at locations where otter regularly come out from the water to territorially spraint-mark or access holt sites. Nevertheless, we found that whilst otters were frequently captured by trail cameras, they were not vocal. Through further work, we found that the collection of otter acoustic data was biased towards females with cubs, and that vocalisations were most likely to be recorded during the winter and spring, where acoustic recorders were installed close to open water and in lake habitats. In particular, high levels of calling within these habitats were associated with the spawning of anurans, which provide an important food resource for otters during the spring.

At the time of writing, we are not aware of any published studies to date in Britain and Ireland that have used bioacoustics for monitoring carnivores. However, the approach has been used very successfully for monitoring carnivores in other parts of the world. For example, bioacoustics has been implemented in surveying grey wolf *Canis lupus* (Passilongo *et al.*, 2015; Papin *et al.*, 2018; Ražen *et al.*, 2020) where individual animals can be identified from within the 'howling chorus' of wolf packs, which has provided information on the status and movement of these packs. Similarly, individual variation in the 'hoo' calls produced by wild dogs have been used to identify pack members. Work is now underway to use acoustic collars to monitor the behaviour and status of endangered African wild dogs and African lions (https://arribada.org/wilddogs/). Acoustic methods have also been used to monitor and examine the behaviour of solitary carnivores such as maned wolves (Ferreira *et al.*, 2019) and Canadian lynx *Lynx canadensis* (Studd *et al.*, 2021). For Canadian lynx, combining acoustic recorders and tri-axial accelerometers has allowed individual hunting behaviour, including prey selection and kill rates, to be documented.

An indication of the possibilities for achieving similar outcomes for some of our carnivore species comes from research carried out on European otter by Castillo (2019), where distinct individual signatures within the 'whistle' call of 11 captive individuals were identified. It is plausible that these techniques could be applied to other carnivore species, for example in mating calls, potentially combined with playback recordings, for monitoring red fox and some of our rarer carnivores, such as pine marten and wildcat.

So far in this section, we have considered acoustics and the use of trail cameras separately, but there is also scope for integrating these approaches. In addition, as the quality of affordable trail cameras continues to improve, the analysis of audio associated with video footage may enable sound libraries to be accompanied by behavioural references, such that descriptions of an animal's behaviour may be attributed to different call types. Indeed, many of the vocalisations for otter (Gordon, 2015), pine marten (D. Bagur) and badger (M. Ingham) provided in this book are evidence of this.

Brown hare – Sandra Graham

CHAPTER 7

Lagomorphs – Hares & Rabbits

7.1 Introduction to this group

This chapter covers the three species of lagomorphs (hares and rabbits) found in the British Isles. All of our species belong to the family Leporidae, represented here as two species of hare (Lepus) and a single species of rabbit (Oryctolagus) (see Table 7.1).

By targeting suitable habitat and taking account of distribution, the presence of each of these species can usually be quite easily established through visual sightings, the use of camera traps and/or finding field-sign evidence (e.g. evidence of faeces and footprints). As a group they are not known for being particularly vocal, and as such the effort involved in recording their acoustic behaviour is likely to be substantial. Having said this, it may be useful to have an appreciation of their vocalisations when seeking to understand their behavioural status and population dynamics (discussed later within this chapter – Section 7.4).

7.2 Overview and comparisons of acoustic behaviour

The species in this group will seldom be encountered acoustically. Instead, olfaction appears to be their primary means of intraspecific communication with scent produced by the anal, inguinal submandibular and Harderian glands being used to convey important information on sex, sexual status and territoriality. The limited use of vocal communication is likely to help avoid detection by predators. Visual and auditory senses will have greater significance for detecting and avoiding predators.

Nevertheless, there are a small number of scenarios where sound could be heard, especially during distress (e.g. in response to an intruder/predator) when all three species emit squeals or screams. Also, in response to a perceived threat, female hares will emit clicking sounds, comparable to teeth chattering used by other small mammals, which likely serves to warn leverets of danger. Rabbits more typically thump their feet when alarmed, possibly signalling to underground nestlings, as well as other above-ground adults (Mullan and Saunders, 2019).

In addition, vocalisations or other non-vocal sounds may occur, for example whilst engaged in mating-related activities (e.g. female hare boxing off unwanted male attention).

The species in this group have good hearing and are able to perceive sounds at both very low frequencies (< 1 kHz) and high frequencies (up to 60 kHz) (Heffner, 1980; Heffner *et al.*, 2020). It is therefore plausible that acoustic communication may be via low frequency or infrasound and/or ultrasound (i.e. using sounds that are outside the hearing range of humans), as well as audible vocalisations.

In the following sections, we will describe a limited number of vocalisations produced by each of the three species and highlight where these species could be confused. Finally, we explore whether or not bioacoustics may be useful for the monitoring and conservation of lagomorphs in the British Isles.

Table 7.1 Lagomorphs occurring in Britain and Ireland, including status, distribution and relative abundance. The relevant chapter sections that describe species-specific vocalisations are also listed.

Order	Family	Species Common name	Genus	Species Scientific name	Status	Distribution range	Abundance within distribution range	Sub-chapter reference
Lagomorpha	Leporidae	Brown hare	*Lepus*	*L. europaeus*	Naturalised	Widely distributed throughout Britain. In Ireland, distribution restricted to north-west and Northern Ireland.	Common	7.3.1
		Mountain hare/ Irish hare		*L. timidus/ L. timidus hibernicus*	Native	Widely distributed throughout Britain and Ireland	Uncommon	7.3.2
		Rabbit	*Oryctolagus*	*O. cuniculus*	Naturalised	Widely distributed throughout Britain and Ireland	Common	7.3.3

Table 7.2 Acoustic spectrum within which you would usually expect to encounter most useful sound for identification purposes within this group

Audible range		Ultrasonic range							
< 10 kHz	10–20 kHz	20–30 kHz	30–40 kHz	40–50 kHz	50–60 kHz	60–70 kHz	70–80 kHz	90–100 kHz	> 100 kHz

Table 7.3 Detectable distance for useful acoustic encounters with lagomorphs (hares and rabbits)

Distance (in metres)							
< 5 m	5–10 m	10–20 m	20–50 m	50–100 m	Up to 500 m	Up to 1,000 m	> 1,000 m
Brown hare/Mountain hare/Irish hare/Rabbit							
Ear/microphone							<<<source of sound

Table 7.4 Typical mating season (shaded in blue) for lagomorphs (hares and rabbits) occurring within Britain and Ireland

Species / Month	Jan	Feb	Mar	April	May	June	July	Aug	Sept	Oct	Nov	Dec
Brown hare												
Mountain hare/ Irish hare												
Rabbit												

7.3 Hare and Rabbit vocalisations

In the following sections, a brief summary of the key characteristics of each of the three lagomorphs will be described, together with details on species-specific vocalisations and, where known, the behavioural function of these sounds.

7.3.1 Brown hare *Lepus europaeus*

Length/Weight	Habitat preferences	Distribution
Brown hare are larger than mountain/Irish hare and rabbit. Body length up to 55 cm. Adult females (c.3.7 kg) are heavier than males (c.3.3 kg).	Grassland Meadows Farmland	Occurs throughout the British Isles, but more thinly populated in the far north-west of Scotland. A small population occurs in north-western Ireland and they are also present on a number of our offshore islands.
Diet		
Herbivorous, typically grazing on grasses and agricultural crops.		

Daily activity pattern	Typically most active overnight, although they may be seen during daylight hours in some locations.
Seasonal activity pattern	Active throughout the year.
Mating activity pattern	Mating typically takes place during the period February to August. Young (leverets) are usually weaned at c.35 days.
Other notes	Typically found at altitudes up to 300 m, above which you would expect to encounter mountain hare. At rest they occupy a form, which can be described as a well-used depression in the ground that helps to conceal the animal from potential predators, as well as offering some protection from wind. During springtime, 'boxing' between males and females may be noticed, as a female attempts to reject a male's advances.

References
Jennings, 2008; Tapper and Yalden, 2010; Lysaght and Marnell, 2016; Mathews *et al.*, 2018; Crawley *et al.*, 2020.

Acoustic behaviour including spectrogram examples

Very little information exists regarding the vocalisations of this species, other than in distress where calls have been described as screaming (Macdonald and Barrett, 1995) or a 'harsh shriek' (Jennings, 2008).

During the breeding season, encounters between individuals will occur that could involve vocalisations, especially during boxing episodes as females fend off unwanted attention from males. One such example was witnessed by one of the authors (SN), who came across a pair of boxing brown hares in May at Hickling Broad (England). From a distance of c.20 feet he heard ferocious growling, described at the time as 'like young bears having a scrap'. Unfortunately, he was unable to record the sound. Separate to this encounter, Wim Jacobs very kindly provided us with a recording of boxing activity 🔊 without any vocalisations, which can be accessed from the sound library.

An additional example we were able to source came to us from Simon Elliott (Figure 7.2) who managed to record a low-frequency call whilst two hares passed by his recording equipment. The purpose behind this call, described by Simon as a 'very quiet woop', is unknown. The animal went on to produce the same sound again some seconds later.

Close-range communicative vocalisations between mothers and young would be anticipated to occur, and tooth-grinding, quiet grunts, chucks and snorts have been reported (Holley, 1992).

Figures 7.1 to 7.3 provide specific examples of spectrograms relating to this species, and where the 🔊 symbol appears, a recording of what is shown within the spectrogram is available to download from the species library. In addition, a single QR code is provided, and can be accessed immediately for listening purposes from most mobile devices.

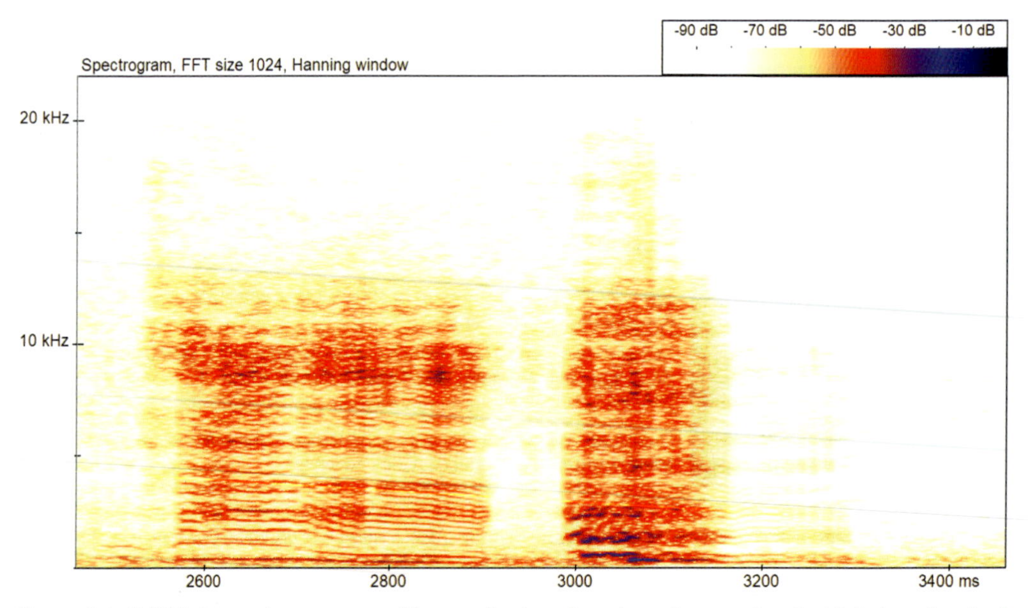

Figure 7.1 🔊 **(QR)** Brown hare – sneeze-like vocalisation, thought to be associated with being disturbed (courtesy of D. Mellor) (frame width: c.1.0 sec)

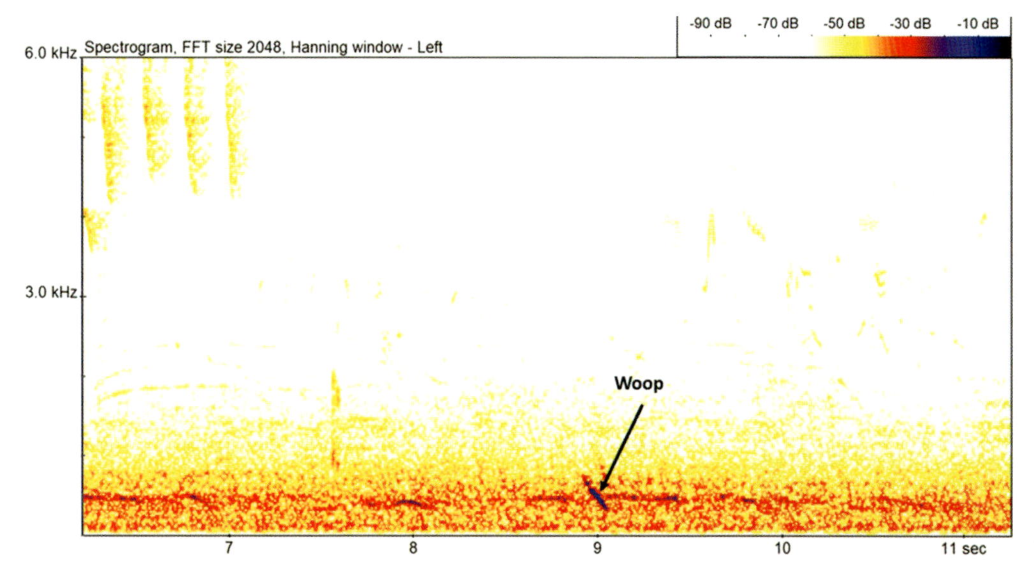

Figure 7.2 🔊 Brown hare – a single 'quiet woop' vocalisation, in amongst bird sounds (courtesy of S. Elliott) (frequency scale: 0 to 6 kHz/frame width: c.5 sec)

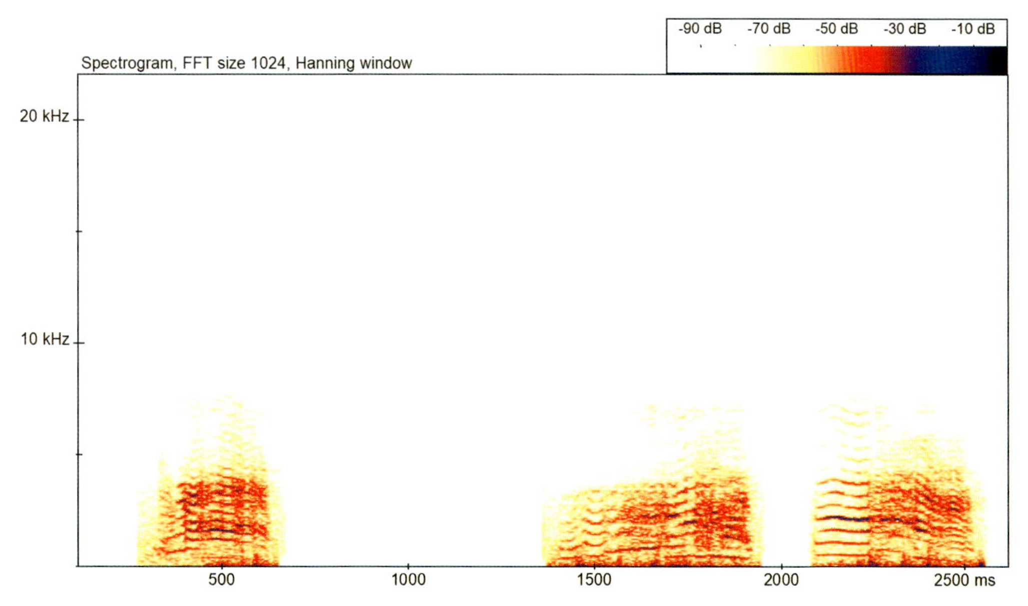

Figure 7.3 Brown hare distress calls (frame width: c.2.5 sec)

QR codes relating to presented figures

Figure 7.1: Brown hare

Potential confusion between brown hare and other species

Table 7.5 Summary of potential confusion between brown hare and other species

Confusion group	Potential confusion species In the first instance, always consider habitat and distribution. Caution is recommended when making a diagnostic identification. Associated visual/video evidence would be beneficial.
Other Lagomorpha species	Distress calls may sound very similar to other hare species or rabbit.
Species in other groups	Squirrel species.

7.3.2 Mountain hare *Lepus timidus*/Irish hare *Lepus timidus hibernicus*

<table>
<tr><th>Length/Weight</th><th>Habitat preferences</th><th>Distribution</th></tr>
<tr><td>Smaller than brown hare, with a conspicuous white tail (above and below).
Body length up to 50 cm.
Adult females (c.2.9 kg) are heavier than males (c.2.6 kg).
In Ireland the weight and size of this species tends to be greater than shown.</td><td rowspan="3">Heather moorland
Upland grassland
Rocky hillsides/mountain habitat</td><td rowspan="3">Occurs within suitable habitat (e.g. heather moorland) throughout the higher altitudes of Scotland (e.g. north-west Scotland, Highlands) as well as throughout Ireland, and within the Peak District (England). Also occurs on many of the Western Isles, the Northern Isles and the Isle of Man.
Combined map for mountain/Irish hare</td></tr>
<tr><th>Diet</th></tr>
<tr><td>Herbivorous, typically grazing on grasses, heather shoots, bilberry, gorse and other vegetation associated with the habitat.</td></tr>
<tr><th>Daily activity pattern</th><td colspan="2">Typically most active from dusk through to dawn, although may be seen during daylight hours in some locations.</td></tr>
<tr><th>Seasonal activity pattern</th><td colspan="2">Active throughout the year.</td></tr>
<tr><th>Mating activity pattern</th><td colspan="2">Not territorial, and mating can take place at any time during the period January to June.
Females can have several litters of one to four offspring per season, with young (leverets) usually weaned within a matter of weeks.</td></tr>
<tr><th>Other notes</th><td colspan="2">At rest they occupy a form, which can be described as a well-used depression in the ground that helps to conceal the animal from potential predators, as well as offering some protection from wind.
During spring males will be in the vicinity of foraging females, and if they get too close the female may eventually attempt to fend off the male with a fighting episode (boxing), using their front feet.</td></tr>
<tr><td colspan="3">References
Iason et al., 2008; Tapper and Yalden, 2010; Lysaght and Marnell, 2016; Mathews et al., 2018; Crawley et al., 2020.</td></tr>
</table>

Acoustic behaviour including spectrogram examples

The most useful study of acoustic behaviour for this species relates to work carried out by Rehnus *et al.* (2019) who conducted research on the vocal emission rate of a small number of captive animals. Demonstrating how silent they were, it was shown that the vocal emission rate *'corresponded to an average rate of one vocalisation every 33 night hours'*. The authors discussed how this low rate may help to minimise the risk of predation, as well as being a symptom of hares not being highly sociable.

In the main, when calls were noted, they related to aggressive behaviour, contact calling or relatively higher-frequency (2 to 4 kHz) isolation calls, emitted in a series, produced by leverets (Rehnus *et al.*, 2019).

Distress calls have been described as screaming. An example of a distress call, provided by Dan Bagur, is shown in Figure 7.4. This distress calling (screaming) was produced by an adult hare being pursued by a fox.

Females have been documented as delivering a clicking danger warning sound to leverets (Macdonald and Barrett, 1995). In addition, other close-range communicative vocalisations between mothers and young would be anticipated to occur.

Figure 7.4 provides one of the few examples that we have of a spectrogram relating to this species, and a recording of what is shown within the spectrogram is available to download from the species library. In addition, a QR code is provided, and this can be accessed immediately for listening purposes from most mobile devices.

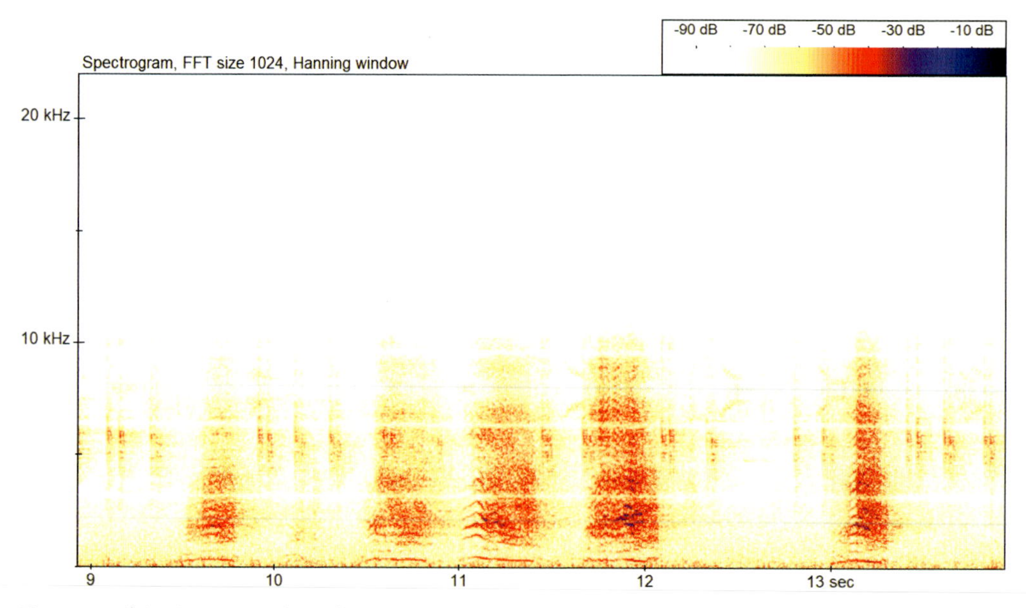

Figure 7.4 🔊 (QR) Mountain hare distress calling whilst being pursued by a fox (courtesy of D. Bagur) (frame width: c.5 sec)

QR codes relating to presented figures

Figure 7.4: Mountain hare

Potential confusion between mountain hare and other species

Table 7.6 Summary of potential confusion between mountain hare and other species

Confusion group	Potential confusion species In the first instance, always consider habitat and distribution. Caution is recommended when making a diagnostic identification. Associated visual/video evidence would be beneficial.
Other Lagomorpha species	Distress calls may sound very similar to other hare species or rabbit.
Species in other groups	Squirrel species.

7.3.3 Rabbit *Oryctolagus cuniculus*

<table>
<tr><th>Length/Weight</th><th rowspan="1">Habitat preferences</th><th>Distribution</th></tr>
<tr><td>Smaller than both of the hare species.
Body length up to 40 cm.
Weight c.1.6 kg.</td><td rowspan="3">Grassland
Meadows
Heathland
Farmland</td><td rowspan="3">Occurring throughout Britain and Ireland, including most offshore islands.</td></tr>
<tr><th>Diet</th></tr>
<tr><td>Herbivorous, typically grazing on a variety of plants, grasses and agricultural crops.</td></tr>
<tr><th>Daily activity pattern</th><td colspan="2">Typically most active at dusk and dawn, as well as overnight.
May be seen during daylight hours in less disturbed areas.</td></tr>
<tr><th>Seasonal activity pattern</th><td colspan="2">Active throughout the year.</td></tr>
<tr><th>Mating activity pattern</th><td colspan="2">Male ranges can be larger and overlapping with female ranges.
Mating can take place at any time, but typically January to August, with potentially a number of litters per female per season.
Young (typically three to seven) are born in nests within a burrow, and are usually weaned at c.21 to 25 days.</td></tr>
<tr><th>Other notes</th><td colspan="2">Reside in a series of burrows (warren) that can be easily recognised in areas where larger colonies are present.</td></tr>
<tr><td colspan="3">References
Cowan and Hartley, 2008; Tapper and Yalden, 2010; Lysaght and Marnell, 2016; Mathews et al., 2018; Crawley et al., 2020.</td></tr>
</table>

Acoustic behaviour including spectrogram examples

The alarm-raising thumping of hind feet may occur in order to warn other members of the group when potential danger exists (e.g. a predator nearby).

Loud vocalisations are extremely rare, and squealing distress calls (Figure 7.5) usually only occur if captured by a predator (Cowan and Hartley, 2008; Mullan and Saunders, 2019). Otherwise, this species is not known for being particularly vocal, though sounds emitted during the breeding season by males (described as grunting), as well as growling during disputes between individuals have been noted (Barrett-Hamilton *et al.*, 1910).

Other documented vocalisations include honing and purring (Mullan and Saunders, 2019).

We did try ourselves, with the assistance of Leonie Washington, to record rabbit vocalisations using trail cameras and audible/ultrasonic static recorders. After several weeks we had very little to show by way of diagnostic results, with only a couple of sounds that we could not confidently assign to species level (hence not included here). The trail camera footage did, however, reveal clear scent-marking activity where several animals rubbed the underside of their chin (submandibular gland) on the same shrub, dead branches and prominent vegetation within the camera's field of view.

Fortunately, however, we were able to obtain a recording from Simon Elliott of a series of low-frequency vocalisations (described by Simon as 'grunts') recorded from an animal close to a microphone, next to a burrow entrance (Figure 7.6). From the same location, Simon was also able to provide us with a foot thump, preceded by what can best be described as a low-frequency 'ooop' sound (Figure 7.7). This is one of two such recordings he has obtained from the same location.

Communicative vocalisations between mothers and young, probably whilst underground, would also be anticipated to occur. Low-frequency sounds (374–667 Hz) (soft-wheeking) have been reported by Schuh *et al.* (2004) prior to nursing. Pre-weaned pup distress calls have also been reported and playback experiments found that these vocalisations evoked a response by the mother shortly after pups were born (when infanticide is most likely) and before young start to disperse from the breeding burrows (Rödel *et al.*, 2013).

Figures 7.5 to 7.7 provide specific examples of spectrograms relating to this species, and where the 🔊 symbol appears, a recording of what is shown within the spectrogram is available to download from the species library.

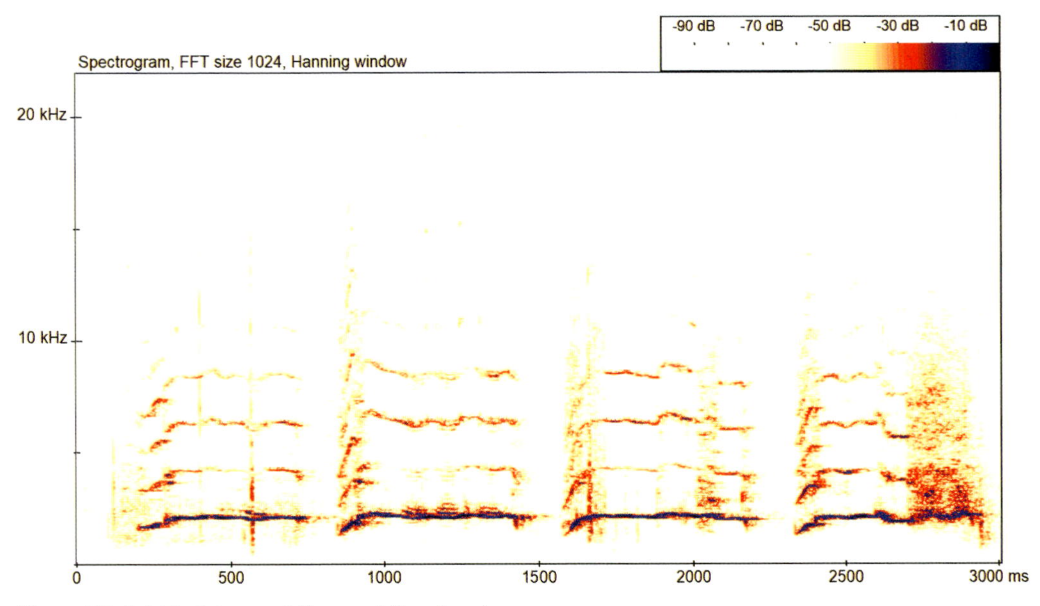

Figure 7.5 Rabbit distress call (frame width: c.3 sec)

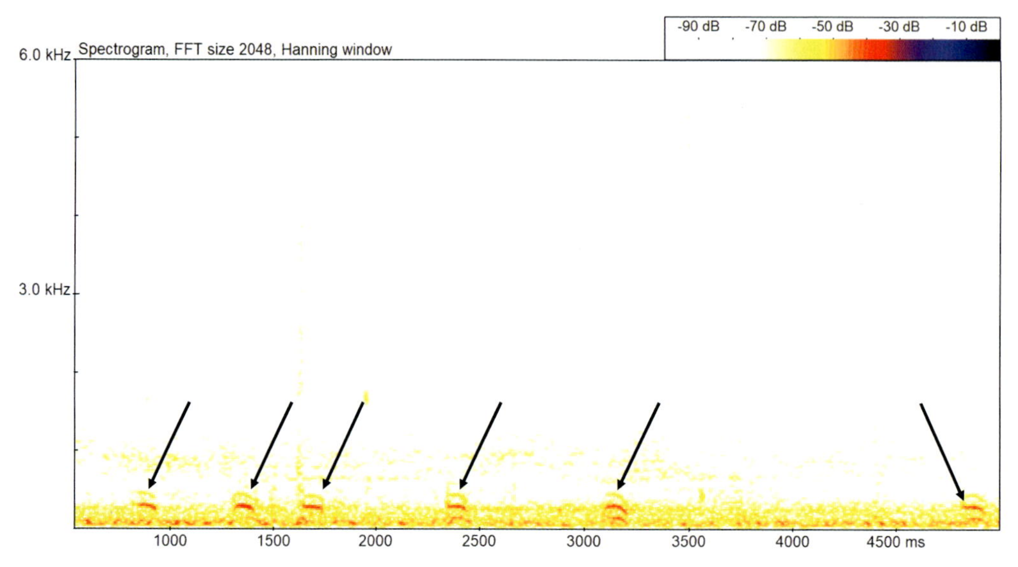

Figure 7.6 Rabbit low-frequency calls (frequency scale: 0 to 6 kHz/frame width: c.5 sec)

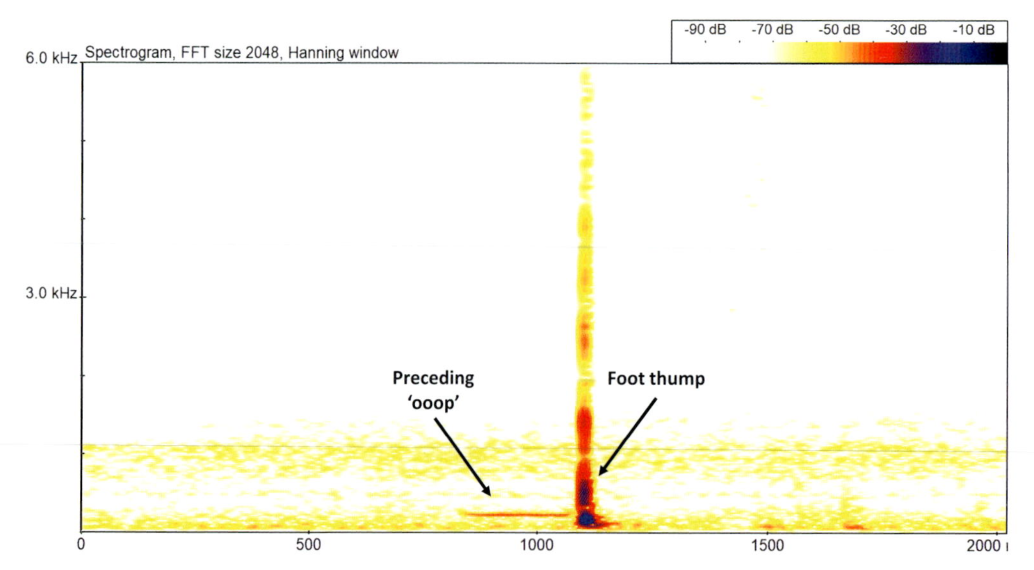

Figure 7.7 Rabbit low-frequency 'ooop' call followed by foot thump (frequency scale: 0 to 6 kHz/frame width: c.2 sec)

Potential confusion between rabbit and other species

Table 7.7 Summary of potential confusion between rabbit and other species

Confusion group	Potential confusion species In the first instance, always consider habitat and distribution. Caution is recommended when making a diagnostic identification. Associated visual/video evidence would be beneficial.
Other Lagomorpha species	Distress calls may sound very similar to hare species.
Species in other groups	Squirrel species. Some sounds may at first sound bird-like.

7.4 Potential application of bioacoustics for monitoring lagomorphs

This group is not making sound on a regular or consistent basis, and as such, a large survey effort over long periods of time, during appropriate seasons (e.g. mating season) would be needed to match anticipated acoustic behaviour. Whilst these mammals could potentially be detected over a larger distance than smaller terrestrial mammals, the number of acoustic encounters is expected to be small.

In conclusion it is probably fair to say that as far as we know at this stage, the application of purely acoustic methods to the study of this group is limited, other than research that specifically focuses upon a better understanding of their bioacoustic behaviour – for example during mating, behaviours during mother/young interactions (e.g. contact and isolation calls), as well as communications associated with distress. A deeper understanding of their vocalisations may in time prove more valuable than is suggested here.

Red squirrel – Sandra Graham

CHAPTER 8

Rodents – Squirrels & Beavers

8.1 Introduction to this group

This chapter covers our three largest species of rodent: red and grey squirrel, and beaver (see Table 8.1).

These species have a separate chapter since, being larger than most other rodent species present in Britain and Ireland (Chapter 9), we would expect them to produce louder and lower-frequency vocalisations, rather than depend on ultrasound for vocal communication. In addition, their presence is comparatively easy to establish through visual sightings, the use of trail cameras and/or finding field-sign evidence (e.g. evidence of foraging, faeces and places of shelter). Nevertheless, as with many of our other larger mammals, there are occasions when having some knowledge of the repertoire of sounds made by these species can help confirm presence, and may be useful for providing further information on behavioural and population status.

8.2 Overview and comparisons of acoustic behaviour

Broadly speaking, the species in this group will be encountered acoustically through the production of audible vocalisations (see Table 8.2) consisting of sounds that can carry over a relatively long distance (see Table 8.3). However, some of the squirrel calls, particularly those produced by neonates, have ultrasonic harmonics (Lishak, 1982a) which may also have a communicative function.

Red squirrels and grey squirrels emit a diversity of calls and it is possible to distinguish between the two species. Squirrel calls can be broadly categorised as alarm, agonistic, discomfort, mating, affiliative and neonatal (Diggins, 2021). Comparing their vocal activity, grey squirrels vocalise more often and are much louder and more shrill than red squirrels (Cresswell, 2023). Vocalisations made by adult grey squirrels are well documented by Lishak (1982b, 1984) who characterised them as 'muk muk', 'buzz', 'quaa' and 'kuk-kuk', and 'sneeze-like' calls. Nestling calls were described as 'squeaks', 'growls', 'screams', 'tooth-chattering' and 'lip smacking' (Lishak 1982a). Comparatively, the vocalisations made by European red squirrel have been less well researched and only 'chak/chuck calls', 'chattering' and 'squeals' are reported in the literature (Sample, 2006; Gurnell *et al.*, 2008a; Cresswell, 2023). Both squirrel species produce non-vocal sounds, most notably adults stamping on a tree to warn off an intruder.

Sound appears to provide a subsidiary form of communication for European beavers. Olfaction seems to be more significant: scent mounds serve to warn off transient beavers and are most prominent in spring when many two-year-olds begin to disperse. Also, oily secretions from the anal glands, which are rubbed onto the fur to provide water-proofing, convey information on sex and family membership (Müller-Schwarze, 2011). Nevertheless, a range of beaver vocalisations have been reported including 'whines', 'growls', 'grunts', 'sighs', 'tooth-chattering' and 'hissing'. As they are group animals, many vocalisations seem to relate to social dominance, with most calls reported from inside the lodge. Young beavers are particularly vocal, emitting a soft high-pitched whine that relates to soliciting food and play. Gargling and bubbling noises have also been reported when beavers leave and enter the lodge. When alarmed or to warn off other animals such as deer feeding on beaver food caches, beavers will typically hiss and grind their

Table 8.1 Squirrel species and beaver occurring in Britain and Ireland, including status, distribution and abundance. The relevant chapter sections that describe species-specific vocalisations are also listed.

Order	Family	Species common name	Genus	Species scientific name	Status	Distribution range	Abundance within distribution range	Sub-chapter reference
Rodentia Rodents	Sciuridae (Squirrels)	Red squirrel	*Sciurus*	*S. vulgaris*	Native	Localised in England, Wales and Ireland; more widespread in Scotland	Scarce in England, Wales and Ireland; locally common in Scotland	8.3.1
		Grey squirrel		*S. carolinensis*	Non-native	Widely distributed throughout much of Britain and Ireland	Common	8.3.2
	Castoridae (Beaver)	Beaver	*Castor*	*C. fiber*	Native Reintroduced	Scotland and south-west England	Uncommon	8.4.1

teeth in defence (Hodgdon and Larson, 1973; Müller-Schwarze, 2011; Pollock *et al.*, 2017). Non-vocal sounds produced by beavers include gnawing associated with wood-cutting, which can be heard over considerable distances. They also make a distinctive tail-slapping alarm noise in water as well as on the ground, if disturbed, which can be heard several hundred metres away (Wilsson, 1971; Hodgdon and Larson, 1973; Thomsen *et al.*, 2007; Müller-Schwarze, 2011; Pollock *et al.*, 2017).

In the following sections, we will describe the sounds produced by each of these three species.

Table 8.2 Acoustic spectrum within which you would usually expect to encounter most useful sound for identification purposes within this group

Audible range		Ultrasonic range							
< 10 kHz	10–20 kHz	20–30 kHz	30–40 kHz	40–50 kHz	50–60 kHz	60–70 kHz	70–80 kHz	90–100 kHz	> 100 kHz
↑	↑								

Table 8.3 Detectable distance for useful acoustic encounters with squirrel species and beaver

Distance (in metres)							
< 5 m	5–10 m	10–20 m	20–50 m	50–100 m	Up to 500 m	Up to 1,000 m	> 1,000 m
Red squirrel/Grey squirrel/Beaver ←	←	←	←				
Ear/microphone							<<<source of sound

Table 8.4 Typical mating season (shaded in blue) for squirrel species and beaver occurring within Britain and Ireland

Species / Month	Jan	Feb	Mar	April	May	June	July	Aug	Sept	Oct	Nov	Dec
Red squirrel	■	■	■	■	■	■	■	■	■	■		■
Grey squirrel	■	■	■	■	■	■	■	■	■	■		■
Beaver	■	■	■	■								■

8.3 Squirrel vocalisations

In the following sections, a brief summary of the key characteristics of each of the two squirrel species will be described, together with details on the species-specific vocalisations and, where known, the behavioural function of these sounds.

8.3.1 Red squirrel *Sciurus vulgaris*

Length/Weight	Habitat preferences	Distribution
Smaller than grey squirrel. Body/tail length of 320 to 435 mm. Weight 220 to 435 g.	Coniferous woodland Broadleaved woodland Parkland	Stronghold occurring throughout northern and western Scotland, the Borders area and Ireland. More localised, patchy distribution occurs within England and Wales. Also found on the Isle of Wight and Brownsea Island.
Diet		
Notable food sources are pine cones, hazelnuts, beech mast and acorns. Shoots and flowers in the spring and a range of nuts, fruits and seeds in autumn and winter, as well as fungi.		

Daily activity pattern	Typically active during daylight hours.
Seasonal activity pattern	Active throughout the year.
Mating activity pattern	Mating takes place during winter or spring, with births occurring during the period February to July. Young, usually one to six, are weaned at c.8 to 10 weeks.
Other notes	Red squirrels are our only native squirrel species, and their range is viewed as contracting due to being displaced by grey squirrels and from the 'squirrel pox' virus, which is carried by the greys and can cause entire populations to die out. Usually solitary, but may share dreys during winter and spring. Dreys (balls of interwoven twigs lined with leaves and moss) situated near tree trunks or in branch forks. Characteristic feeding remains of hazelnuts and cones.

References
Gurnell *et al.*, 2008b; Gurnell *et al.*, 2012; Lysaght and Marnell, 2016; Mathews *et al.*, 2018; Crawley *et al.*, 2020.

Acoustic behaviour including spectrogram examples

Red squirrels are known for producing loud 'chuck-like' noises (Figures 8.1 and 8.2) during aggressive encounters. Softer chucking may also occur, as well as other sounds: for example 'explosive "wrruhh" (Figures 8.2, 8.3, 8.6 and 8.7), various moans and teeth chattering' (Gurnell *et al.*, 2008a). In addition, 'scream-like' calls have also been recorded (Sample, 2006; Cresswell, 2023) (Figures 8.4, 8.5 and 8.6). An additional call we have encountered ourselves (courtesy of M. Smith, Banchory, Aberdeenshire) is provided in Figure 8.8, described here as 'murmurs', but it could very well fall into the previously discussed 'moans' category as described by Gurnell *et al.*, 2008a.

Unlike grey squirrels and North American red squirrels, there have been very few studies on the vocalisations of European red squirrels. Nevertheless, based on comparisons with other tree squirrels (McRae, 2020), we would expect that different calls and/or combinations of calls elicit specific information and responses – for example, whether a perceived threat is terrestrial or aerial. Similarly, a variety of calls would be anticipated from youngsters within the drey, in a similar way that such behaviour has been described for grey squirrel (Lishak, 1982a).

Figures 8.1 to 8.8 provide specific examples of spectrograms relating to this species, and where the 🔊 symbol appears, a recording of what is shown within the spectrogram is available to download from the species library. In addition, QR codes are provided for selected examples, and these can be accessed immediately for listening purposes from most mobile devices.

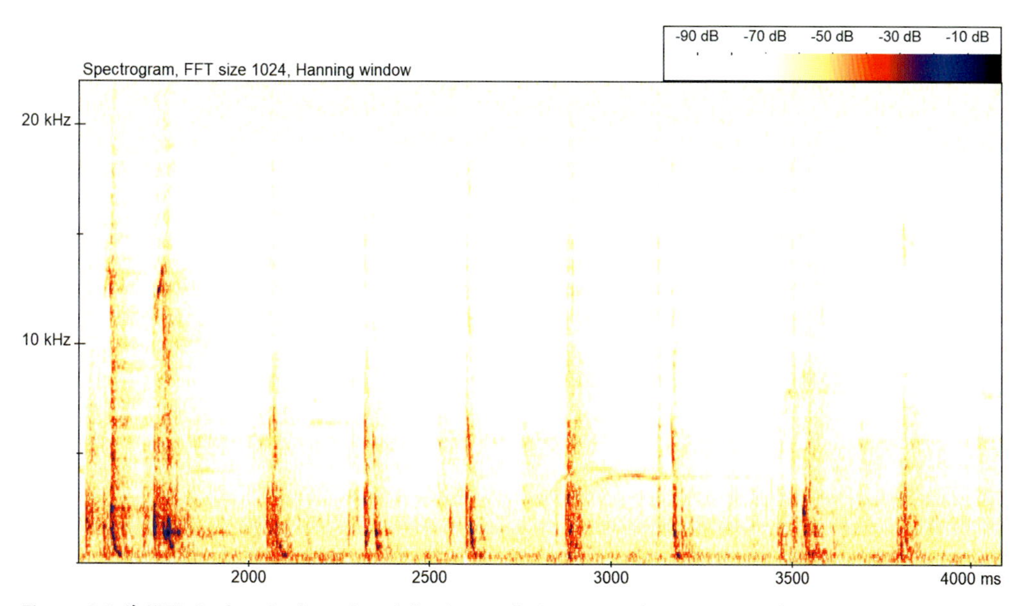

Figure 8.1 🔊 **(QR)** Red squirrel – series of chucking calls (courtesy of M. Ferguson) (frame width: c.2.5 sec)

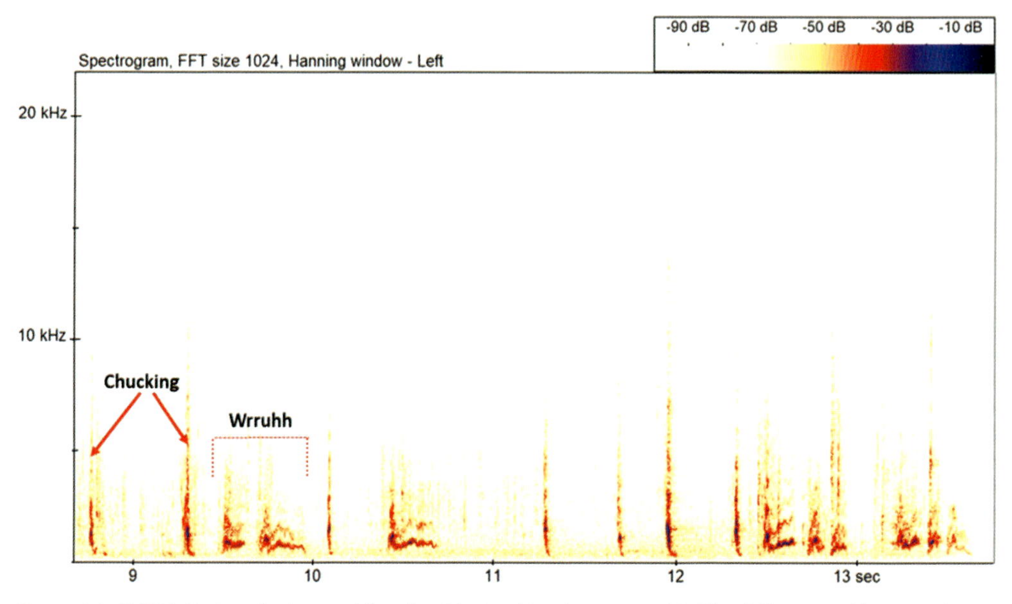

Figure 8.2 🔊 **(QR)** Red squirrel – wrruhh calls with chucking (courtesy of S. Elliott) (frame width: c.5 sec)

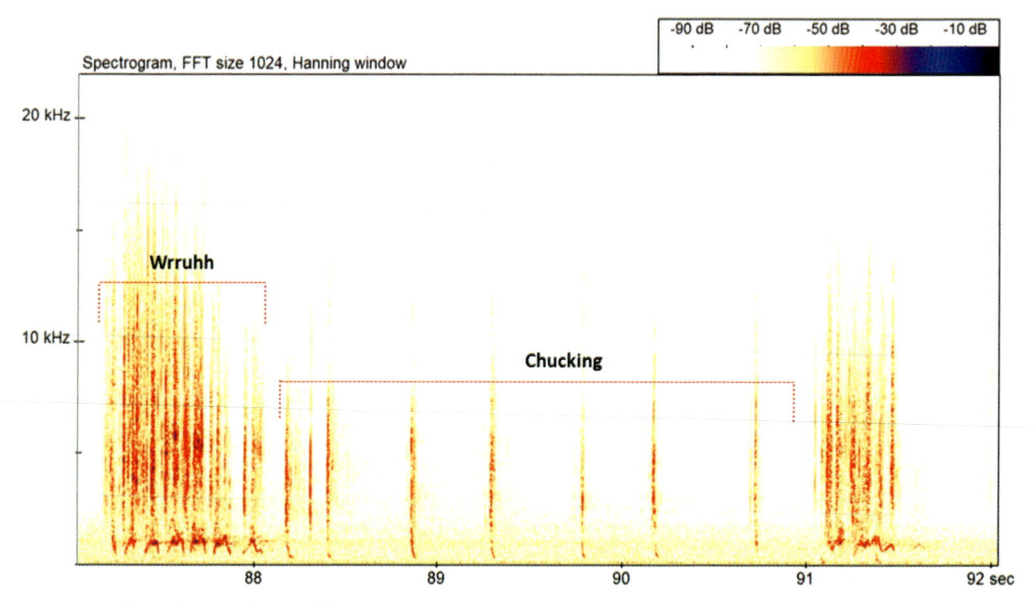

Figure 8.3 🔊 Red squirrel – wrruhhs at start and end, with chucking in between (courtesy of S. Elliott) (frame width: c.5 sec)

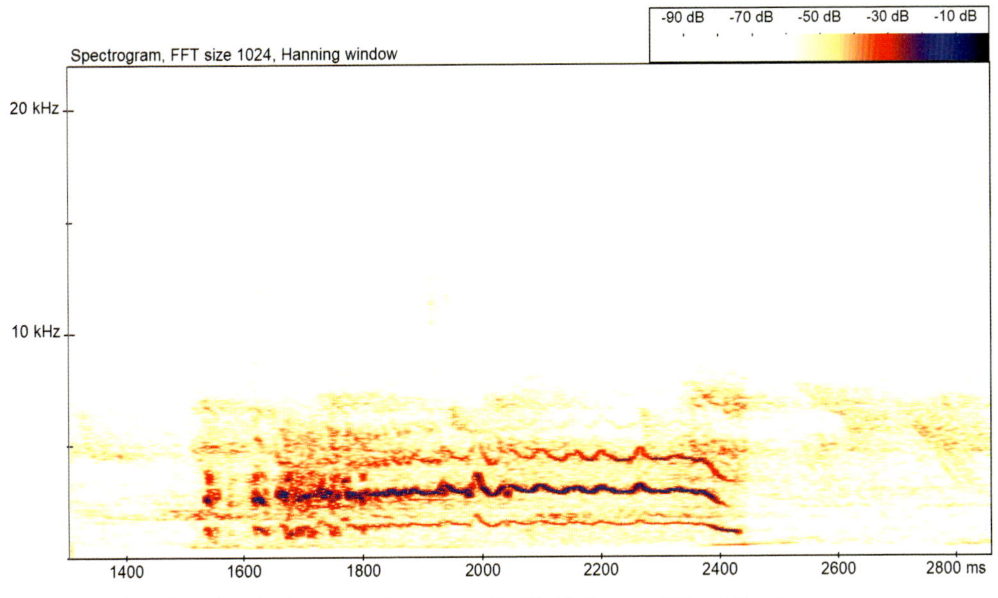

Figure 8.4 🔊 **(QR)** Red squirrel – scream (courtesy of S. Elliott) (frame width: c.1.5 sec)

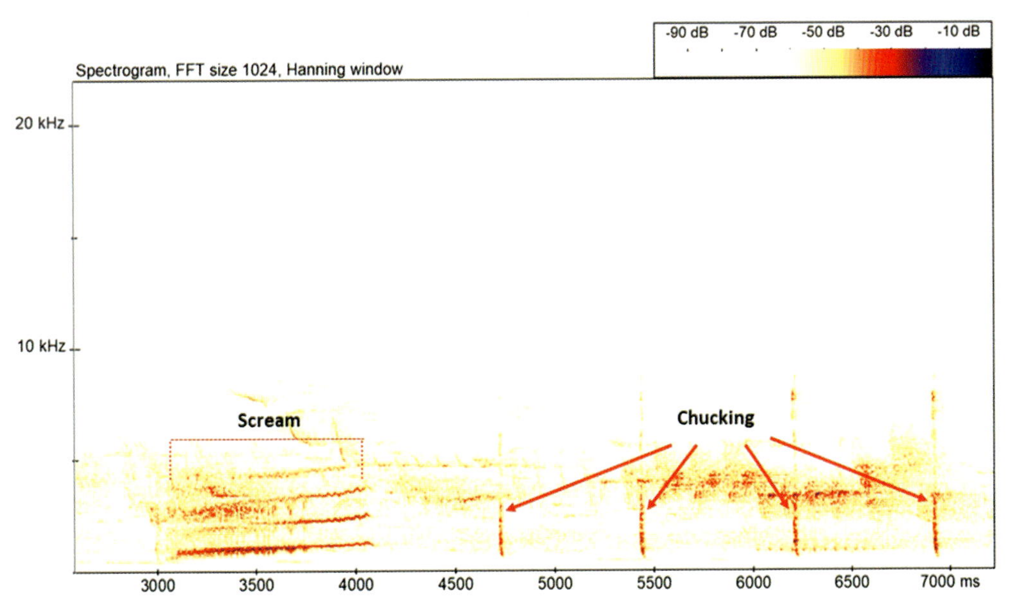

Figure 8.5 🔊 Red squirrel – scream with chucking, and also some birdsong at higher frequencies (courtesy of S. Elliott) (frame width: c.5 sec)

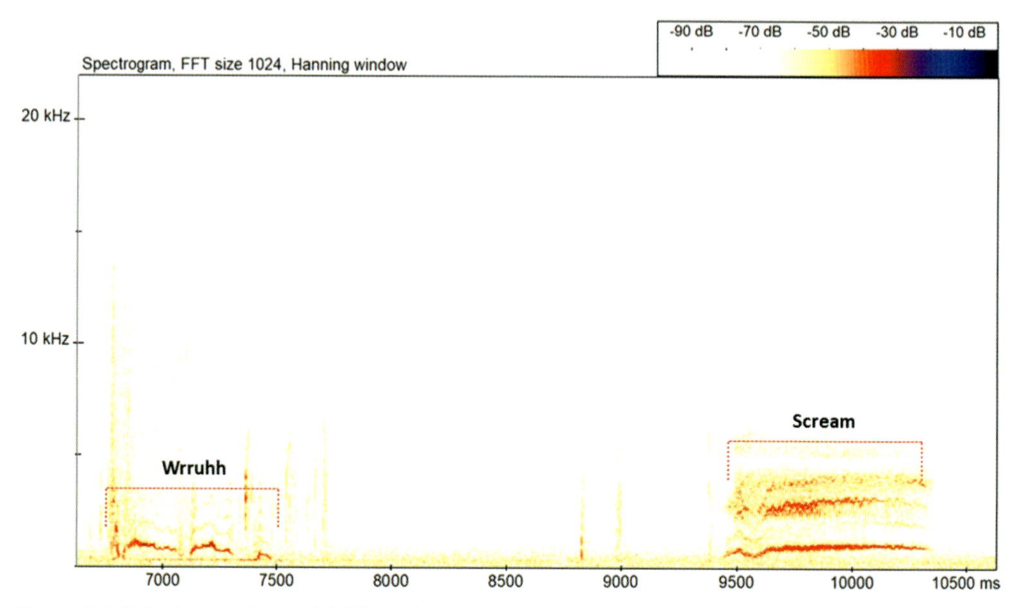

Figure 8.6 Red squirrel – wrruhh followed by a scream (courtesy of S. Elliott) (frame width: c.4 sec)

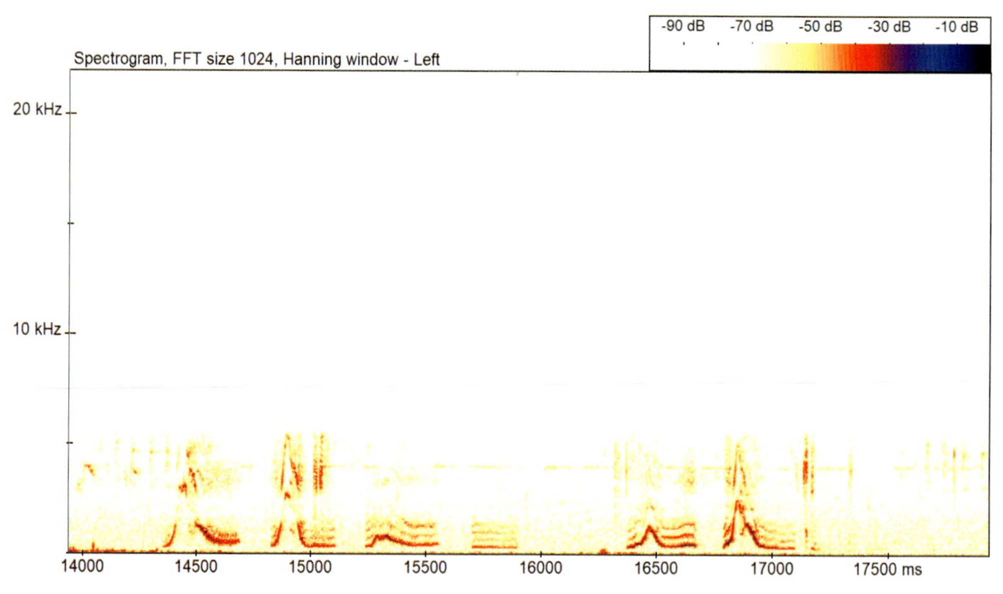

Figure 8.7 Red squirrel – wrruhhs (courtesy of M. Smith) (frame width: c.4 sec)

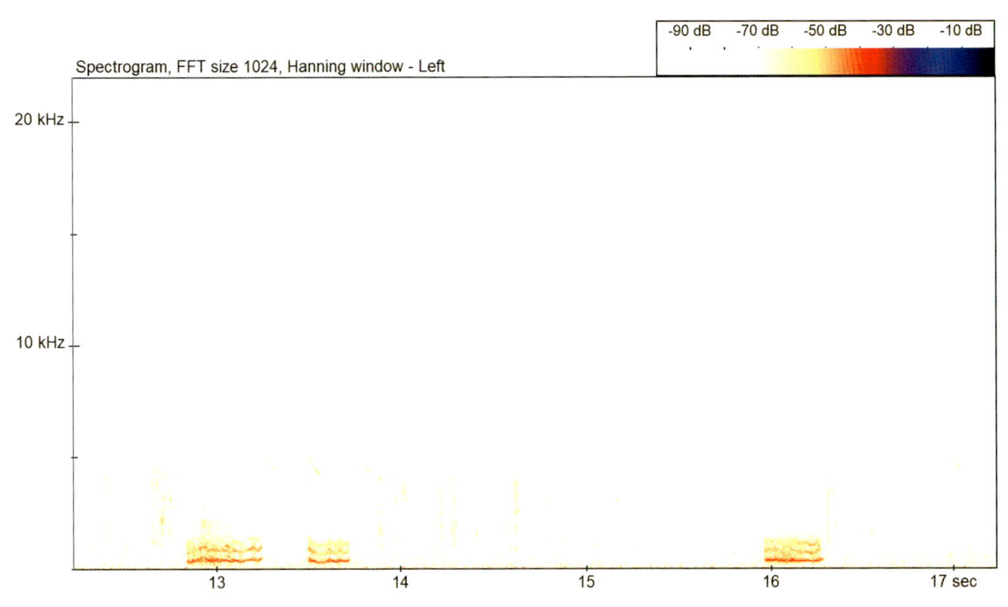

Figure 8.8 🔊 Red squirrel – murmurs (moans!) (courtesy of M. Smith) (frame width: c.4 sec)

QR codes relating to presented figures

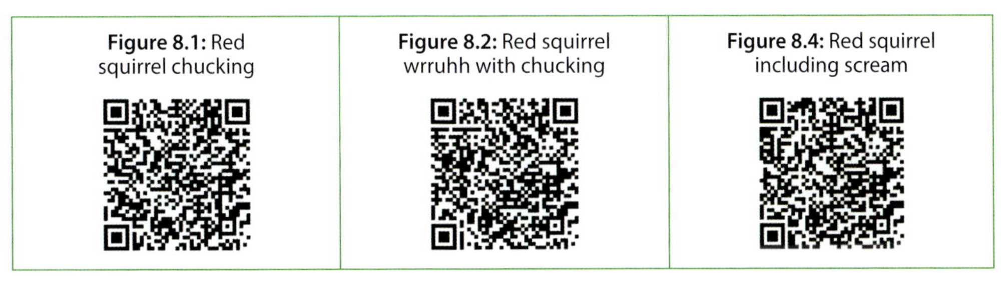

Figure 8.1: Red squirrel chucking	**Figure 8.2:** Red squirrel wrruhh with chucking	**Figure 8.4:** Red squirrel including scream

Potential confusion between red squirrel and other species

Table 8.5 Summary of potential confusion between red squirrel and other species

	Potential confusion species
Confusion group	In the first instance, always consider habitat and distribution.
Other squirrel species	Grey squirrel.
Species in other groups	Unlikely to be confused with other mammals, but it could be confused with a number of bird species.

8.3.2 Grey squirrel *Sciurus carolinensis*

<table>
<tr><th>Length/Weight</th><th>Habitat preferences</th><th>Distribution</th></tr>
<tr><td>Larger than red squirrel.
Body/tail length of 435 to 425 mm.
Weight 400 to 710 g.</td><td rowspan="4">Coniferous woodland
Broadleaved woodland
Parkland
Gardens</td><td rowspan="4">Occurs throughout much of Britain and Ireland, with the exception of north-west Scotland, western Ireland, the Isle of Wight and Brownsea Island.</td></tr>
<tr><th>Diet</th></tr>
<tr><td>Acorns, beechnuts and chestnuts during the autumn and winter; they may also take other food sources, for example fruits, shoots, buds, bark, lichen and flowers during the spring and summer. They have also been reported to feed on carrion and bird's eggs.</td></tr>
<tr></tr>
<tr><th>Daily activity pattern</th><td colspan="2">Typically active during daylight hours.</td></tr>
<tr><th>Seasonal activity pattern</th><td colspan="2">Active throughout the year.</td></tr>
<tr><th>Mating activity pattern</th><td colspan="2">Mating activity can occur at any time, but mainly during spring and summer, with births occurring during the periods March to April, and July to October.
Young, usually two or three, tend to be born either in spring or early summer, being fully weaned at c.10–12 weeks.</td></tr>
<tr><th>Other notes</th><td colspan="2">Grey squirrels were introduced into the British Isles during the late 1800s/early 1900s.
Usually solitary.
Dreys situated near tree trunks or in branch forks, but may also use tree cavities. In urbanised areas, they may nest within roof spaces of buildings.
Characteristic feeding remains of hazelnuts and cones.</td></tr>
<tr><td colspan="3">References
Gurnell et al., 2008a; Gurnell et al., 2012; Lysaght and Marnell, 2016; Mathews et al., 2018; Crawley et al., 2020.</td></tr>
</table>

Acoustic behaviour including spectrogram examples

A range of vocalisations have been documented for this species, with calls defined as 'buzz', 'kuk', 'quaa', 'moan', 'squeak', 'growl', 'scream', 'tooth-chatter', 'lip-smacking', 'muk-muk' and 'sneeze-like' mating calls, emitted by the males when chasing a female (Lishak, 1982a, 1982b and 1984).

The repetitive quaa call (Figures 8.9, 8.10 and 8.11) is associated with warning behaviour when threatened (Gurnell *et al.*, 2008a). Alarm calling has been described as buzz, kuk, quaa and moan, all of which can be produced in pure or mixed sequences or as one-off single-note calls (Lishak, 1984; McRae and Green, 2014).

Buzz calls are a rapid series (ranging from 1 to 30) of nasal low-intensity FM swept notes, with fmax being as high as 12 kHz and fmin being at c.0.25 kHz. FmaxE occurs within the range of 3 to 6 kHz.

Research by McRae and Green (2015) found that information about a perceived threat was associated with call rate and the type of calls/call combinations (kuks, quaa and/ or moans) emitted. Kuk and quaa calls are usually emitted in response to a terrestrial predator. Moans are emitted in response to aerial predators. If the initial 30 seconds of calling contains kuks, then the elicited threat is terrestrial in about 69% of cases, and the ambiguity of the threat is lessened when the rate of kuk calls is more rapid. The likely presence of a terrestrial threat is increased when quaa calls are also emitted (89% of the time when quaa calls were used, the threat was terrestrial). Conversely, if the first 30 seconds of the call contains moans, the threat is always aerial. Moan tonality is influenced by the anatomical features of the vocal airway and consequently these calls may also provide additional information about the identity of the caller.

Quaas and kuks are similar in frequency, with an FmaxE in the region of 2 to 5 kHz, but differing in call length and time interval between emissions. Components with a duration of 0.15 seconds or less can be described as kuks, whereas longer calls are described as quaas (Lishak, 1984). The production of quaas may appear to be slower (i.e, less components per unit of time) than kuks.

Quite often shorter kuks are produced, preceding fewer but longer quaa sounds, as shown in Figures 8.9 to 8.11. Figure 8.11 has an additional element included just before the longer calls, this additional feature being a shudder (rumble) of quick notes. Based on the characteristics of these call sequences, these vocalisations are likely a response to a terrestrial predator/threat.

Figure 8.12 shows a series of low-amplitude growls, whilst Figure 8.13 gives examples of squeak-type sounds as well as the moan call, which has a longer flatter structure (i.e. relatively constant frequency), often with a minimum frequency range of 1.2 to 2 kHz. The latter call is likely a response to an aerial threat.

Calls produced by young grey squirrels were first fully documented as part of a focused effort by Robert Lishak (Lishak, 1982b), within which a variety of vocalisations were described (squeaks, growls, screams, teeth chattering and lip smacking), including the development of sound with maturity. Ultrasonic harmonics ranging from 16 to 50 kHz were also noted but their communicative value remains unclear. The paper also describes the challenges associated with coming across such vocalisations because of the inaccessibility (both visually and audibly) of dreys to humans, and also the infrequency and low intensity of any sounds produced by young squirrels.

Figures 8.9 to 8.13 provide specific examples of spectrograms relating to this species, and where the 🔊 symbol appears, a recording of what is shown within the spectrogram is available to download from the species library. In addition, QR codes are provided for selected examples, and these can be accessed immediately for listening purposes from most mobile devices.

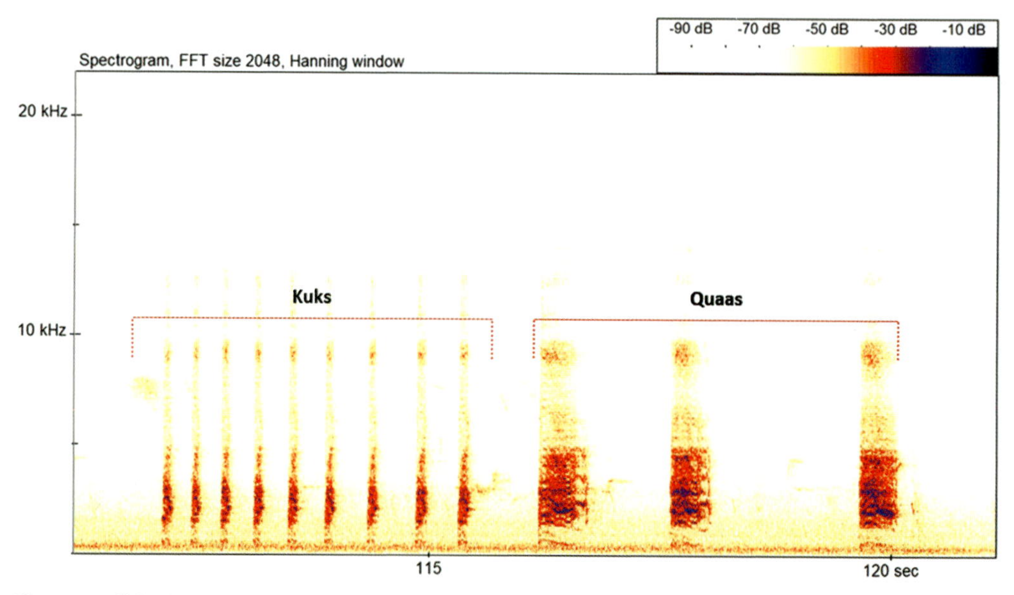

Figure 8.9 🔊 **(QR)** Grey squirrel – alarm calling, a series of kuks followed by three quaa emissions (courtesy of M. Ferguson) (frame width: c.10 sec)

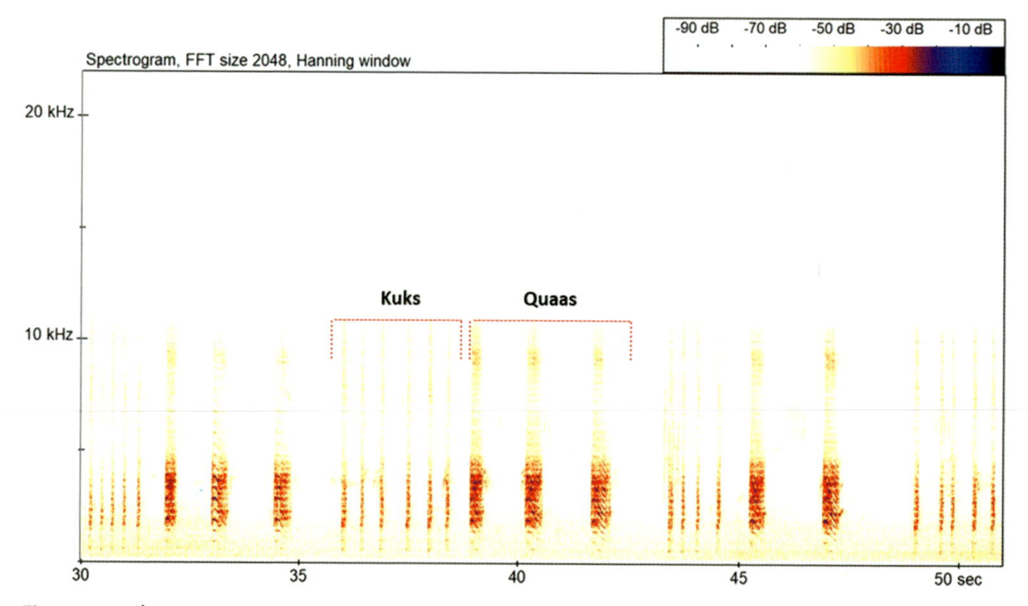

Figure 8.10 🔊 Grey squirrel – a longer series of kuks and quaas showing the emission rate and relative lengths of each (courtesy of S. Elliott) (frame width: c.26 sec)

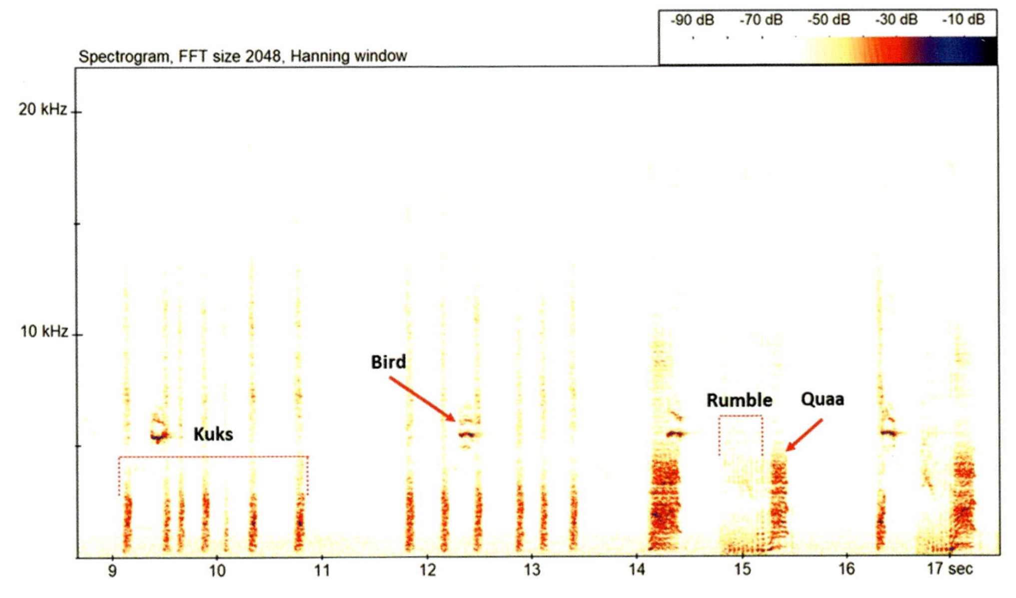

Figure 8.11 🔊 Grey squirrel – showing kuks and rumbles occurring immediately before quaas (courtesy of S. Elliott) (frame width: c.9 sec)

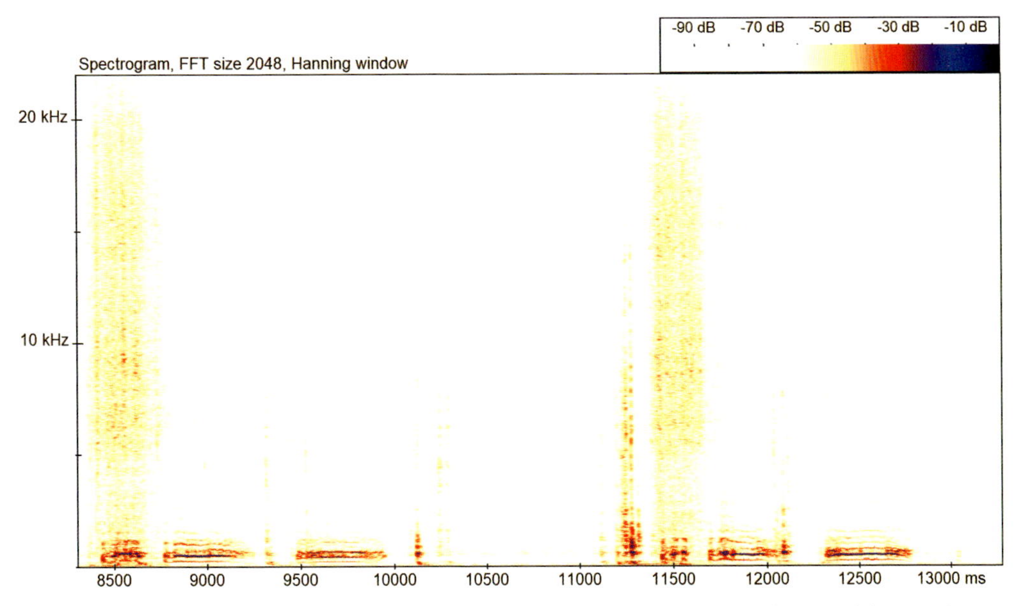

Figure 8.12 🔊 **(QR)** Grey squirrel – low-amplitude growling (courtesy of S. Elliott) (frame width: c.6 sec)

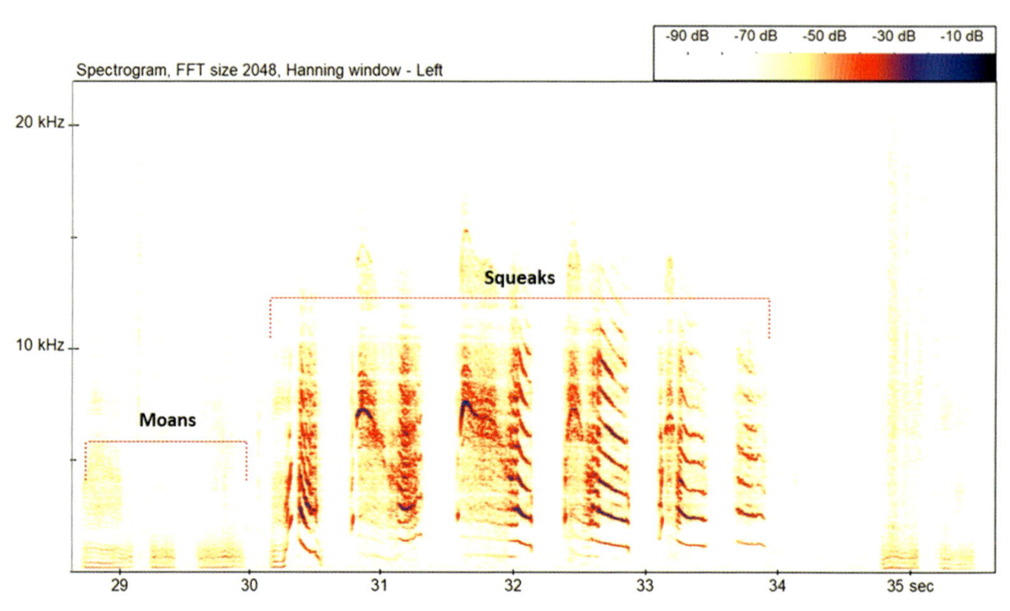

Figure 8.13 🔊 **(QR)** Grey squirrel – a combination of moans and squeaks (courtesy of S. Elliott) (frame width: c.7 sec)

QR codes relating to presented figures

Figure 8.9: Grey squirrel kuks and quass	**Figure 8.12:** Grey squirrel growls	**Figure 8.13:** Grey squirrel medley squeaks and moans

Potential confusion between grey squirrel and other species

Table 8.6 Summary of potential confusion between grey squirrel and other species

	Potential confusion species
Confusion group	In the first instance, always consider habitat and distribution.
Other squirrel species	Red squirrel.
Species in other groups	Unlikely to be confused with other mammals. May be confused with some bird species, especially jay.

8.4 Beaver vocalisations and associated sounds

In the following section, a brief summary of the key characteristics of the single species of beaver present within Britain and Ireland, the Eurasian beaver, will be described, together with details on the species-specific vocalisations and, where known, the behavioural function of these sounds.

8.4.1 Beaver *Castor fiber*

Length/Weight	Habitat preferences	Distribution
Body length 74 to 81 cm. Weight 12.5 to 30 kg.	Rivers Lakes Wetlands Broadleaved woodland	Beyond a reintroduction programme in Scotland, other pockets of free-ranging activity occur due to escapees and illegal introduction to areas.
Diet		
Herbivorous, feeding mainly on vegetation and bark, remaining close to water (rarely more than 100 m from the water's edge). Herbaceous species are favoured in the summer months. In autumn, broadleaved woody plants especially aspen *Populus tremula* and willows *Salix* species are preferred.		

Daily activity pattern	Typically most active from dusk through to dawn.
Seasonal activity pattern	Active throughout the year, living as a family group comprising of adult male and female, along with this year's youngsters as well as potentially young from the previous breeding season.
Mating activity pattern	Mating takes place during December to April. A single litter per annum (usually two to four young) are born during the period May to June, and weaned within two to three months.
Other notes	Webbed tracks with dragging tail imprint often noticed on bank sides or trails between water bodies. Characteristic evidence due to fallen trees and remaining gnawed stumps. Trees with trunk diameters of 3 to 10 cm are felled and bark, twigs and leaves are consumed or cached. Wood may be used for engineering works (lodges, dams). Constructs dams with logs, branches, rocks and mud. Dens/nests often within naturally occurring or burrowed holes in bankside.

References
Cole *et al.*, 2008; Galbraith and Gaywood, 2008; Mathews *et al.*, 2018; Crawley *et al.*, 2020.

Acoustic behaviour including spectrogram examples

Beaver are not known for being particularly vocal, but have been recorded making growling sounds, as well as scream-like vocalisations and hisses (Cole *et al.*, 2008). When family members greet each other sometimes a whining noise can be heard (Campbell-Palmer *et al.*, 2016). Examples of calls recorded by ourselves and others are provided in Figures 8.14 to 8.19.

Youngsters are more vocal, with a range of mews, squeaking and crying noises, thought to be related to soliciting food and during play (Wilsson, 1971). The frequency of whine calls and the pitch of the whine decreases with age (Hodgdon and Larson, 1973).

A non-vocal characteristic sound is that of tail-slapping when alarmed, both in the water or on the ground, which can be heard several metres away (Wilsson, 1971; Hodgdon and Larson, 1973; Thomsen *et al.*, 2007) – a recording of which we initially thought we'd be unlikely to obtain, but thanks to a focused effort from Paul Howden-Leach we are delighted to share with you here (Figure 8.17). Beavers seem to be able to discriminate between tail slaps from different individuals, and there is variability in the response to beaver tail-slapping between conspecifics (Müller-Schwarze, 2011; Pollock *et al.*, 2017). This behaviour may therefore have a social function in terms of group cohesion, as well as serving to alert other group members of a potential threat.

Feeding sounds, although not true vocalisation, can also be quite loud and distinctive, with quick chewing shown in Figure 8.18, and a much slower gnawing presented in Figure 8.19.

Figures 8.14 to 8.19 provide specific examples of spectrograms relating to this species, and where the 🔊 symbol appears, a recording of what is shown within the spectrogram is available to download from the species library. In addition, QR codes are provided for selected examples, and these can be accessed immediately for listening purposes from most mobile devices.

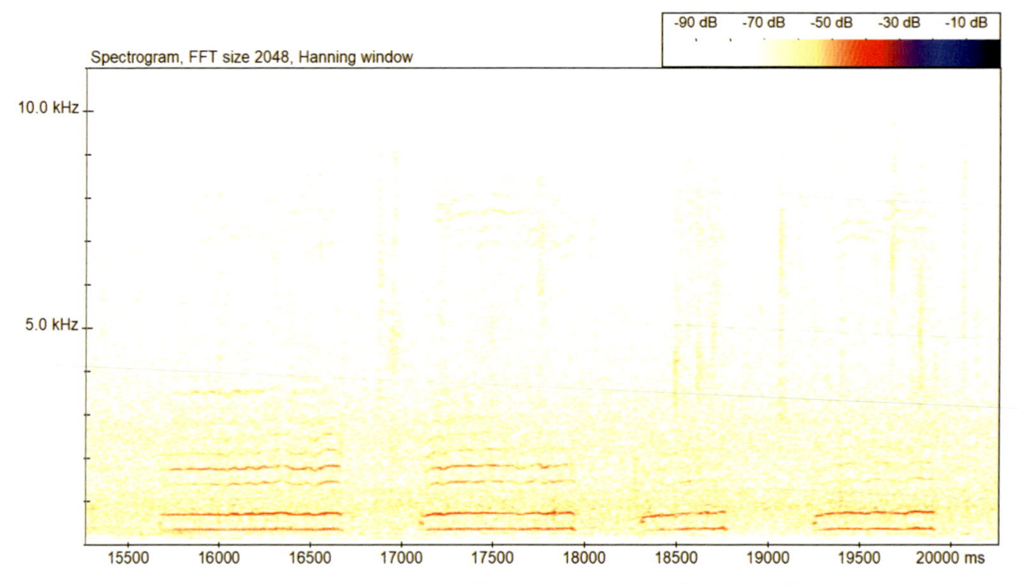

Figure 8.14 🔊 Beaver – a series of long whining calls (courtesy of P. Howden-Leach) (frame width: c.5 sec)

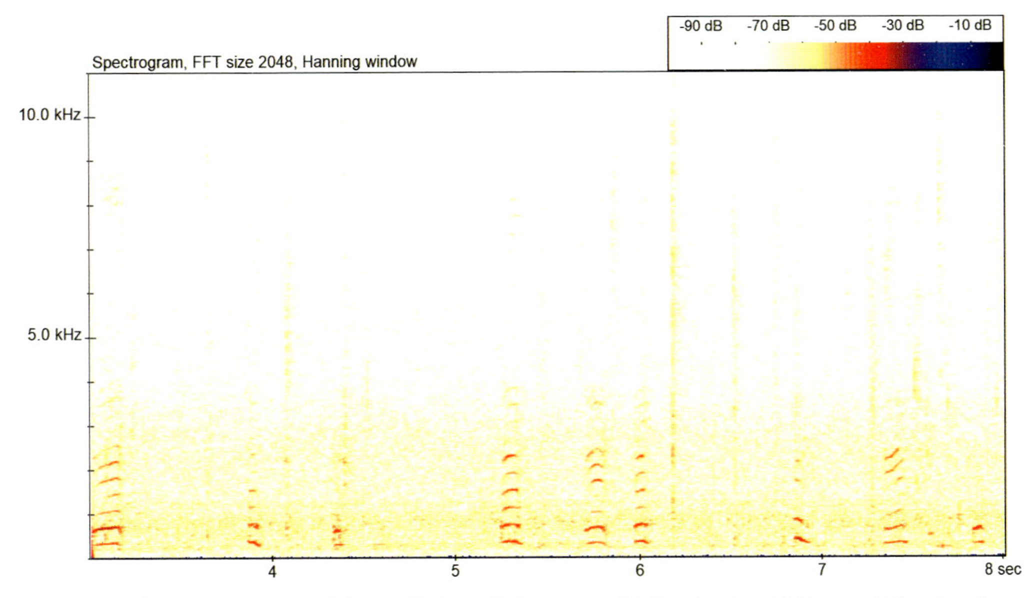

Figure 8.15 🔊 Beaver – a series of short whining calls (courtesy of P. Howden-Leach) (frame width: c.5 sec)

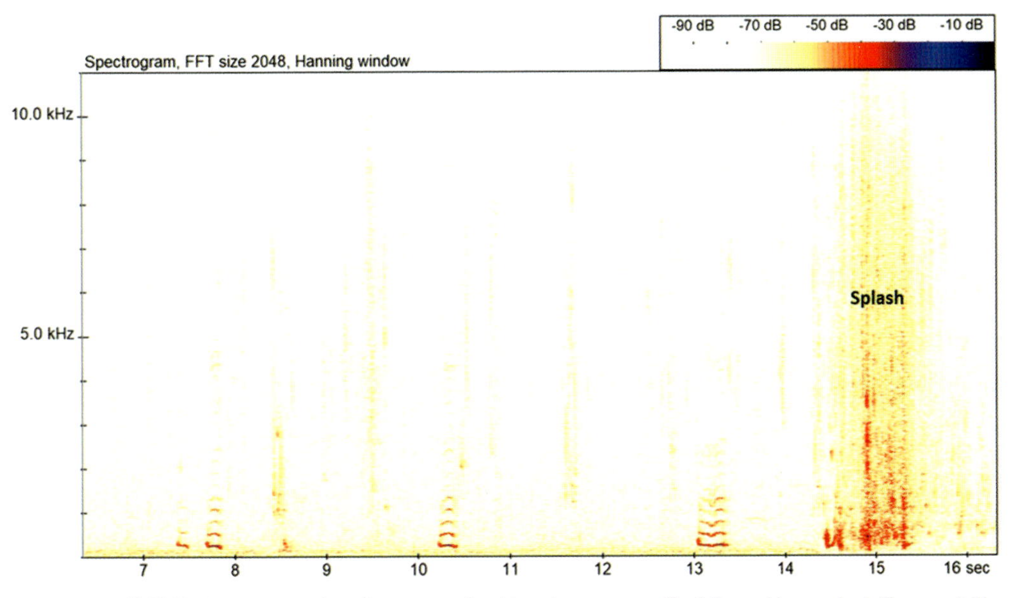

Figure 8.16 🔊 **(QR)** Beaver – a series of uncategorised low-frequency calls, followed by a splash (frame width: c.10 sec)

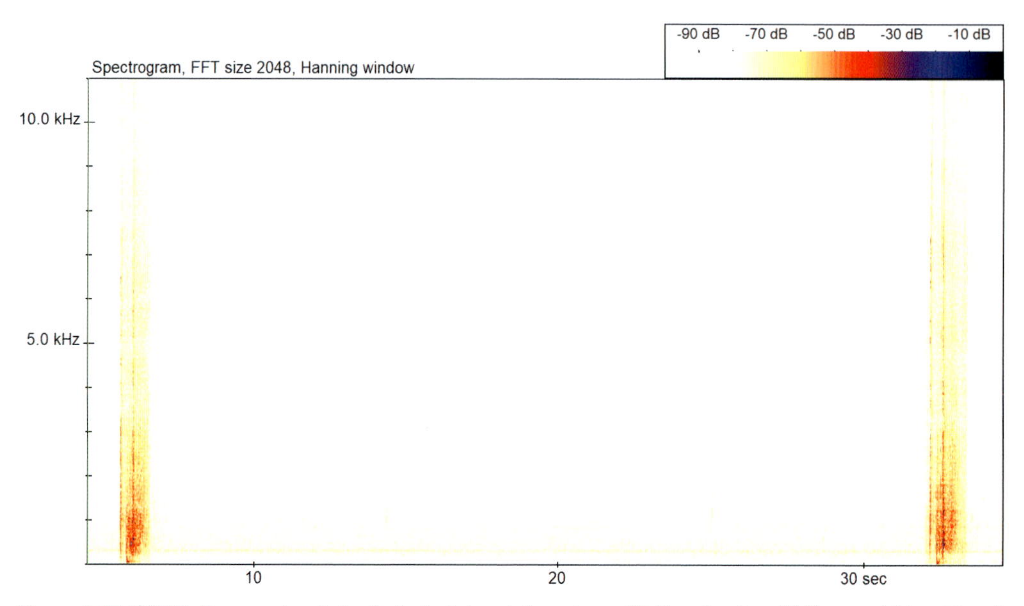

Figure 8.17 🔊 (QR) Beaver – two tail splash alert signals (courtesy of P. Howden-Leach) (frame width: c.30 sec)

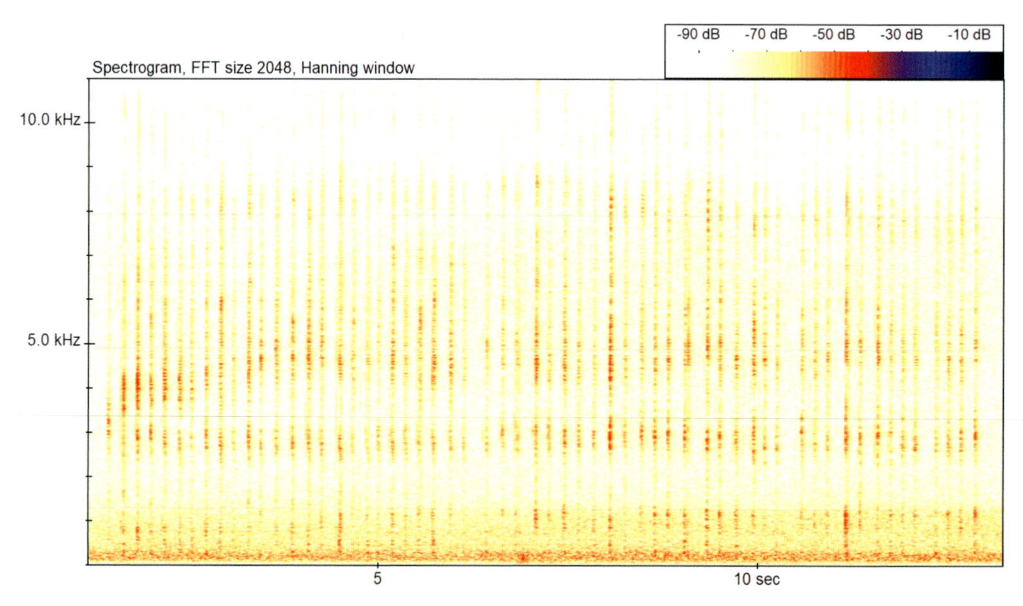

Figure 8.18 🔊 (QR) Beaver – a long series of fast chewing sounds (frame width: c.12 sec)

-90 dB -70 dB -50 dB -30 dB -10 dB
Spectrogram, FFT size 2048, Hanning window
10.0 kHz
5.0 kHz
5
10 sec

Figure 8.19 Beaver – a series of longer gnawing sounds produced whilst feeding (frame width: c.12 sec)

QR codes relating to presented figures

Figure 8.16: Beaver low-frequency calls	**Figure 8.17:** Beaver tail slap	**Figure 8.18:** Beaver chewing

Potential confusion between beaver and other species

Table 8.7 Summary of potential confusion between beaver and other species.

Confusion group	Potential confusion species
	In the first instance, always consider habitat and distribution.
Other rodent species	Unlikely to be confused with other rodent species.
Species in other groups	Unlikely to be confused with other species.

8.5 Potential application of bioacoustics for monitoring squirrels and beavers

Traditional methods to estimate population densities of red squirrel include: counting individuals, dreys and feeding sites (in coniferous forests); monitoring activity at artificial corncob baiting stations (Kopij, 2014); and the use of hair tube surveys (Mortelliti and Biotani, 2008). Since red squirrels often reside high in the canopy, visual detection may be problematic.

For drey counts, when both red and grey squirrels are present in an area, it is difficult to accurately discern to which of the squirrel species dreys should be attributed without direct sightings. The use of feeding signs also has limitations, as discriminating feeding activity between the squirrel species is largely reliant on pine cones and therefore coniferous forest habitats. Corncob baiting, although reliable for species identification, also has the disadvantage of altering the animals' natural behaviour by adding a food source into their environments. Hair tubes carry a risk of underestimating the number of individuals since one event may contain hairs from several individuals, or overestimating numbers due to 'tube-happy' animals repeatedly visiting (Mortelliti and Biotani, 2008). Furthermore, the cost (financial and manpower) of accurately analysing hair samples may impact the effectiveness of applying this survey method at a landscape scale.

Studies have been carried out in Europe to explore the effectiveness of Small Mammal Monitoring Units that utilise trail cameras (and sound recordings) (Di Cerbo and Biancardi, 2012) for monitoring red squirrel populations. Although effective in enabling some level of individual identification, and therefore population estimates, the key disadvantage of these methods is that monitoring stations are usually baited and consequently most behavioural activity is related to feeding. For areas that support both squirrel species, there are also biosecurity implications whereby red squirrels are inadvertently encouraged to feed at the same sites as the greys, which could facilitate transmission of the squirrel pox virus, if present in the population. There is also the potential for animals to become 'trap-happy', and, also, regular visitation by greys may act to exclude reds from entering the monitoring stations.

Based on these factors, acoustic surveys may offer a more suitable non-intrusive method for surveying red squirrel populations.

The application of manned acoustic surveys will frequently elicit alarm calls in response to an intruder, and/or be biased towards the breeding season when squirrels will be emitting mate-chasing calls. Although such surveys may have value, more significant will be the application of remote recorders, which will enable vocalisations to be recorded without any observer influence on behaviour. Similarly, baiting stations should not be required to gather data, thus eliminating any increased risk of viral transmission. Nevertheless, acoustic surveys will result in large datasets, meaning algorithms that can accurately identify the species from the surrounding soundscape would be beneficial, if not essential, for this method to be efficient.

Pioneering research is currently underway by Will Cresswell (University of Bristol) in collaboration with The Mammal Society to assess the potential application of acoustic monitoring surveys for red squirrels. Information on the sounds produced by red squirrels is still deficient and further investigations into the repertoire of calls they produce are needed before field testing acoustic methods for identifying the presence of red squirrel and/or population monitoring (Abass, 2021; Millman, 2021; Cresswell, 2023: www.mammal.org.uk/science-research/the-red-squirrel-acoustic-monitoring-project/).

For beavers, traditional methods that involve counting beavers at dusk and dawn at active lodges, and in core areas to record the number of beaver kits, would seem to remain the most effective method of monitoring numbers (Elmeros *et al.*, 2003). There are however opportunities for acoustics to be applied in behavioural studies, to better understand the species ecology. In particular, microphones could be deployed inside lodges, as well as within the surrounding environment, including the potential use of hydrophones to monitor acoustics during underwater activities. Such studies may benefit from combining acoustic surveys with trail camera or video surveillance systems. Following the collection of acoustic data and a better understanding of beaver vocalisations, it may in the future be possible to differentiate between individuals in a colony (e.g. by comparing the pitch and frequency of the most common 'whine' calls), with the potential application for wider species monitoring, aided by the automated extraction of calls.

Wood mouse – Sandra Graham

CHAPTER 9

Rodents – Rats, Mice, Voles & Dormice

9.1 Introduction to this group

This chapter covers the smaller rodents, within which we consider two species of rat, four species of mouse, four species of vole (one of which, common vole, is represented by two subspecies) and two species of dormouse (see Table 9.1).

Being small and usually most active at night, the presence of these species (and shrews as covered in Chapter 10) is often difficult to establish through visual sightings, despite being considerably more abundant than other mammals that are seen more regularly (to demonstrate this point, see Table 9.2).

Consequently, a variety of survey techniques have been designed to facilitate the monitoring of these species: some methods are more invasive, for example live trapping (Gurnell and Flowerdew, 2019) and nest box/nest tube surveys (Bright *et al.*, 2006), and others less invasive, for example the use of trail cameras (Littlewood *et al.*, 2021), owl pellet analysis (Yalden and Morris, 1993) and/or finding field-sign evidence (e.g. evidence of foraging, faeces and footprints).

Despite the array of monitoring procedures already available and the comparatively new application of bioacoustics for this species group, recent studies have highlighted the benefits of using acoustics for establishing species presence for some rodents and for shrews (see Chapter 10) (Newson and Pearce, 2022; Newson *et al.*, 2023). Accordingly, knowing the more regularly encountered and in some cases easily identifiable sounds made by the species in this group can help confirm species presence, and with this, there is the opportunity for acoustics to be used for informing on species distribution and status, potentially at any spatial scale (e.g. regionally, nationally, internationally). We will explore this later within this chapter (Section 9.7).

9.2 Overview and comparisons of acoustic behaviour

Broadly speaking the species in this group will be encountered through the production of vocalisations that are made ultrasonically (see Table 9.3) and consist of sounds that can carry over a relatively short distance (see Table 9.4 and Figure 9.1). The exception is edible dormouse which vocalises, comparatively, more often using audible sounds.

There is variability between the detectable distance of the species in this group. Some factors influencing their acoustic detection that should be considered before embarking on acoustic surveys are outlined below.

- Specialist ultrasound microphones (e.g. bat detectors) that are able to detect sounds above human hearing range (>20 kHz) are required to record many of their vocalisations.
- Species that vocalise at a lower sound frequency (e.g. brown rat, hazel dormouse) will typically be more detectable than those that vocalise at a much higher frequency (e.g. house mouse).
- For some species (e.g. voles), olfaction is the primary means of communication, and these species of small mammal tend not to vocalise often, so as to avoid detection by predators.
- Many of these species are often active within dense vegetation: in the forest undergrowth, amongst leaf litter, in tall grassy tussocks and within the leafy tree canopy, all of which acts to absorb sound, and thus hinders successful acoustic (ultrasound) detection.

Table 9.1 Small rodents occurring in Britain and Ireland, including status, distribution and relative abundance. The relevant chapter sections that describe species-specific vocalisations are also listed.

Order	Family	Species common name	Genus	Species scientific name	Status	Distribution range	Abundance within distribution range	Sub-chapter reference
Rodentia	Muridae	Brown rat	*Rattus*	*R. norvegicus*	Naturalised	Throughout Britain and Ireland	Common	9.3.1
		Black rat		*R. rattus*	Naturalised	Restricted, probably only occurring on some offshore islands	Rare, possibly no longer present	9.3.2
		Wood mouse	*Apodemus*	*A. sylvaticus*	Native	Throughout Britain and Ireland	Common	9.4.1
		Yellow-necked mouse		*A. flavicollis*	Native	Southern England and Wales	Common	9.4.2
		Harvest mouse	*Micromys*	*M. minutus*	Native	Throughout England and Wales, absent from Ireland	Common	9.4.3
		House mouse	*Mus*	*M. musculus*	Native	Throughout Britain and Ireland	Common	9.4.4
	Cricetidae	Water vole	*Arvicola*	*A. amphibius*	Native	Throughout mainland Britain, absent from Ireland	Uncommon	9.5.1
		Bank vole	*Myodes*	*M. glareolus*	Native	Throughout mainland Britain, with more localised distribution in Ireland towards the south-west	Common	9.5.2
		Field vole	*Microtus*	*M. agrestis*	Native	Throughout Britain, including some of the Western Isles, absent from Ireland	Common	9.5.3
		Orkney/Guernsey (Common) vole		*M. arvalis orcadensis/ M. arvalis sarnius*	Naturalised	Orkney/Guernsey	Common where present	9.5.4
	Gliridae	Hazel dormouse	*Muscardinus*	*M. avellanarius*	Native Reintroduced	Wales, southern England and Cumbria	Uncommon	9.6.1
		Edible dormouse	*Glis*	*G. glis*	Non-native	Southern England, mostly focused in the Chilterns area	Uncommon	9.6.2

- Animals that live in colonies (e.g. brown rat) or small social groups (e.g. house mouse) will be more vocal than those species that are only seasonally gregarious (e.g. water vole), communal (e.g. wood mouse) or territorially polygamous (e.g. field vole), or those that are territorially monogamous (e.g. yellow-necked mouse), due to the incidence/likelihood of intraspecific interactions.
- For species that occupy burrows, notably the voles (and to a lesser extent wood mouse and yellow-necked mouse), social interactions and associated vocal activity will occur more often inside burrow systems, rather than above ground, making acoustic detection less likely.
- There will be daily/nightly differences in detectability according to periods of activity. For example, some species such as wood mouse may be active throughout the night, whereas others, such as voles, are crepuscular in their habits, with acoustic detections expected more often around dusk and dawn.
- There will be seasonal variations in detectability according to the extent of vegetation cover (i.e. they may be easier to detect in late autumn to early spring); changes in species social structures and the degree of interactions (e.g. species are more vocal when breeding and/or with young); and when the availability of food resources affects seasonal habits (e.g. whether a species is active throughout the year or hibernates).

As a consequence, to successfully encounter these species acoustically, the recording equipment often needs to be deployed in close proximity to the animal in order to have a chance of picking sound up (Table 9.4). Also to be considered is the appropriate time period (e.g. within the day–night cycle and/or within the year, according to their periods of activity). In most cases, a significant amount of survey effort will need to be carried out to record a target species, and the acoustic results will often amount to only low numbers of vocal contacts.

Species such as brown rat and hazel dormouse will be more easily detected since they produce louder, lower-frequency calls, and have greater sociality (i.e. live in colonies or small groups). Of these species, brown rat will be the easiest to detect since they are ubiquitous and active throughout the year, whereas hazel dormouse has a localised range, occurs at only low population densities, is mainly found within dense habitats (broadleaved woodlands/ancient hedgerows) and hibernates (i.e. is inactive for up to six months of the year).

Our research to date suggests that the voles are significantly harder to detect than the other small rodents (Newson and Pearce, 2022), and successful detections (albeit in very low numbers) were mostly recorded when they occurred at high population densities (e.g. during mast years). This could be a consequence of them spending a significant amount of time within burrow systems where they may be more vocal but are beyond the range of the acoustic equipment, compared to above ground where olfactory cues (latrines, and marking of runways with faeces and urine) are the preferred communication method. Additionally, we speculate that, similar to other burrowing/subterranean rodent species such as mole-rats, prairie dogs and gophers (Heffner *et al.*, 1994; Buffenstein, 1996; Bednářová *et al.*, 2012; Narins *et al.*, 2016; Heffner *et al.*, 2020), it is conceivable that they can hear and are able to communicate using low-frequency sound/infrasound and/or seismic signalling – sounds which are undetectable using audible or ultrasonic microphones.

Within the following species accounts we tend to focus mostly on identifiable (i.e. most useful) sounds, which in the main are vocalisations emitted ultrasonically (a notable exception being edible dormouse), and so require the use of specialist recording equipment (e.g. bat detectors). In addition to the examples provided, we have many others which have not been included, usually because they fall into one of the following categories:

- Audible sounds – usually considered to be distress-related or threat calls (e.g. animal being attacked).
- Mother/young interactions – contact calls, begging calls etc.
- Examples only encountered on a very small number of occasions where it is not yet possible to be confident as to either their origin or significance.

So, please bear in mind that what we are showing you here is by no means the full vocal repertoire for any of the species involved. There are examples that we have seen that have not been included, and we expect that there will be other examples of calls that we have not yet encountered.

Whilst it might be expected that different species of rodents in Britain and Ireland produce fairly similar calls, there are some big differences between species when it comes to their vocal emissions. There are undoubtedly some vocalisations that are fairly nondescript and/or difficult to describe because of the level of variation, or because the calls do not follow a precise structure or order of events. However, there are also sounds which do follow a set of 'rules' at a species-specific level. Differentiating most of the species in this chapter from one another can be achieved, though with different degrees of confidence, at least on occasions when you have recorded one of the more distinctive call structures.

In conclusion, some calls are more identifiable than others, and some species are more likely to provide you with more acoustic clues than others (i.e. they are, relatively speaking, more vocal).

Table 9.2 Comparison of population estimates of widespread small rodent/shrew species with larger more commonly seen mammals (Mathews *et al.*, 2018)

Common name	Scientific name	Population estimate Great Britain
Red fox	*Vulpes vulpes*	357,000
Roe deer	*Capreolus capreolus*	265,000
Badger	*Meles meles*	562,000
Brown rat	*Rattus norvegicus*	7,070,000
Wood mouse	*Apodemus sylvaticus*	39,600,000
House mouse	*Mus musculus*	5,203,000
Field vole	*Microtus agrestis*	59,900,000
Common shrew	*Sorex araneus*	21,100,000
Pygmy shrew	*Sorex minutus*	6,300,000

Table 9.3 Acoustic spectrum within which you would usually expect to encounter most useful sound for identification purposes within this group

Audible range		Ultrasonic range							
< 10 kHz	10–20 kHz	20–30 kHz	30–40 kHz	40–50 kHz	50–60 kHz	60–70 kHz	70–80 kHz	90–100 kHz	> 100 kHz
		⟷	⟷	⟷	⟷	⟷	⟷	⟷	⟷

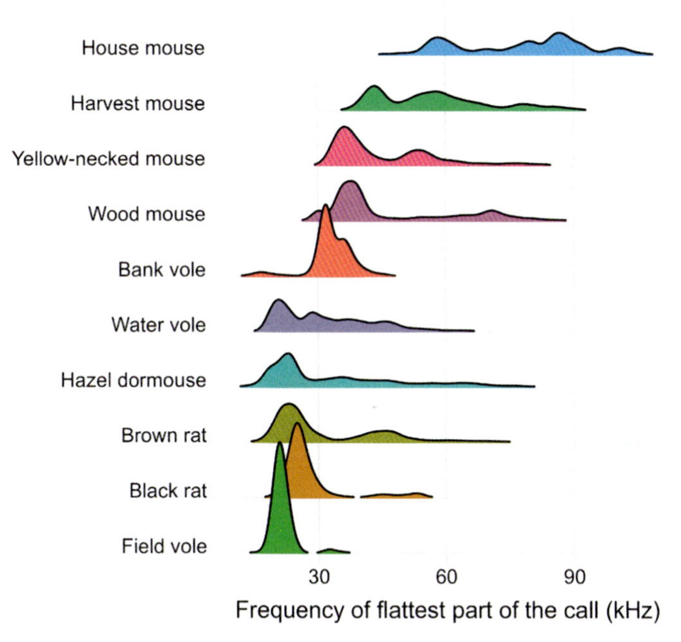

Figure 9.1 Frequency distribution of rodent calls (some selected species) occurring within Britain and Ireland (updated from Newson *et al.*, 2020 with new data)

Table 9.4 Detectable distance for useful acoustic encounters with small rodents

Distance (in metres)							
< 5 m	5–10 m	10–20 m	20–50 m	50–100 m	Up to 500 m	Up to 1,000 m	> 1,000 m
Brown rat							
Ear/microphone							<<<source of sound
Edible dormouse							
Ear/microphone							<<<source of sound
Mouse spp./Vole spp.							
Ear/microphone							<<<source of sound

Table 9.5 Typical mating season (shaded in blue) for small rodents occurring within Britain and Ireland

Species / Month	Jan	Feb	Mar	April	May	June	July	Aug	Sept	Oct	Nov	Dec
Brown rat												
Wood mouse												
Yellow-necked mouse												
Harvest mouse												
House mouse												
Water vole												
Bank vole												
Field vole												
Orkney vole/ Guernsey vole												
Hazel dormouse												
Edible dormouse												

9.3 Rat vocalisations

In the following section, a brief summary of the key characteristics of brown rat and black rat will be discussed. Details on the species-specific vocalisations and, where known, the behavioural function of these sounds will be described.

Ultrasonic sound made by rats has been researched and reported upon extensively, with many studies available for reference, including an overview for identification purposes within Britain, published in *British Wildlife* in 2020 (Newson *et al.*, 2020). The occurrence of emitted sound is dependent upon the animal's age and sex, but also the presence or absence of conspecifics, and their age and sex (Sales and Pye, 1974; Quy and Macdonald, 2008; Wöhr and Schwarting, 2013).

9.3.1 Brown rat *Rattus norvegicus*

<table>
<tr><th>Length/Weight</th><th>Habitat preferences</th><th>Distribution</th></tr>
<tr><td>Head/body length up to 280 mm.
Weight c.600 g.</td><td rowspan="3">Ubiquitous, opportunistic species that is closely associated with humans and has colonised the urban and rural environment including buildings, agricultural areas and areas associated with human activities, rubbish tips, sewage systems, urban water courses etc.
Also found in woodlands, along water courses and in coastal areas.</td><td rowspan="3">Common and widely distributed throughout Britain and Ireland.</td></tr>
<tr><th>Diet</th></tr>
<tr><td>Omnivorous, feeding across a vast range of food sources, varying according to availability, season and locality.</td></tr>
<tr><th>Daily activity pattern</th><td colspan="2">Typically most active from dusk through to dawn.</td></tr>
<tr><th>Seasonal activity pattern</th><td colspan="2">Active throughout the year.</td></tr>
<tr><th>Mating activity pattern</th><td colspan="2">Promiscuous mating can take place throughout the year, with mean litter size of nine. Most births occur during spring through to late summer.</td></tr>
<tr><th>Other notes</th><td colspan="2">Ultrasonic sounds are more frequent than anything audible, so usually not heard audibly.
Inhabits burrow systems with clearly defined runways connecting burrow entrances, but will also inhabit built structures, making use of any suitable place of shelter (e.g. sewage systems).</td></tr>
<tr><td colspan="3">References
Quy and Macdonald, 2008; Lysaght and Marnell, 2016; Mathews et al., 2018.</td></tr>
</table>

Acoustic behaviour including spectrogram examples

Of all of the small mammal species that we receive correspondence about, brown rat is one of the most regular to appear in our inboxes. This species is more likely than many others to be picked up during bat detector surveys, for a number of reasons. Firstly, they are widely distributed and abundant. They are also relatively more vocal than most other small mammals, and because they produce loud low-frequency calls, the distance at which they can be detected is greater than for many other small mammal species here (Newson and Pearce, 2022).

Compared to most of the other small mammals discussed in this book, there has been quite a lot of academic research on the vocalisations of the brown rat (e.g. Halls, 1981; Burgdorf *et al.*, 2008; Takahashi *et al.*, 2010; Coffey *et al.*, 2019). Rats are highly social, forming large colonies, meaning that vocalisation is an important form of communication between individuals. Academic work largely refers to two groups of common call types, described as alarm (22 kHz) calls (see Figures 9.2 to 9.5) and appetitive or 'friendly' (50 kHz) calls (Brudzynski, 2009) with examples shown in Figures 9.6 to 9.10. A broad range of calls have been noted during mating activities (Quy and Macdonald, 2008; Sales, 2010).

The 22 kHz call (Figures 9.2 to 9.5) – which is a long-duration, quasi-constant-frequency (QCF) call – serves as an alarm call emitted by a member of the colony when it senses danger or experiences anxiety (Kim *et al.*, 2010). Such calls may also occur during aggressive interactions. Brudzynski (2009) describes these calls and the 50 kHz call types in more detail. The 22 kHz calls last from 100 ms to 3,000 ms, with a frequency of maximum energy (FmaxE) usually within the range 20–23 kHz (bandwidth 1–4 kHz), although frequencies beyond this can occur (Figure 9.4). Normally a series of two to seven calls is emitted, which may be followed immediately afterwards by shorter calls (c.100 ms) at the same or similar frequencies.

The 50 kHz calls (Figures 9.6 to 9.10) appear to be emitted in more positive circumstances (e.g. sexual contact, feeding, non-aggressive interaction with conspecifics). The average duration of these calls is much shorter than the 22 kHz calls, with these higher-frequency emissions being 30–40 ms in duration. Although they are described conveniently as 50 kHz calls, there is a lot of variation in their frequency, with typical FmaxE range being 45–55 kHz, but higher frequencies are also known to occur (up to c.70 kHz). The bandwidth is considerably wider than for the 22 kHz calls, at 5–7 kHz (i.e. these calls are considerably more FM in appearance). The calls that fall into this broad category also vary structurally and can be relatively simple or more complex as shown in Figure 9.10.

One study concluded that the calls traditionally referred to as 50 kHz calls ranged in peak frequency (FmaxE) from 35 to 75 kHz (mean at c.55 kHz) (Burgdorf *et al.*, 2008). This study went on to separate the calls at these levels into two subcategories, named (1) flat (i.e. quasi-CF) 50 kHz, and (2) FM 50 kHz. In differentiating these subcategories, the authors described the flat calls as having only a flat (CF) 50 kHz (or similar) component, while the FM calls could also contain the flat component but always a trill or stepped component. In analysing the behaviour of rats in conjunction with these call types, they concluded that the FM 50 kHz calls were related to *'positively valanced appetitive behaviour'* during mating, play or aggressive encounter (also discussed in Wöhr, 2018). The flat 50 kHz calls were not associated with the same behaviours and are more evident during aggressive interactions.

For identification purposes within Britain (and considered here to be equally applicable to Ireland) the authors produced an article published in *British Wildlife* (Newson *et al.*, 2020) which included brown rat (as well as black rat).

Figures 9.2 to 9.10 provide specific examples of spectrograms relating to this species. Variations of the call types shown are known to exist, and the examples provided here are not intended to be the full repertoire, but instead the call types that are mostly likely to be encountered.

Where the 🔊 symbol appears, a recording of what is shown within the spectrogram is available to download from the species library. In addition, QR codes are provided for selected examples, and these can be accessed immediately for listening purposes from most mobile devices.

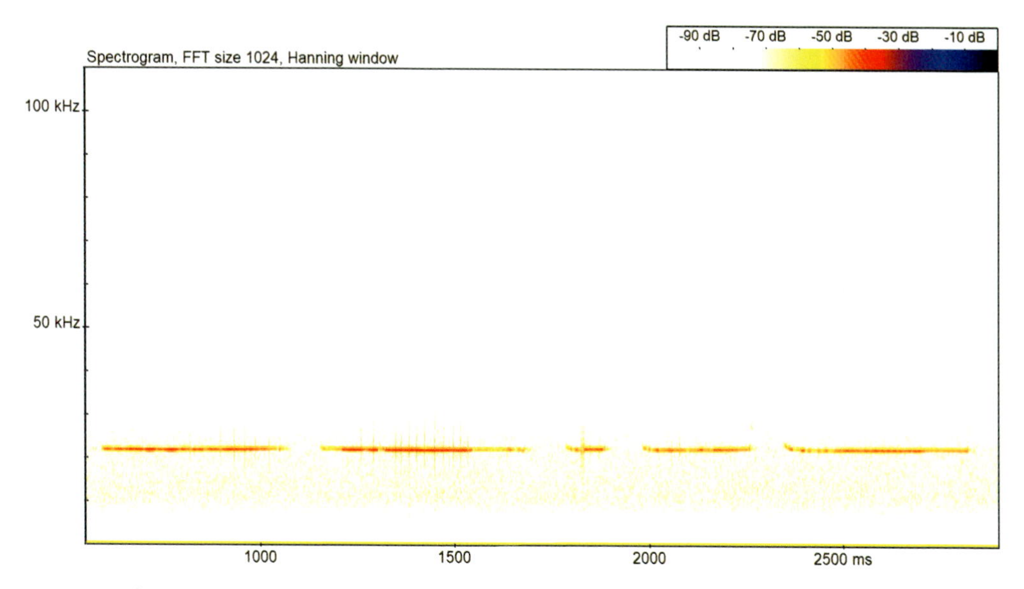

Figure 9.2 🔊 Brown rat – 22 kHz quasi-constant-frequency (QCF) alarm calls (x5) (frame width: c.2.5 sec)

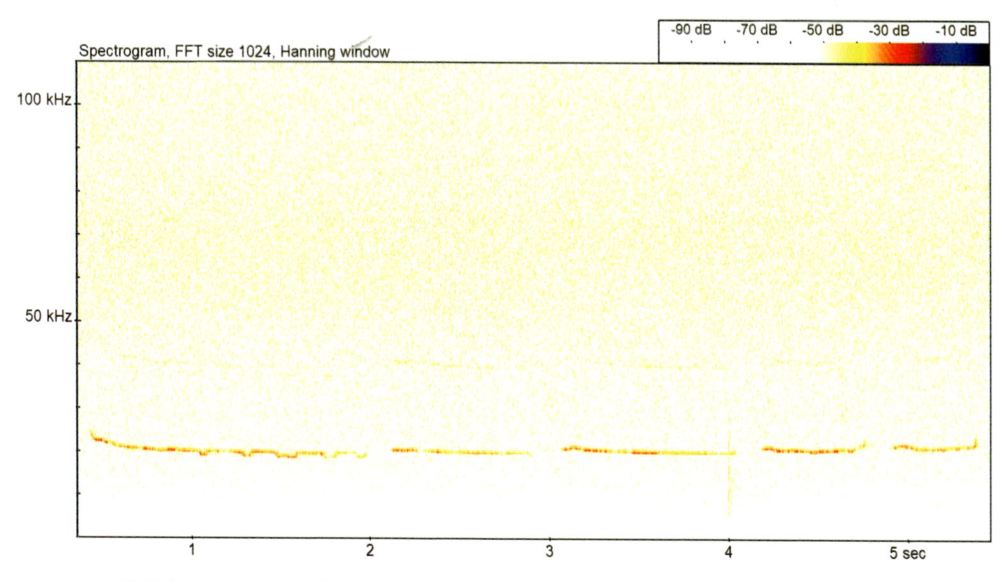

Figure 9.3 🔊 **(QR)** Brown rat – 22 kHz (QCF) alarm calls (x5) showing slight variances in frequency throughout (frame width: c.5 sec)

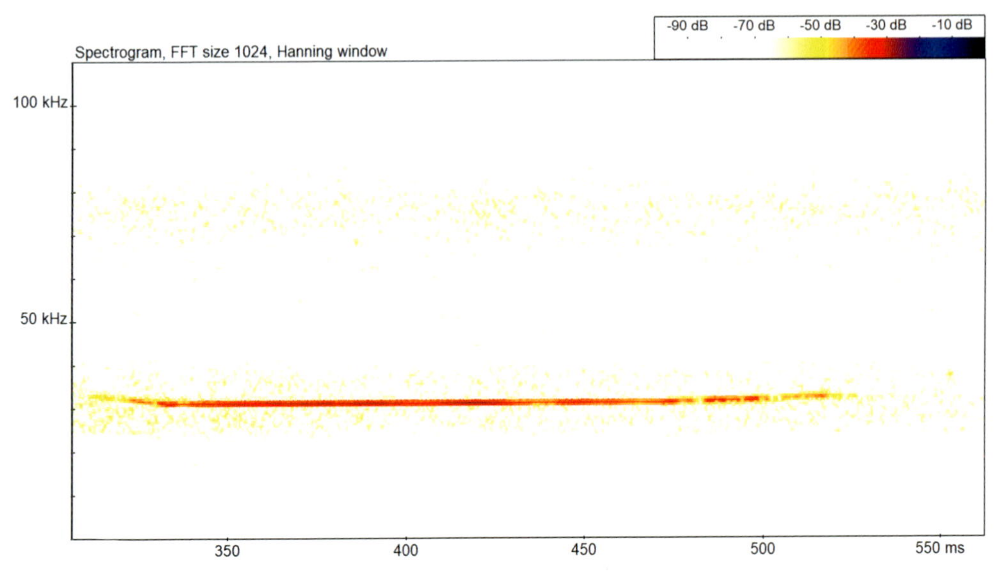

Figure 9.4 🔊 Brown rat – QCF alarm call at c.30 kHz, zoomed in and at a higher frequency than the more typical 22 kHz QCF call types (courtesy of I. Kaergaard) (frame width: c.250 ms)

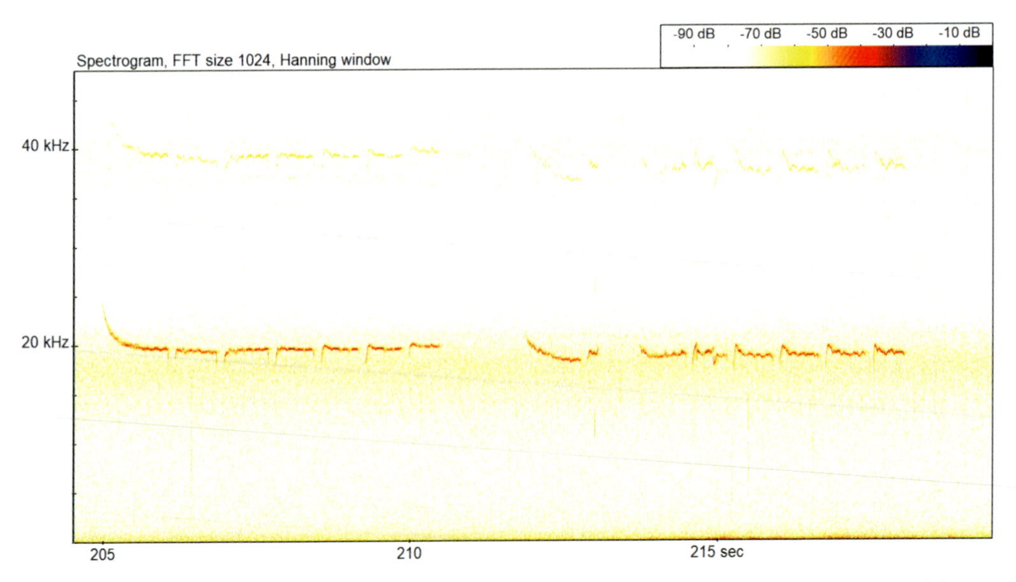

Figure 9.5 🔊 Brown rat – 22 kHz QCF alarm calling (two sequences) but showing more frequency modulation for some components (courtesy of S. Elliott) (frame width: c.20 sec)

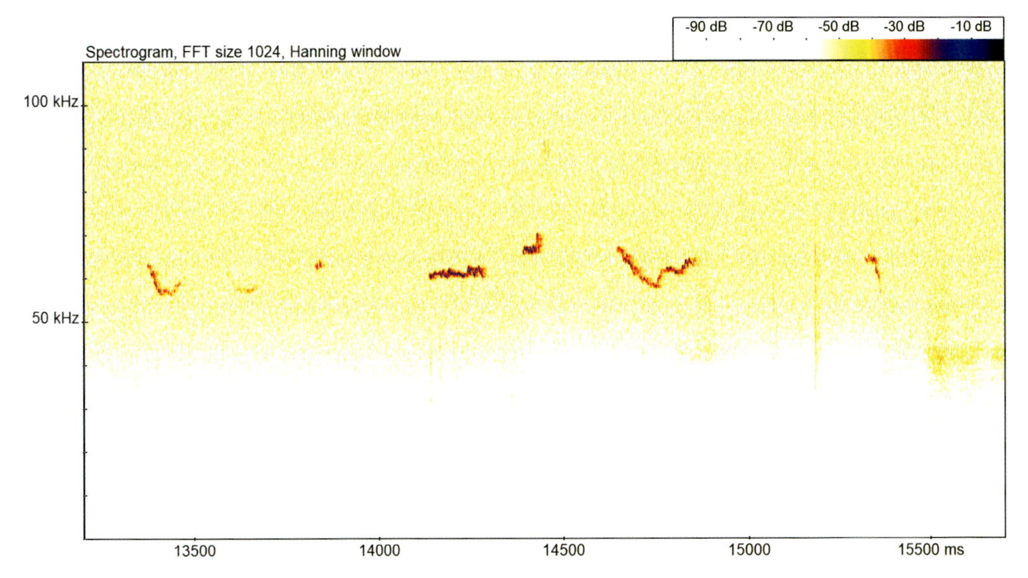

Figure 9.6 🔊 **(QR)** Brown rat – 50 kHz FM calls associated with positive/non-aggressive interactions, showing variation in frequency and call structures (frame width: c.2.5 sec)

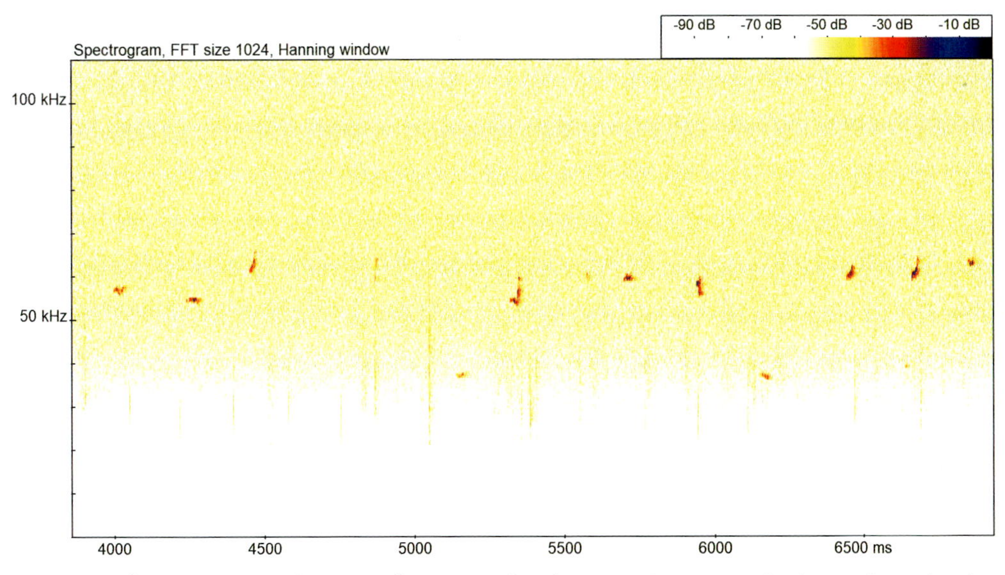

Figure 9.7 🔊 Brown rat – 50 kHz FM calls associated with positive/non-aggressive interactions, showing variation in frequency and call structures (frame width: c.3 sec)

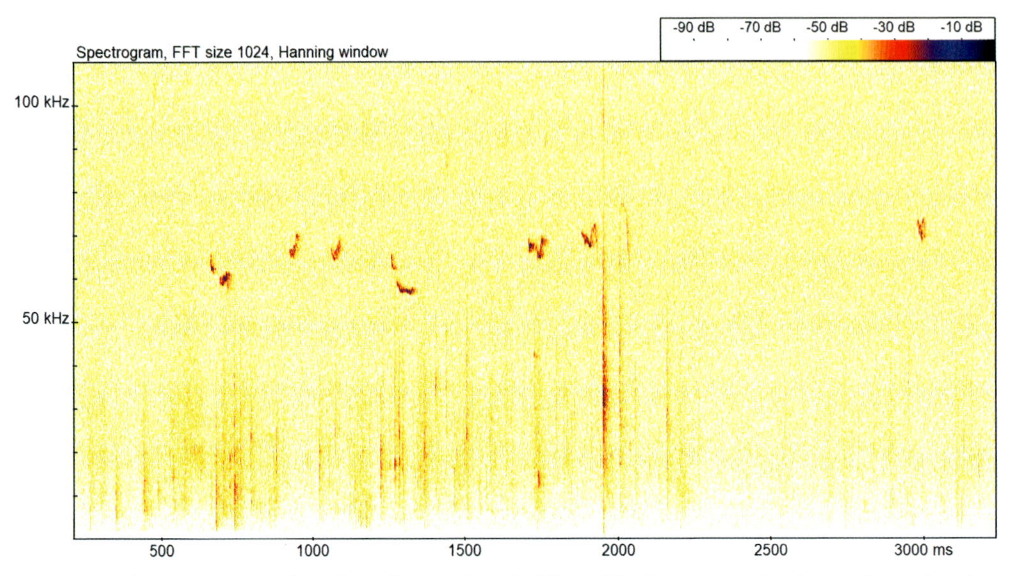

Figure 9.8 🔊 Brown rat – 50 kHz FM calls associated with positive/non-aggressive interactions, showing variation in frequency and structure (lower-frequency background noise also present) (frame width: c.2.5 sec)

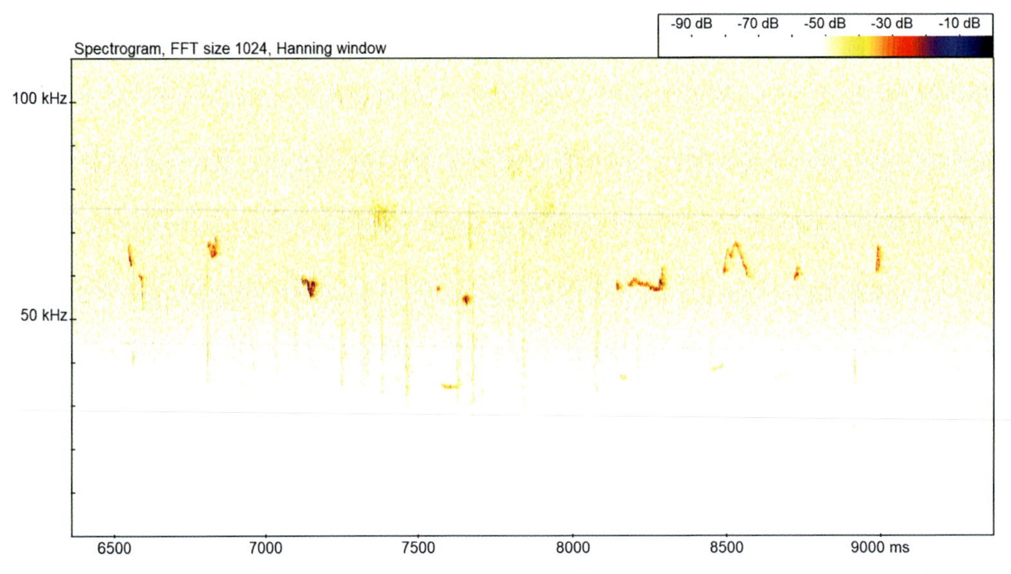

Figure 9.9 🔊 Brown rat – 50 kHz FM calls associated with positive/non-aggressive interactions, showing variation in frequency and call structures (frame width: c.3 sec)

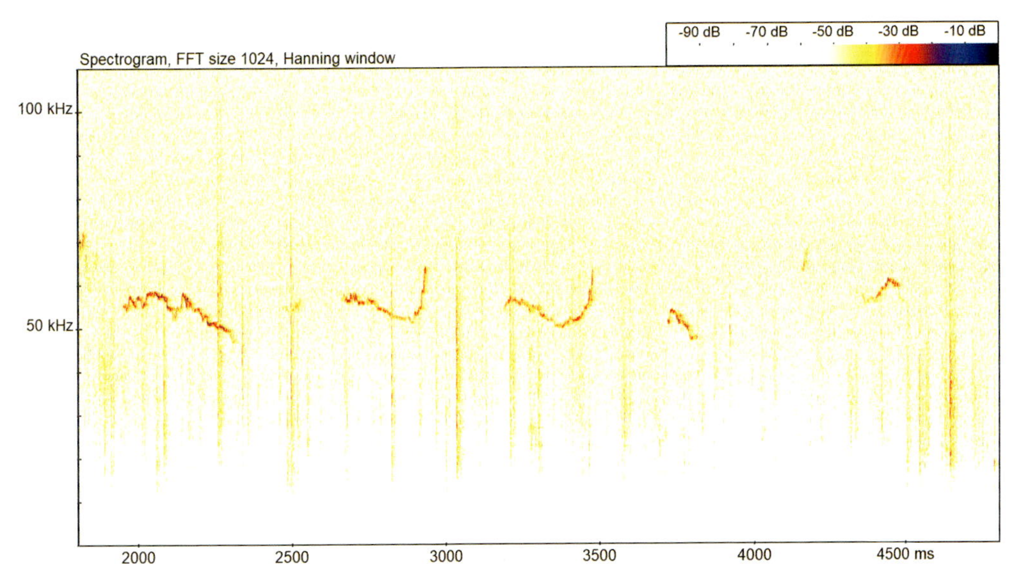

Figure 9.10 🔊 **(QR)** Brown rat – 50 kHz FM calls associated with positive/non-aggressive interactions, showing variation in frequency and call structures (frame width: c.3 sec)

QR codes relating to presented figures

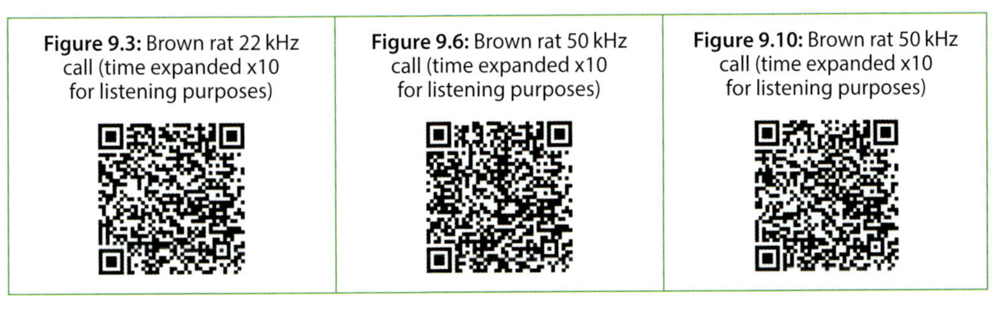

Figure 9.3: Brown rat 22 kHz call (time expanded x10 for listening purposes)	**Figure 9.6:** Brown rat 50 kHz call (time expanded x10 for listening purposes)	**Figure 9.10:** Brown rat 50 kHz call (time expanded x10 for listening purposes)

Potential confusion between brown rat and other species

Table 9.6 Summary of potential confusion between brown rat and other species

Confusion group	Potential confusion species
	In the first instance, always consider habitat and distribution.
Other rat species	Long-duration alarm calls of brown rat tend to be marginally lower in frequency when compared to black rat.
Species in other groups	Hazel dormouse. Lesser white-toothed shrew. Field vole – juvenile calls of brown rat. *Nyctalus* bat species quasi-constant-frequency echolocation in open environment, particularly where the quality of the recording is poor. Single constant-frequency electrical sounds and mechanical sounds, including some sounds produced by cars, may look very similar.

9.3.2 Black (Ship) rat *Rattus rattus*

Length/Weight	Habitat preferences	Distribution
Smaller in size than brown rat, but with slightly bigger ears and eyes, and a longer and thinner tail.	Historically, strongly associated with shipyards (which have since been redeveloped) and buildings.	Introduced during Roman times, black rats were once widespread and closely associated with dockyards. Today they are very rare and possibly extinct from mainland Britain. A population survives on Lambay Island in Ireland, and on Sark and possibly Alderney in the Channel Islands. Previously reported on some offshore islands (e.g. Lundy and Shiants) but have since been eradicated as they posed a threat to nesting birds.
Diet		
Omnivorous, feeding across a vast range of food sources, varying according to availability, season and locality. Feeding preference regarded as leaning more towards vegetarian food (e.g. cereals and fruits) than brown rat.		

Daily activity pattern	Most active 2–3 hours after dusk.
Seasonal activity pattern	Active throughout the year.
Mating activity pattern	Breed mid-March to mid-November, with females having two to three litters during this time, comprising on average seven pups per litter.
Other notes	Unlikely to be encountered outside of Sark and Alderney, since very rare/ extinct from mainland Britain.

References
Twigg *et al.*, 2008; Lysaght and Marnell, 2016; Mathews *et al.*, 2018; Harris, 2022.

Acoustic behaviour including spectrogram examples

Black rat are known to produce a variety of ultrasonic calls (> 20 kHz) during mating activities (Sales, 2010).

During our own observations, which includes collecting recordings from two long-term captive colonies (c.50 individuals), these mammals were very vocal. A range of call structures was recorded, many of which were not obviously dissimilar to those of brown rat, although the long-duration alarm calls of black rat tend to be marginally higher in frequency. They also produce similar higher-frequency calls during positive, non-aggressive encounters.

Figures 9.11 to 9.15 provide specific examples of spectrograms relating to this species, and where the 🔊 symbol appears, a recording of what is shown within the spectrogram is available to download from the species library.

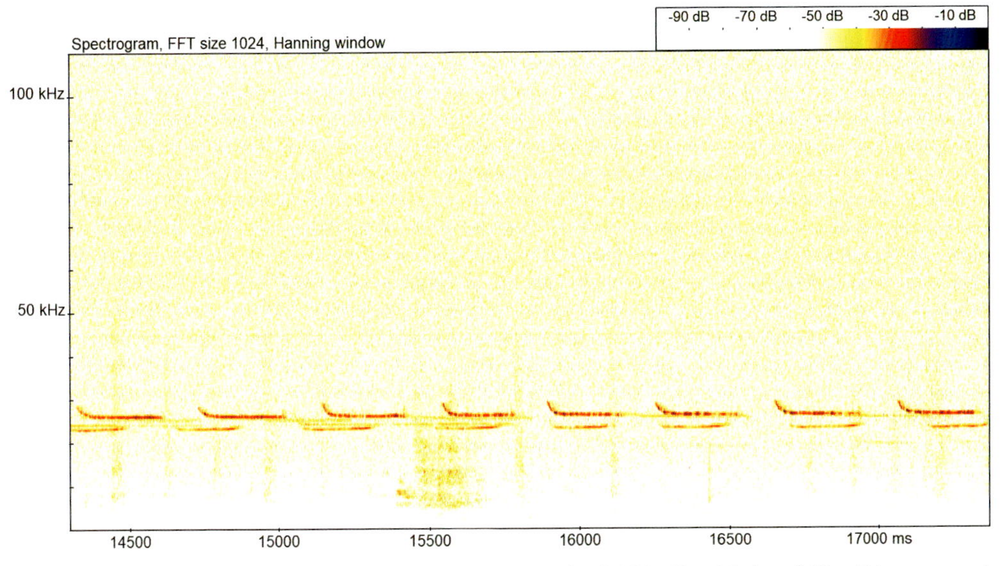

Figure 9.11 Black rat – a sequence of low-frequency (20–30 kHz) QCF calls, with short falling FM component at start, produced by at least two animals (frame width: c.3 sec)

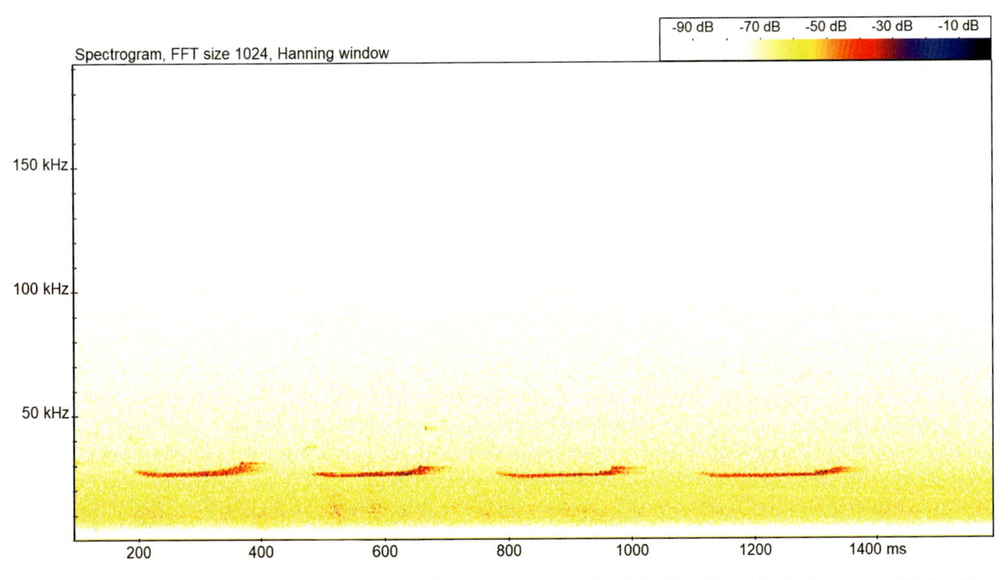

Figure 9.12 Black rat – a sequence of low-frequency (20–30 kHz) QCF calls, with slight upward inflection (rising FM) at end (frequency scale: 0 to 190 kHz/frame width: c.1.4 sec)

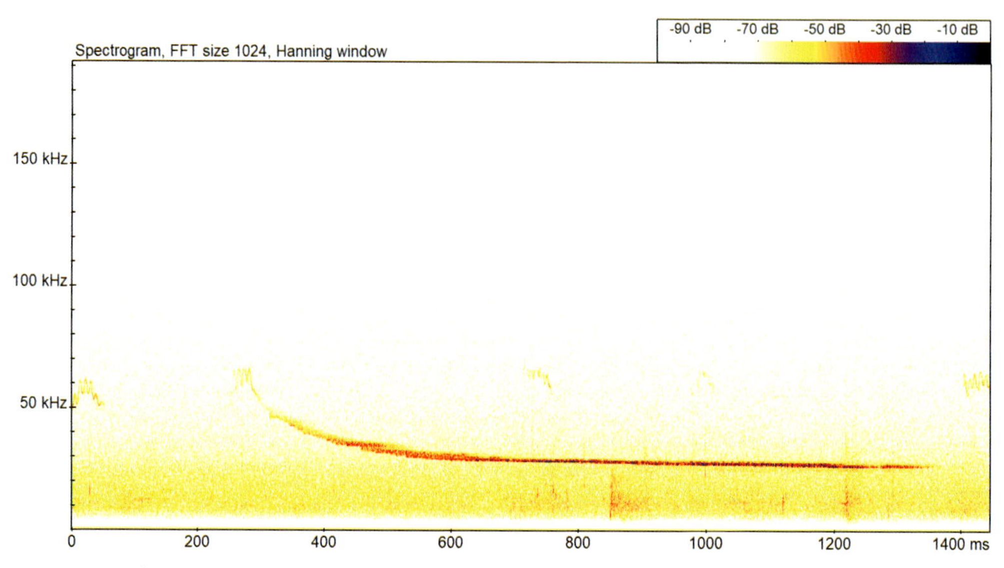

Figure 9.13 🔊 Black rat – low-frequency (20–30 kHz) QCF call with higher-frequency elements at start (frequency scale: 0 to 190 kHz/frame width: c.1.4 sec)

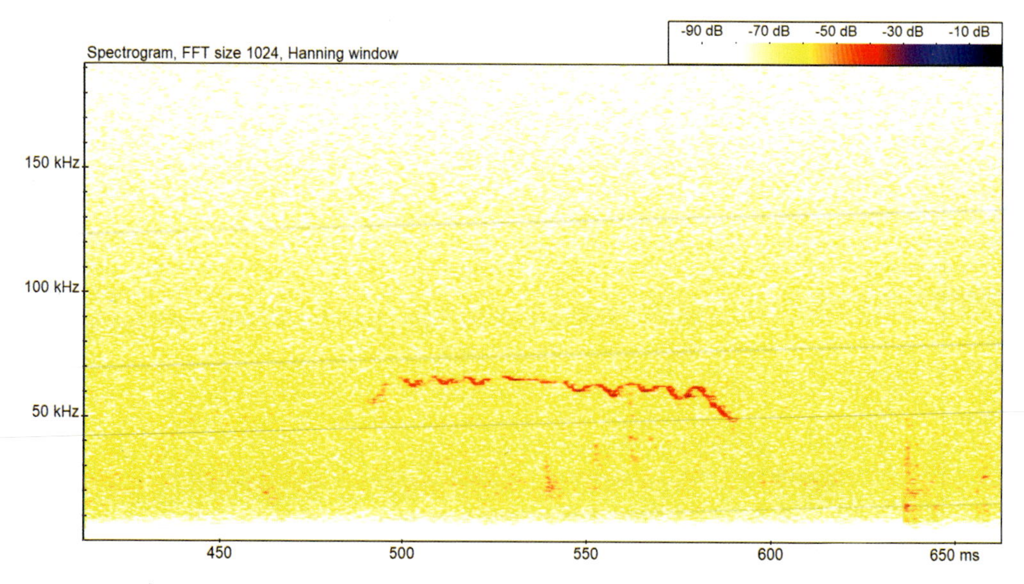

Figure 9.14 🔊 Black rat – a higher-frequency and more complex call likely associated with non-aggressive encounters and appetitive behaviours (frequency scale: 0 to 190 kHz/frame width: c.0.25 sec)

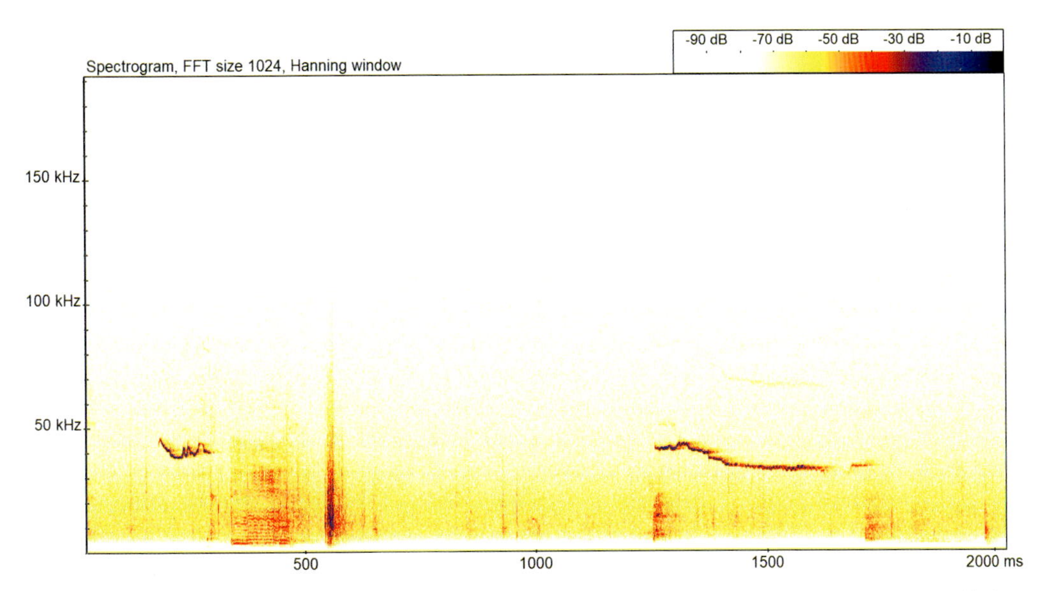

Figure 9.15 🔊 Black rat – higher-frequency, more complex calls (x2), with audible call at c.400 ms, likely associated with non-aggressive encounters, possibly adult–young interaction (frequency scale: 0 to 190 kHz/ frame width: c.2 sec)

Potential confusion between black rat and other species

Table 9.7 Summary of potential confusion between black rat and other species

Confusion group	Potential confusion species In the first instance, always consider habitat and distribution.
Other rat species	Long-duration alarm calls of black rat tend to be marginally higher in frequency when compared to brown rat.
Species in other groups	Hazel dormouse. Lesser white-toothed shrew. Field vole – juvenile calls of black rat. *Nyctalus* bat species quasi-constant-frequency echolocation in open environment, particularly where the quality of the recording is poor. Single constant-frequency electrical sounds and mechanical sounds, including some sounds produced by cars, may look very similar.

9.4 Mouse vocalisations

In the following section, we provide a brief summary of the key characteristics of our four mouse species, including details on species-specific vocalisations and, where known, the behavioural function of these sounds.

It is well known that mice produce ultrasonic sounds (Coffey *et al.*, 2019), which for identification purposes in Britain was previously reported on in *British Wildlife* in 2020 (Newson *et al.*, 2020). As well as ultrasonic sound, all mouse species here make audible noise. As with rats, the occurrence of emitted sound is dependent on the animal's age and sex, but also the presence or absence of conspecifics, and their age and sex (Wöhr and Schwarting, 2013).

The majority of published studies on mice have been on the house mouse (e.g. von Merten *et al.*, 2014; Lupanova and Egorova, 2015). The other mouse species that occur within our region have not had anywhere near the same amount of attention with respect to the study of their vocalisations.

Little previous research relevant to this book exists for harvest mouse, although we have been fortunate enough to obtain data ourselves. For wood mouse and yellow-necked mouse, some early research looked at the ultrasonic behaviour of these species (Hoffmeyer and Sales, 1977; Gyger and Schenk, 1983, 1984). More recently, a study of their distress calls, produced when individuals are handled (carried out in Italy), demonstrated a very high rate (98%) of accuracy in differentiating these two species from each other *'based on a combination of acoustic variables and body mass'* (Ancillotto *et al.*, 2017). This study established that the approach taken by its authors was more reliable than morphology alone.

9.4.1 Wood mouse (Long-tailed field mouse) *Apodemus sylvaticus*

Length/Weight	Habitat preferences	Distribution
Adult males and females very similar in size, with body length (excluding tail) from 81 to 103 mm. Weight from 13 to 27 g. **Diet** Omnivorous feeding across a range of vegetative (seeds, fruits, shoots) and insectivorous food sources as well as fungi, varying according to availability, season and locality.	Broadleaved woodland Coniferous woodland Unimproved grassland Urban and gardens Hedgerow	Occurs throughout Britain and Ireland.

Daily activity pattern	Typically most active from dusk through to dawn.
Seasonal activity pattern	Active throughout the year.
Mating activity pattern	Promiscuous mating occurs throughout the year, usually during the period March to October, with mean litter size of four to seven. Most births occur during spring through to late summer.
Other notes	Due to small size difficult to encounter visually, but often recorded using appropriately positioned trail cameras. Encountered during nest box/nest tube surveys and differentiated from yellow-necked mouse by the absence of a complete 'yellow' collar of fur across the chest. Not particularly audible, with most sounds occurring within the ultrasonic range.

References
Flowerdew and Tattersall, 2008; Lysaght and Marnell, 2016; Mathews *et al.*, 2018.

Acoustic behaviour including spectrogram examples

For this species, ultrasonic calls are known to be emitted during exploration, mating, chase sequences, aggressive behaviour and contact with conspecifics (Stoddart and Sales, 1985; Sales, 2010). Their exploratory calls are usually higher in frequency than in yellow-necked mouse (Newson *et al.*, 2020), usually in the region of 70 kHz (Figures 9.16 to 9.18). These higher-frequency calls are most useful for separating wood mouse from other mouse species in Britain and Ireland, with Figure 9.16 showing a typical example of what to expect reasonably often when encountering this species. Unfortunately, we were unable to source recordings of a subspecies of wood mouse endemic to the Scottish archipelago of St Kilda, the St Kilda field mouse *Apodemus sylvaticus hirtensis*, and as such it is not specifically covered here.

In addition to higher-frequency calls, wood mouse also makes a range of calls at lower frequencies: c.35 kHz to 50 kHz, which show a range of structures (Figures 9.19 to 9.23). These are far harder to differentiate from other species based on our knowledge so far (but see Newson *et al.*, 2020).

With reference to mating behaviour, wood mouse is known to make arched calls (upside-down 'V'-shaped) in a series, with maximum frequency of c.90–100 kHz (Gyger and Schenk, 1983). Unfortunately, we have not encountered such calls ourselves to date.

Audible calls are also known to be produced by distressed animals (Flowerdew and Tattersall, 2008; Ancillotto *et al.*, 2017) but are not covered here, as we are unsure how commonly they are produced under normal field conditions.

Figures 9.16 to 9.23 provide specific examples of spectrograms relating to this species, and where the 🔊 symbol appears, a recording of what is shown within the spectrogram is available to download from the species library. In addition, QR codes are provided for selected examples, and these can be accessed immediately for listening purposes from most mobile devices.

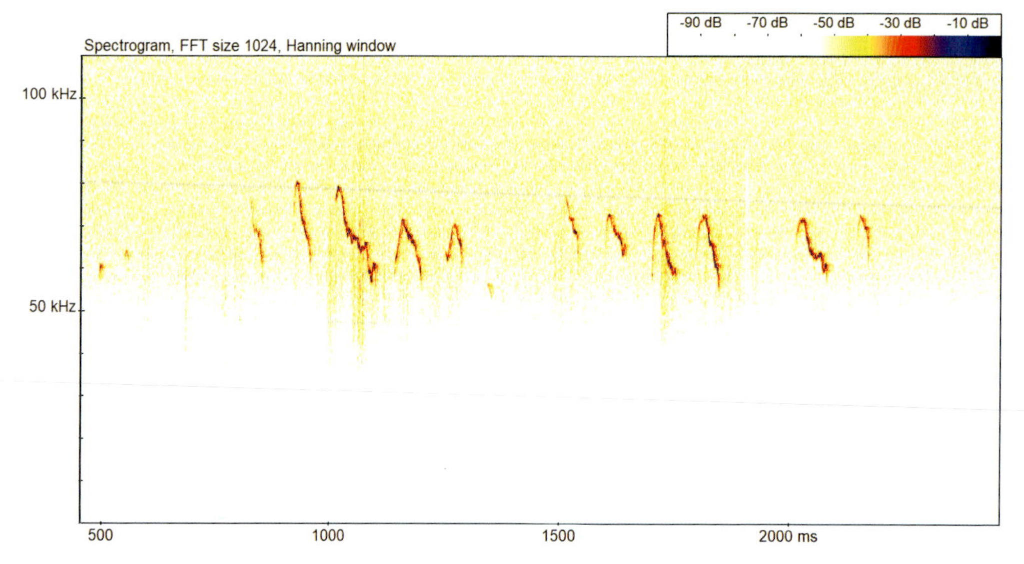

Figure 9.16 🔊 (QR) Wood mouse – higher-frequency hooked call sequence, fairly typically showing an fmax > 70 kHz (frame width: c.2 sec)

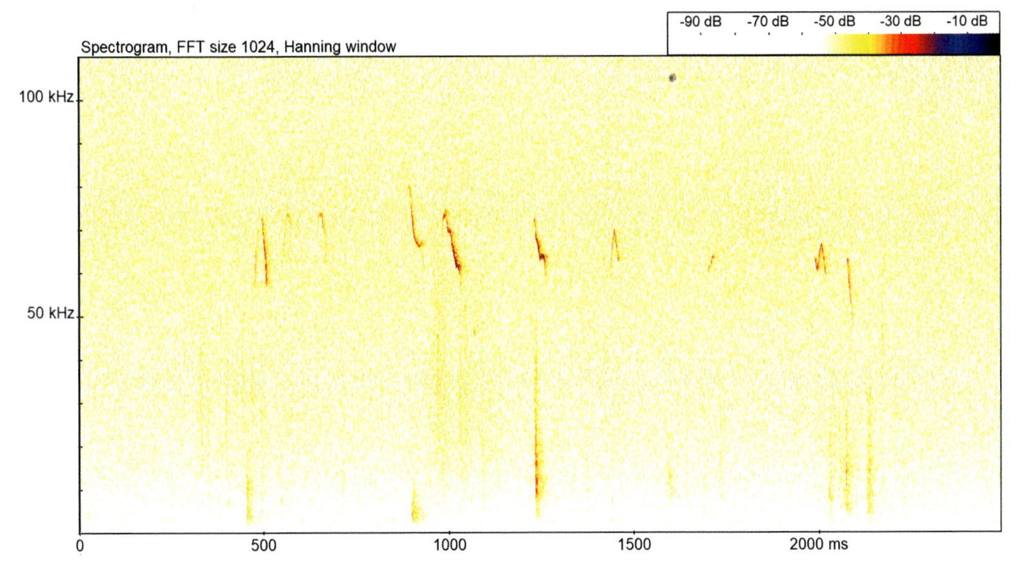

Figure 9.17 🔊 Wood mouse – a series of higher-frequency calls with fmax > 70 kHz (frame width: c.2.5 sec)

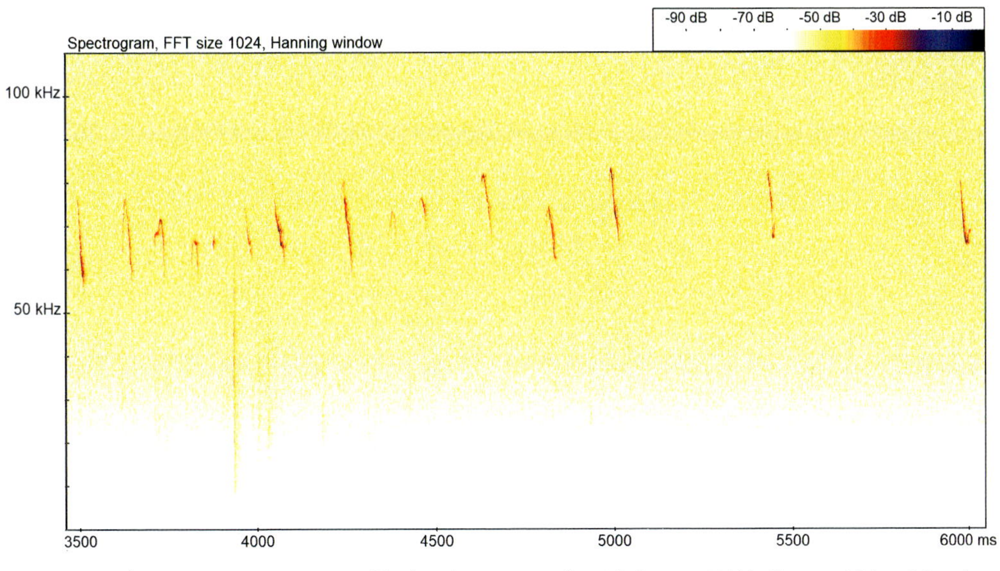

Figure 9.18 🔊 Wood mouse – a series of higher-frequency calls with fmax > 70 kHz (frame width: c.2.5 sec)

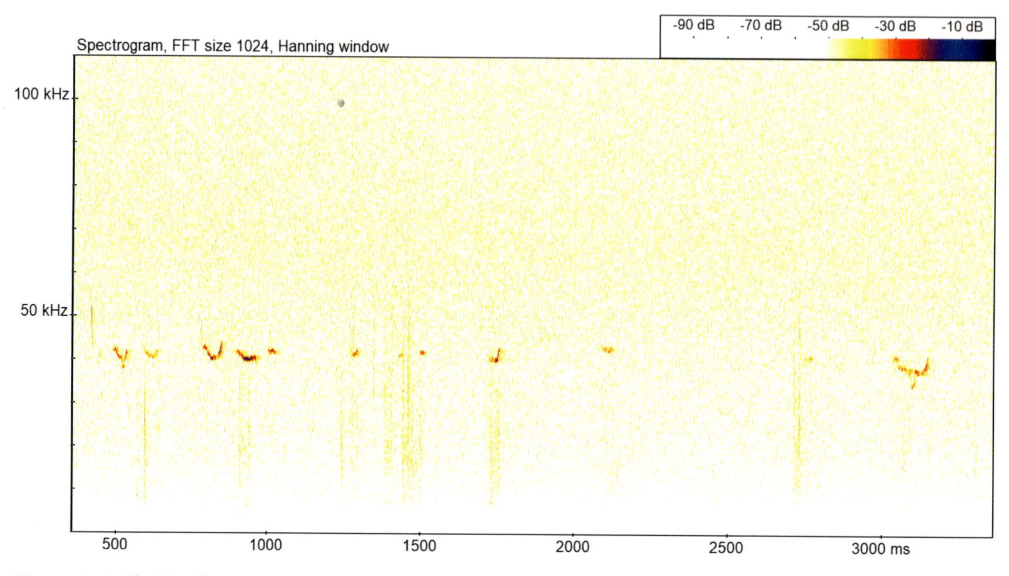

Figure 9.19 🔊 Wood mouse – a series of low-frequency calls (c.40 kHz) showing U-shaped structure (frame width: c.3 sec)

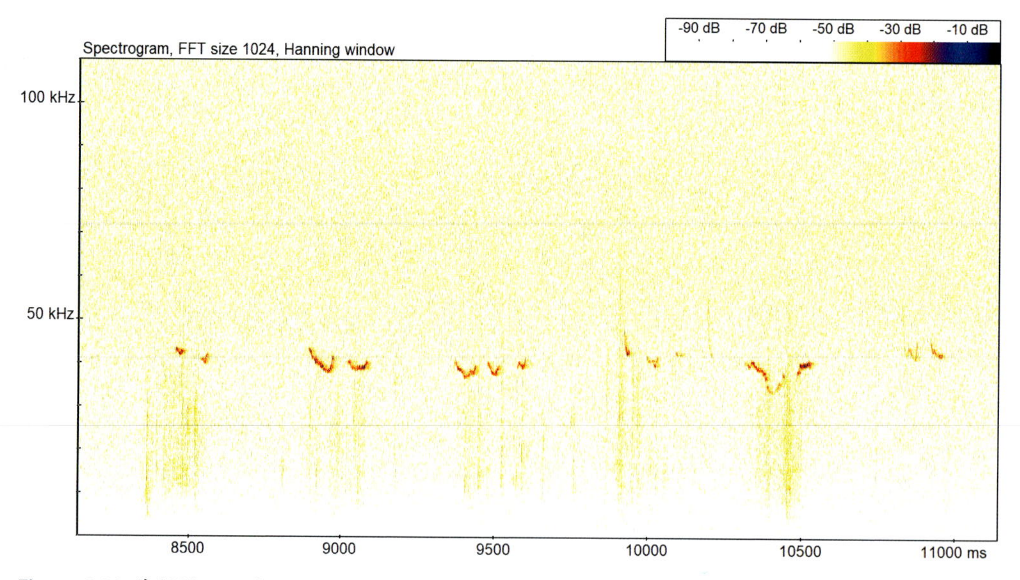

Figure 9.20 🔊 **(QR)** Wood mouse – a series of low-frequency calls (c.40 kHz) showing U-shaped structure (frame width: c.3 sec)

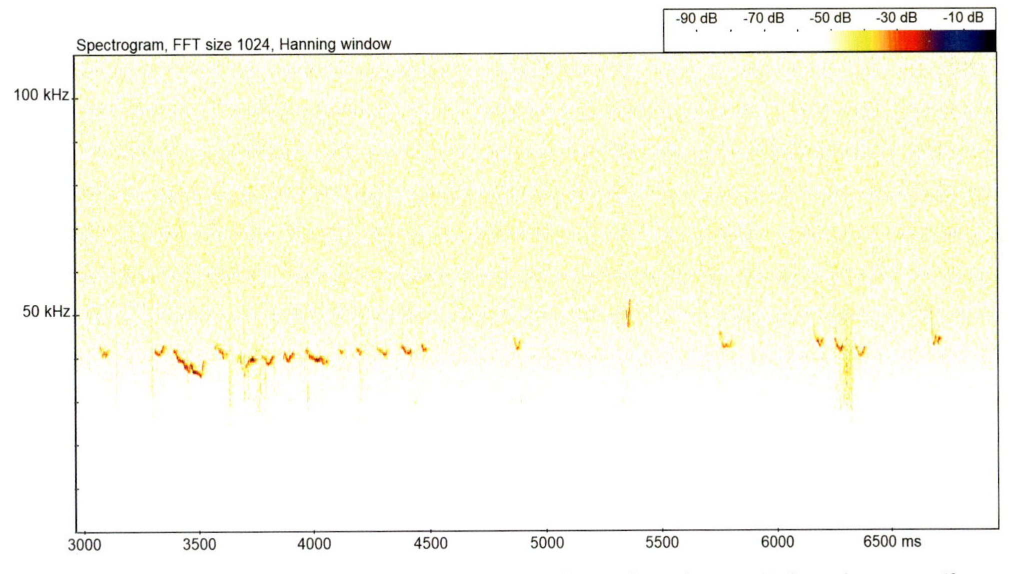

Figure 9.21 🔊 Wood mouse – a series of low-frequency calls (c.40 kHz) showing U-shaped structure (frame width: c.4 sec)

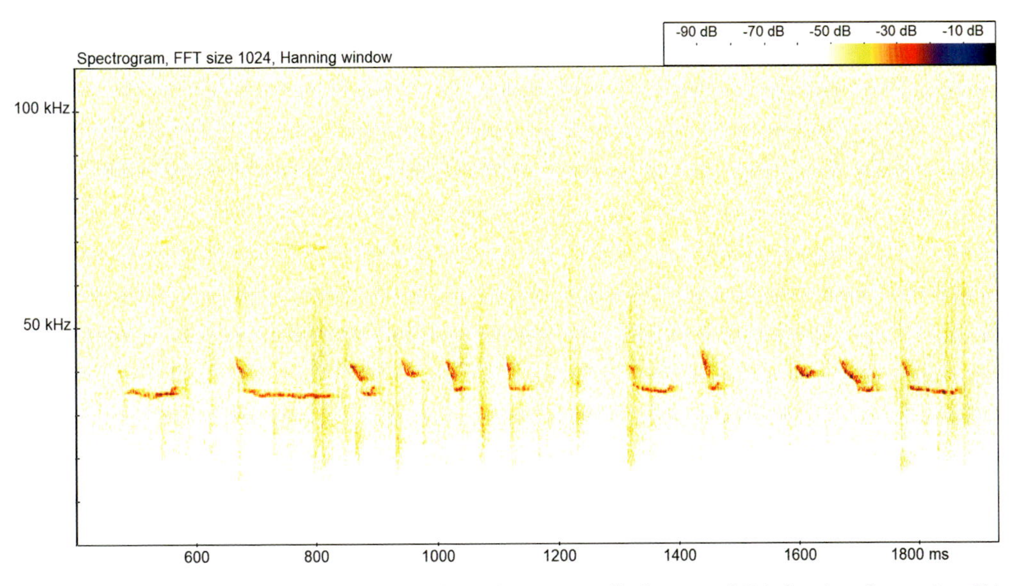

Figure 9.22 🔊 **(QR)** Wood mouse – a series of low-frequency calls (fmin c.35 kHz) showing descending FM followed by longer QCF portion (frame width: c.1.5 sec)

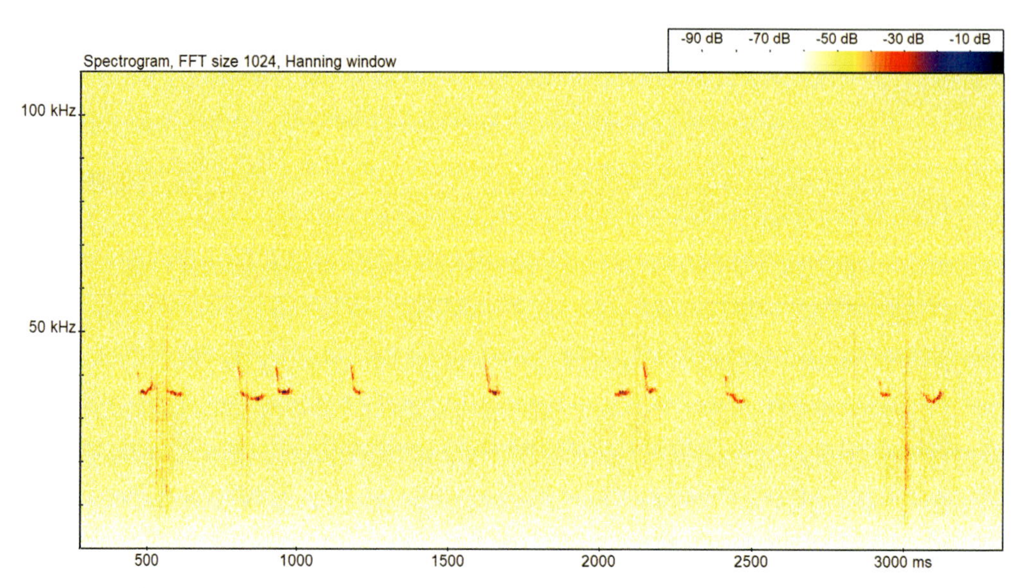

Figure 9.23 Wood mouse – a series of low-frequency calls (fmin c.35 kHz) showing descending FM followed by longer QCF portion (frame width: c.3 sec)

QR codes relating to presented figures

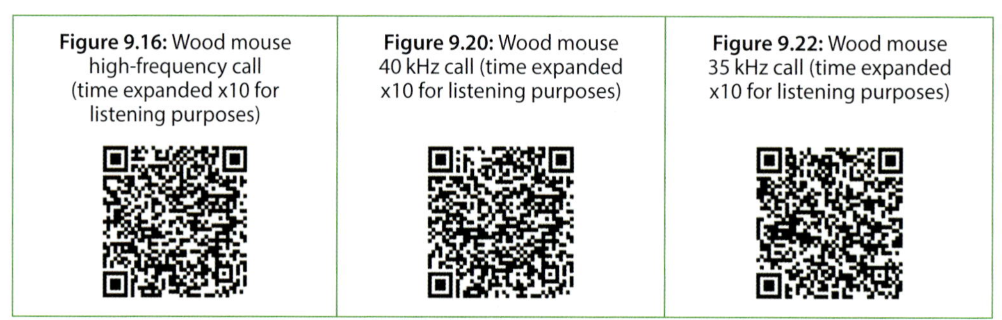

Potential confusion between wood mouse and other species

Table 9.8 Summary of potential confusion between wood mouse and other species

Confusion group	Potential confusion species
	In the first instance, always consider habitat and distribution.
Other mouse species	Yellow-necked mouse. Higher fmax calls from wood mouse could potentially overlap house mouse examples.
Species in other groups	Low-amplitude, 40 kHz common pipistrelle echolocation calls. Hazel dormouse. Nathusius' pipistrelle echolocation calls. Stoat – U-shaped calls.

9.4.2 Yellow-necked mouse *Apodemus flavicollis*

Length/Weight	Habitat preferences	Distribution
Body length (excluding tail) from 95 to 120 mm. Weight from 14 to 45 g.	Broadleaved woodland Urban and gardens Coniferous woodland Hedgerow	Occurs in southern England and Wales. Absent from Scotland and Ireland.
Diet		
Mainly herbivorous feeding across a range of vegetative food sources (e.g. seeds (particularly beech mast), buds, fruit and some insects and insect larvae), varying according to availability, season and locality.		

Daily activity pattern	Typically most active from dusk through to dawn.
Seasonal activity pattern	Active throughout the year.
Mating activity pattern	Promiscuous mating occurs usually during the period February to October, with litter size of two to eleven. Most births occur during spring through to late summer.
Other notes	Due to small size and more arboreal habits, they are difficult to encounter visually, but may be recorded using appropriately positioned trail cameras. Sometimes use nest boxes/nest tubes. Not particularly audible, with most sounds occurring within the ultrasonic range.

References
Marsh and Montgomery, 2008; Lysaght and Marnell, 2016; Mathews *et al.*, 2018.

Acoustic behaviour including spectrogram examples

Yellow-necked mouse is known to make ultrasonic sounds in the range of 30 to 60 kHz (Stoddart and Sales, 1985; Marsh and Montgomery, 2008; Newson *et al.*, 2020), and these calls are typically lower in frequency than the higher calls (i.e. > 70 kHz fmax) that can be produced by wood mouse. Wood mouse also make calls in the 30 to 60 kHz range (Newson *et al.*, 2020), and care should be taken to consider both species, in areas where both are potentially present. Audible calls are also known to be produced by distressed animals (Ancillotto *et al.*, 2017).

Figures 9.24 to 9.32 provide specific examples of spectrograms relating to this species, and where the 🔊 symbol appears, a recording of what is shown within the spectrogram is available to download from the species library. In addition, QR codes are provided for selected examples, and these can be accessed immediately for listening purposes from most mobile devices.

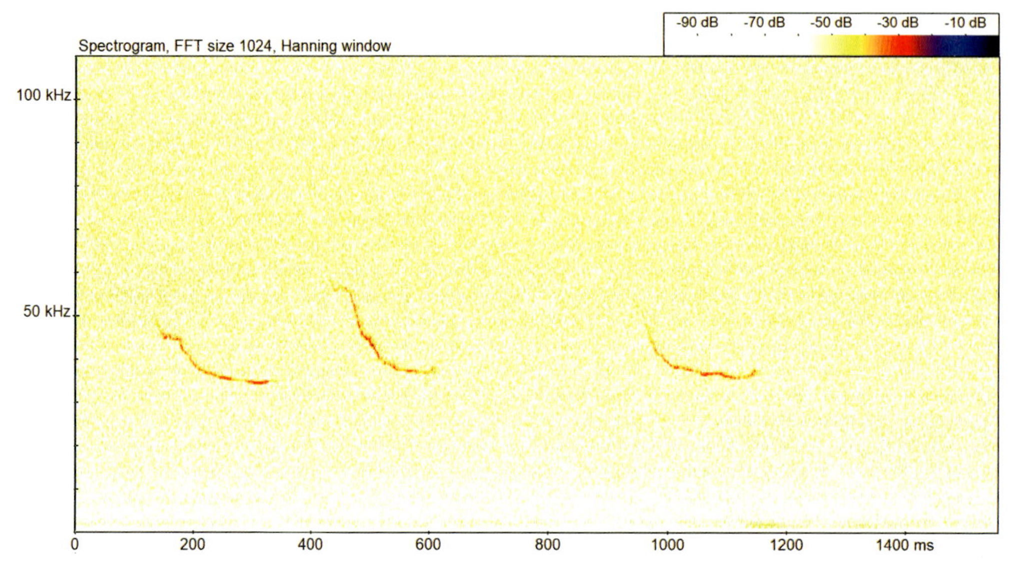

Figure 9.24 🔊 **(QR)** Yellow-necked mouse – FM to QCF call structures (frame width: c.1.5 sec)

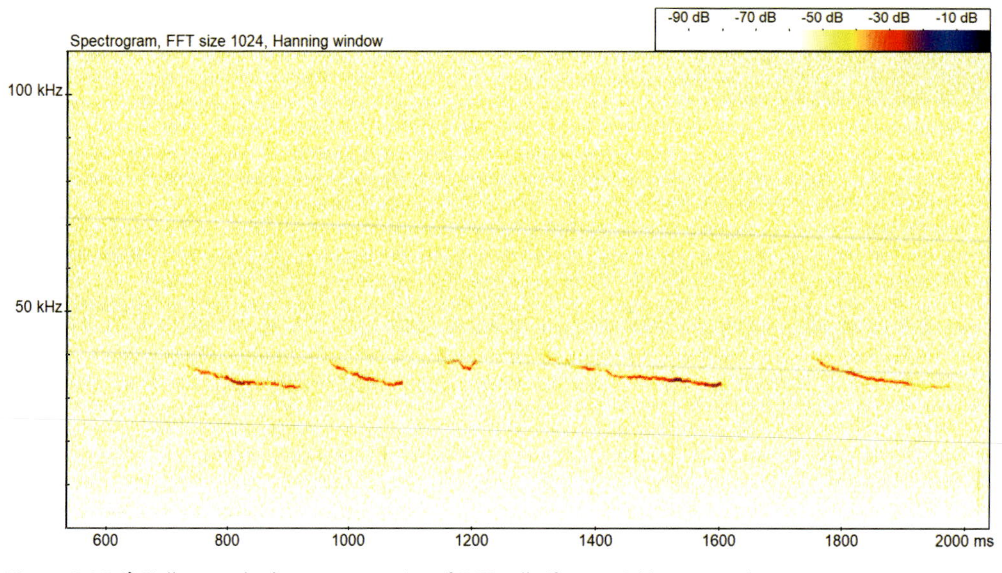

Figure 9.25 🔊 Yellow-necked mouse – a series of QCF calls (frame width: c.1.5 sec)

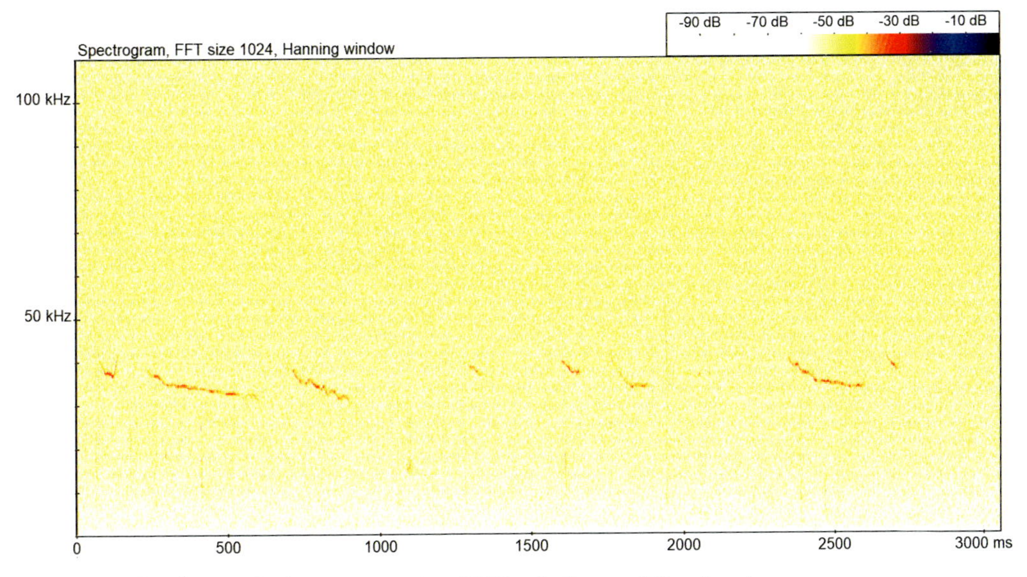

Figure 9.26 🔈 Yellow-necked mouse – a series of QCF calls (frame width: c.3 sec)

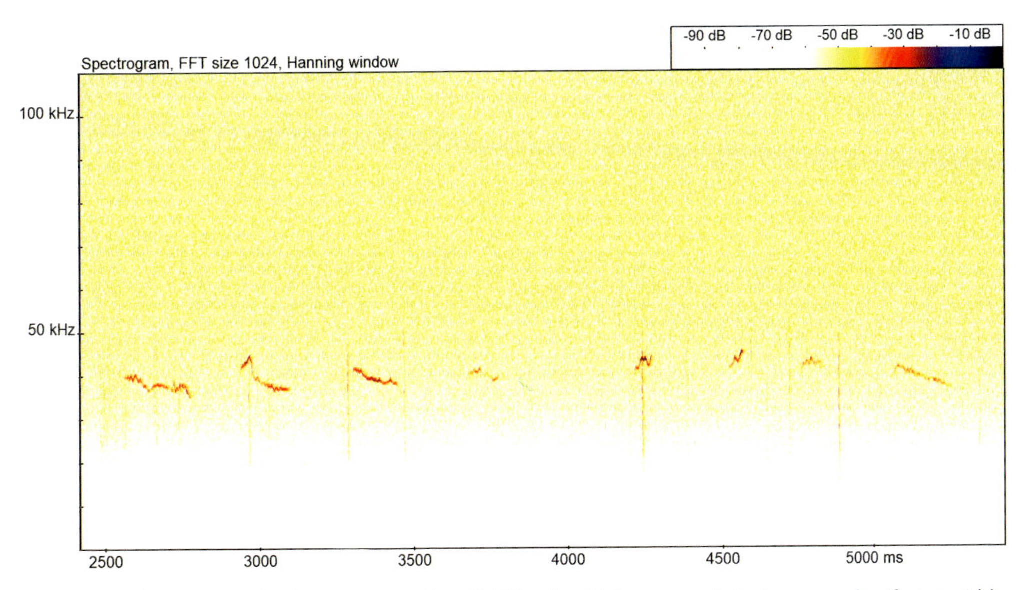

Figure 9.27 🔈 Yellow-necked mouse – a series of QCF calls with longer and shorter examples (frame width: c.3 sec)

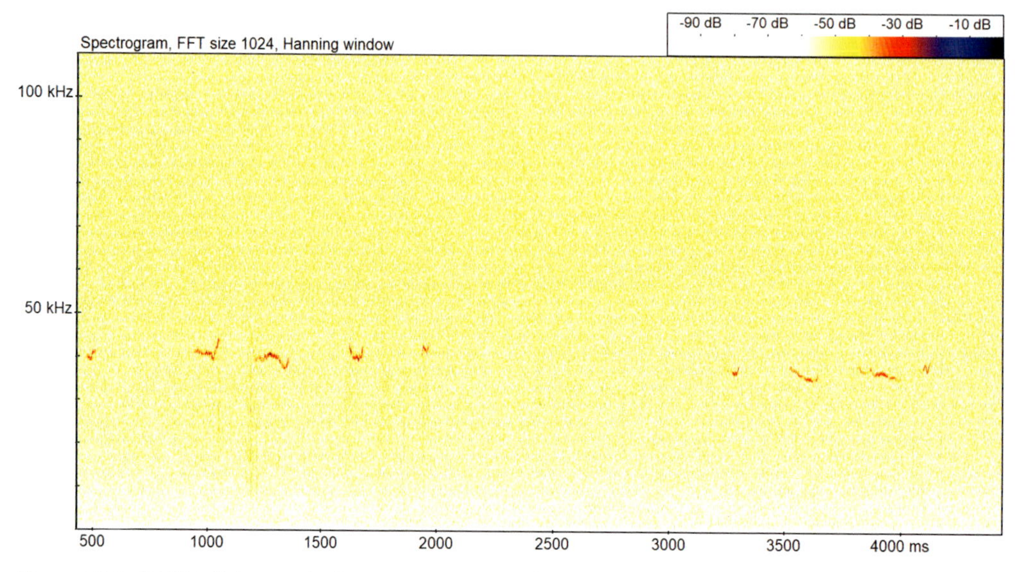

Figure 9.28 🔊 **(QR)** Yellow-necked mouse – a series of calls including U-shaped and humped examples (frame width: c.4 sec)

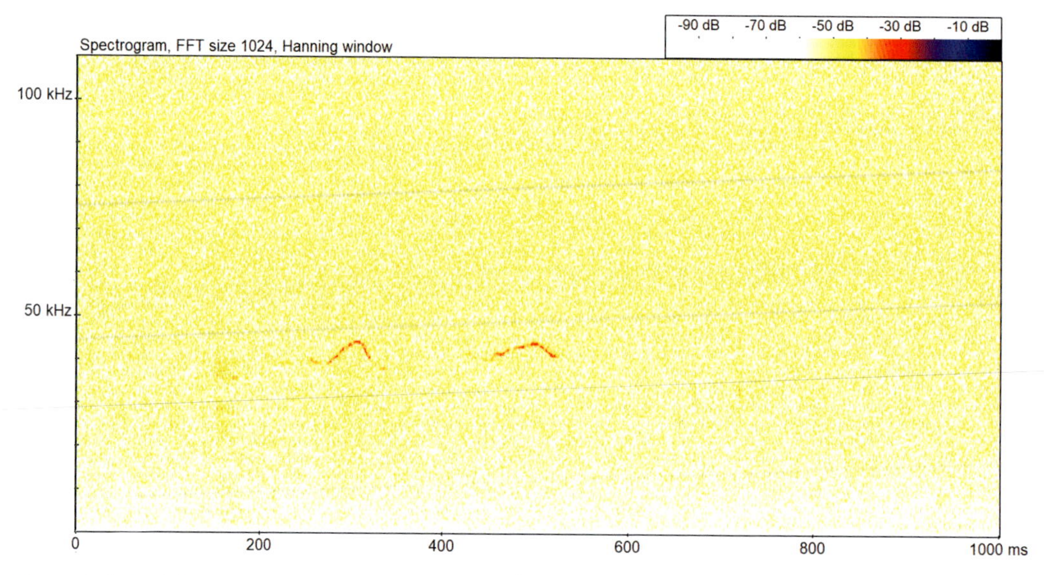

Figure 9.29 🔊 Yellow-necked mouse – two humped call structures (frame width: c.1 sec)

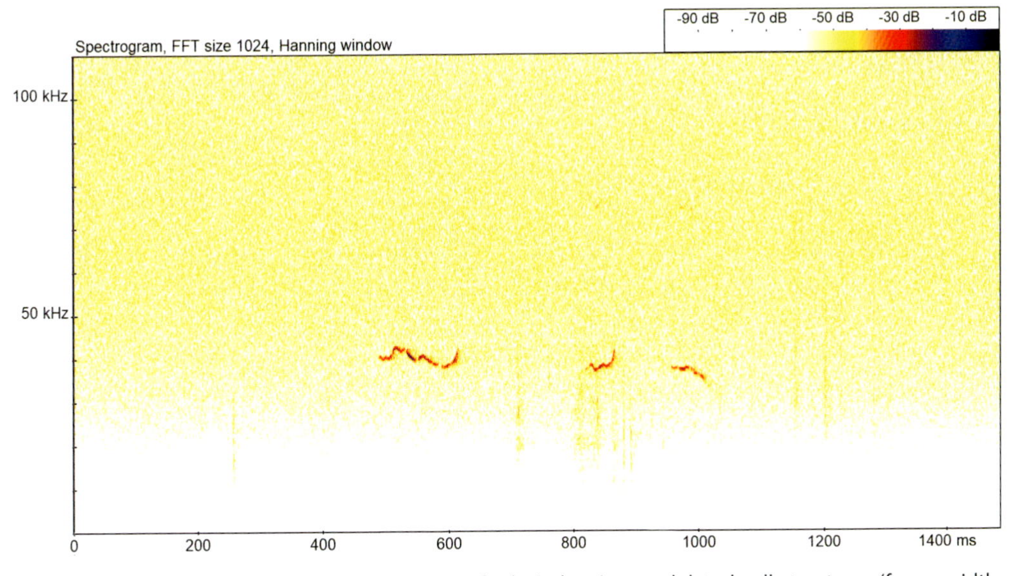

Figure 9.30 🔊 Yellow-necked mouse – a series of relatively more undulated call structures (frame width: c.1.5 sec)

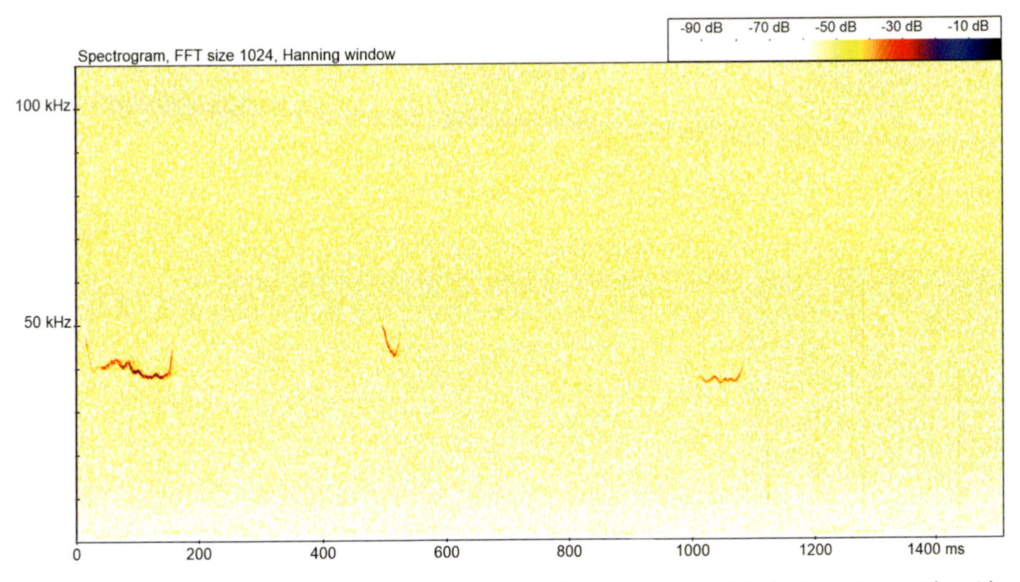

Figure 9.31 🔊 Yellow-necked mouse – a series including relatively more undulated call structures either side of a U-shaped call (frame width: c.1.5 sec)

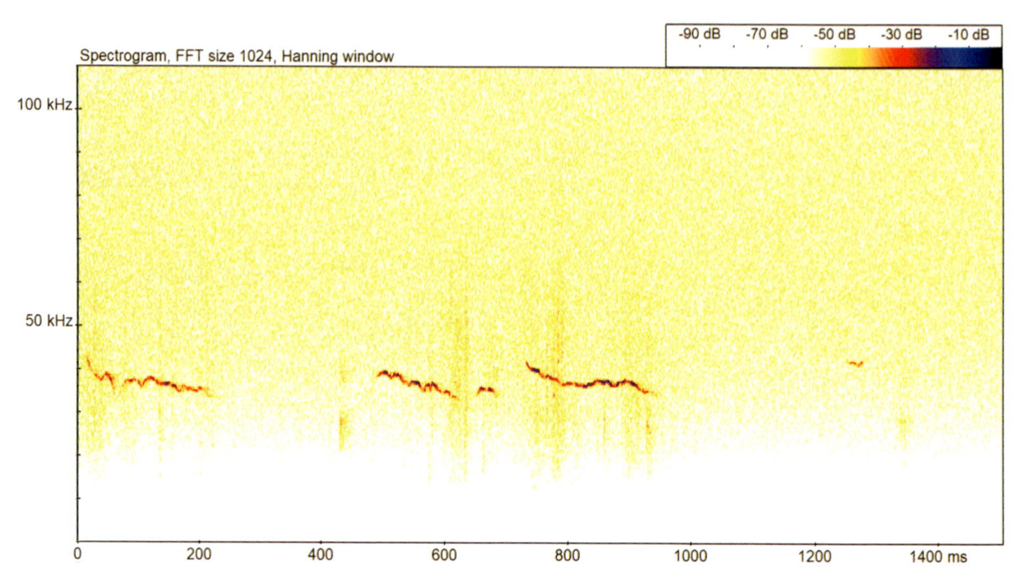

Figure 9.32 🔊 **(QR)** Yellow-necked mouse – a series of undulated call structures (frame width: c.1.5 sec)

QR codes relating to presented figures

Potential confusion between yellow-necked mouse and other species

Table 9.9 Summary of potential confusion between yellow-necked mouse and other species

	Potential confusion species
Confusion group	In the first instance, always consider habitat and distribution.
Other mouse species	Wood mouse – 30–50 kHz and U-shaped calls.
Species in other groups	Low-amplitude, 40 kHz common pipistrelle calls. Hazel dormouse.

9.4.3 Harvest mouse *Micromys minutus*

Length/Weight	Habitat preferences	Distribution
Our smallest rodent. Body length (excluding tail) from 50 to 70 mm. Weight c.6 g.	Arable and horticulture Unimproved grassland Hedgerow Wetlands, particularly reedbed and wet meadows	Occurs throughout much of England and Wales, and also southern Scotland. Absent from Ireland.
Diet		
Omnivorous feeding across a range of vegetative (seeds, fruits, berries, young shoots) and insectivorous food sources, varying according to availability, season and locality.		

Daily activity pattern	Often active from beyond dusk, but can also be active during daytime hours especially in the height of summer.
Seasonal activity pattern	Active throughout the year.
Mating activity pattern	Promiscuous mating occurs usually during the period May to October, with litter size of two to eight. Most births occur during summer through to early autumn. Distinctive breeding nests occur at height within tall grass, reeds, hedges and scrub.
Other notes	Due to small size difficult to encounter visually, but may be recorded from nest surveys or using trail cameras positioned onto purpose-built feeding stations. Not particularly audible, with most sounds occurring within the ultrasonic range.

References
Trout and Harris, 2008; Morris *et al.*, 2013; Lysaght and Marnell, 2016; Mathews *et al.*, 2018.

Acoustic behaviour including spectrogram examples

Compared to other small mammal species here, there have been few studies on the bioacoustics of harvest mouse. Ultrasonic calls have been described in previous publications produced by the authors (Middleton, 2020; Newson *et al.*, 2020).

Notable call structures that appear to differ from other mouse species in the region fall into four types:

(1) A rapid series of FM calls occurring over two frequency ranges produced almost simultaneously, where the higher-frequency call shows an fmin at c.50 kHz, and the lower call shows an fmin at c.35 kHz (Figures 9.33 to 9.35).

(2) Narrowband FM calls that are slightly humped (concave) at start (40–50 kHz) and with an fmin that can fall below 30 kHz (Figure 9.36).

(3) A series of numerous arch-shaped calls produced in succession, with an fmax falling within the range of 25 to 40 kHz (Figures 9.37 and 9.38).

(4) A higher-frequency U-shaped call (fmax c.60–70 kHz), as described in Figure 9.39.

Although we cannot be certain that other mouse species do not produce anything similar to these call structures, thus far we have failed to encounter these structures for any of the other species covered in this book.

Audible calls have been noted during courtship and mating behaviour, as well as between conspecifics (Trout and Harris, 2008).

Figures 9.33 to 9.39 provide specific examples of spectrograms relating to this species, and where the 🔊 symbol appears, a recording of what is shown within the spectrogram is available to download from the species library. In addition, QR codes are provided for selected examples, and these can be accessed immediately for listening purposes from most mobile devices.

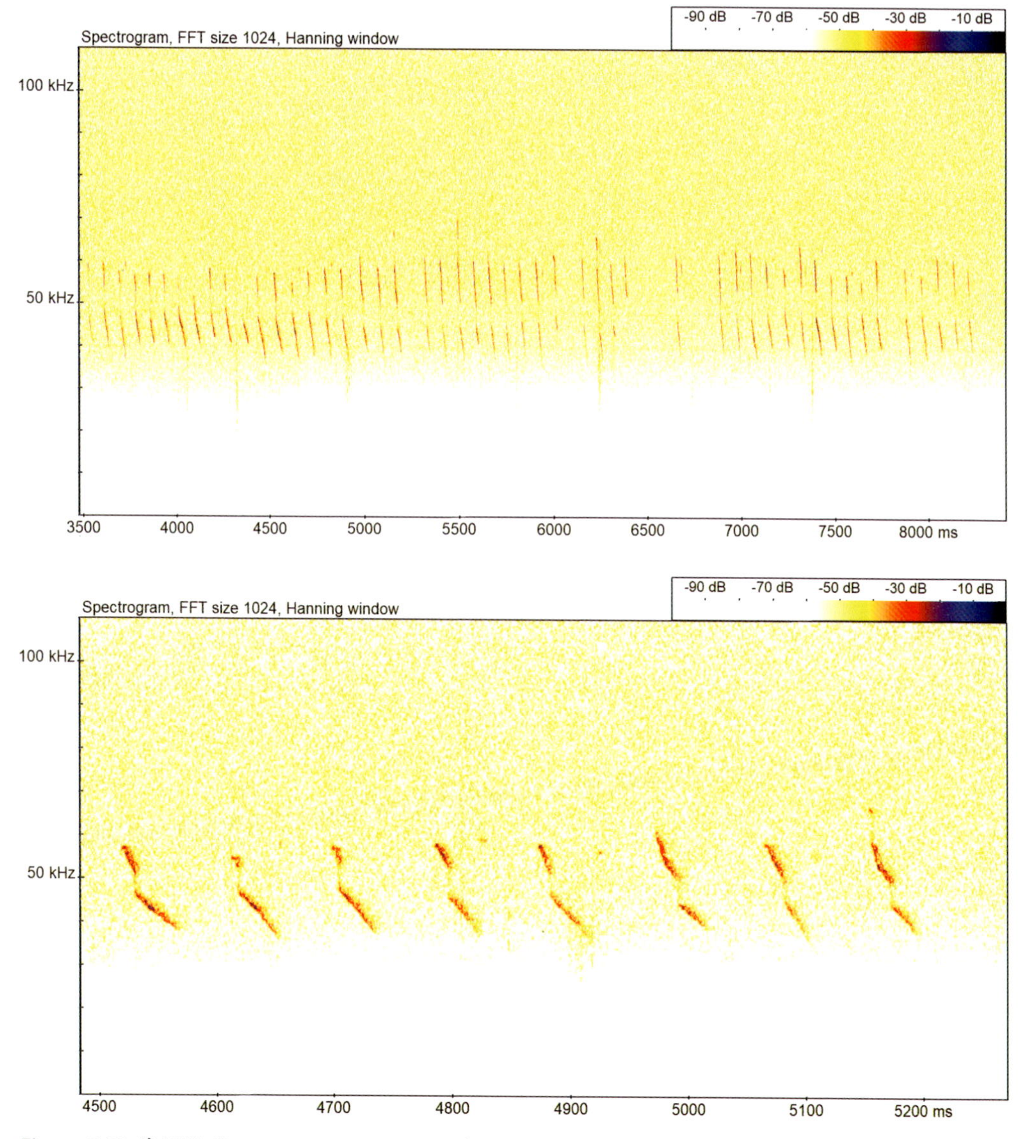

Figure 9.33 🔊 **(QR)** Harvest mouse – a series of rapid FM emissions produced over two frequency ranges, with lower graphic zoomed in to give more detail regarding call structure (frame width: c.5 sec, zoomed to c.1 sec)

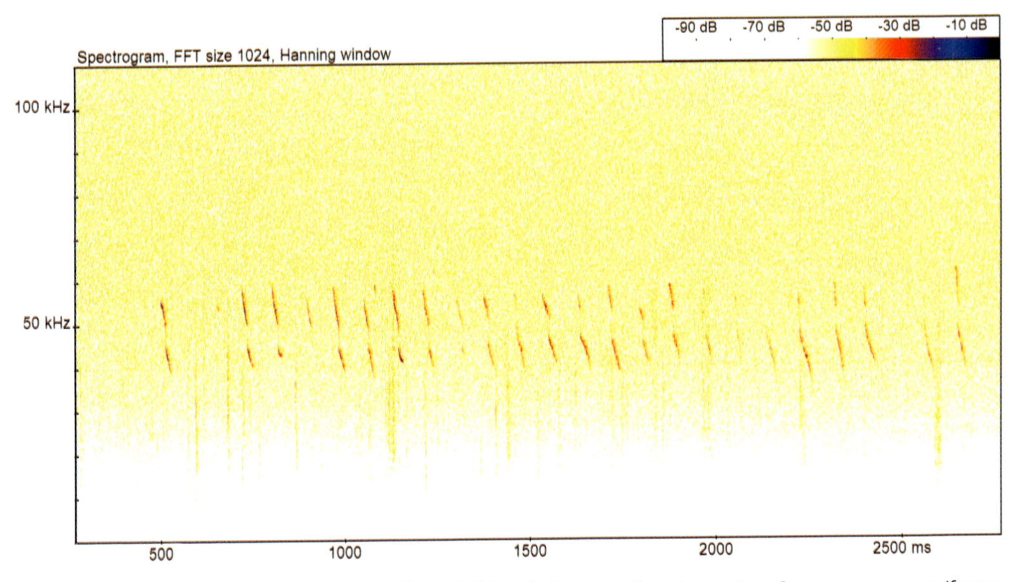

Figure 9.34 🔊 Harvest mouse – a series of rapid FM emissions produced over two frequency ranges (frame width: c.2.5 sec)

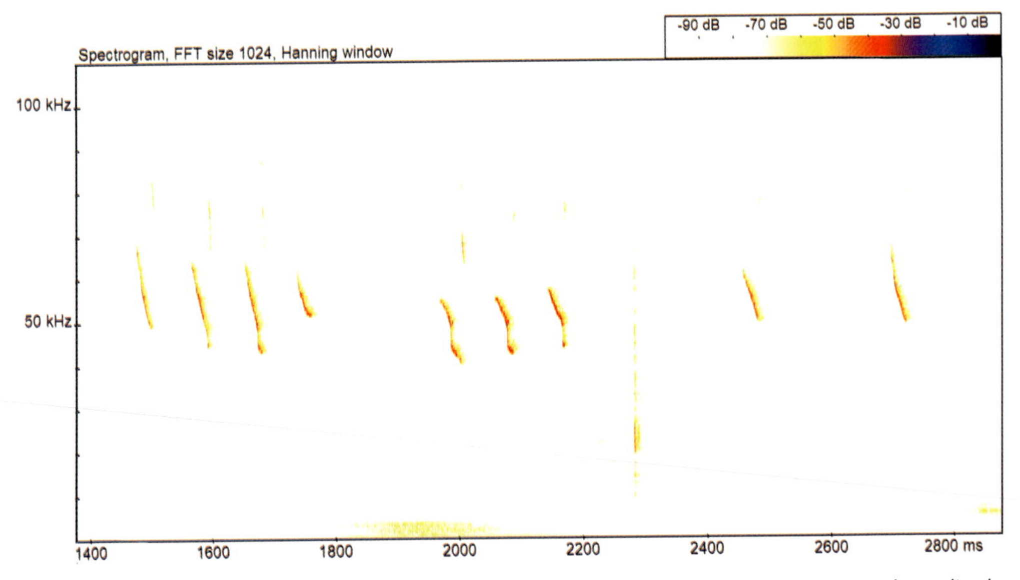

Figure 9.35 🔊 Harvest mouse – a variation of structure to that shown in Figures 9.33 and 9.34, with amplitude more evenly distributed, showing call structure as being more continuous (frame width: c.1.5 sec)

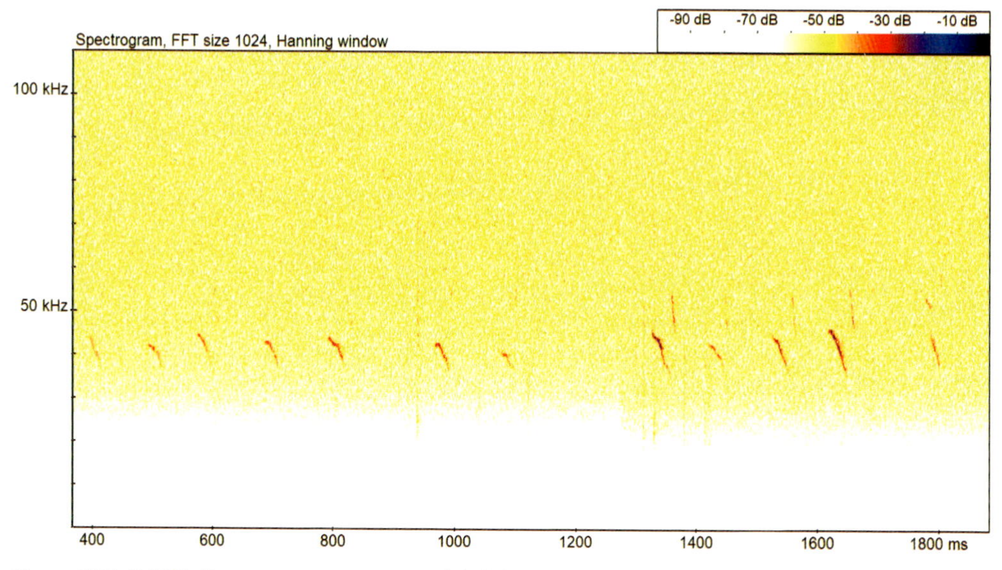

Figure 9.36 🔊 (QR) Harvest mouse – a series of slightly humped (i.e. concave) narrowband FM calls with an fmin at c.35 kHz (frame width: c.1.5 sec)

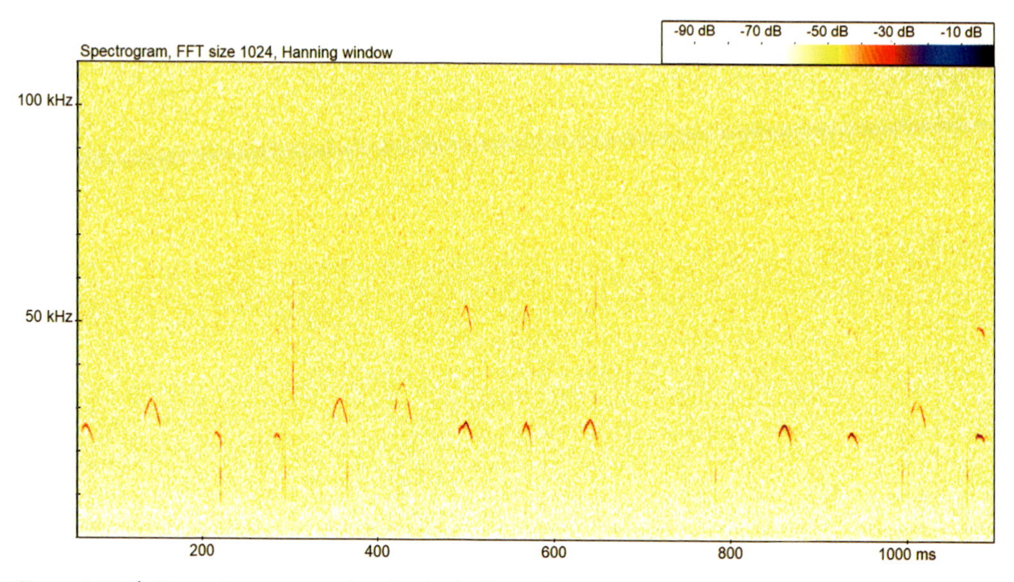

Figure 9.37 🔊 Harvest mouse – a series of arched calls produced at various frequencies, with some harmonics visible (frame width: c.1 sec)

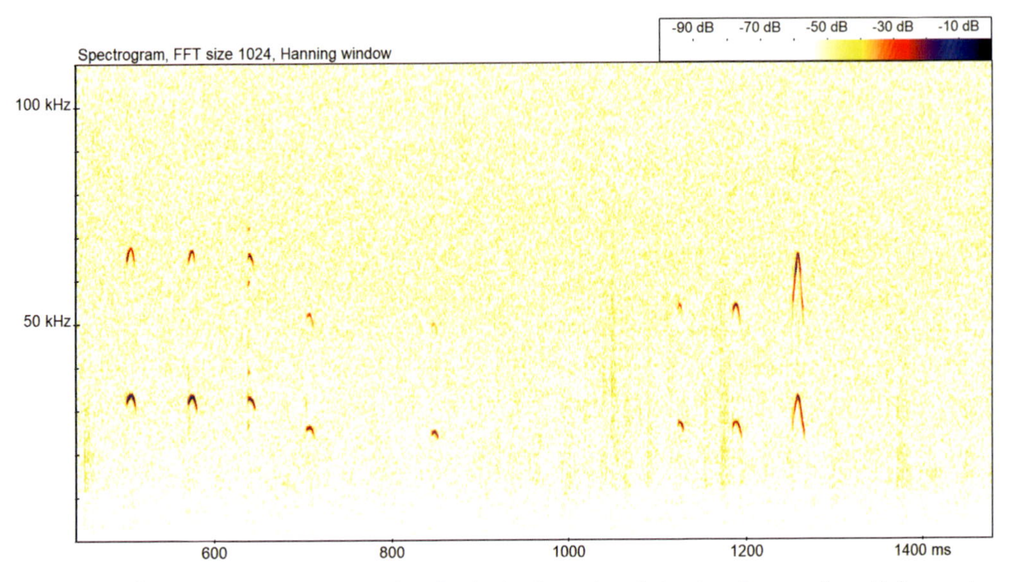

Figure 9.38 🔊 **(QR)** Harvest mouse – a series of arched calls produced at various frequencies, with harmonics visible (frame width: c.1 sec)

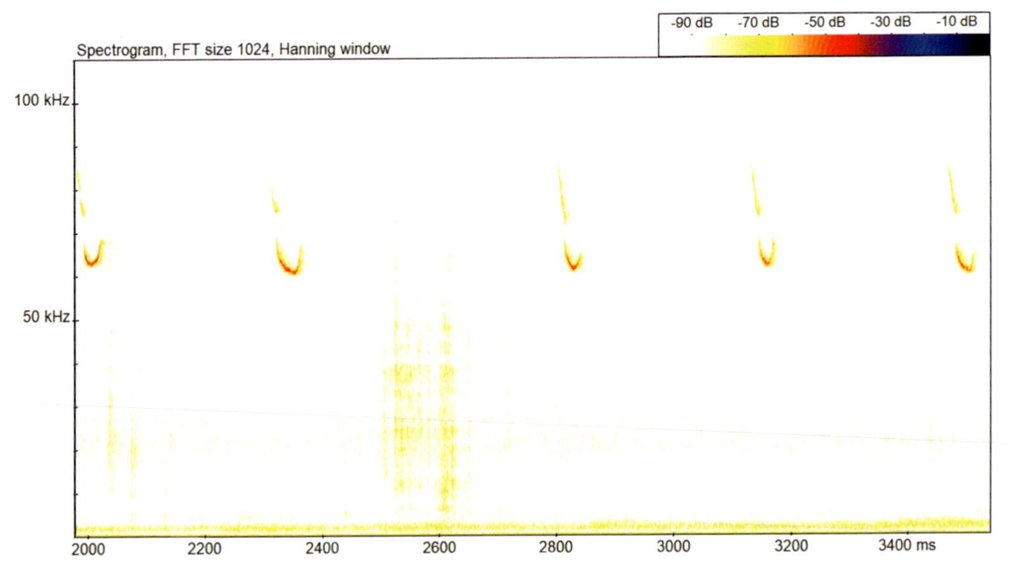

Figure 9.39 🔊 Harvest mouse – a series of U-shaped calls produced at various frequencies (frame width: c.1.5 sec)

QR codes relating to presented figures

Figure 9.33: Harvest mouse (time expanded x10 for listening purposes)

Figure 9.36: Harvest mouse (time expanded x10 for listening purposes)

Figure 9.38: Harvest mouse (time expanded x10 for listening purposes)

Potential confusion between harvest mouse and other species

Table 9.10 Summary of potential confusion between harvest mouse and other species

	Potential confusion species
Confusion group	In the first instance, always consider habitat and distribution.
Other mouse species	Unlikely to be confused with other mouse species, with the exception of wood mouse – arched calls.
Species in other groups	Hazel dormouse. Weak echolocation calls of common and soprano pipistrelle together could be confused with split calls. Nathusius' pipistrelle echolocation. Barbastelle echolocation. Daubenton's bat 'walking stick' social calls.

9.4.4 House mouse *Mus musculus*

Length/Weight	Habitat preferences	Distribution
Adult males are slightly smaller than females, with head/body length (excluding tail) from 70 to 90 mm. Weight described as variable.	Urban and rural environments; closely associated with human activities. Buildings and other human-made structures (domestic, commercial, industrial and agricultural).	Occurs throughout Britain and Ireland, although less so in northern Scottish mainland.
Diet		
Omnivorous, feeding across a range of vegetative and insectivorous food sources, varying according to availability, season and locality.		

Daily activity pattern	Largely nocturnal, therefore most often active from beyond dusk.
Seasonal activity pattern	Active throughout the year.
Mating activity pattern	Promiscuous mating occurs throughout the year.
Other notes	Due to small size, difficult to encounter visually, but may be recorded using appropriately positioned trail cameras.

References
Berry *et al.*, 2008; Lysaght and Marnell, 2016; Mathews *et al.*, 2018.

Acoustic behaviour including spectrogram examples

The house mouse is well studied when it comes to ultrasonic vocalisations, which are produced for many social contexts, including during mating and contact with conspecifics (Berry *et al.*, 2008; Lahvis *et al.*, 2011; von Merten *et al.*, 2014; Lupanova and Egorova, 2015). Call structure can be complex and variable (Lahvis *et al.*, 2011; Hammerschmidt *et al.*, 2012; von Merten *et al.*, 2014). Calls used by males during courtship may also serve a territorial function (Hammerschmidt *et al.*, 2012).

Audible sounds have also been noted, albeit less often, and seem to relate mostly to animals in distress or agonistic interactions.

In Table 9.11 we provide schematic examples of adult-generated ultrasonic calls as described by Zala *et al.* (2020). It should be borne in mind that the call parameters and examples shown are approximate and are also taken from a study of a small group of animals. Further variations to calls shown here are likely amongst wild populations, but these examples highlight that house mouse produces calls across a broad range of frequencies and comprise of a wide range of call shapes.

In our own study of house mouse, we have recorded many of the call types in Zala *et al.* (2020) and found that some are more common than others. Figure 9.40 shows a typical example of the relatively high ultrasonic calls that are regularly recorded by this species. Based on data collected by others and our own recordings, house mouse vocalisations usually occur at higher frequencies (Figures 9.40 to 9.42) than the other

Table 9.11 Typical ultrasonic call structures of house mouse (adapted from Zala *et al.*, 2020)

Call type (as described by Zala *et al.*, 2020)	**Description of structure**	**Schematic representation** (not to scale)
Ultra-short Short	Vocalisations below 91 kHz and less than 10 ms duration	
Flat	Quasi-constant frequency (bandwidth less than 5 kHz) Vocalisations below 91 kHz, longer than 10 ms in duration	
Down	FM descending Vocalisations below 91 kHz, decreasing in frequency, with a bandwidth greater than 5 kHz	
Up	FM ascending Vocalisations below 91 kHz, increasing in frequency, with a bandwidth greater than 5 kHz	
U-shaped	FM descending/ascending (U-shaped) Vocalisations below 91 kHz, decreasing and then increasing in frequency, with a bandwidth greater than 5 kHz	
U-shaped inverted	FM ascending/descending (arched) Vocalisations below 91 kHz, increasing and then decreasing in frequency, with a bandwidth greater than 5 kHz	
Complex	Vocalisations below 91 kHz, variable in frequency (i.e. with two or more directional shifts), with a bandwidth greater than 5 kHz	91 kHz
Complex 2 Complex 3 Complex 4 Complex 5	Vocalisations below 91 kHz, consisting of two or more separate components (variable in structure), with two or more frequency jumps (not separated by time)	
Ultra-high	Vocalisations of any structure, with an fmax greater than 91 kHz	91 kHz

mouse species encountered in Britain and Ireland, although there is still some potential for overlap (see Figures 9.43 to 9.45) with, for example, the higher maximum frequency (fmax) wood mouse vocalisations. Based on our recordings to date, however, we would not expect wood mouse to be achieving an fmax of 100 kHz or more.

Bat workers should be aware that when listening to house mouse in time expansion (x10), the calls may sound similar to what you would expect to hear from a horseshoe species, albeit structurally the calls are quite different.

House mouse also make calls at lower frequencies, which show a range of structures (e.g. Figure 9.46), with arch-shaped calls that end between c.50 kHz and 60 kHz being the most commonly recorded.

Compared with wood and yellow-necked mouse, audible calls were also quite often recorded, particularly during agonistic interactions over food items (Figure 9.47).

Finally, we provide an example of an easily audible call (Figure 9.48).

Figures 9.40 to 9.48 provide specific examples of spectrograms relating to this species, and where the 🔊 symbol appears, a recording of what is shown within the spectrogram is available to download from the species library. In addition, QR codes are provided for selected examples, and these can be accessed immediately for listening purposes from most mobile devices.

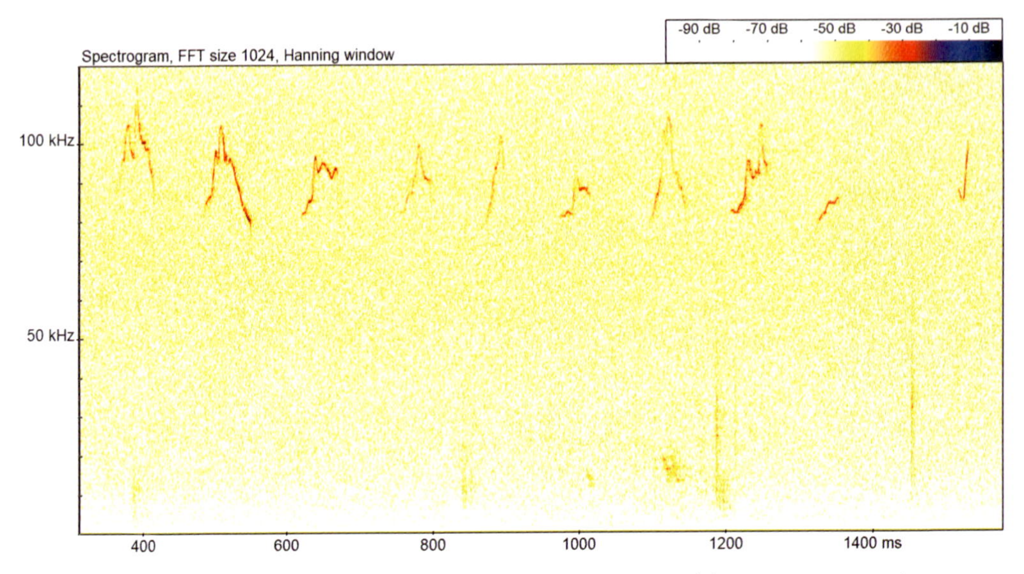

Figure 9.40 🔊 **(QR)** House mouse – a series of ten calls showing typical frequency range and structures (frequency scale: 0 to 120 kHz/frame width: c.1.25 sec)

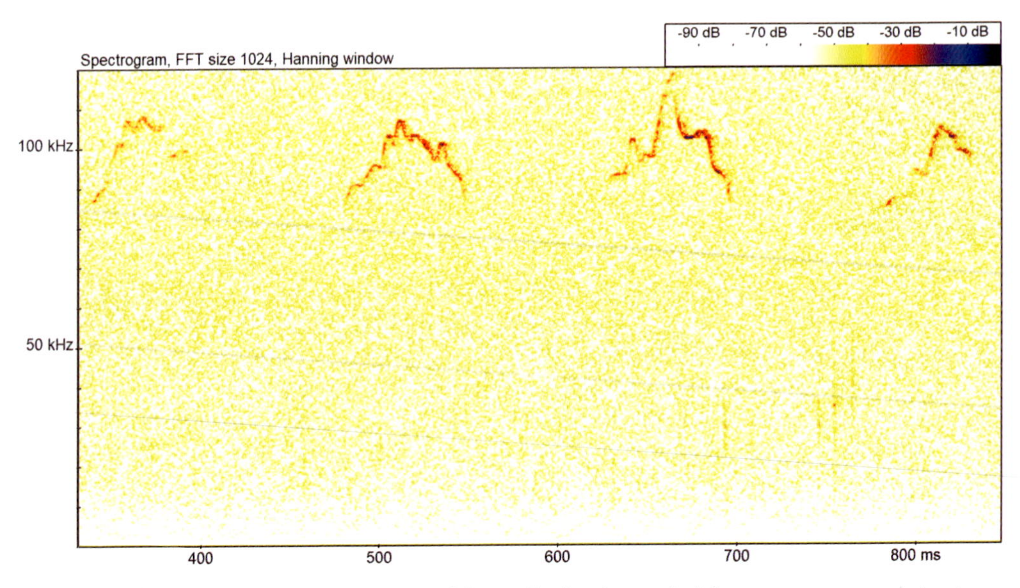

Figure 9.41 🔊 **(QR)** House mouse – a series of four calls showing typical frequency range and structures (frequency scale: 0 to 120 kHz/frame width: c.0.5 sec)

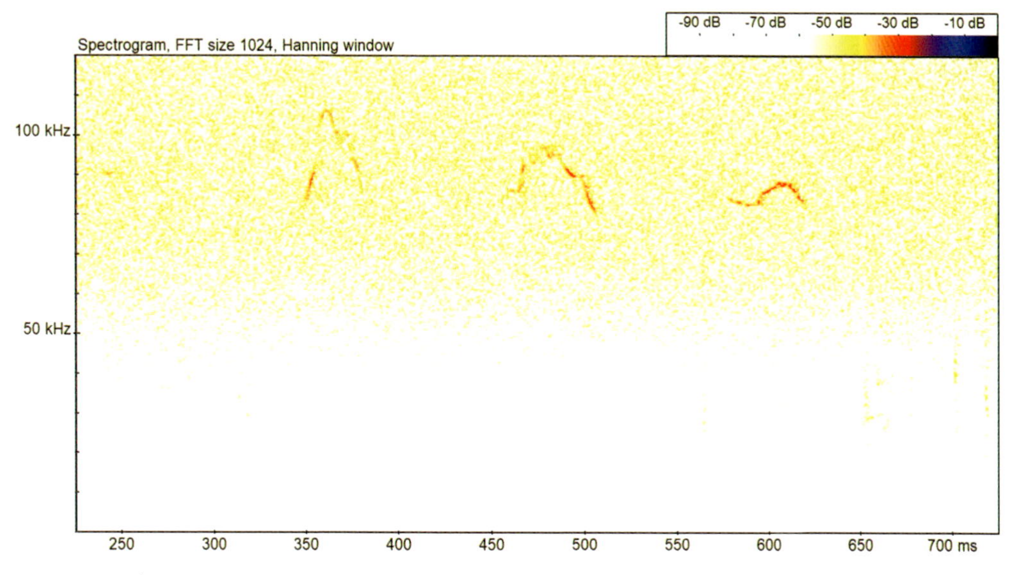

Figure 9.42 House mouse – a series of three calls showing a degree of variation in structure and maximum frequencies (frequency scale: 0 to 120 kHz/frame width: c.0.5 sec)

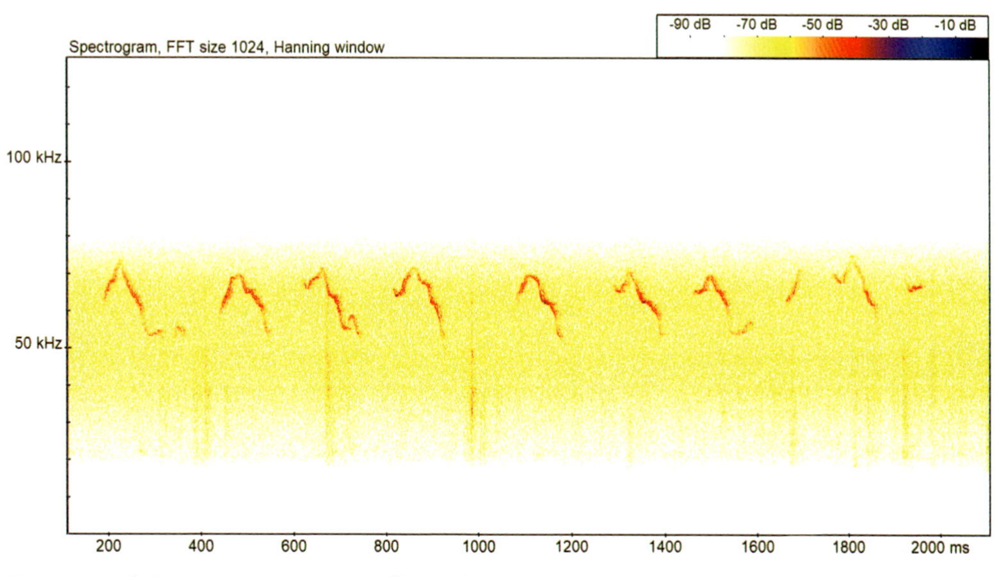

Figure 9.43 House mouse – a series of lower-frequency calls showing a degree of variation in structure (frequency scale: 0 to 130 kHz/frame width: c.2 sec)

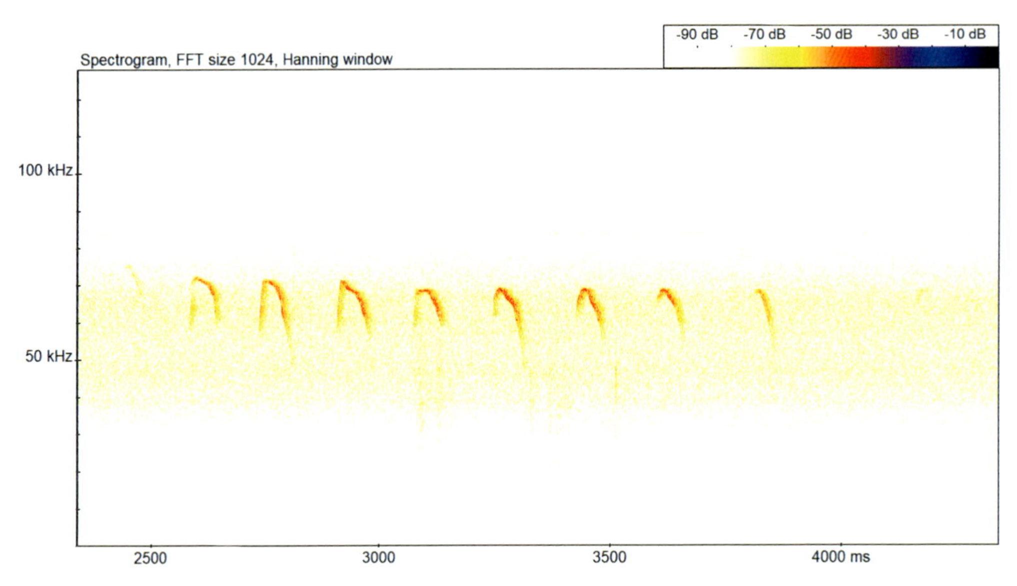

Figure 9.44 House mouse – a series of lower-frequency calls showing a degree of variation in structure (frequency scale: 0 to 130 kHz/frame width: c.2 sec)

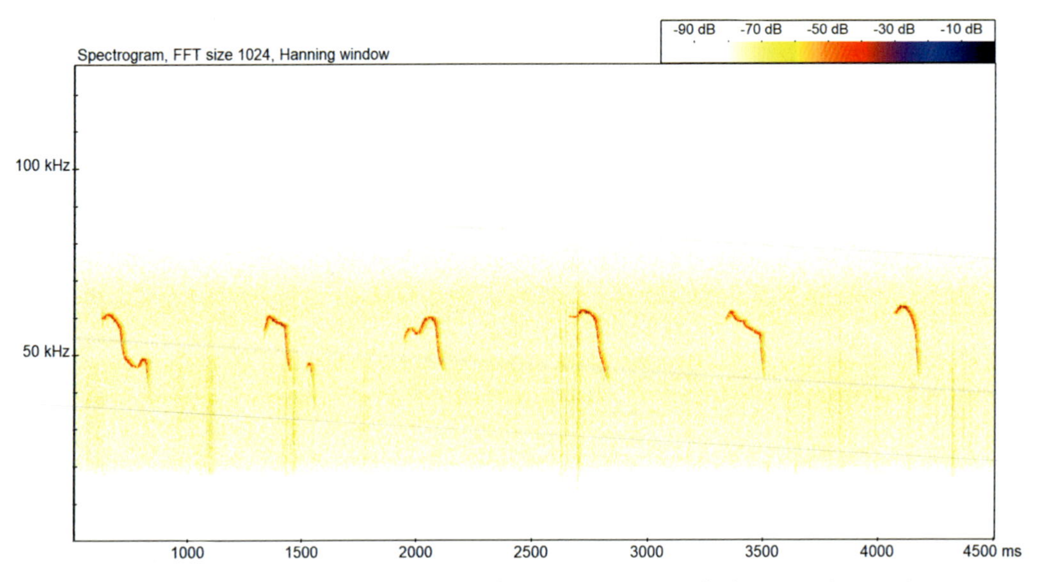

Figure 9.45 **(QR)** House mouse – a series of lower-frequency, complex calls showing a degree of variation in structure (frequency scale: 0 to 130 kHz/frame width: c.4 sec)

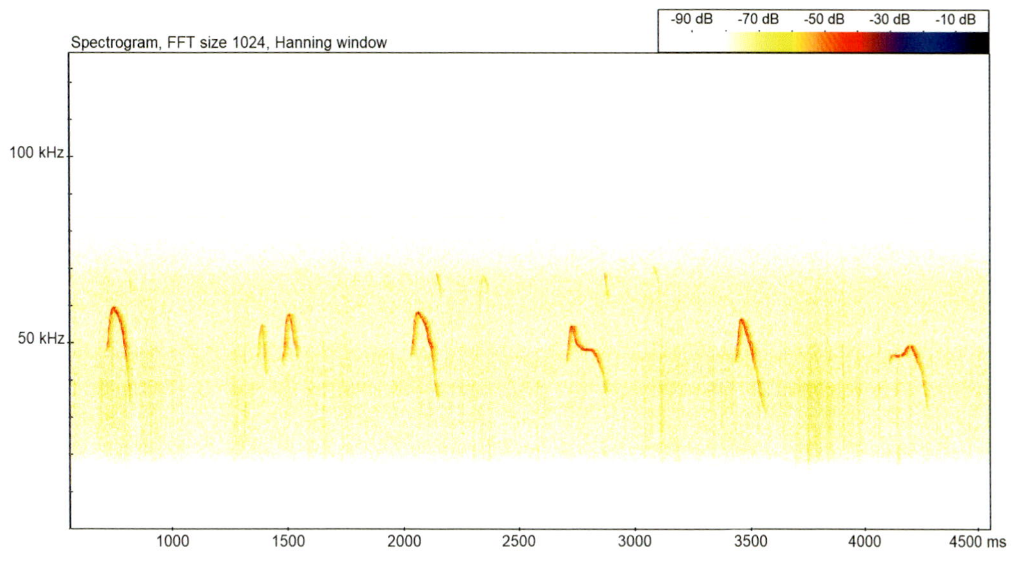

Figure 9.46 🔊 House mouse – a series of lower-frequency calls, more evenly arched (frequency scale: 0 to 130 kHz/frame width: c.4 sec)

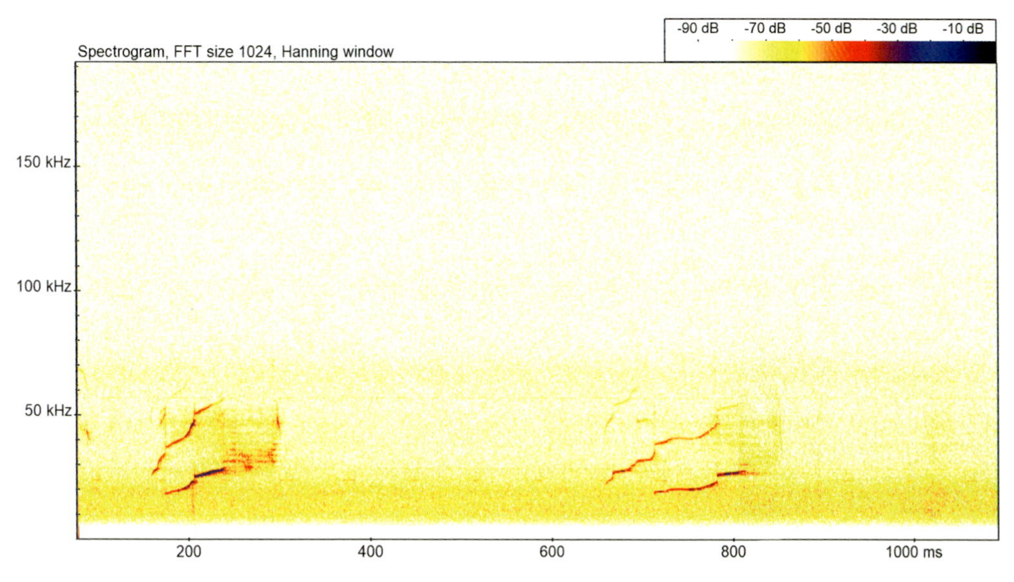

Figure 9.47 🔊 House mouse – low-frequency calls, ascending FM (frequency scale: 0 to 140 kHz/frame width: c.1 sec)

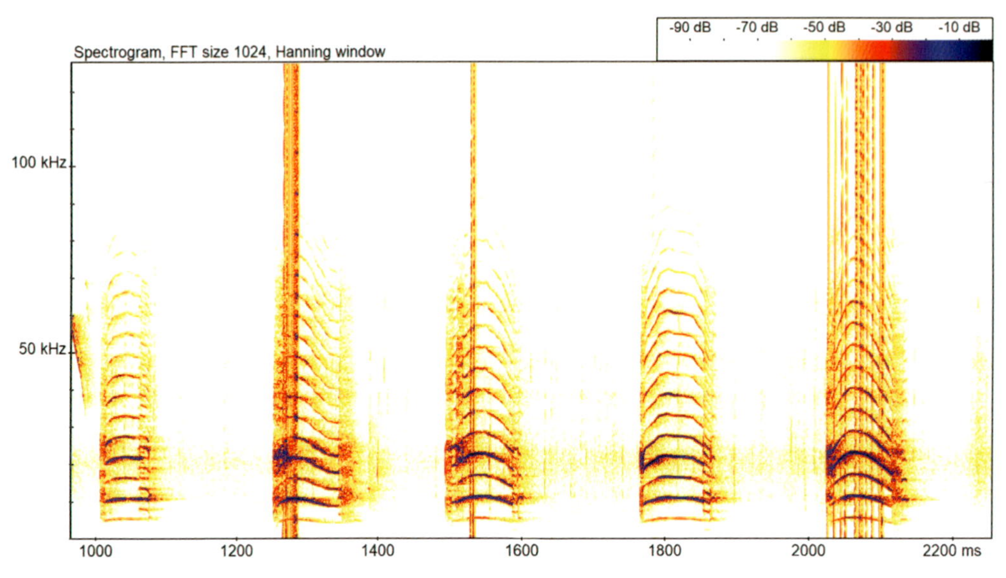

Figure 9.48 🔊 House mouse – a series of audible calls (frequency scale: 0 to 130 kHz/frame width: c.1.25 sec)

QR codes relating to presented figures

Potential confusion between house mouse and other species

Table 9.12 Summary of potential confusion between house mouse and other species

	Potential confusion species
Confusion group	In the first instance, always consider habitat and distribution.
Other mouse species	Lower fmax calls from house mouse could potentially overlap calls from wood and yellow-necked mouse.
Species in other groups	When listening to calls in time expansion (x10) they sound similar to horseshoe bat examples (confusion perhaps most likely with horseshoe bat social calls).

9.5 Vole vocalisations

In the following section, a brief summary of the key characteristics of the four species of vole (one of which, common vole, is represented by two subspecies) will be described, along with details on the species-specific vocalisations and, where known, the behavioural function of these sounds.

Three species of vole occur within mainland Britain (voles are absent from Northern Ireland). Within the Republic of Ireland, only bank vole occurs (in the south-west). In addition a further species, common vole *Microtus arvalis*, is represented in the region by two subspecies, Orkney vole and Guernsey vole, which occur in Orkney and Guernsey respectively. Unfortunately, we were not able to source recordings of the two subspecies of bank vole (i.e. Jersey vole *Myodes glareolus caesarius* and Skomer vole *Myodes glareolus skomerensis*), and as such, these have not been included.

Acoustic studies on voles have been fairly limited, although they were included in an overview for identification purposes within Britain, published in *British Wildlife* in 2020 (Newson *et al.*, 2020). Voles are burrowing species and their social interactions, and consequently vocalisations, occur most often underground within their burrow systems. As such, most vocal activity may be outside the range of the microphone, which could explain the low call rate for voles compared to other small mammal species. Above ground, communication is, in the main, via olfactory cues (e.g. scent-marking latrines and runways).

9.5.1 Water vole *Arvicola amphibius*

<table>
<tr><th>Length/Weight</th><th rowspan="4">Habitat preferences

Closely associated with rivers, streams and wetlands, favouring steep banks for burrowing, although will also build nests above ground with vegetation.
In some parts of Scotland they are non-amphibious, favouring pasture/meadow.
Rivers
Streams
Wetlands</th><th rowspan="4">Distribution

Occurs throughout Britain, including many islands. Absent from Ireland.</th></tr>
<tr><td>Our largest vole species.
Males are larger than females. Body length from 145 to 220 mm.
Weight from 150 to 300 g.</td></tr>
<tr><td>Diet</td></tr>
<tr><td>Vegetarian, feeding primarily upon grasses, sedges and reeds.</td></tr>
<tr><td>Daily activity pattern</td><td colspan="2">Can be active throughout daytime and overnight.</td></tr>
<tr><td>Seasonal activity pattern</td><td colspan="2">Active throughout the year.</td></tr>
<tr><td>Mating activity pattern</td><td colspan="2">Mating can occur from February through to October, with young (around six) being born typically during the period April to September.</td></tr>
<tr><td>Other notes</td><td colspan="2">Distinctive 'plop' sound may be heard when diving.
Usually monitored by searching for characteristic latrine sites and feeding remains. They are also easily captured using trail cameras, especially if baited feeding stations are provided.</td></tr>
<tr><td colspan="3">References
Woodroffe et al., 2008; Lysaght and Marnell, 2016; Mathews et al., 2018; Mammal Society, 2022.</td></tr>
</table>

Acoustic behaviour including spectrogram examples

Compared to other small mammals, we have found that the call rate of captive and wild voles, including water vole, is much lower than other species groups. Given that voles are burrowing animals, vocal interactions may occur more often within burrow systems, and therefore vocal activity is usually outside of the range of where microphones are usually deployed. Since they are gregarious, particularly during the breeding season, it is likely that sound does have some communicative function.

Most available literature refers to audible sound in connection with this species. For example, during agonistic encounters water vole will make irregular calls at a frequency range from 2.5 kHz to 4.4 kHz, with duration of c.100 ms (Woodroffe *et al.*, 2008). Audible sound also features when animals are in a state of distress.

Although not a vocalisation, the audible 'plop' noise heard when this species dives is thought to act as a warning to conspecifics (Strachan, 1999) and, usefully, can also alert us to the presence of this species.

Based on our own data, ultrasonic sounds are made, some of which appear to be distinctive. The best examples appear as stepped structured calls either typically progressing upwards as in Figures 9.49 to 9.53, or occasionally downwards as in the first call shown in Figure 9.50. The frequencies of calls are similar to brown rat, black rat and hazel dormouse, although in these other species, the call structures usually appear to show a smoother transition through frequencies (i.e. not typically stepped as in water vole). Occasionally we have encountered stepped water vole calls at higher frequencies, for example as shown in Figure 9.53 where the minimum frequency (fmin) is c.40 kHz. The only other example of stepped call we have encountered in a vole species to date involves a single example from Orkney vole (see Figure 9.76) although this occurred at a higher frequency (c.10 kHz higher) than is typical for water vole. Since Guernsey vole is also a subspecies of common vole (as is Orkney vole), it also may in time be shown to produce similar calls.

A range of other call types have also been recorded by ourselves and others, and Figures 9.54 to 9.57 provide examples of what may also be encountered when recording this species.

Figures 9.49 to 9.57 provide specific examples of spectrograms relating to this species, and where the 🔊 symbol appears, a recording of what is shown within the spectrogram is available to download from the species library. In addition, QR codes are provided for selected examples, and these can be accessed immediately for listening purposes from most mobile devices.

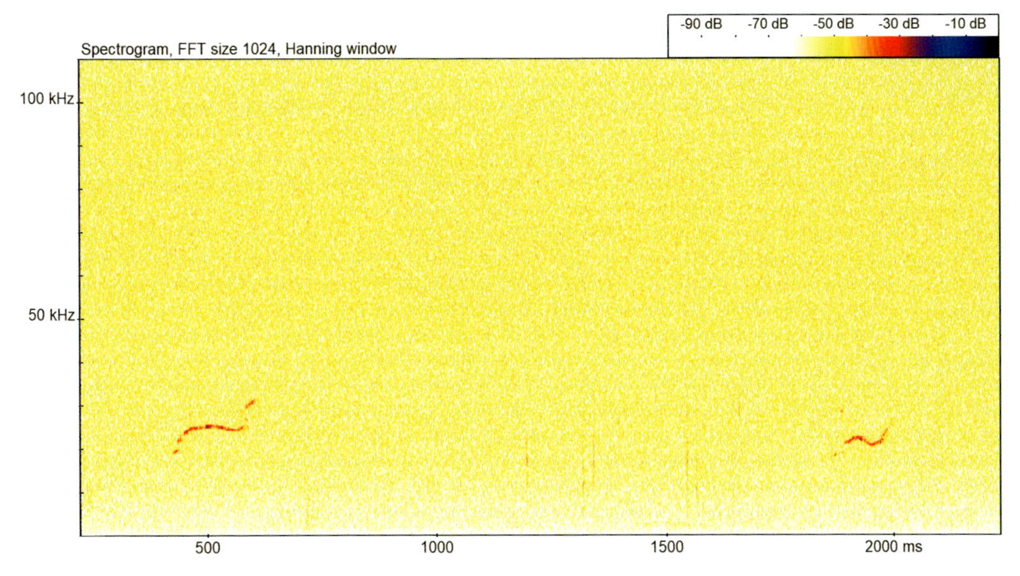

Figure 9.49 🔊 **(QR)** Water vole – two typical rising FM stepped calls (frame width: c.2 sec)

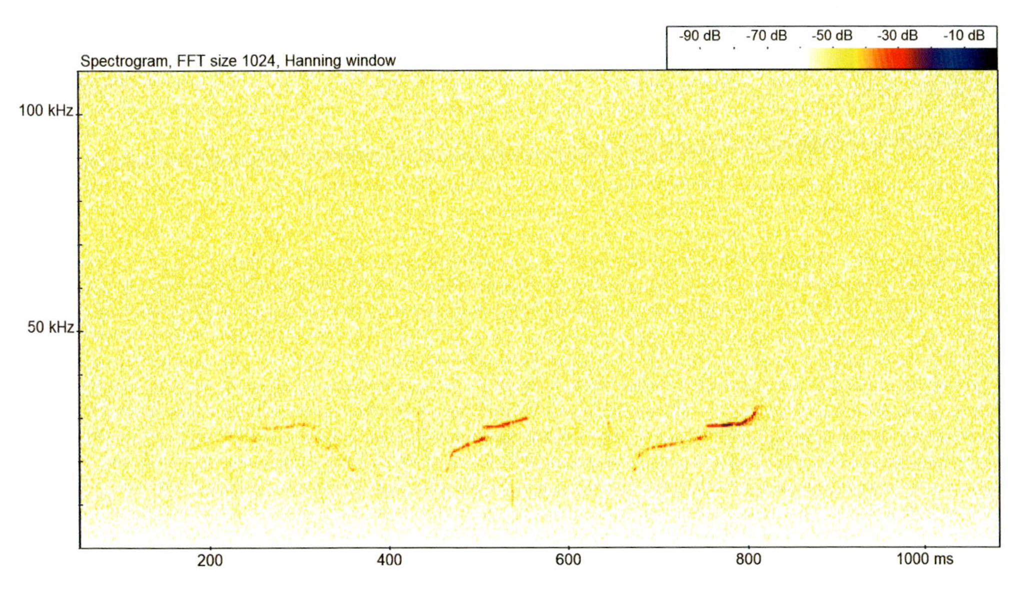

Figure 9.50 🔊 Water vole – three calls, the first of which is very faint but shows an initial rise in frequency followed by a drop. The remaining two calls (rising FM stepped) are more typical (frame width: c.1 sec).

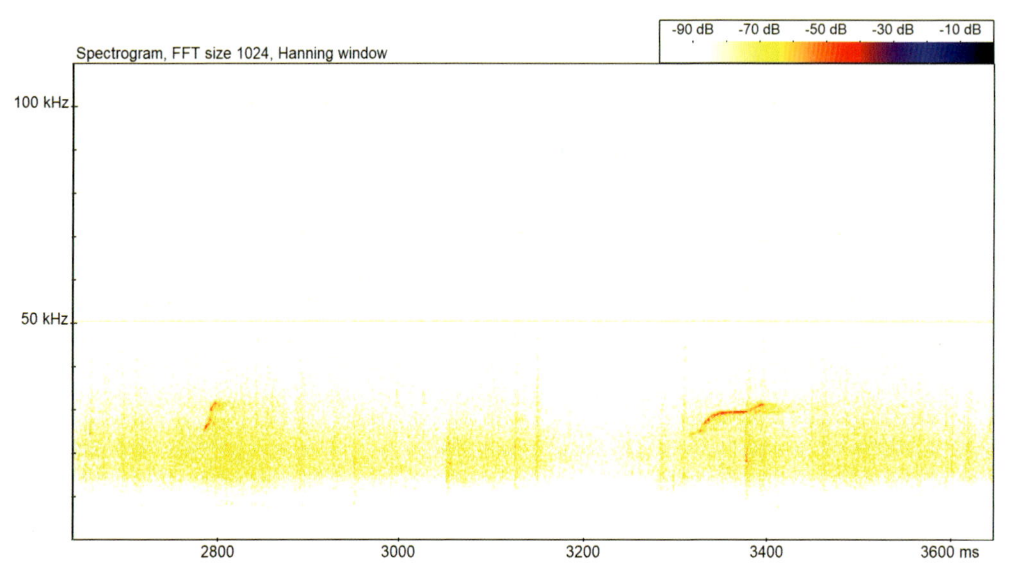

Figure 9.51 🔊 Water vole – two calls, the second being typically rising FM stepped (courtesy of C. Macdonald) (frame width: c.1 sec)

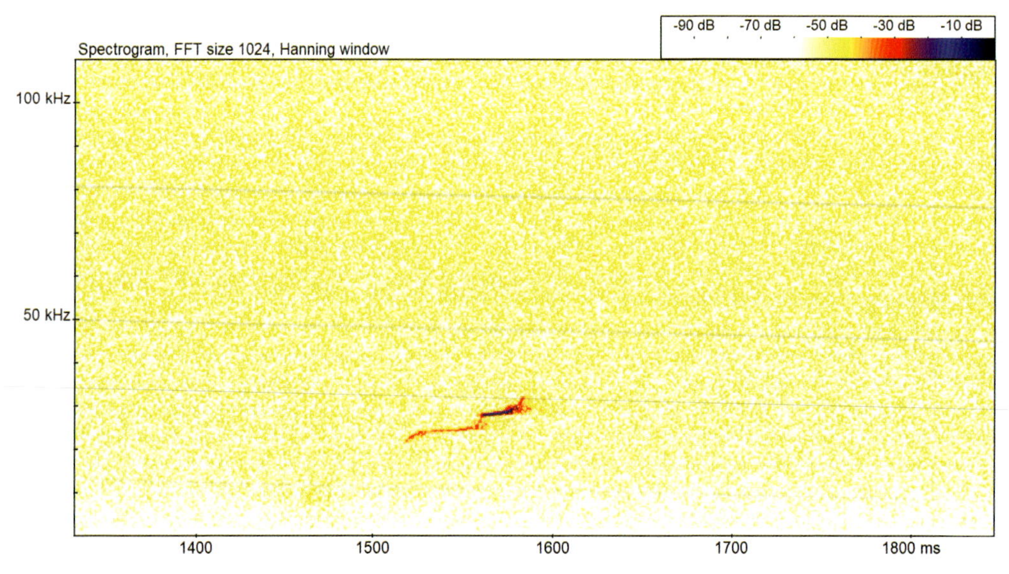

Figure 9.52 🔊 Water vole – a single rising FM stepped example (frame width: c.0.5 sec)

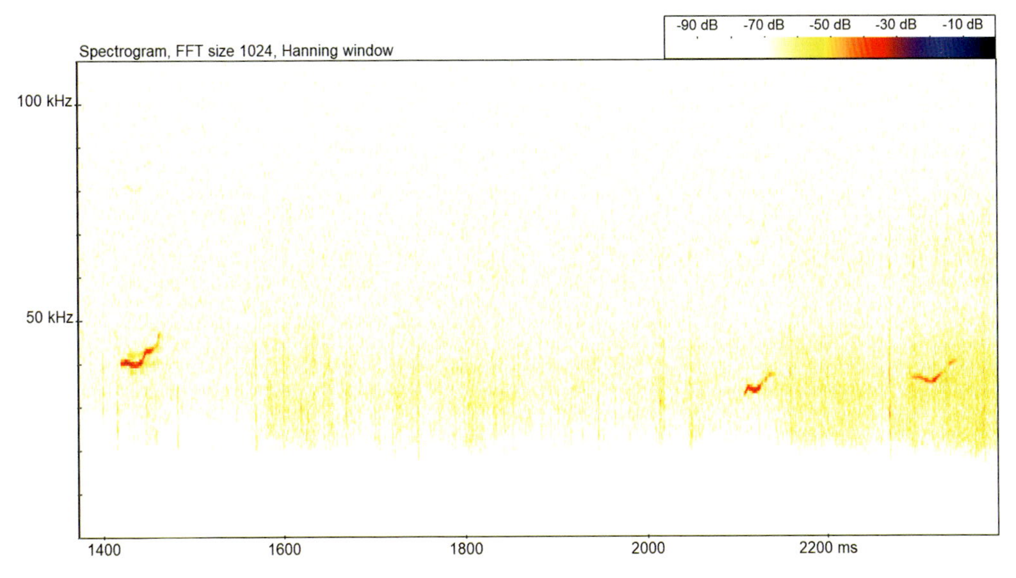

Figure 9.53 🔊 Water vole – a variation of rising FM stepped calls (x3) at a much higher frequency than typical (frame width: c.1 sec)

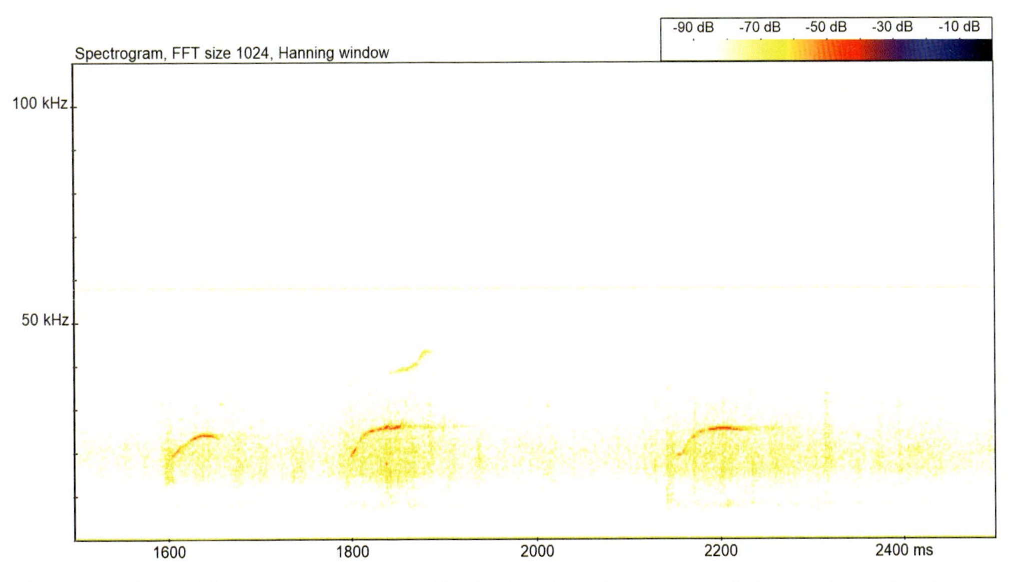

Figure 9.54 🔊 **(QR)** Water vole – three rising FM calls (courtesy of C. Macdonald) (frame width: c.1 sec)

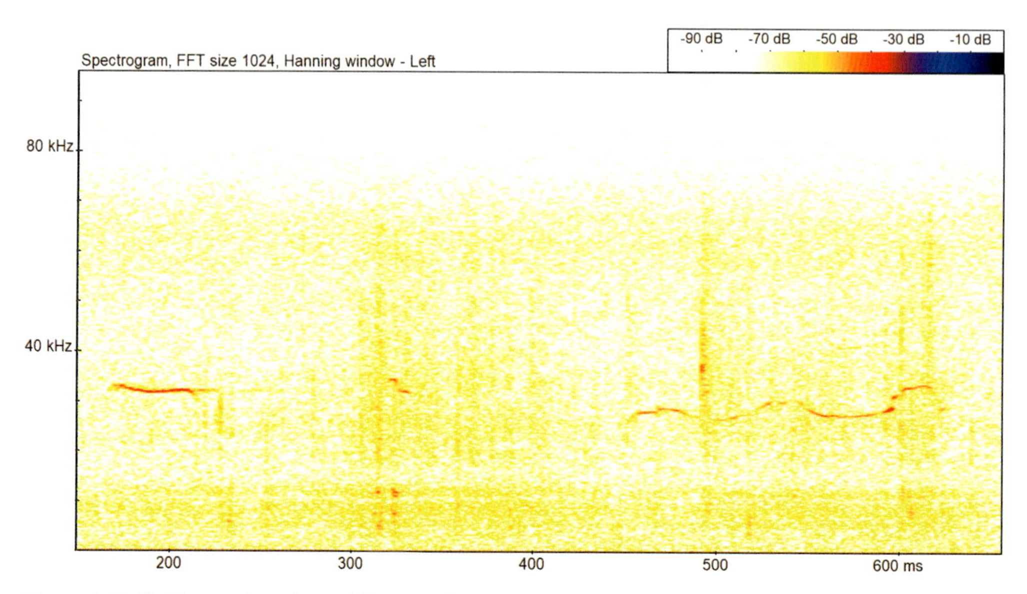

Figure 9.55 🔊 Water vole – three different call types in one sequence (frequency scale: 0 to 95 kHz/frame width: c.0.5 sec)

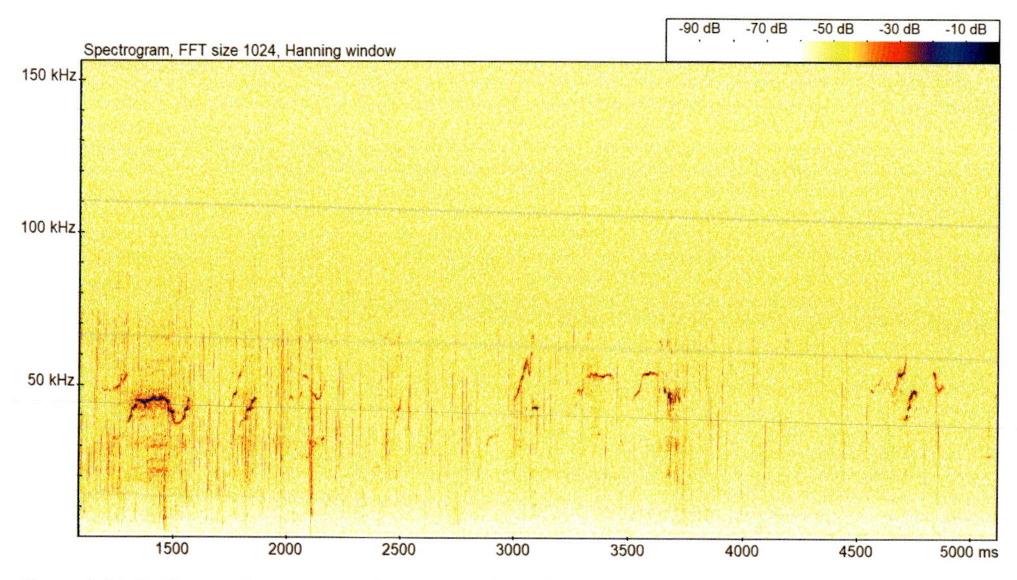

Figure 9.56 🔊 Water vole – a series of more complex call structures (frequency scale: 0 to 150 kHz/frame width: c.4 sec)

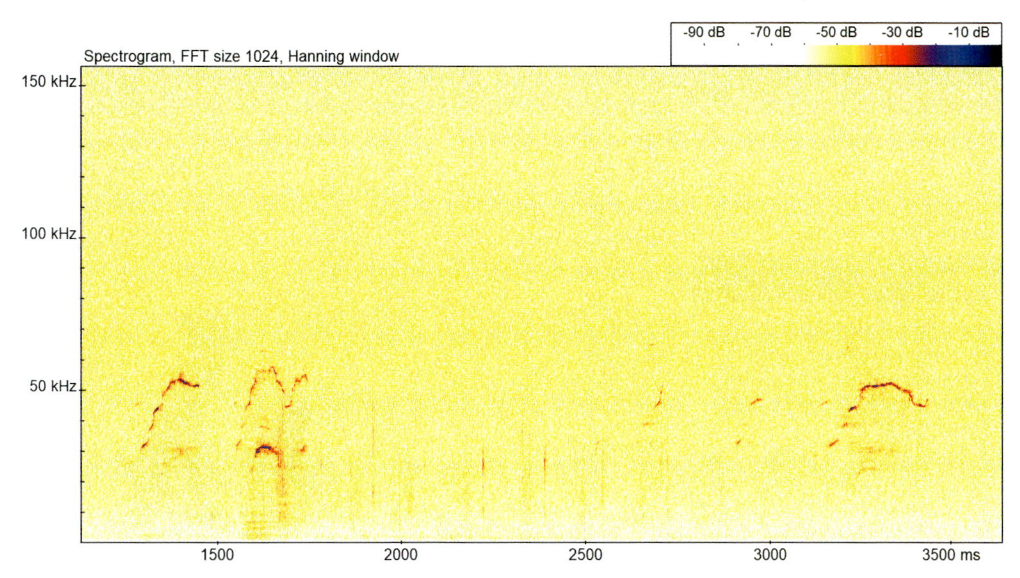

Figure 9.57 🔊 (QR) Water vole – a series of more complex call structures (frequency scale: 0 to 150 kHz/frame width: c.2.5 sec)

QR codes relating to presented figures

Figure 9.49: Water vole call (time expanded x10 for listening purposes)	**Figure 9.54:** Water vole call (time expanded x10 for listening purposes)	**Figure 9.57:** Water vole call (time expanded x10 for listening purposes)

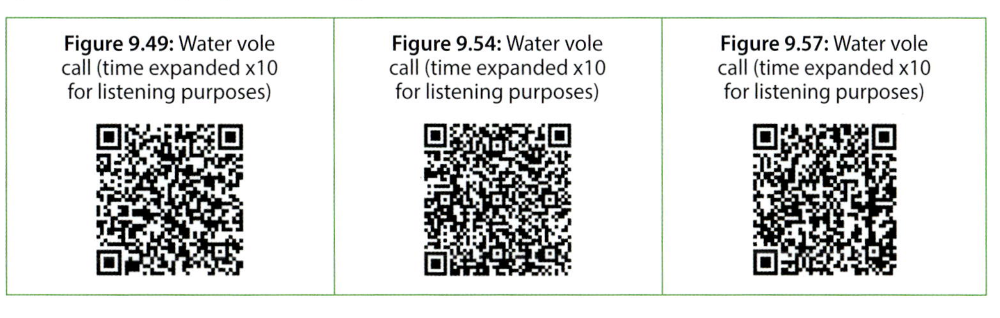

Potential confusion between water vole and other species

Table 9.13 Summary of potential confusion between water vole and other species

	Potential confusion species
Confusion group	In the first instance, always consider habitat and distribution.
Other vole species	Orkney vole occasionally produces stepped call structure, and this may also transpire to be the case for Guernsey vole.
Species in other groups	Brown rat, black rat and hazel dormouse produce calls within similar frequency ranges, but these tend to be much smoother in structure compared to the stepped water vole emissions.

9.5.2 Bank vole *Myodes glareolus*

Length/Weight	Habitat preferences	Distribution
Adult males are larger than females. Body length from 90 to 110 mm. Weight from 20 to 40 g.	Broadleaved woodland Urban and gardens Coniferous woodland Dwarf shrub heath Hedgerows	Occurs throughout Britain, including many surrounding islands. Present in south-west of Ireland.
Diet		
Mainly vegetarian, feeding primarily upon leaves, seeds and fruits, and to a lesser extent fungi, roots, flowers, moss and invertebrates.		

Daily activity pattern	Can be active during day and night, but peaking more at dusk and dawn. Less active overnight during winter period.
Seasonal activity pattern	Active throughout the year.
Mating activity pattern	Promiscuous mating usually takes place from March to October. Young (around four) are normally born during the period April to September, with seasonal peak in birth rate occurring during May to June.
Other notes	Easily captured using baited small mammal trail camera boxes.

References
Shore and Hare, 2008; Lysaght and Marnell, 2016; Mathews *et al.*, 2018; Littlewood *et al.*, 2021; Mammal Society, 2022.

Acoustic behaviour including spectrogram examples

Bank voles produce a range of audible and ultrasonic sounds (Newson *et al.*, 2020) but, as with all the vole species, their call rate, at least above ground, is much lower than that of other small mammals.

Audible sounds can have a duration of c.50 to 113 ms, occurring at a minimum frequency (fmin) of 1.5 to 2.7 kHz. Examples of audible calls are shown in Figures 9.64 and 9.65, these calls being very distinctive when listened to in time expansion (x10). Infants are known to make noise while in their nest, these sounds being both audible (< 8 kHz) and ultrasonic (c.20 to 35 kHz) (Stoddart and Sales, 1985; Marchlewska-Koj, 2000).

High-frequency calls have also been described during mating (Kapusta and Sales, 2009). In other studies, adults have been described as producing both ultrasonic and audible sounds (Kapusta *et al.*, 2007; Osipova and Rutovskaya, 2000; Sales, 2010). Adults are also known to produce high-pitched sounds, for example between males during aggressive encounters (Miska-Schramm *et al.*, 2018), as well as a sound described as teeth chattering (Stoddart and Sales, 1985).

Kapusta and Sales (2009) found the ultrasonic calls of bank vole had a minimum frequency of 24 to 30 kHz, lasting approximately 61 to 70 ms. Their experience was that the calls of bank vole were lower in frequency than those produced by field vole. We have examples of higher-frequency field vole recordings (see section on field vole below), but when recording captive field voles ourselves, we recorded more lower-frequency calls of about 20 kHz. Kapustra and Sales (2009) also found that bank vole were generally less vocal than field vole, which is also our experience.

In our examples, those shown in Figures 9.58 to 9.61 seem fairly typical, although compared with other small mammal species, we have fewer examples to draw conclusions from, and hence we say this with a degree of caution. Notice how these calls are fairly constant in frequency, with minimum frequencies being c.35 kHz. Figures 9.62 and 9.63 show more complex arrangements emitted over a longer period of time.

The examples provided show a range of call structures, none of which we can say typically occur more often than others, as our experience in recording bank vole tells us that obtaining any calls at all involves lengthy periods of recording with little to show for the effort. When calls are successfully encountered, they tended to be brief, compared to other small mammals.

Figures 9.58 to 9.65 provide specific examples of spectrograms relating to this species, and where the 🔊 symbol appears, a recording of what is shown within the spectrogram is available to download from the species library. In addition, QR codes are provided for selected examples, and these can be accessed immediately for listening purposes from most mobile devices.

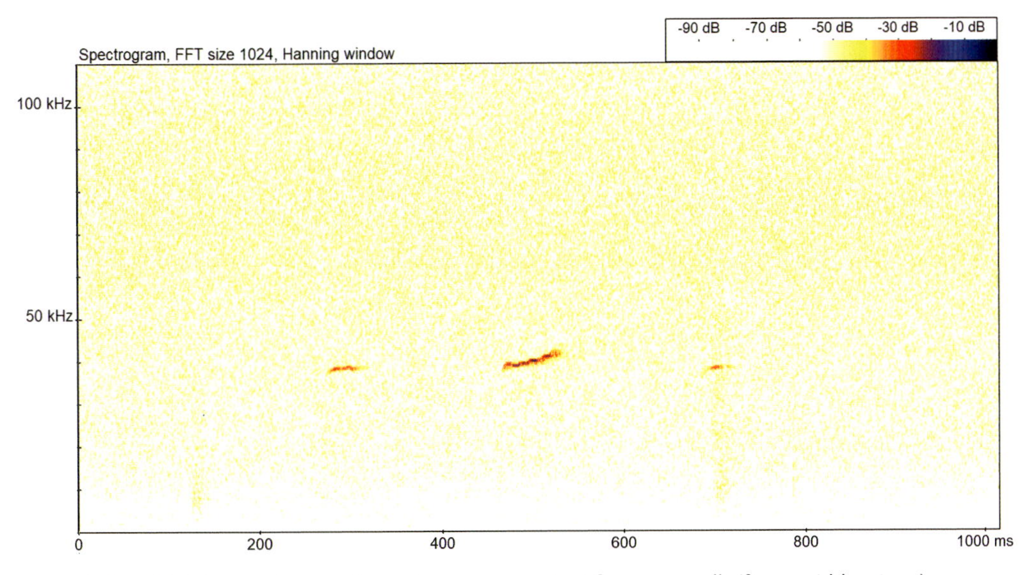

Figure 9.58 🔊 (QR) Bank vole – a series of three near-constant-frequency calls (frame width: c.1 sec)

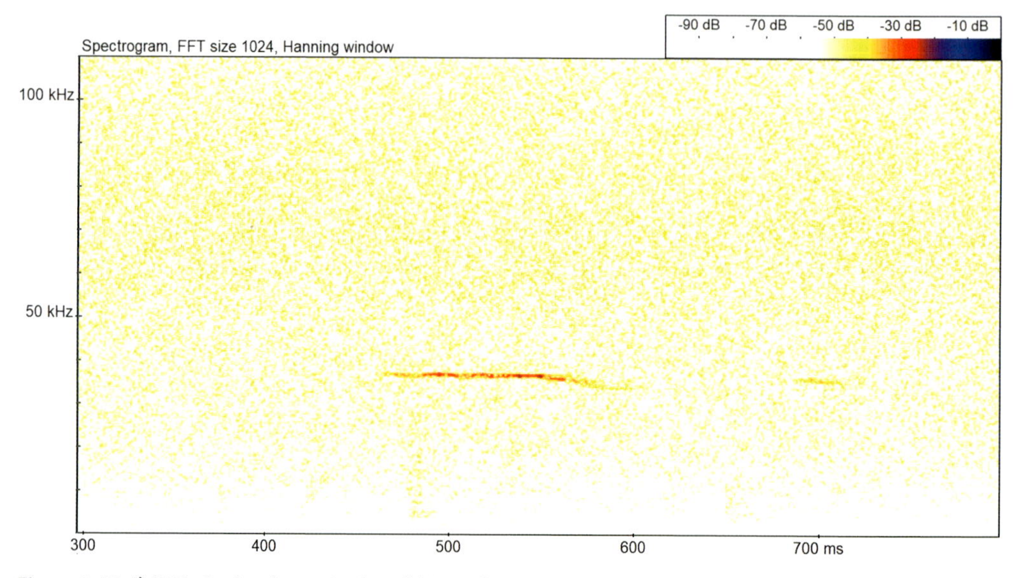

Figure 9.59 🔊 (QR) Bank vole – a single call (QCF) (frame width: c.0.5 sec)

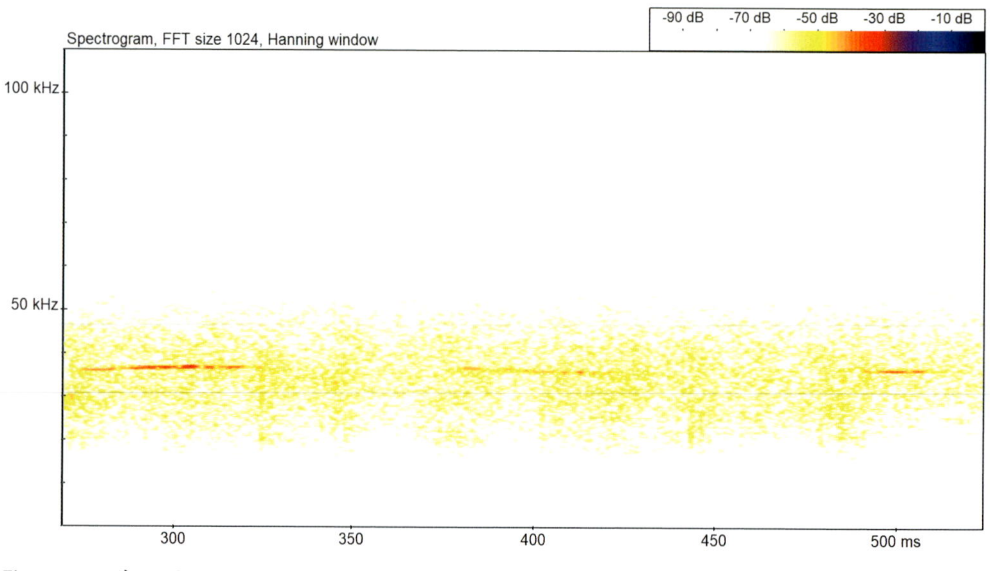

Figure 9.60 🔊 Bank vole – a series of three QCF calls (frame width: c.0.25 sec)

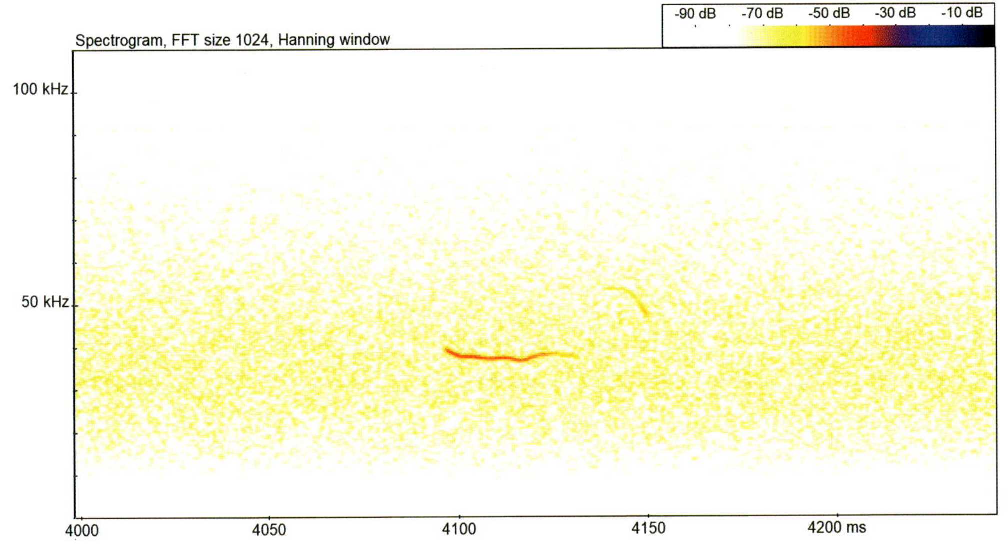

Figure 9.61 🔊 Bank vole – a QCF call (frame width: c.0.25 sec)

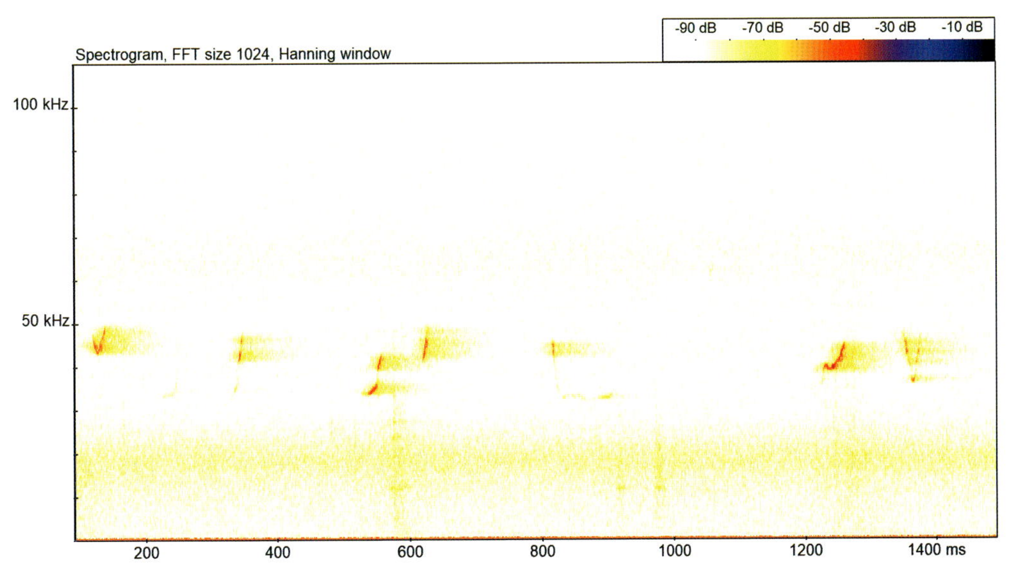

Figure 9.62 🔊 Bank vole – a series of variable calls (frame width: c.1.5 sec)

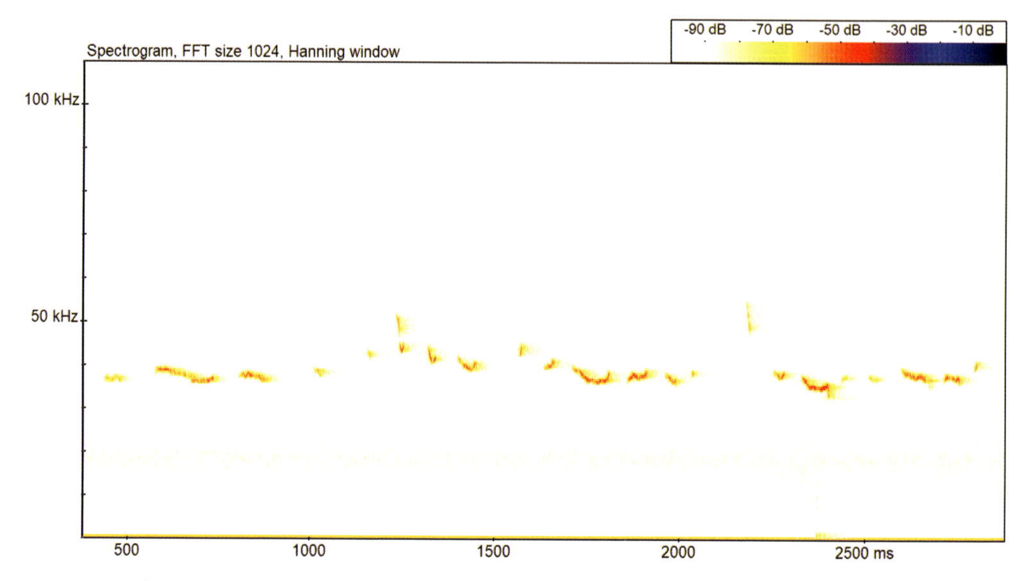

Figure 9.63 🔊 Bank vole – a series of variable calls (frame width: c.2.5 sec)

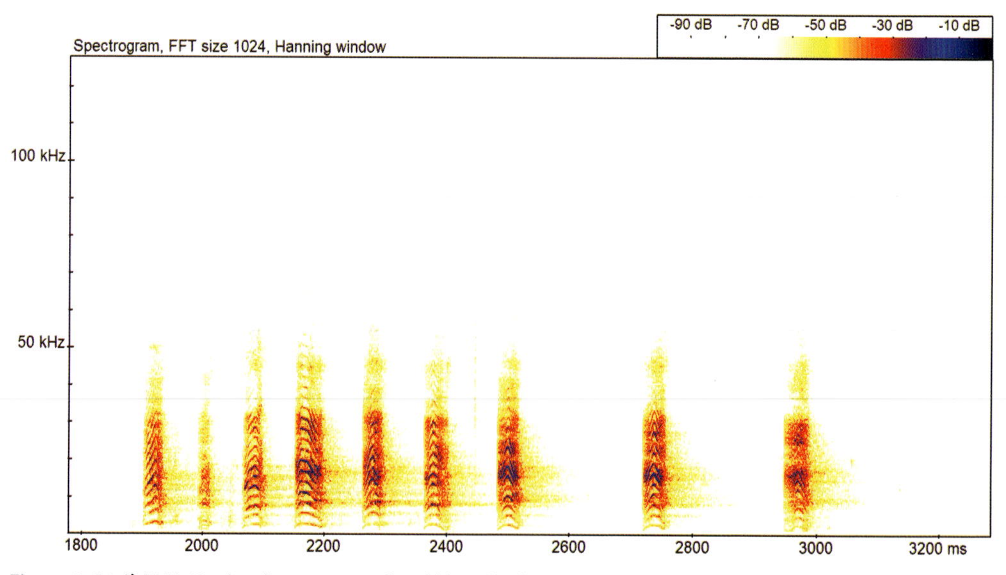

Figure 9.64 🔊 **(QR)** Bank vole – a series of audible calls (frequency scale: 0 to 130 kHz/frame width: c.1.5 sec)

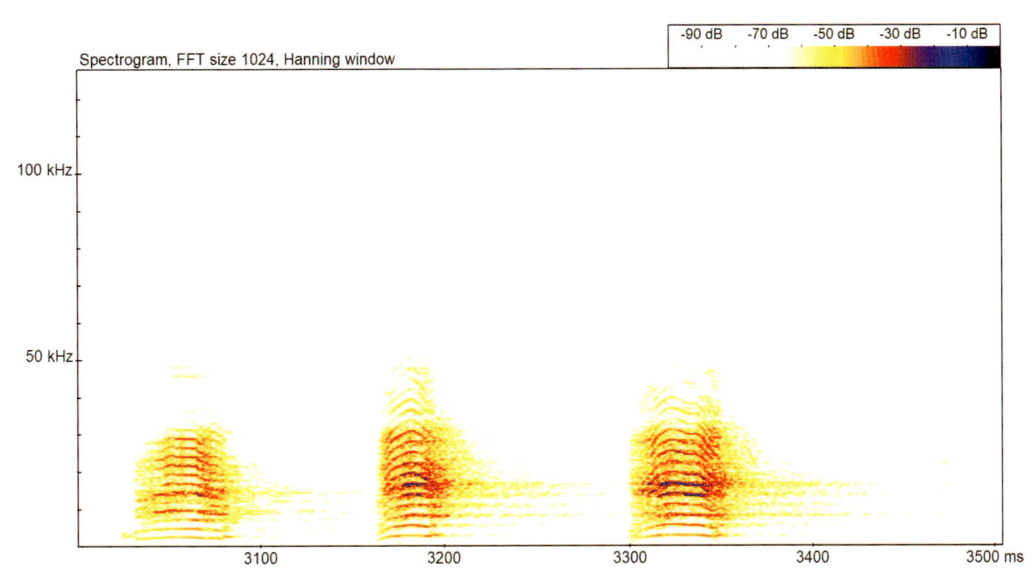

Figure 9.65 🔊 Bank vole – a series of audible calls (frequency scale: 0 to 130 kHz/frame width: c.0.5 sec)

QR codes relating to presented figures

Figure 9.58: Bank vole call (time expanded x10 for listening purposes)	**Figure 9.59:** Bank vole call (time expanded x10 for listening purposes)	**Figure 9.64:** Bank vole audible call

Potential confusion between bank vole and other species

Table 9.14 Summary of potential confusion between bank vole and other species

	Potential confusion species
Confusion group	In the first instance, always consider habitat and distribution.
Other vole species	Field vole, although frequency range may differ.
Species in other groups	Can be confused with juvenile rat calls, as well as lower frequency *Apodemus* calls.

9.5.3 Field vole *Microtus agrestis*

<table>
<tr><th>Length/Weight</th><th>Habitat preferences</th><th>Distribution</th></tr>
<tr><td>Males are larger than females.
Body length from 90 to 120 mm.
Weight from 20 to 40 g.</td><td rowspan="3">Unimproved grassland
Urban and gardens
Arable
Broadleaved woodland
Coniferous woodland
Hedgerows</td><td rowspan="3">Occurs throughout Britain, including many surrounding islands. Absent from Ireland.</td></tr>
<tr><th>Diet</th></tr>
<tr><td>Mainly vegetarian, feeding primarily upon leaves, seeds and fruits, and to a lesser extent fungi, roots, flowers, moss and invertebrates.</td></tr>
<tr><th>Daily activity pattern</th><td colspan="2">Can be active during day and night, but generally nocturnal during summer months and diurnal during winter.</td></tr>
<tr><th>Seasonal activity pattern</th><td colspan="2">Active throughout the year, but less so overnight during winter period.</td></tr>
<tr><th>Mating activity pattern</th><td colspan="2">Promiscuous mating usually takes place from March to October.
Young (typically four to five) are normally born during the period April to October, with seasonal peak in birth rate occurring during summer.</td></tr>
<tr><th>Other notes</th><td colspan="2">Easily captured using baited small mammal trail camera box.</td></tr>
<tr><td colspan="3">References
Lambin, 2008; Lysaght and Marnell, 2016; Mathews et al., 2018; Littlewood et al., 2021; Mammal Society, 2022.</td></tr>
</table>

Acoustic behaviour including spectrogram examples

Adults have been described as producing both ultrasonic and audible sounds (Stoddart and Sales, 1985; Sales, 2010), but are generally not considered as being particularly audible. Audible sounds have a longer duration (82 to 138 ms), occurring at a minimum frequency of 2.2 to 3.2 kHz.

A range of ultrasonic sounds have been documented (Middleton, 2020; Newson *et al.*, 2020). Compared to other small mammals however, voles, including this species, have a much lower call rate. It is plausible that social and vocal encounters occur more often within burrows, rather than above ground, and are therefore outside the range of microphones.

Ultrasonic calls can occur at a minimum frequency of 42 to 46 kHz, lasting approximately 61 to 73 ms. Higher-frequency calls have also been described during mating (Kapusta and Sales, 2009) and are considerably higher than those recorded from bank voles. Also, field vole is generally more vocal than bank vole (Kapusta and Sales, 2009).

Figure 9.66 shows a quasi-constant-frequency (QCF) call occurring at c.20 kHz, whereas Figures 9.67 to 9.69 show examples which vary slightly more in frequency, but at higher levels (c.35 kHz and 50 kHz respectively).

The example shown in Figure 9.70 differs in that there is a slight hook at the start as frequency increases, before dropping down through a bandwidth of c.10 kHz. Figures 9.71 and 9.72 show a more complex series of emissions. Finally, Figure 9.73 provides an example of an audible sound, which when listened to in time expansion (x 10) is quite distinctive.

The examples provided show a range of call structures, none of which we can say typically occur more often than others, as our experience in recording this species tells us that obtaining any calls involves lengthy periods of recording with little to show for the effort. When calls are successfully encountered, they tend to be brief engagements, compared to other small mammals.

Figures 9.66 to 9.73 provide specific examples of spectrograms relating to this species, and where the 🔊 symbol appears, a recording of what is shown within the spectrogram is available to download from the species library. In addition, QR codes are provided for selected examples, and these can be accessed immediately for listening purposes from most mobile devices.

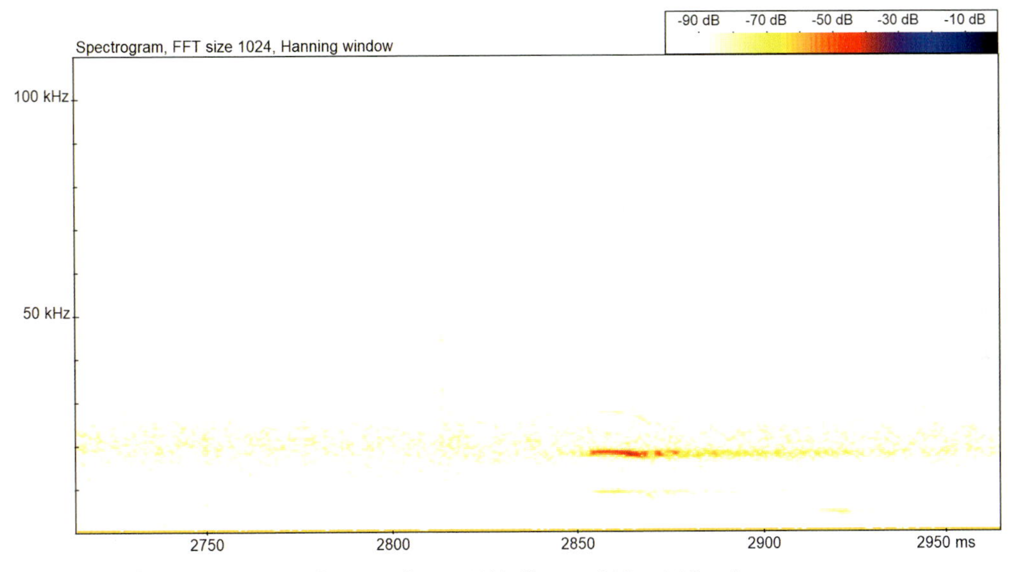

Figure 9.66 🔊 Field vole – a single QCF call at c.20 kHz (frame width: c.0.25 sec)

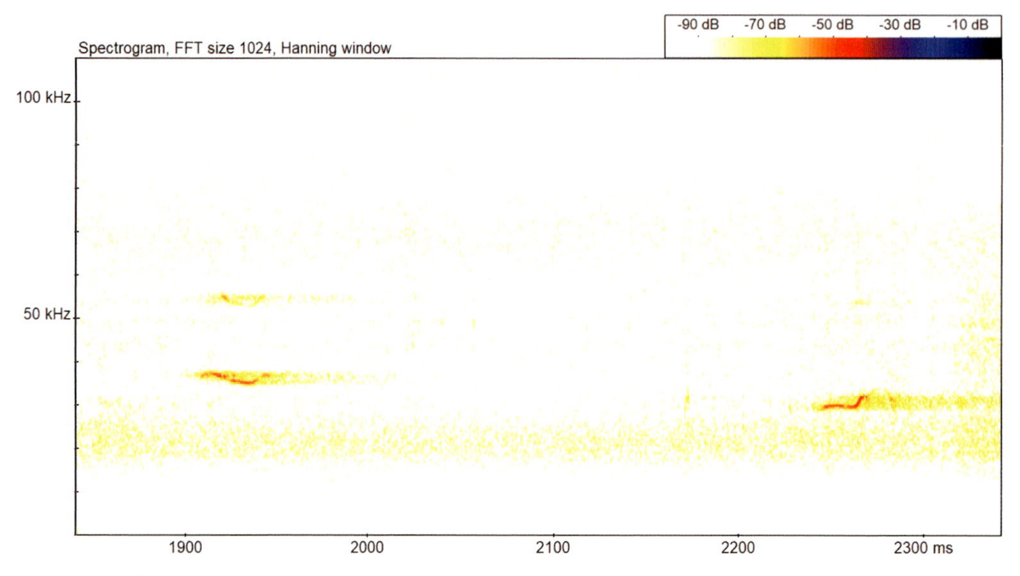

Figure 9.67 🔊 **(QR)** Field vole – two calls fluctuating slightly in frequency at c.35 kHz (frame width: c.0.5 sec)

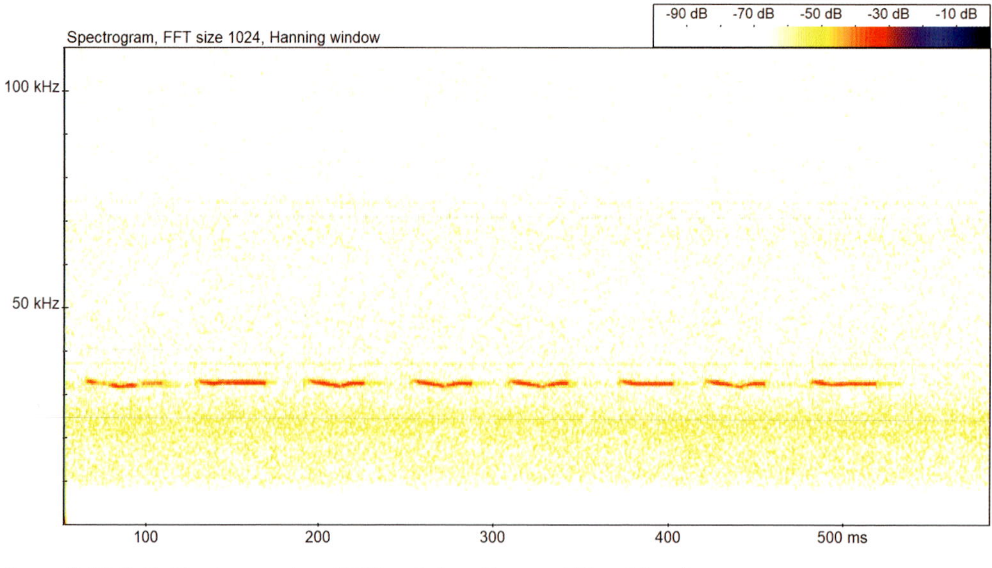

Figure 9.68 🔊 Field vole – a series of QCF emissions (frame width: c.0.5 sec)

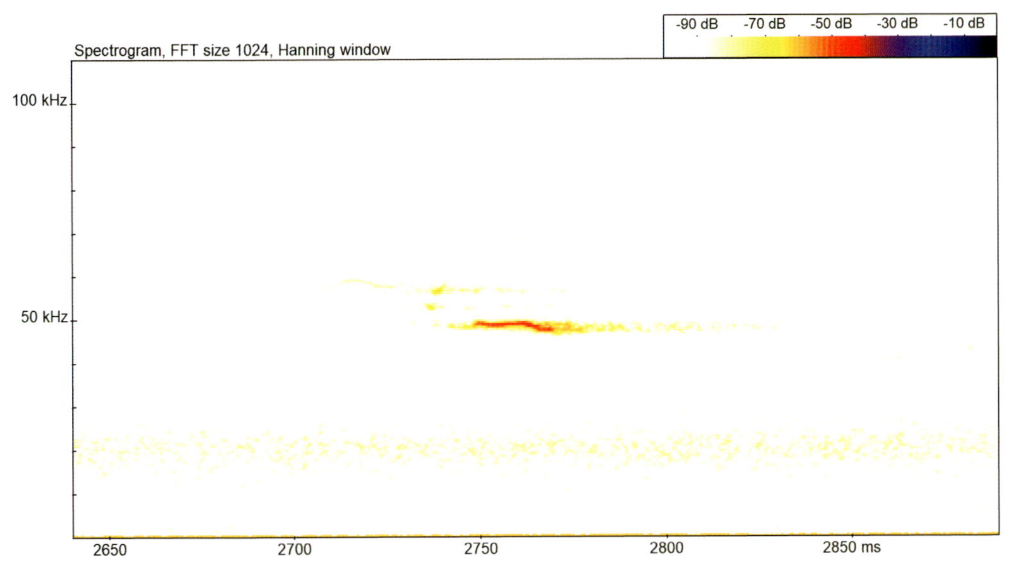

Figure 9.69 🔊 Field vole – a single call at c.50 kHz (frame width: c.0.25 sec)

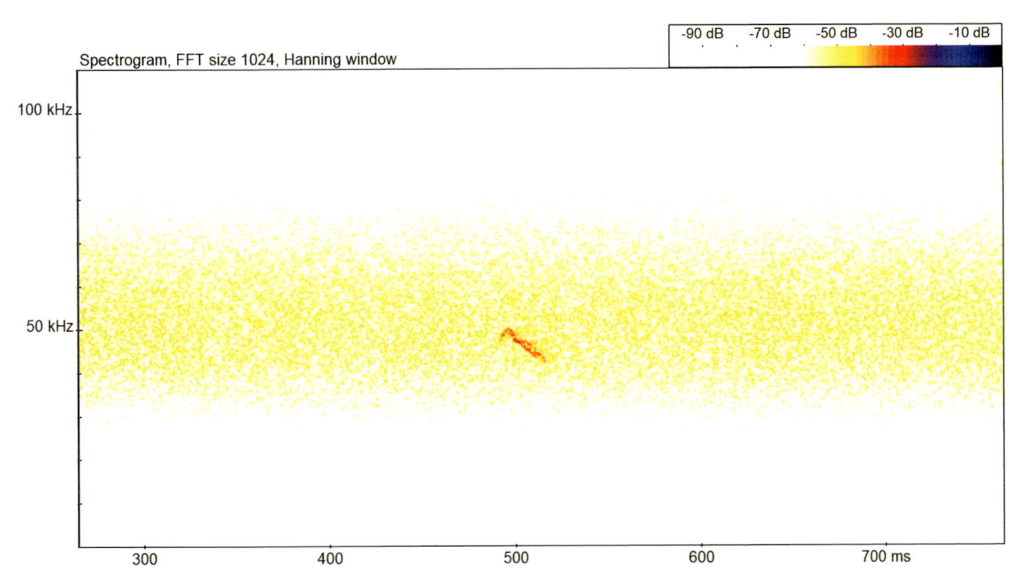

Figure 9.70 🔊 **(QR)** Field vole – a hooked call, FM ascending, then descending, with a bandwidth of c.10 kHz (frame width: c.0.5 sec)

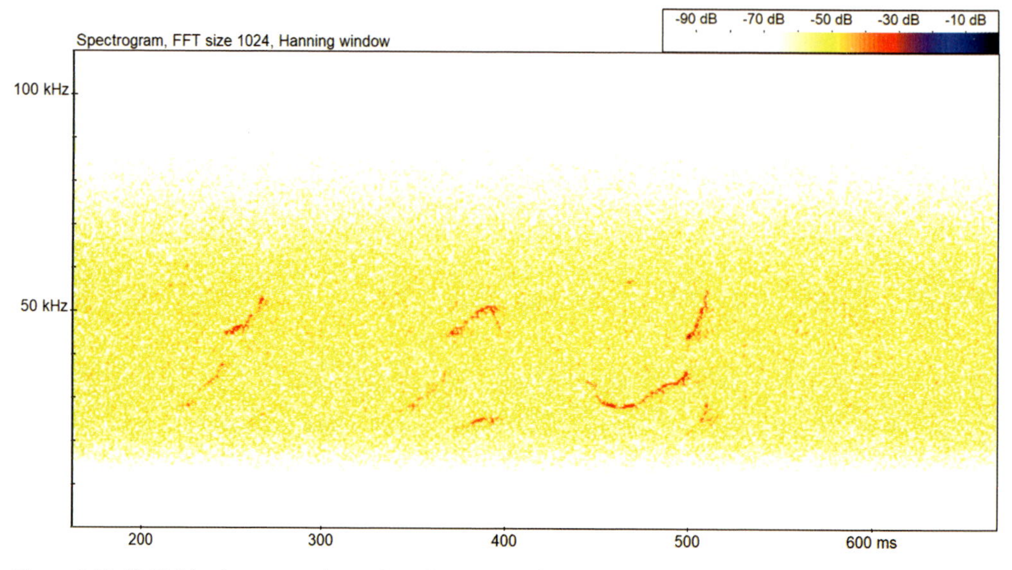

Figure 9.71 🔊 Field vole – a complex series of emissions (frame width: c.0.5 sec)

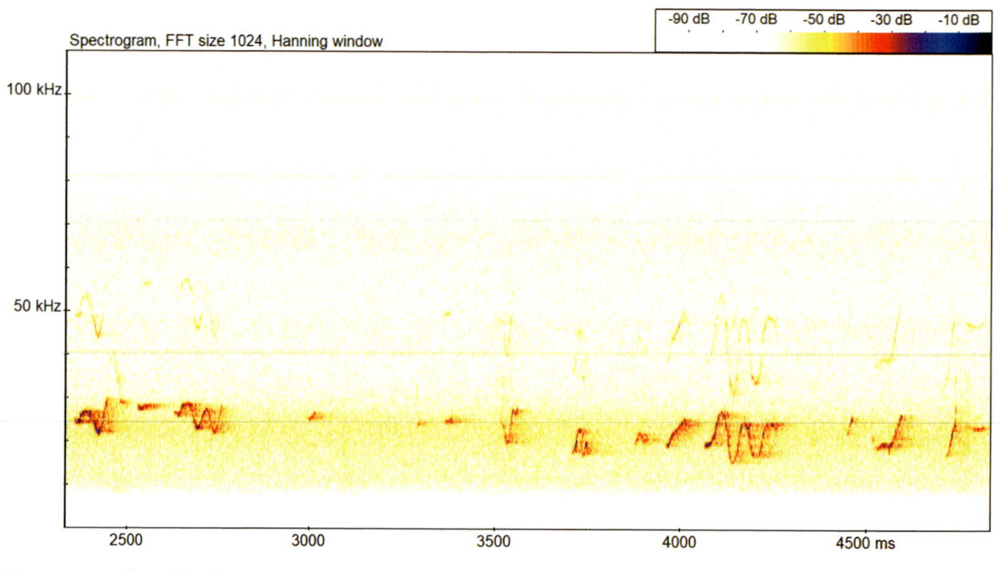

Figure 9.72 🔊 Field vole – a complex series of emissions (frame width: c.2.5 sec)

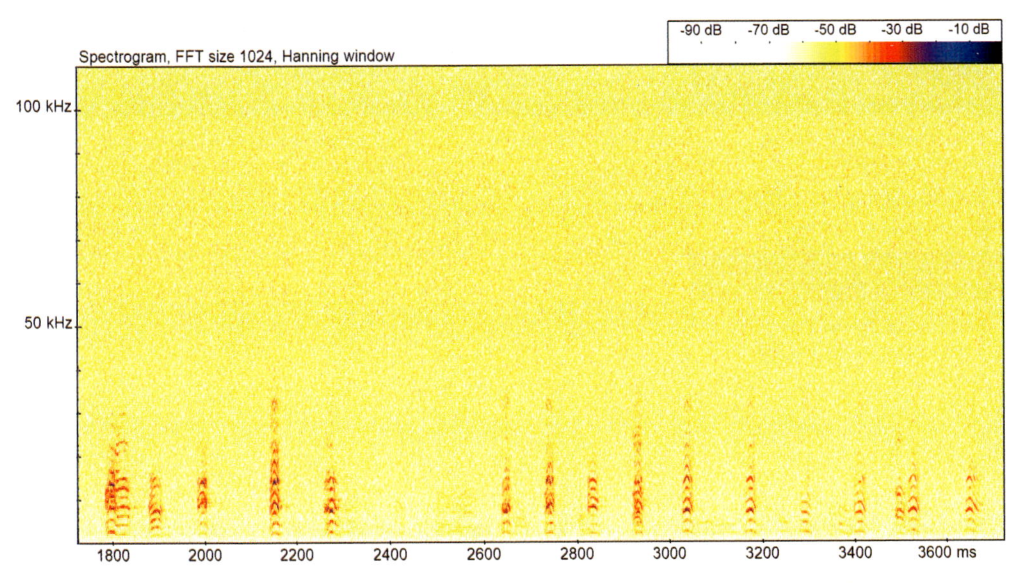

Figure 9.73 🔊 **(QR)** Field vole – a series of audible calls (frame width: c.0.5 sec)

QR codes relating to presented figures

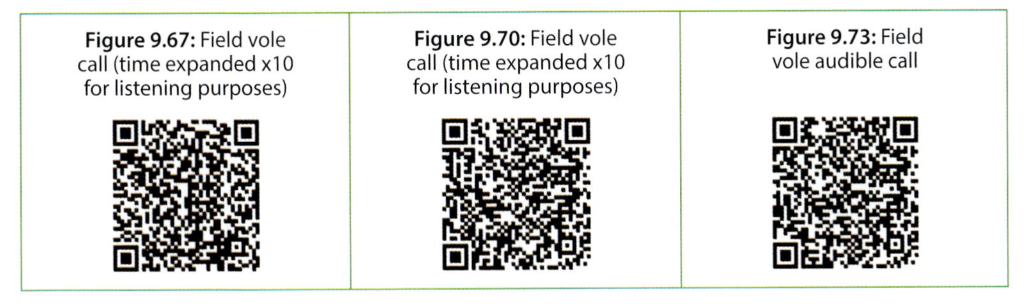

Figure 9.67: Field vole call (time expanded x10 for listening purposes)	**Figure 9.70:** Field vole call (time expanded x10 for listening purposes)	**Figure 9.73:** Field vole audible call

Potential confusion between field vole and other species

Table 9.15 Summary of potential confusion between field vole and other species

	Potential confusion species
Confusion group	In the first instance, always consider habitat and distribution.
Other vole species	Bank vole, although frequency range may differ.
Species in other groups	Unlikely to be confused with other small mammals.

9.5.4 Orkney vole *Microtus arvalis orcadensis*/Guernsey vole *Microtus arvalis sarnius*

Length/Weight	Habitat preferences	Distribution
Adult males can be slightly larger than females. Body length from 97 to 134 mm. Weight from 22 to 67 g.	Coniferous woodland Deciduous woodland Wetlands Moorland Rough grassland Arable farmland	Orkney vole occurs in Orkney, and Guernsey vole occurs on Guernsey (Channel Islands).
Diet		
Herbivorous, feeding primarily upon leaves, stems, heather and roots.		

Daily activity pattern	Can be active during day and night, but generally more diurnal during winter.
Seasonal activity pattern	Active throughout the year.
Mating activity pattern	Mating usually takes place from February to October. Young (one to six) are normally born during the period March to November.
Other notes	Due to small size, difficult to encounter visually, but may be recorded using appropriately positioned camera traps.

References
Gorman and Reynolds, 2008; Lysaght and Marnell, 2016; Mathews *et al.*, 2018; Littlewood *et al.*, 2021; Mammal Society, 2022.

Acoustic behaviour including spectrogram examples

These two discretely localised vole subspecies are considered historically to have evolved from the common vole (*Microtus arvalis*) which occurs in continental Europe, and in each case (Orkney and Guernsey) they are the only vole species present within their locality.

Information on the most distinctive sounds produced by these species is lacking, but in the absence of other vole species on the islands where they occur, any vocalisations are useful when attempting to assign sounds recorded in the field to species.

A range of ultrasonic sounds have been encountered by ourselves but, as for the other vole species, we found that the call rate was much lower than the other small mammal species in this group.

The examples provided (Figures 9.74 to 9.79) show a range of call structures, none of which we can say typically occur more often than others. Nevertheless, many of the examples have similarities with the call structure of other vole species, in that fairly constant frequency emissions occur. In Figure 9.76 we have a stepped rising FM structure similar to that which occurs in water vole, but with the frequency being c.10 kHz higher.

Figures 9.74 to 9.79 provide specific examples of spectrograms relating to this species, and where the 🔊 symbol appears, a recording of what is shown within the spectrogram is available to download from the species library. In addition, QR codes are provided for selected examples, and these can be accessed immediately for listening purposes from most mobile devices.

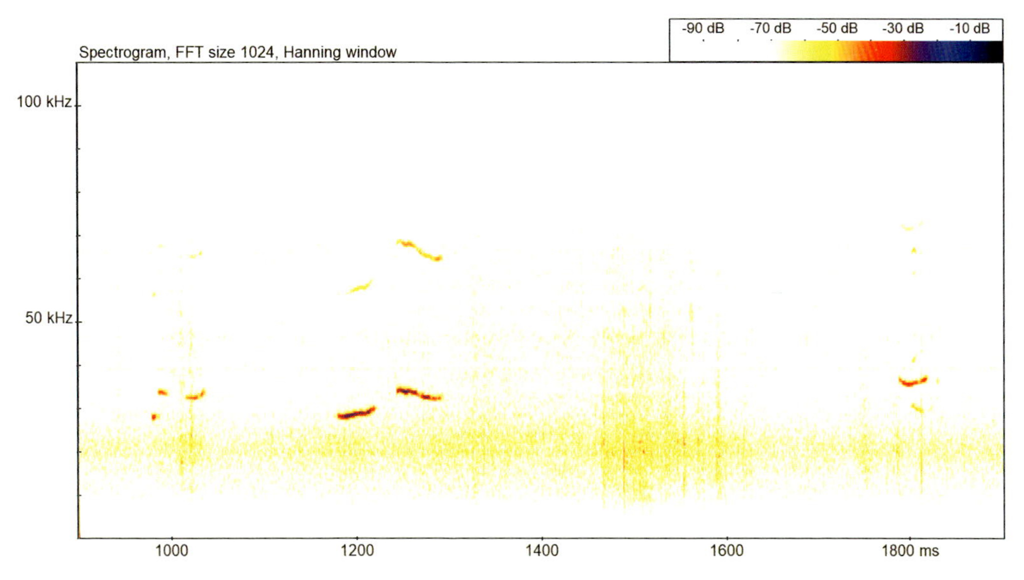

Figure 9.74 🔊 (QR) Orkney vole – a series of variable call structures (courtesy of A. Hough) (frame width: c.1 sec)

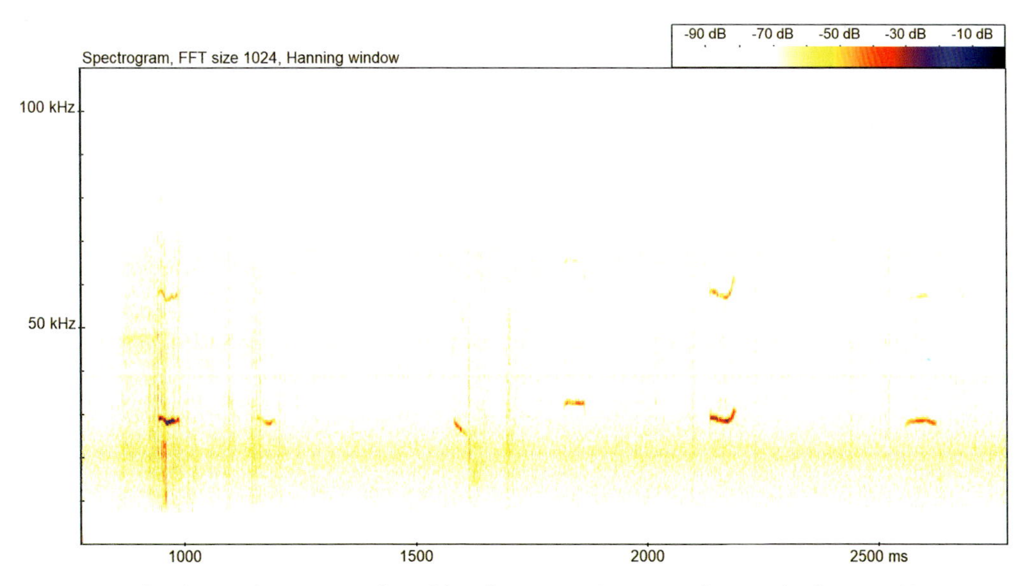

Figure 9.75 🔊 Orkney vole – a series of variable call structures (courtesy of A. Hough) (frame width: c.2 sec)

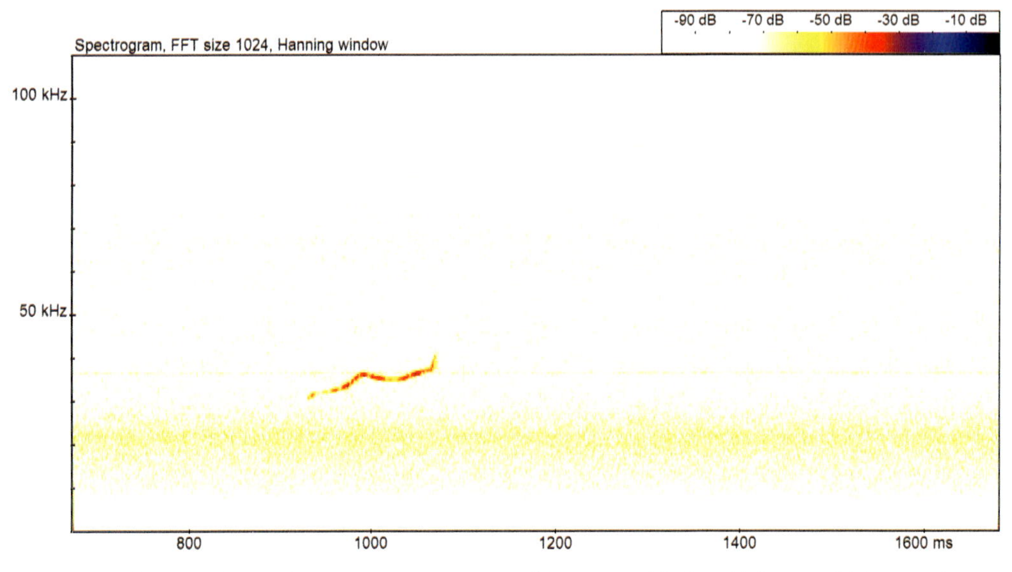

Figure 9.76 🔊 (QR) Orkney vole – a rising FM stepped call (courtesy of A. Hough) (frame width: c.1 sec)

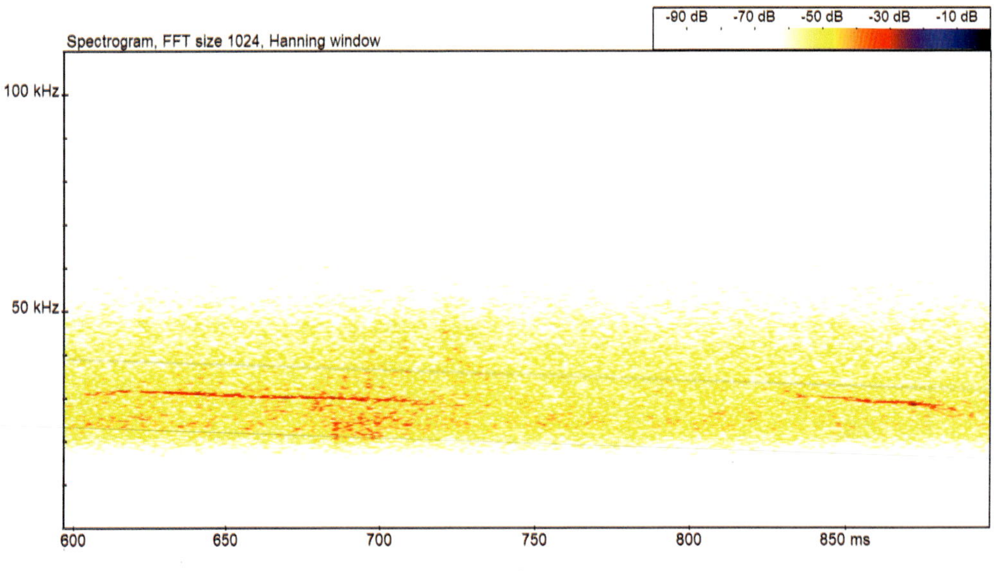

Figure 9.77 🔊 Guernsey vole – two QCF calls (courtesy of N. Jee) (frame width: c.0.3 sec)

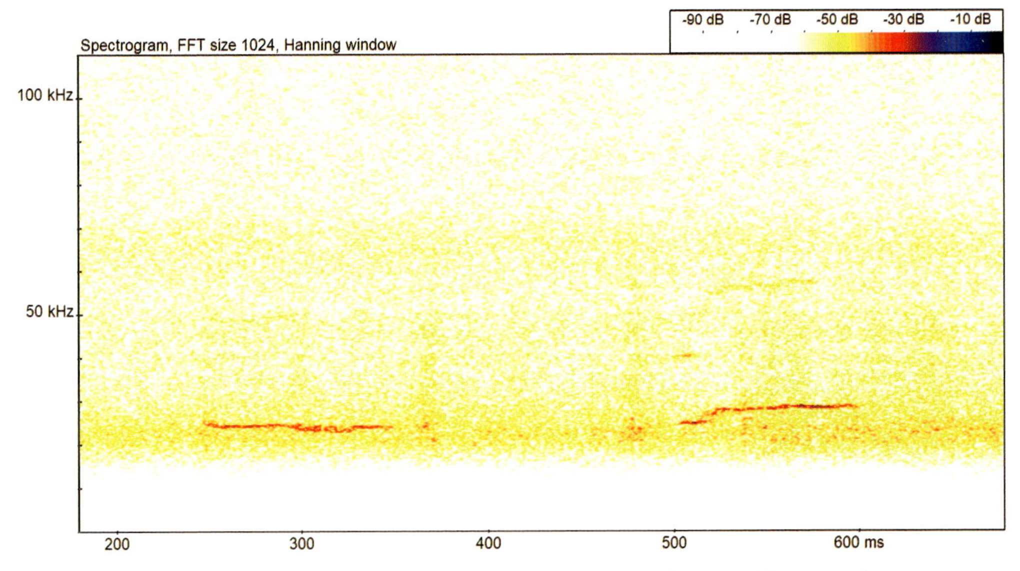

Figure 9.78 🔊 **(QR)** Guernsey vole – two QCF calls (courtesy of N. Jee) (frame width: c.0.5 sec)

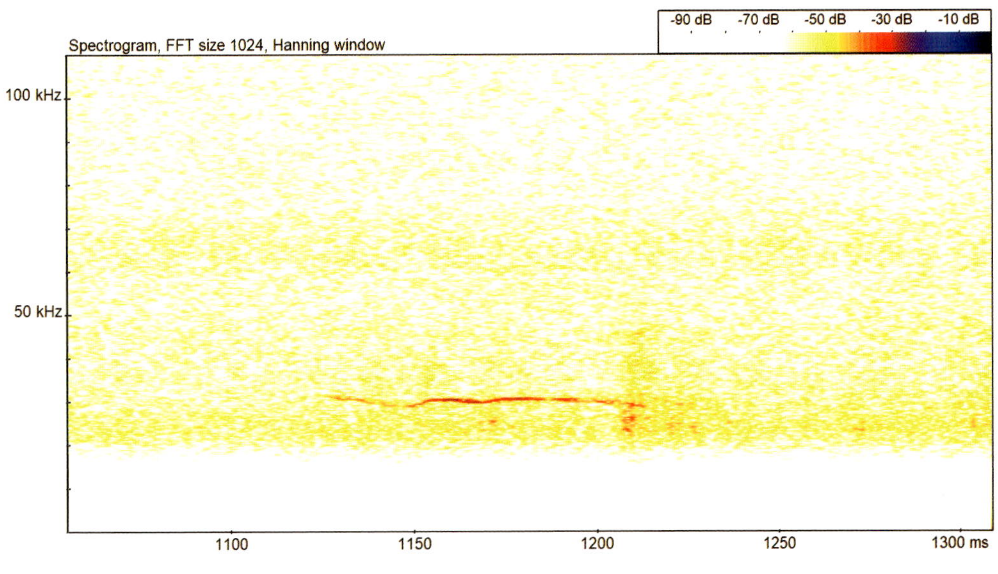

Figure 9.79 🔊 Guernsey vole – a single QCF call (courtesy of N. Jee) (frame width: c.0.25 sec)

QR codes relating to presented figures

Figure 9.74: Orkney vole call (time expanded x10 for listening purposes)

Figure 9.76: Orkney vole call (time expanded x10 for listening purposes)

Figure 9.78: Guernsey vole call (time expanded x10 for listening purposes)

Potential confusion between Orkney/Guernsey vole and other species

Table 9.16 Summary of potential confusion between Orkney/Guernsey vole and other species

Confusion group	Potential confusion species
	In the first instance, always consider habitat and distribution.
Other vole species	Bank vole and field vole. Water vole stepped calls.
Species in other groups	Unlikely to be confused with other small mammals.

9.6 Dormouse vocalisations

In the following section, a brief summary of the key characteristics of the two species of dormouse will be described, along with details on the species-specific vocalisations and, where known, the behavioural function of these sounds.

Both of our species of dormouse are normally nocturnal and arboreal when active. During colder months, they are inactive and enter into a long period of hibernation, normally from October through to May. When resources are limited, for example during non-mast years, edible dormouse has been found to remain in torpor for an extended period beyond the typical hibernation season and delay breeding until the following year (Roger Trout, 2023, pers. comm.).

9.6.1 Hazel dormouse *Muscardinus avellanarius*

Length/Weight	Habitat preferences	Distribution
Body length (excluding tail) of 50 to 90 mm. Weight from 15 to 25 g, heavier in weight immediately prior to hibernation.	Broadleaved woodland Coniferous woodland Hedgerows	Patchily distributed across southern England and Wales. Isolated populations in northern England (Cumbria). Recently found in Ireland (County Kildare) and regarded as non-native (and potentially invasive) there.
Diet		
Mainly herbivorous, feeding across a range of food sources, varying according to availability, including flowers, pollen, fruits, berries and nuts. Insects may also be consumed.		

Daily activity pattern	Nocturnal.
Seasonal activity pattern	Active throughout warmer months. Typically hibernates from October through to May.
Mating activity pattern	When breeding, males are strongly territorial. Polygynous mating mostly takes place during June to August, with births occurring as late as September. Young (usually two to six) are born during the period June to September.
Other notes	Nocturnal and arboreal, they are difficult to encounter visually, although appropriately positioned trail cameras may prove successful. Feeding signs can be used for surveying purposes, whereby hazelnut shells are opened distinctively leaving smooth round holes. Monitoring can be carried out using nest boxes/nest tubes and footprint tunnels.

References
Bright *et al.*, 2006; Bright and Morris, 2008; Glass *et al.*, 2015; Lysaght and Marnell, 2016; Mathews *et al.*, 2018; Bullion and Looser, 2019; Melcore *et al.*, 2020; Mammal Society, 2022.

Acoustic behaviour including spectrogram examples

Whilst hazel dormice have been known to make audible calls during aggressive/chase encounters, this species is not regarded as being particularly vocal in the audible range, and is certainly much quieter than the larger edible dormouse (Bright and Morris, 2008).

As well as some audible sounds, ultrasonic sounds (Figures 9.80 to 9.93) have been recorded, occurring in the frequency range of 25 to 100 kHz including harmonics. This includes calls produced by young in the nest, which can emit ultrasonic sounds within the range of 20 to 30 kHz (Juškaitis and Büchner, 2013).

A study carried out on a small group (n = 5 adults + 3 pups) of captive individuals (Ancillotto *et al.*, 2014) described six types of vocalisations, five of which typically occurred within the frequency range of 18 to 48 kHz. The sixth call type related to mother/infant 'clucking' communication, occurring within the audible range of 6 to 8 kHz.

In Table 9.17 we provide schematic examples of the five adult-generated ultrasonic calls described by Ancillotto *et al.* (2014). It should be borne in mind that the call parameters and examples shown are approximate, and are also taken from a study of a small group of animals, over a short period of time (10 days). It should be anticipated that further variation to what is shown here could occur, and the authors of the study themselves say that *'further analysis relying on a larger sample size would be beneficial in order to create a more complete assessment'*.

Ultrasonic sound examples for this species were also looked at in a published overview of vocalisations for the purpose of identification within Britain, published in *British Wildlife* in 2020 (Newson *et al.*, 2020).

In the examples described here, we start off with typical ascending FM calls, commencing and ending at different frequencies (Figures 9.80 to 9.83). These examples broadly fit into

Table 9.17 Typical call structure of hazel dormouse (adapted from Ancillotto *et al.*, 2014).

Call type and context	Description of structure	Schematic representation
'A', produced by solitary males and females	Described as an 'arising stepped call', or shallow FM stepped rising, commencing at c.18 kHz and rising to c.34 kHz	18 kHz < c.600 ms >
'B', produced by solitary males and females	Described as an 'arising, smooth call', or shallow FM smooth, commencing at c.19 kHz and rising to c.25 kHz	20 kHz < c.600 ms >
'C', produced by male with female present	'Arising sloped call', giving a slightly humped impression, commencing at c.35 kHz and rising to c.43 kHz	35 kHz < c.400 ms >
'D', courtship song, produced by male with female present	'Down-sweep' call, comprising up to 62 individual notes, split over two phases. Phase 1: numerous steep FM sweep calls of short duration, commencing at c.40 kHz, dropping steeply to c.18 kHz, followed by Phase 2: more complex variety of call structures, occurring within a similar frequency range to Phase 1	40 kHz 20 kHz Phase 1 → Phase 2 < up to c.3,200 ms >
'E', produced by solitary males	'Flat wiggled' call, occurring within the frequency range c.44 kHz to c.53 kHz	50 kHz < 500 ms >

what Ancillotto *et al.* (2014) described as Type A, B and C calls. Figures 9.84 to 9.86 show slightly more complex arrangements whereby both the duration and frequency of the calls differ across the call sequence.

An important consideration when analysing sounds such as those shown in Figures 9.80 to 9.86 would be to always keep in mind that brown rat calls have similar structure and can occur at similar frequencies to hazel dormouse. Nevertheless, the constant-frequency calls produced by brown rat, which do not noticeably alter in frequency during their entirety, do not appear to occur in hazel dormouse. As such, if you are seeing this 'very constant frequency', it is more likely to relate to brown rat. Furthermore, when listening (time expansion x10) to these low-frequency calls for either species, there is often a clear distinction in the sound. The brown rat call is very clean and almost electrical in sound, whereas the hazel dormouse call is more of a vibrato approach, as the sound quickly fluctuates, and to our ears sounds similar to a whistling kettle as it achieves boiling point. Finally, when looking at brown rat calls you will quite often see that at the start of the call the frequency drops, before becoming more constant (i.e. FM descending). In hazel dormouse it is usually the case that the frequency increases throughout the duration of the call (i.e. FM ascending).

Another regular feature we have encountered for hazel dormouse relates to the examples shown in Figures 9.87 to 9.89. Here you see totally different types of call structure occurring, in that staple-shaped calls are evident, as well as ascending/descending arched calls.

Figures 9.90 and 9.91 show different structures again, broadly fitting in to Type D calls (Ancillotto *et al.*, 2014). The final examples (Figures 9.92 and 9.93) relate to calls from sub-adult hazel dormouse whilst in captivity.

Figures 9.80 to 9.93 provide specific examples of spectrograms relating to this species, and where the 🔊 symbol appears, a recording of what is shown within the spectrogram is available to download from the species library. In addition, QR codes are provided for selected examples, and these can be accessed immediately for listening purposes from most mobile devices.

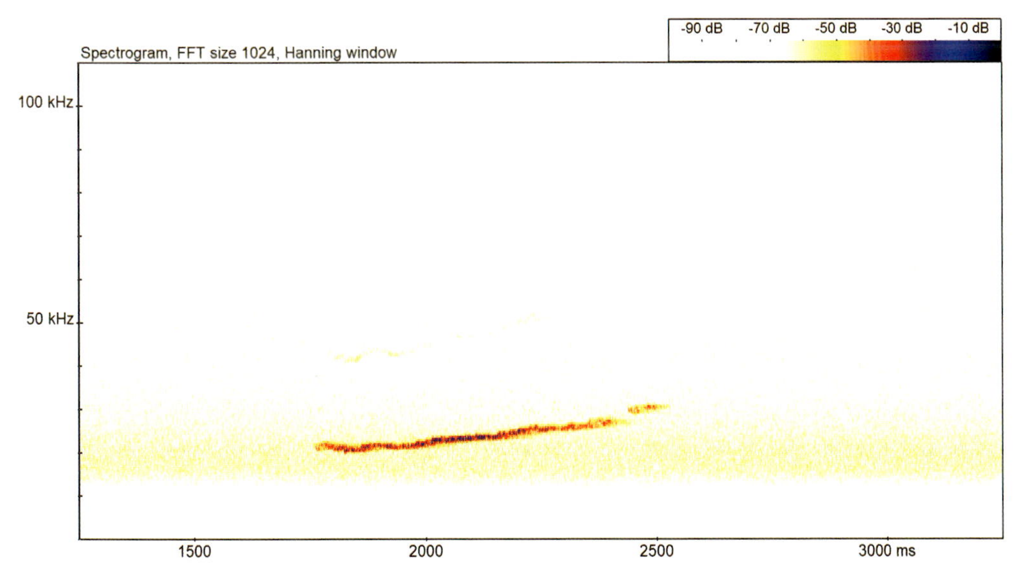

Figure 9.80 🔊 (QR) Hazel dormouse – ascending FM call (courtesy of L. Griffiths, Nottinghamshire Dormouse Group) (frame width: c.2 sec)

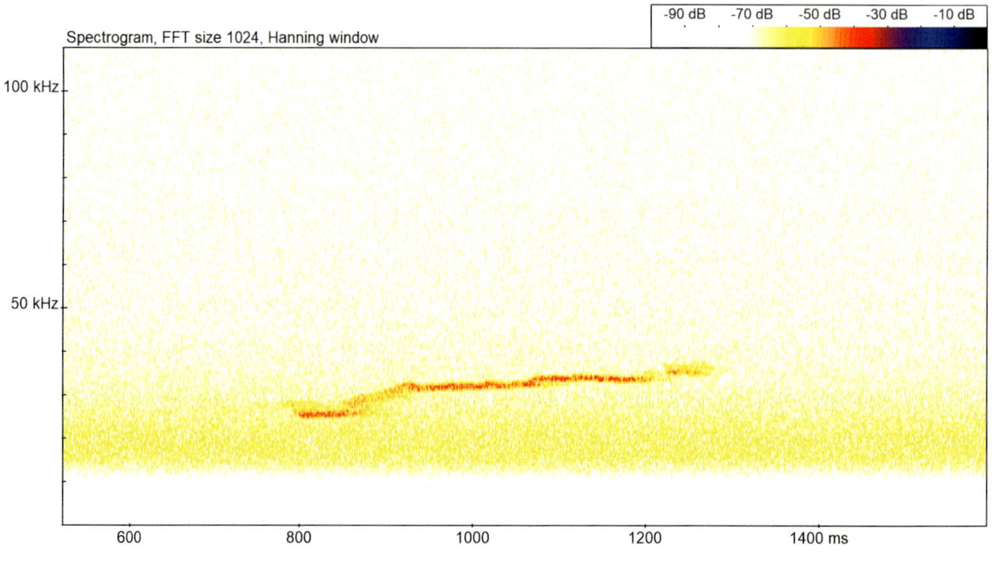

Figure 9.81 🔊 Hazel dormouse – ascending stepped FM call (courtesy of L. Griffiths, Nottinghamshire Dormouse Group) (frame width: c.1 sec)

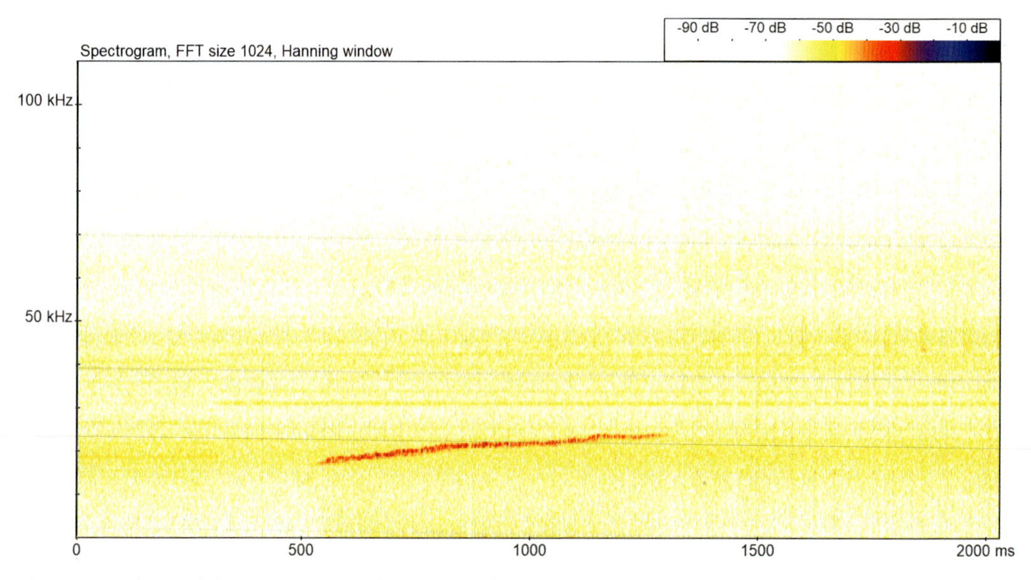

Figure 9.82 🔊 Hazel dormouse – ascending FM call (lower frequency) (courtesy of J.-F. Godeau) (frame width: c.2 sec)

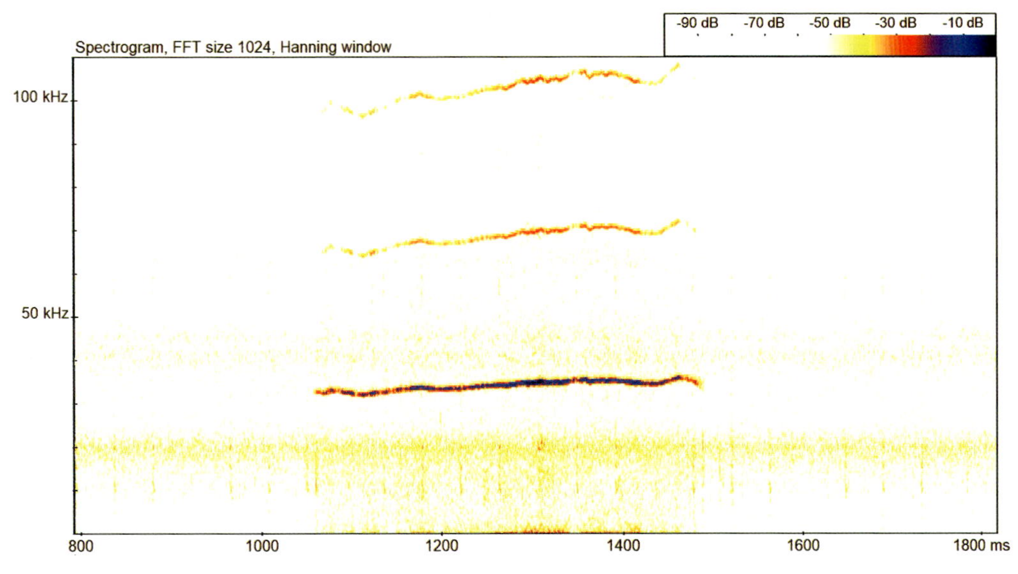

Figure 9.83 🔊 Hazel dormouse – ascending FM call (higher frequency) (courtesy of J.-F. Godeau) (frame width: c.1 sec)

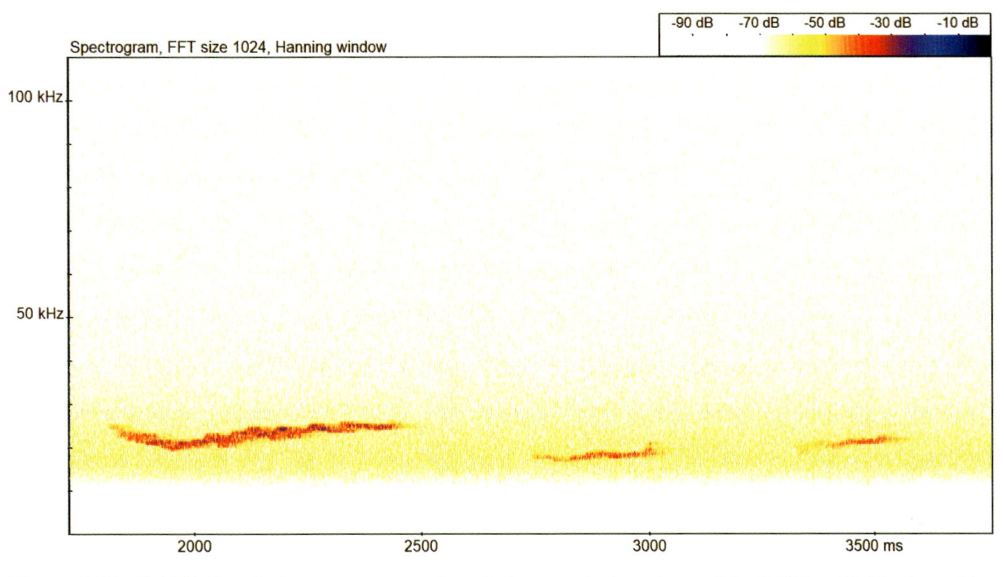

Figure 9.84 🔊 **(QR)** Hazel dormouse – a series of three calls (courtesy of L. Griffiths, Nottinghamshire Dormouse Group) (frame width: c.2 sec)

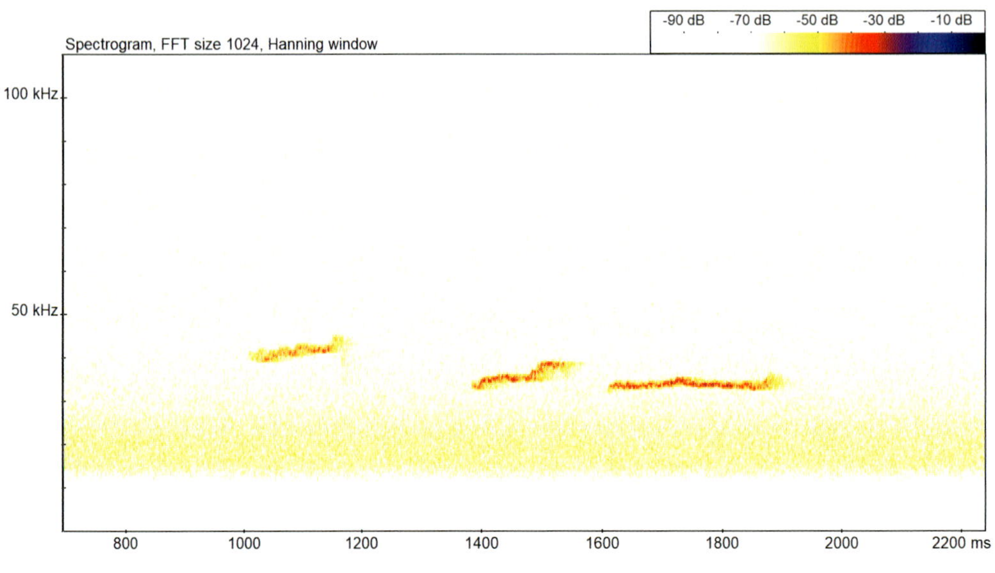

Figure 9.85 Hazel dormouse – a series of three calls (courtesy of L. Griffiths, Nottinghamshire Dormouse Group) (frame width: c.1.5 sec)

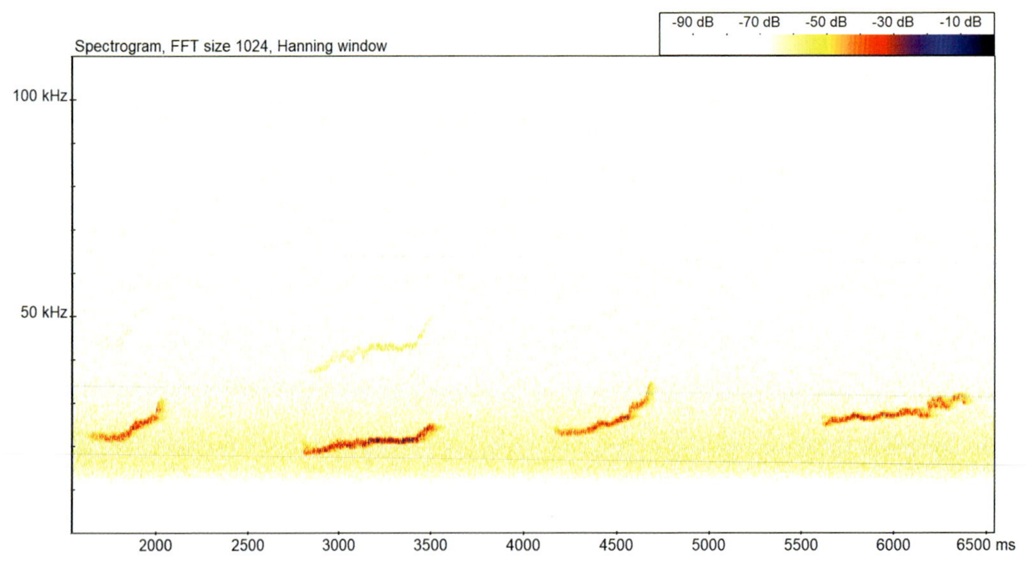

Figure 9.86 Hazel dormouse – a series of four calls (courtesy of L. Griffiths, Nottinghamshire Dormouse Group) (frame width: c.5 sec)

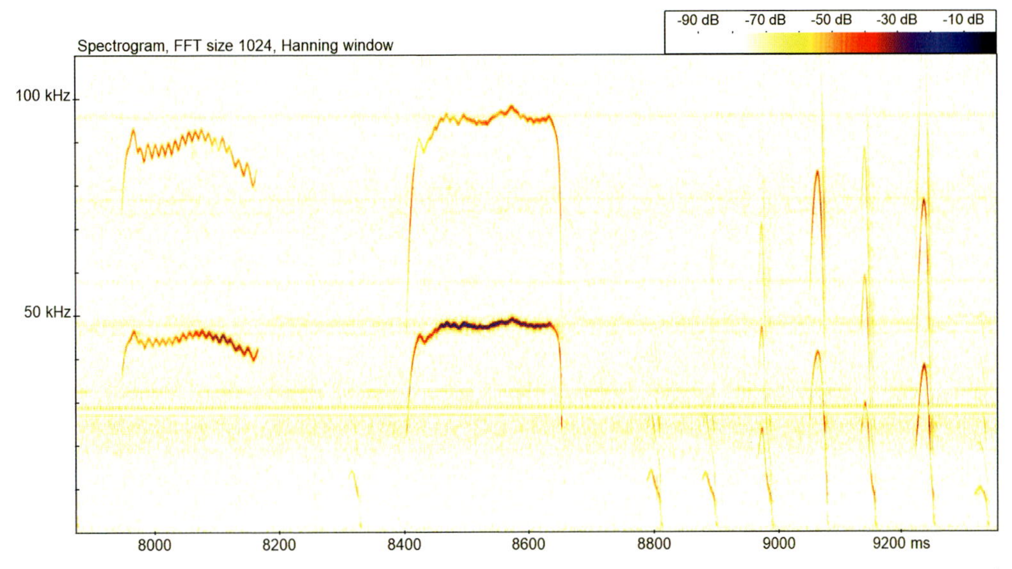

Figure 9.87 🔊 **(QR)** Hazel dormouse (sub-adult) – a series of calls showing variation in structure (courtesy of K. Bettoney, Mid Devon Bat Rescue Centre) (frame width: c.1.5 sec)

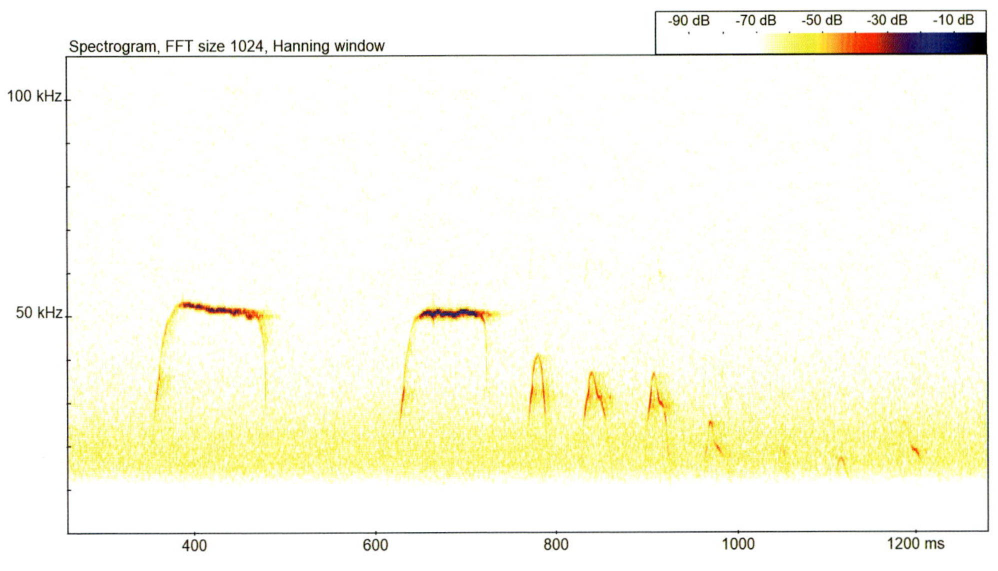

Figure 9.88 🔊 Hazel dormouse – a series of calls showing variation in structure (courtesy of L. Griffiths, Nottinghamshire Dormouse Group) (frame width: c.1 sec)

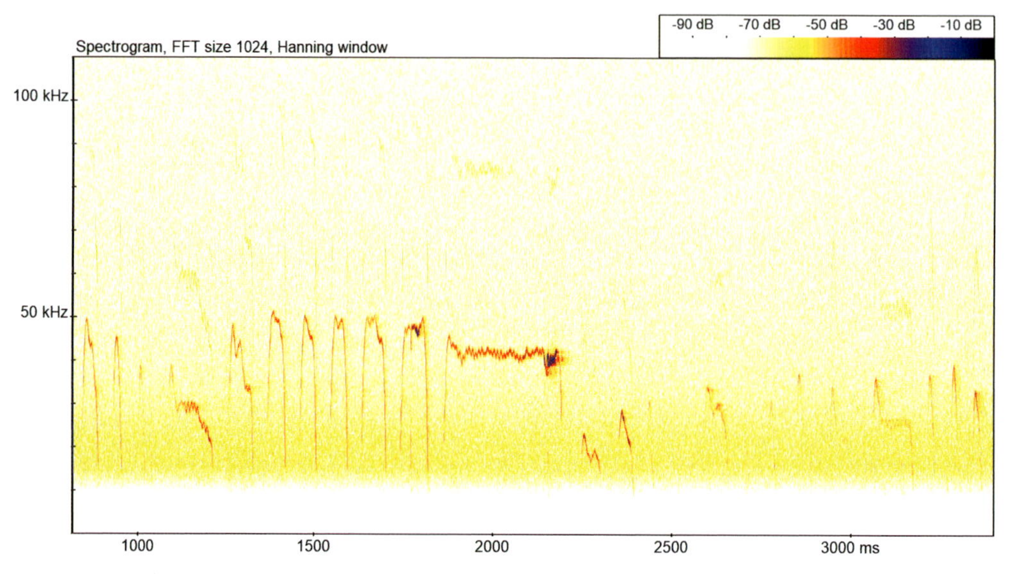

Figure 9.89 Hazel dormouse – a series of calls showing variation in structure (courtesy of L. Griffiths, Nottinghamshire Dormouse Group) (frame width: c.2.5 sec)

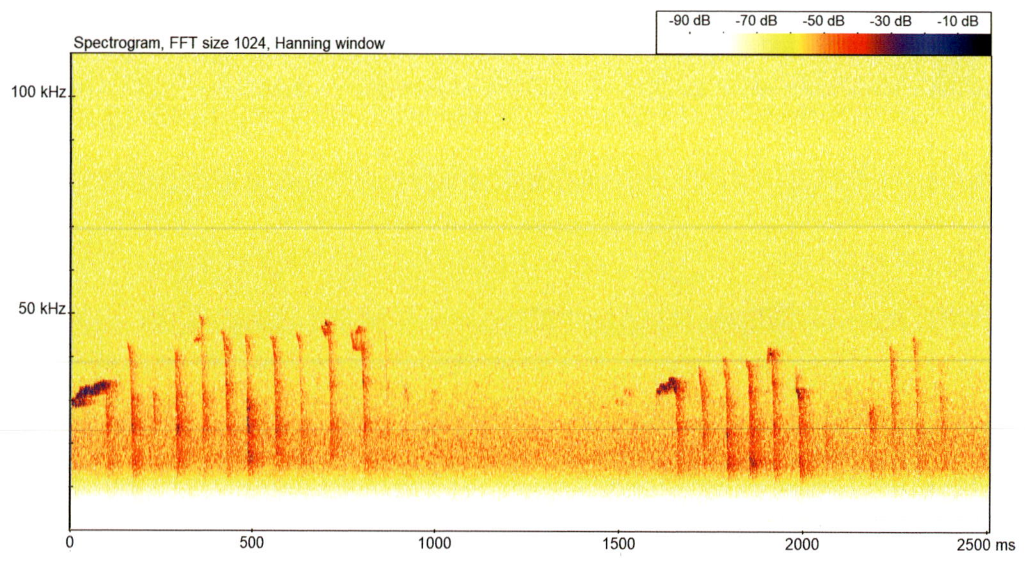

Figure 9.90 Hazel dormouse – a series of steep FM 'Type D' calls (courtesy of L. Griffiths, Nottinghamshire Dormouse Group) (frame width: c.2.5 sec)

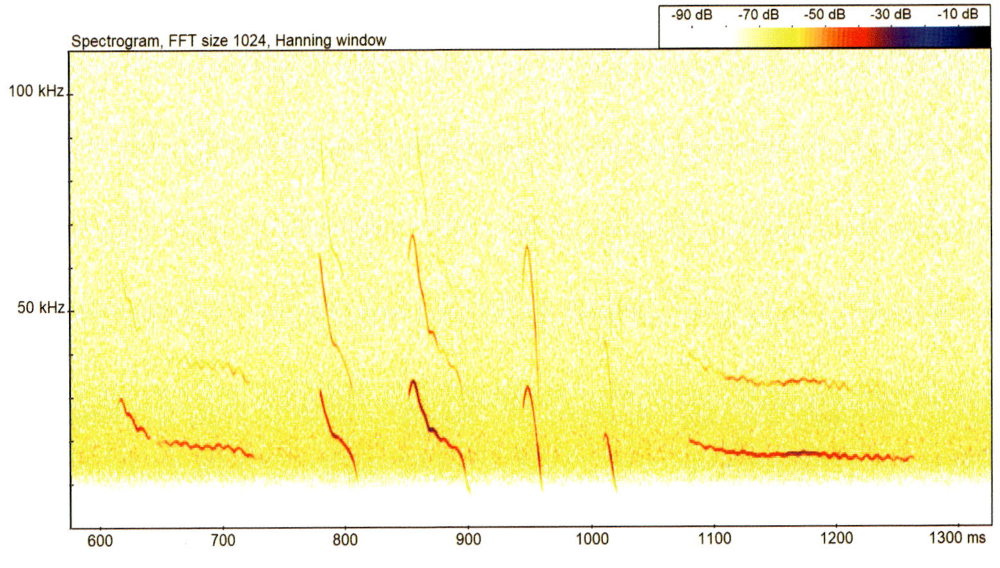

Figure 9.91 Hazel dormouse – a series of calls including steep FM hooked structures (courtesy of L. Griffiths, Nottinghamshire Dormouse Group) (frame width: c.0.75 sec)

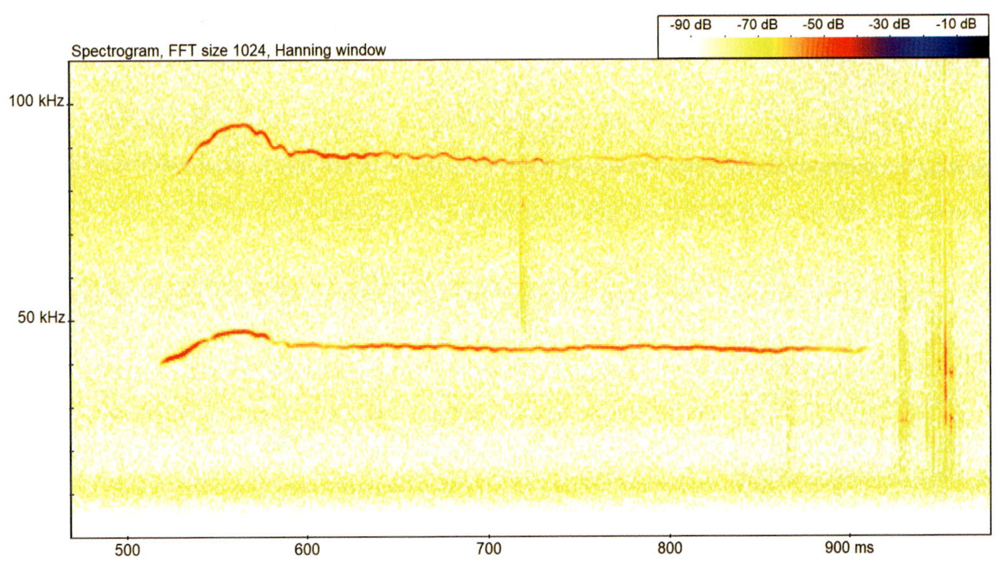

Figure 9.92 Hazel dormouse (sub-adult) – ascending FM, descending FM, QCF call (courtesy of K. Bettoney, Mid Devon Bat Rescue Centre) (frame width: c.0.5 sec)

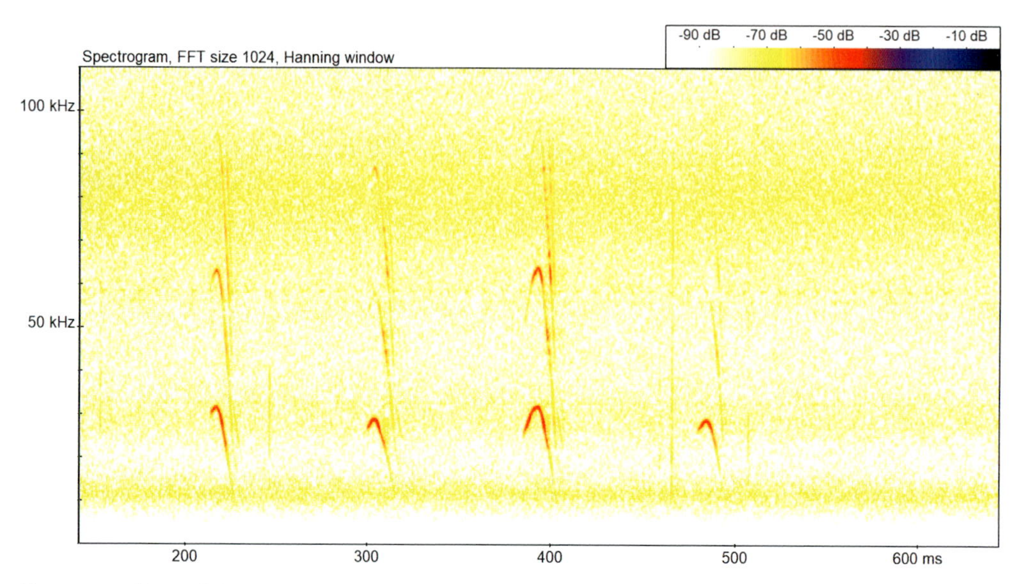

Figure 9.93 🔊 Hazel dormouse (sub-adult) – ascending FM, descending FM call (courtesy of K. Bettoney, Mid Devon Bat Rescue Centre) (frame width: c.0.5 sec)

QR codes relating to presented figures

Potential confusion between hazel dormouse and other species

Table 9.18 Summary of potential confusion between hazel dormouse and other species

	Potential confusion species
Confusion group	In the first instance, always consider habitat and distribution.
Other dormouse species	Due to the restricted range of edible dormouse, and its far more audible behaviour, confusion is unlikely, taking all matters into consideration.
Species in other groups	Brown rat, black rat, water vole and harvest mouse. *Apodemus* calls, but dormouse calls tend to be of longer duration. Shrew species. Polecat – QCF calls. Daubenton's bat 'walking stick' social call. *Nyctalus* species echolocation calls. Brown long-eared bat Type D1 social calls (Middleton *et al.*, 2022).

9.6.2 Edible dormouse *Glis glis*

<table>
<tr><th>Length/Weight</th><th>Habitat preferences</th><th>Distribution</th></tr>
<tr><td>Body length (excluding tail) of 120 to 175 mm.
Weight c.140 g, heavier in weight immediately prior to hibernation.</td><td rowspan="4">Mixed woodland
Can be found living within buildings, including occupied houses.</td><td rowspan="4">Occurs primarily in an area centred around Tring (Chilterns). Absent from Wales, Scotland and Ireland.</td></tr>
<tr><th>Diet</th></tr>
<tr><td>Mainly herbivorous, feeding across a range of food sources, varying according to availability and season, including beech mast, nuts, buds, fruit and berries. Also known to eat insects, fungi, eggs and carrion.</td></tr>
<tr></tr>
<tr><th>Daily activity pattern</th><td colspan="2">Nocturnal.</td></tr>
<tr><th>Seasonal activity pattern</th><td colspan="2">Active throughout warmer months. Typically hibernates from October through to May.</td></tr>
<tr><th>Mating activity pattern</th><td colspan="2">When breeding, males are strongly territorial. Mating mostly takes place during June to August, with births usually occurring from August to September.</td></tr>
<tr><th>Other notes</th><td colspan="2">Difficult to encounter visually, but very audibly vocal at dusk.</td></tr>
<tr><td colspan="3">References
Morris, 2008; Lysaght and Marnell, 2016; Mathews et al., 2018.</td></tr>
</table>

Acoustic behaviour including spectrogram examples

Edible dormouse is far more vocal within the audible frequency range, compared to the smaller and audibly quieter hazel dormouse.

Edible dormouse will call from trees and can be particularly vocal during their mating season (June to August). A variety of sounds are known to be emitted (Hutterer and Peters, 2001) including a threat call described as an explosive churring, and a distinctive raucous squealing 'chirp' call (Figures 9.94 to 9.96) emitted from a high vantage point, which may be territorial (Morris, 2008).

A study of captive edible dormice in a stable colony (Iesari *et al.*, 2017) distinguished seven types of audible calls (ultrasonic vocalisations were not studied), that were characterised as: chatter, chirp, hissing, squeak, twitter, volley and whistle. The 'chirp' call (Figures 9.94 to 9.96) was described as being 'high-pitched, acute, short and recurring sounds with constant frequency modulation' delivered in a 'long series'. The context of these 'chirp' calls is thought to be related to agonistic territorial behaviour.

These audible calls are particularly useful for detecting the presence of this species at night (Rodolfi, 1994; Ciechanowski and Sachanowicz, 2014; Iesari *et al.*, 2017). The calls described in Figure 9.97 were produced by an animal (one in a group of four) which was believed to have been defending a food resource from an external threat. To our ears, it sounds very similar to the examples provided in Figures 9.94 to 9.96.

Figure 9.98 shows a call emitted as a one-off during what would otherwise seem to have been territorial 'chirping' behaviour.

Figure 9.99 shows a different style, with chirp calls being immediately preceded by a series of shorter calls.

A more unusual audible example is described in Figure 9.100. This is best described as a rattling-type sound, in this instance produced by an adult, but similar sounds have also been recorded from sub-adults. After corresponding with Stefanie Kruse (who drew these particular sounds to our attention) we asked her for some context that could perhaps explain such behaviour. Stefanie felt as follows: *'I'd say the sound is a kind of threat. They always produce it when you come near their boxes. Kind of "don't you dare to disturb me!" They do it not only in my aviaries, but also naturally outside, for example when checking bird nest boxes, you sometimes have a Glis glis inside, which you immediately know when you hear this noise.'*

At this stage, we do not have examples of ultrasonic sounds produced by edible dormouse, and are unsure how important ultrasound is for this species. We have however provided a couple of examples of higher-frequency vocalisations (not quite ultrasonic) thought to have been emitted by this species (Figures 9.101 and 9.102).

Figures 9.94 to 9.102 provide specific examples of spectrograms relating to this species, and where the 🔊 symbol appears, a recording of what is shown within the spectrogram is available to download from the species library. In addition, QR codes are provided for selected examples, and these can be accessed immediately for listening purposes from most mobile devices.

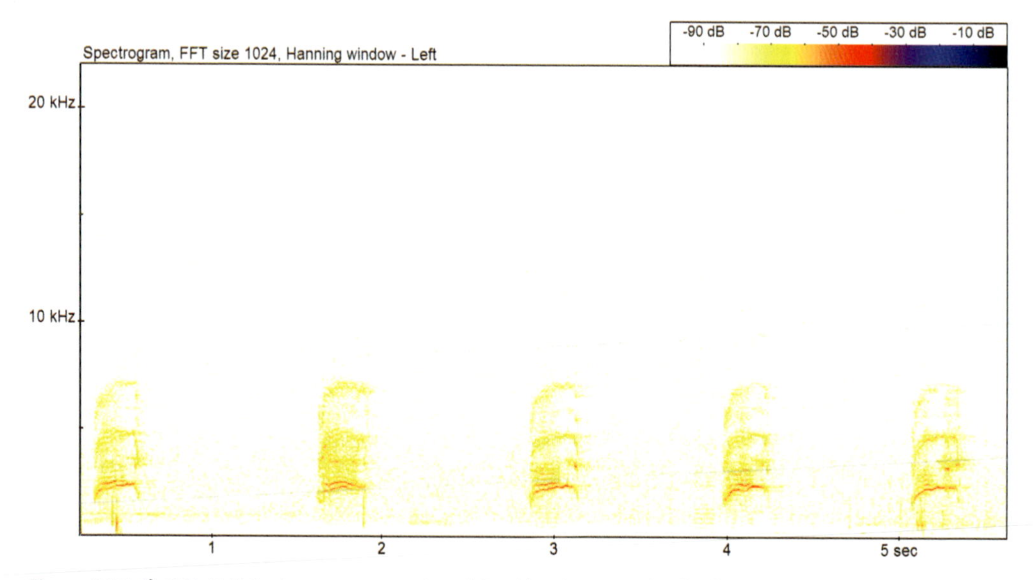

Figure 9.94 🔊 (QR) Edible dormouse – a series of five 'chirp' territorial calls (frame width: c.5 sec)

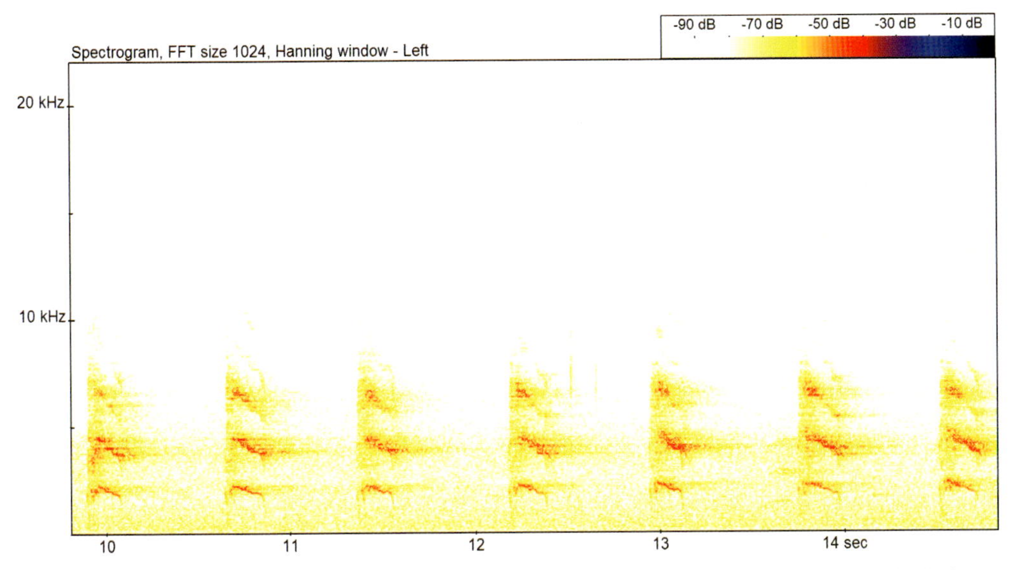

Figure 9.95 🔊 **(QR)** Edible dormouse – a series of seven 'chirp' territorial calls (courtesy of T. Fulford) (frame width: c.5 sec)

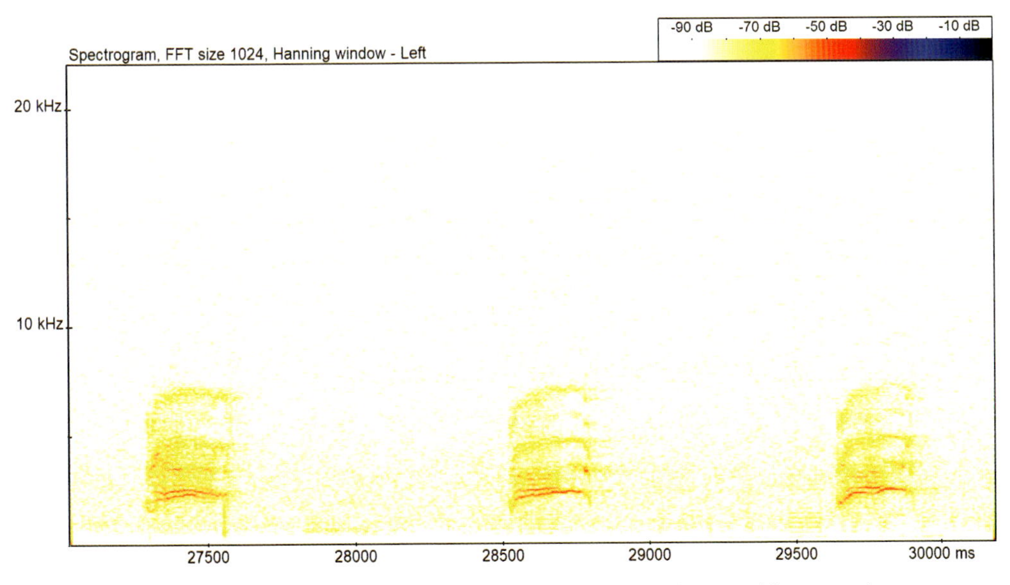

Figure 9.96 🔊 Edible dormouse – a series of three 'chirp' territorial calls (frame width: c.2.5 sec)

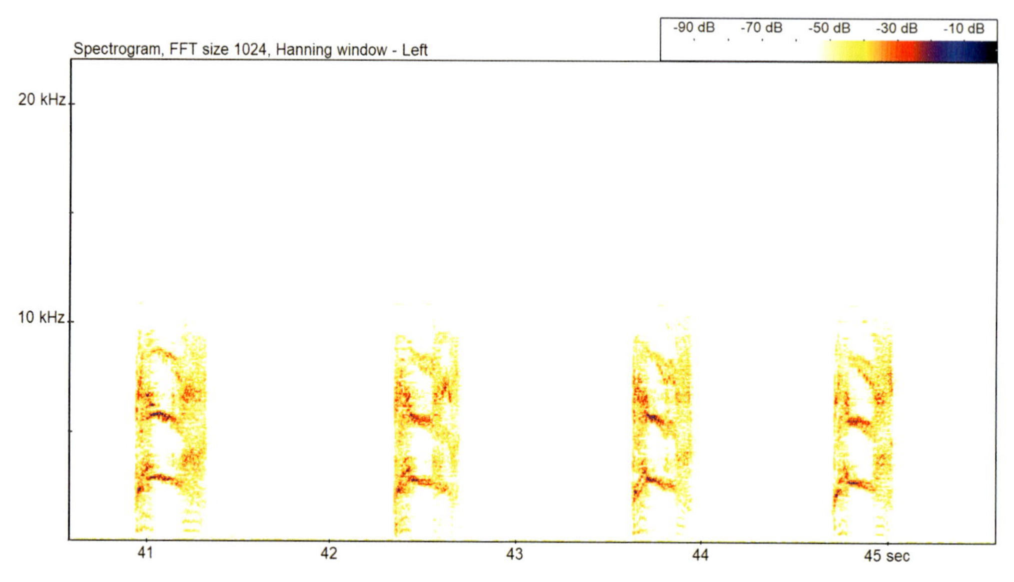

Figure 9.97 Edible dormouse – a series of four threat-type calls (courtesy of H. French) (frame width: c.5 sec)

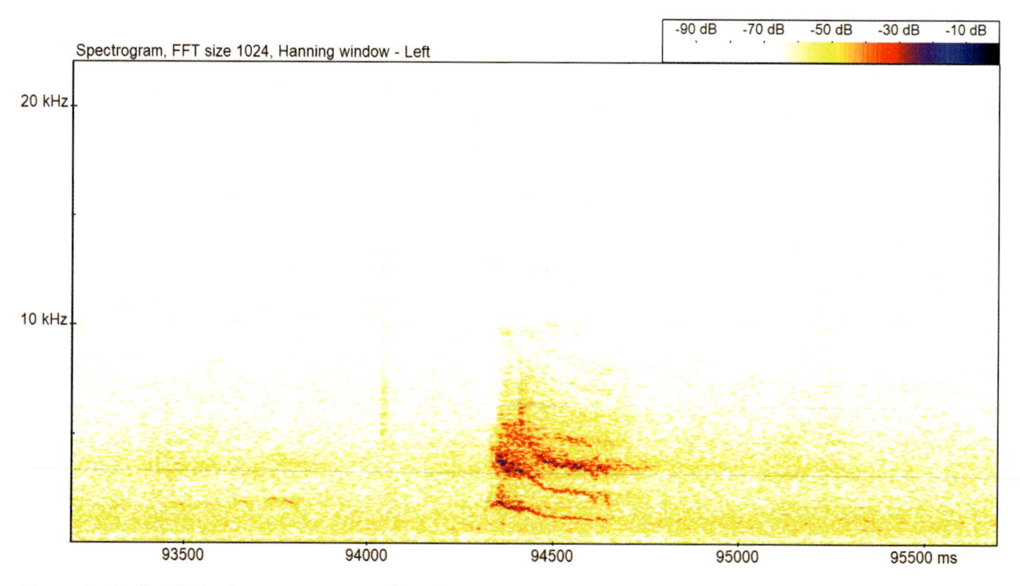

Figure 9.98 Edible dormouse – a single call variation emitted during normal 'chirp' territorial behaviour (courtesy of T. Fulford) (frame width: c.2.5 sec)

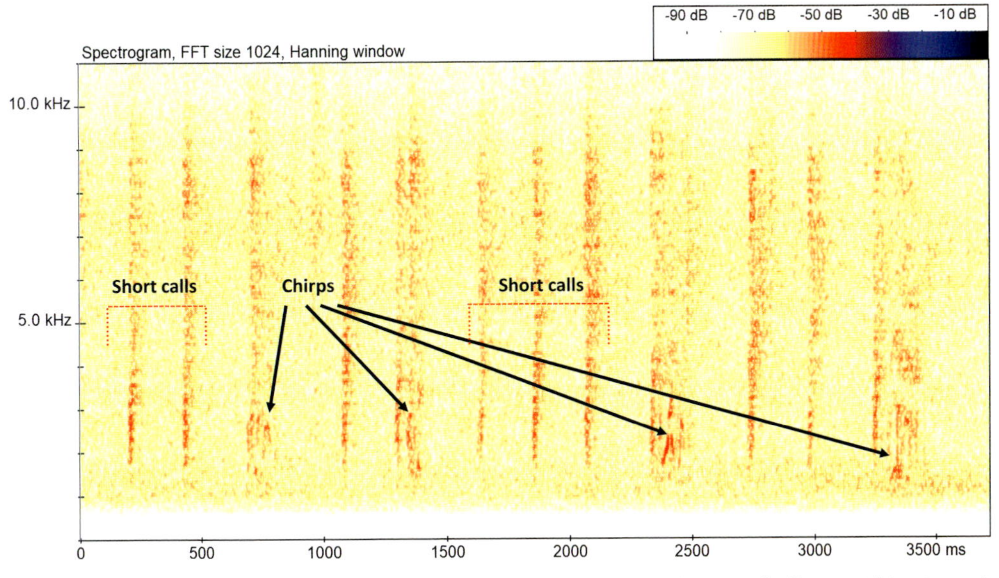

Figure 9.99 🔊 Edible dormouse – chirp calls (black arrows) preceded by shorter calls (frame width: c.3.5 sec)

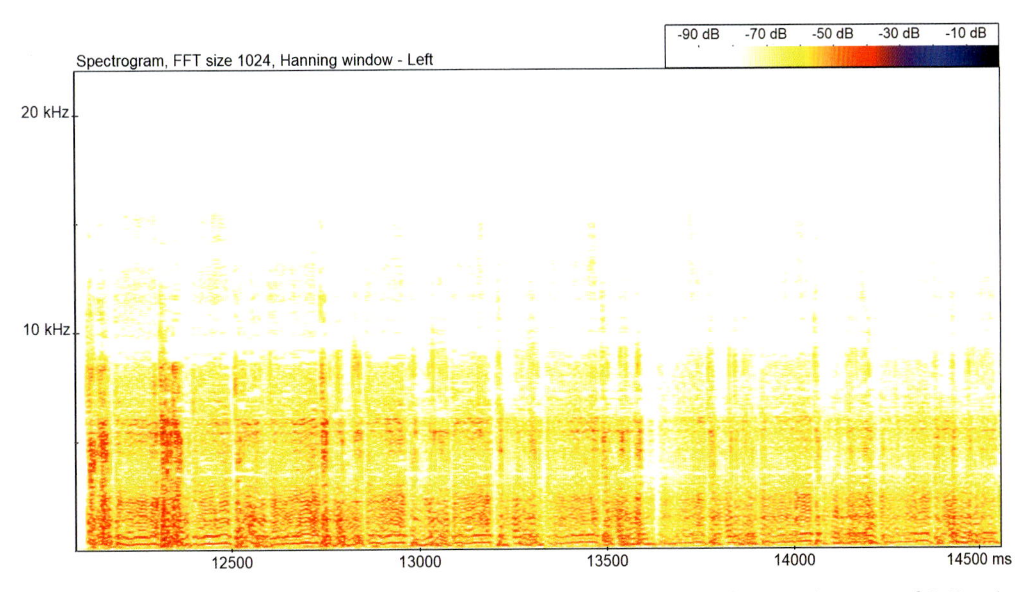

Figure 9.100 🔊 **(QR)** Edible dormouse – a series of rattling threat-type vocalisations (courtesy of S. Kruse) (frame width: c.2.5 sec)

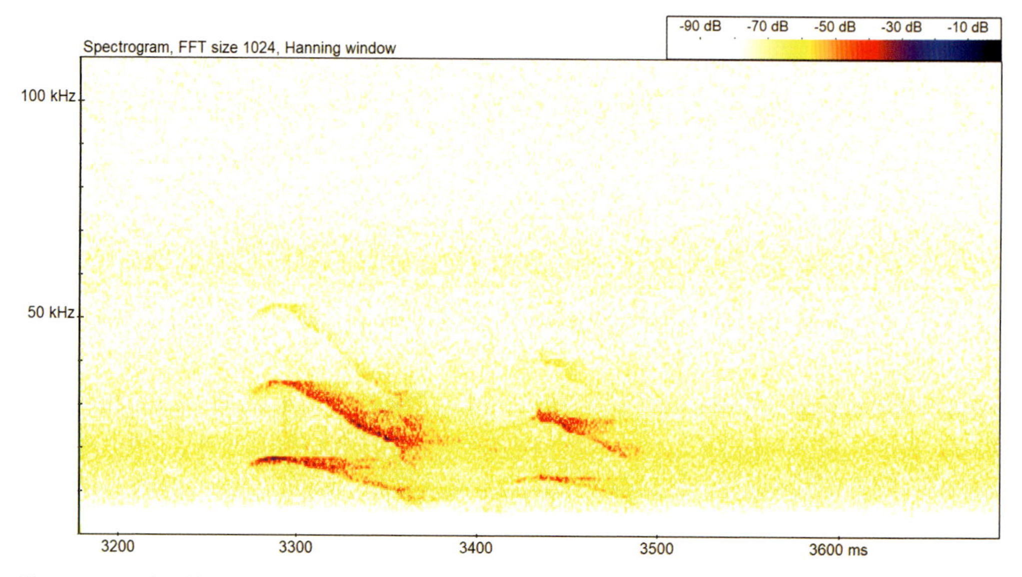

Figure 9.101 Edible dormouse – a couple of higher-frequency emissions (frame width: c.0.5 sec)

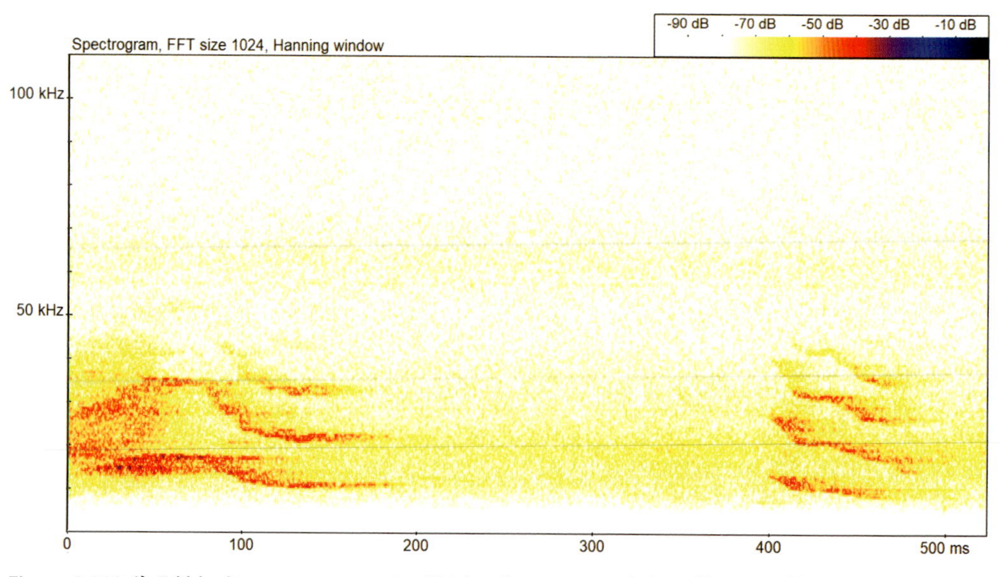

Figure 9.102 Edible dormouse – a couple of higher-frequency emissions (frame width: c.0.5 sec)

QR codes relating to presented figures

Figure 9.94: Edible dormouse audible call

Figure 9.95: Edible dormouse audible call

Figure 9.100: Edible dormouse audible call

Potential confusion between edible dormouse and other species

Table 9.19 Summary of potential confusion between edible dormouse and other species

Confusion group	Potential confusion species
	In the first instance, always consider habitat and distribution.
Other dormouse species	Due to the restricted range of edible dormouse, and its far more audible behaviour, confusion is less likely with hazel dormouse, taking all matters into consideration.
Species in other groups	Unlikely to be confused with other small mammals. Bird calls (e.g. contact call from tawny owl, and begging calls from tawny owl chick).

9.7 Potential application of bioacoustics for monitoring small rodents

Acoustic studies offer a number of opportunities over other survey approaches for rodents (as well as for shrews – see Chapter 10): they can be carried out without directly impacting on an animal's behaviour, and the techniques can be deployed by many surveyors across much larger areas than would be possible using more conventional invasive methods, such as live trapping. This is a novel field of study, where there are important questions still to address, but for an exploratory look at some of the possibilities that acoustics could offer for rodents see Newson and Pearce (2022). This includes case studies that illustrate in which situations and (with the support of the BTO Acoustic Pipeline for processing recordings) how acoustics could inform our understanding of small mammal populations at local, regional or national scales. With this, there are clear opportunities for acoustics to complement existing monitoring and recording methods.

In a Britain and Ireland context, it is worth mentioning that a huge amount of acoustic survey work is already conducted annually for bats, where small mammals including rodents are recorded as by-catch. This offers a potentially cost-effective opportunity for making use of existing or planned bat surveys to also improve our knowledge of small mammals. As an illustration, whilst verification of these records/recordings is still needed, the first two years of the BTO Acoustic Pipeline alone has resulted in over 100,000 small mammal identifications. If it were possible to make use of the by-catch from more bat-related studies in Britain and Ireland, the amount of data available for small mammals including rodents would be immense, and far greater than is likely to be collected through more intensive and intrusive live trapping.

Looking at the rodent groups in more detail, rats are highly vocal species that are relatively easy to detect using ultrasonic microphones and are regularly recorded incidentally during static bat detector surveys (Newson *et al.*, 2022; Newson, 2023). Acoustics may therefore provide a cost-effective approach for monitoring the presence of brown rats or black rats on seabird islands where conservation efforts involve predator detection and eradication practices (Newson and Pearce, 2022). There are also important opportunities for monitoring or detecting the presence of licensed species/species of conservation interest, potentially in combination with other traditional monitoring approaches. This includes for climbing and arboreal species, like the harvest mouse and hazel dormouse, that can be difficult to monitor.

Lastly, acoustics could provide opportunities for monitoring mice and voles, but the lower detection distance for both groups and the low call rate for voles mean that bat detectors would need to be positioned closer to the ground, or be more targeted to where the animal is likely to be found. For voles, the time of day when they are potentially most active does not necessarily tie in with recording bats at the same time. The survey effort needed to record bank and field voles in particular is likely to be much higher than would be needed for other small mammal species, and therefore acoustics is perhaps more likely to provide only a supplementary function to conventional survey methods. Table 9.20 gives some examples of where acoustic studies could be deployed, either stand-alone or in conjunction with other survey techniques such as live trapping. Some preliminary investigations into the application of acoustics for monitoring small mammals are provided in Appendix II (Case Studies).

Table 9.20 Usefulness of acoustic studies – small rodent examples

Species	Potential applications/benefits
All species	An opportunity to add considerably more data where information on any specific species is lacking (e.g. less well-covered parts of Britain and Ireland) or out of date. This may be particularly cost-effective if bat surveys are already being carried out in these areas, and the recordings can be mined for small mammals (e.g. using the BTO Acoustic Pipeline).
All species	A wider awareness of acoustics for this group means there is greater chance of someone realising that something unknown or unexpected is occurring in their area (e.g. presence of a native species or new invasive species, previously thought to be absent).
Brown rat	A cost-effective approach for detecting the presence of brown rats, potentially during eradication work on seabird islands.
Black rat	Continued monitoring to confirm absence from areas where the presence of black rat has in the past caused issues (e.g. offshore island seabird colonies).
Harvest mouse	Monitoring species presence at local/national level.
House mouse	Monitoring for presence/absence in areas where distribution knowledge is poor.
Water vole	Supporting the survey work relating to this species, which is a species that is under severe pressure in many areas.
Vole species	Additional information to show health of local populations.
Hazel dormouse	Determining species presence in areas where other methods may be difficult, time-consuming or expensive. Carrying out of non-intrusive acoustic studies by a greater number of people, where licensing may not be required in order to deploy acoustic surveys.
Edible dormouse	Monitoring local populations to help confirm species presence or number of territorial males in a given area.

Hedgehog – Matt Errington

CHAPTER 10

Insectivores – Hedgehogs, Moles & Shrews

10.1 Introduction to this group

This chapter covers the insectivorous mammal species that occur across the geographic range of this book, within which we have a single species each of hedgehog and mole, along with six species of shrew (see Table 10.1).

Hedgehogs are a widely distributed species; however, their numbers have declined considerably in recent years. Urban hedgehog populations appear to be stable or recovering and they are often seen at dusk opportunistically feeding around bird feeders within gardens. Comparatively, rural populations have declined by up to 75% in the last 20 years (Wembridge *et al.*, 2022) and are encountered less often. A variety of survey techniques have been developed for monitoring hedgehogs from scats, using trail cameras and using baited footprint tunnels (Thomas and Wilson, 2018). Although acoustics is not a recognised survey method, hedgehogs are known to produce a variety of vocalisations within the audible range, with some extending into the ultrasonic spectrum. Loud huffing sounds displayed during courtship are probably the most recognisable.

Moles, again, are widely distributed, and although rarely seen, their presence is evident from their visible molehills, which offer a means of estimating the number of mole territories (Fellowes *et al.*, 2020). Due to their fossorial lifestyle, they are considerably harder to encounter acoustically, and far less is known about their vocalisations. Similar to other burrowing/subterranean species such as mole-rats, prairie dogs and gophers (Heffner *et al.*, 1994; Buffenstein, 1996; Bednářová 2012; Nairns *et al.*, 2016; Heffner *et al.*, 2020), it is conceivable that they communicate using very low-frequency sound/infrasound (< 10 kHz) and/or seismic signalling – sound frequencies that are not covered in this book.

Being small, the presence of shrews is usually difficult to establish through visual sightings and field signs. The most frequently used methods are live trapping (which requires a licence), analysis of scats from bait tubes (Churchfield *et al.*, 2000), hair tubes (Pocock and Jennings, 2006) and more recently using small mammal trail camera boxes (Littlewood *et al.*, 2021). Shrews, despite their size, are very vocal within the ultrasonic range and research indicates that they use sound to echo-orientate within their environment (Tomasi, 1979; Siemers *et al.*, 2009; Sanchez *et al.*, 2019) as well as for communication between conspecifics.

The higher level of vocalisations produced by shrews, compared to other small mammals, make them extremely good candidates for acoustic study. There are proven benefits of using acoustics for establishing the presence of shrew species (Newson and Pearce, 2022; Newson *et al.*, 2023) and the potential for acoustics to help inform our understanding of shrew species distribution and status, across spatial scales (e.g. regionally, nationally, internationally) (discussed later within this chapter – Section 10.6).

For moles and hedgehogs, bioacoustics data is likely to be supplementary to existing monitoring techniques, but knowing the more regularly encountered sounds made by these species can help confirm species presence and may offer value for interpreting behaviours.

10.2 Overview and comparisons of acoustic behaviour

Broadly speaking the species in this group will be encountered through the production of vocalisations that are made audibly for hedgehogs and moles, and ultrasonically for shrews (see Table 10.2), and consist of sounds that carry over a relatively short distance (see Table 10.3).

A diversity of audible sounds have been noted for hedgehogs, many of which have been linked to specific behaviours including dominance interactions, courtship (huffing), alarm calls, responses to a threat or distress (hissing), agonistic behaviours (e.g. territorial disputes; bark or scream) and appetitive behaviour (e.g. searching for and finding food; sniffing detected in the ultrasonic range). Comparatively, very little is known about the vocalisations made by moles, and although we are only able to provide two examples of audible vocalisations, both made by an animal when above ground (courtesy of D. Mellor), given their fossorial lifestyle, it is likely that moles produce a greater diversity of sounds below ground, some of which may fall within the range of infrasound and/or seismic signalling. Vocalisations above ground are likely to be associated with agonistic interaction between dispersing individuals or distress, given that this species is predominantly subterranean.

Compared to moles and hedgehogs, shrews vocalise predominantly within the ultrasonic range and are highly vocal, perhaps because they are believed to utilise sound as a means of echo-orientation (Tomasi, 1979; Siemers *et al.*, 2009; Sanchez *et al.*, 2019). They are also active throughout the day and night, with peaks in activity at dusk and dawn, which offers greater opportunity for recording them. Their ultrasonic calls fall within the lower end of the ultrasonic range (below 30 kHz) and therefore travel greater distances than most other small mammal species that call at higher frequencies. They are therefore regularly recorded during bat surveys, even when bat detectors are deployed at height. Whilst it might be expected that different species of shrews in Britain and Ireland produce similar calls to one another, there are some big differences between species that are useful for identification. Undoubtedly some vocalisations of shrews are more complex, and are potentially more difficult to assign to species given these alone, but on the whole shrews do seem to follow a set of 'rules' at a species level that are useful for identification. Differentiating most of the species in this chapter from one another should be possible, at least on occasions when you have recorded one of the more distinctive call structures.

Within the following species accounts we tend to focus mostly on identifiable (i.e. most useful) sounds, which in the main fall into vocalisations emitted audibly by hedgehogs, and ultrasonically by shrews, and hence acquired with the use of specialist recording equipment (e.g. bat detectors). In addition to the examples presented we have many others which have not been included, usually as they fall into one of the following categories:

- Audible sounds – usually considered to be distress-related or threat calls (e.g. animal being attacked).
- Mother/young interactions – contact calls, begging calls etc.
- Examples only encountered on a very small number of occasions where it is not yet possible to be confident as to either their origin or significance.

So, please bear in mind that what we are showing you here is by no means the full vocal repertoire for any of the species involved. There are examples that we have seen that have not been included, and we expect that there will be examples of calls that we have not as yet encountered.

Table 10.1 Insectivores occurring in Britain and Ireland, including status, distribution and relative abundance. The relevant chapter sections that describe species-specific vocalisations are also listed.

Order	Family	Species common name	Genus	Species scientific name	Status	Distribution range	Abundance within distribution range	Sub-chapter reference
Erinaceomorpha	Erinaceidae	Hedgehog	*Erinaceus*	*E. europaeus*	Native	Widely distributed throughout Britain and Ireland	Common, but declining	10.3.1
Soricomorpha	Talpidae	Mole	*Talpa*	*T. europaea*	Native	Widely distributed throughout Britain, absent from Ireland	Common	10.4.1
	Soricidae (Shrews)	Common shrew	*Sorex*	*S. araneus*	Native	Widely distributed throughout Britain, absent from Ireland	Common	10.5.1
		Pygmy shrew		*S. minutus*	Native	Widely distributed throughout Britain and Ireland	Common	10.5.2
		Millet's shrew		*S. coronatus*	Native	Jersey	Locally common	10.5.3
		Water shrew	*Neomys*	*N. fodiens*	Native	Widely distributed throughout Britain, absent from Ireland	Common	10.5.4
		Lesser white-toothed shrew	*Crocidura*	*C. suaveolens*	Non-native (naturalised), but possibly native	Scilly Isles and Jersey and Sark – The Channel Islands	Locally common	10.5.5
		Greater white-toothed shrew		*C. russula*	Non-native (naturalised), but possibly native in the Channel Islands, non-native in Ireland and north-east England.	Guernsey, Alderney and Hern Ireland and north-east England (Sunderland)	Common in Channel Islands, locally uncommon elsewhere	10.5.6

Table 10.2 Acoustic spectrum within which you would usually expect to encounter most useful sound for identification purposes within this group

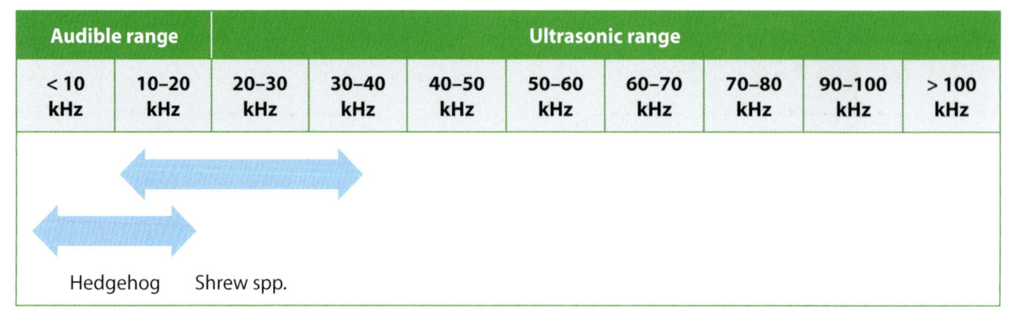

Audible range		Ultrasonic range							
< 10 kHz	10–20 kHz	20–30 kHz	30–40 kHz	40–50 kHz	50–60 kHz	60–70 kHz	70–80 kHz	90–100 kHz	> 100 kHz

Table 10.3 Detectable distance for useful acoustic encounters with insectivorous mammal species

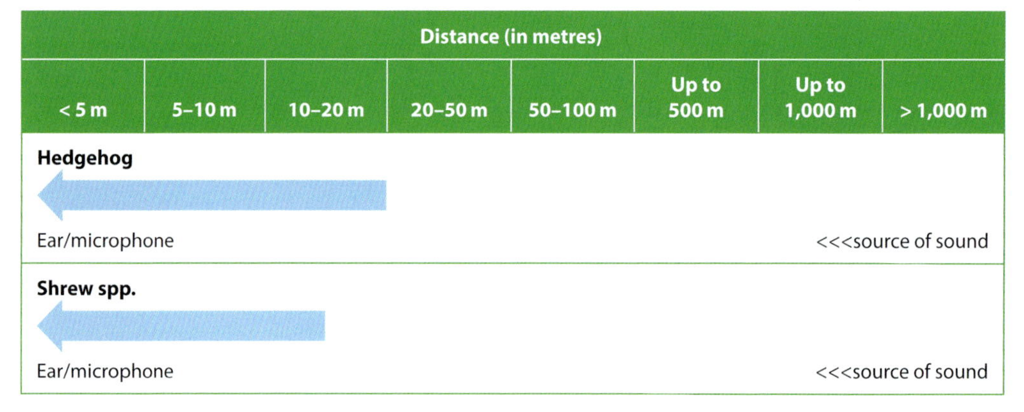

Distance (in metres)							
< 5 m	5–10 m	10–20 m	20–50 m	50–100 m	Up to 500 m	Up to 1,000 m	> 1,000 m

Table 10.4 Typical breeding season (shaded in blue) for insectivorous species within Britain and Ireland

Species \ Month	Jan	Feb	Mar	April	May	June	July	Aug	Sept	Oct	Nov	Dec
Hedgehog				■	■	■	■	■	■			
Mole		■	■	■	■							
Common shrew				■	■	■	■	■	■			
Pygmy shrew				■	■	■	■	■	■			
Millet's shrew					■	■	■	■	■			
Water shrew				■	■	■	■	■	■			
Greater white-toothed shrew		■	■	■	■	■	■	■	■	■		
Lesser white-toothed shrew			■	■	■	■	■	■	■			

10.3 Hedgehog vocalisations

In the following section, a brief summary of the key characteristics of hedgehog will be described, along with details on the species-specific vocalisations and, where known, the behavioural function of these sounds.

10.3.1 Hedgehog *Erinaceus europaeus*

Length/Weight	Habitat preferences	Distribution
Adult males are bigger than females, with a head/body length of c.190 to 265 mm. Female body length ranges from c.180 to 260 mm. Males weigh 900 to 1,200 g and females 800 to 1,025 g. Weights can be much higher than this, especially just prior to entering hibernation.	Grassland habitat associated with woodland edge, scrub and hedgerows Woodland Gardens Parkland Farmland	Our only species of hedgehog, occurring throughout Britain and Ireland, with the exception of some offshore islands.*
Diet		
Insectivorous, mostly feeding across a range of ground-dwelling invertebrates, and also occasionally recorded consuming carrion, eggs and chicks of ground-nesting birds.		

Daily activity pattern	Mainly nocturnal.
Seasonal activity pattern	Active during spring, summer and autumn, going into hibernation during the period November to March within a nest comprising leaf litter.
Mating activity pattern	Mating takes place during April to September, with peak pregnancies occurring in May to July, and September.
Other notes	May be more easily encountered visually at dusk or dawn. Although producing sounds within the audible range, not often encountered acoustically, other than snuffling sounds associated with foraging if seen at close distance, and loud huffing sounds associated with courtship.

References

Morris and Reeve, 2008; Morris, 2011; *Lysaght and Marnell, 2016; *Mathews *et al.*, 2018; Mammal Society, 2022.

Acoustic behaviour including spectrogram examples

Hedgehogs do not appear to have been extensively studied with respect to their vocalisations, although they are known to make a 'loud squealing noise when alarmed' (Morris, 2011). Loud snorting and huffing sounds are often heard during the breeding season when males engage in a circular courtship display with females and frequently attract rival males, evoking agonistic and dominance calls (hissing, snorting, rasping).

Young hedgehogs are known to make a range of sounds, including a chirping/peeping-type sound (K. Bettoney, 2021 and M. Ferguson, 2023, pers. comm.).

Gillian Prince of Hedgehog Bottom, Berkshire (www.hedgehog-rescue.org.uk) very kindly gave us access to a range of recordings that we can share with you here, and most of the descriptions of the various hedgehog sounds are Gillian's interpretations whilst studying hedgehogs over many years.

For a summary of many of the more commonly encountered vocalisations please refer to Table 10.5. There are other sounds that may also be encountered (e.g. made by sub-adults) which are not documented here, with the exception of examples in Figures 10.2 and 10.12.

There may very well be sounds within the ultrasonic range, for example, that we are not yet aware of.

Table 10.5 A summary of adult hedgehog vocalisations with descriptions adapted from those by Gillian Prince (www.hedgehog-rescue.org.uk)

Call description	Structure and considered context of vocalisation
Coughing	Can produce numerous coughing sounds, including a dry rasping version as described within Figure 10.1.
Distress	Produced by animals under stress or in distress. Figure 10.2 is an example emitted by a juvenile which found itself stuck.
Hissing	Probably a threat-type gesture that is produced when disturbed or feeling threatened (Figure 10.3).
Challenge/Warning	A gruff bark-like sound, thought to be a warning call produced during threatening and potentially aggressive behaviour (Figure 10.4). Can also produce a longer scream-like sound (Figure 10.5).
Huffing	May be heard when two hedgehogs encounter each other. Thought to be a show of dominance, as the winner is usually the loudest. Huffing is heard a lot during the mating season when a male is pursuing a female (Figure 10.6).
Screaming	Sounds made when an animal is injured and in pain.
Note: For more information about other hedgehog-related sounds, and indeed anything to do with hedgehogs, visit www.hedgehog-rescue.org.uk.	

In addition to the sounds discussed in Table 10.5 we have also encountered other sounds, at less audible frequencies (Figures 10.7 to 10.11). Many recordings, similar to the examples here, were encountered during our monitoring of hedgehogs over a prolonged period using bat detectors in combination with trail cameras. Figures 10.8, 10.9 and 10.11 show sounds that when slowed down (time expansion x10) sound like sniffing sounds, and as shown in Figures 10.8 and 10.9 these are often accompanied by a whistling sound. Comparatively, Figures 10.7 and 10.10 were associated agonistic interactions between rival animals.

Figure 10.10 shows a short burst of squeaks produced during one of our prolonged recording sessions. It was unusual in that it was the only occasion when such a sound was produced.

Figure 10.11 shows a longer and more varied series of emissions. Such vocalisations were encountered far less often than those shown in Figures 10.7 to 10.9.

Finally, Figure 10.12 shows a call sequence that we felt may be useful to document here, in that it was a noticeable vocalisation made by a sub-adult accompanying a mother whilst appearing to be on a foraging trip. This was recorded by Mark Ferguson (Mark Ferguson Audio) who describes the encounter as follows: *'This seemed to be a contact call from the juvenile hedgehog, following a period of roaming alongside the mum in our suburban garden. All diurnal activity, so not sure what was going on: maybe a food shortage, and foraging by necessity pre-hibernation? Following the recording, both individuals bedded down in their nest.'*

Figures 10.1 to 10.12 provide specific examples of spectrograms relating to this species, and where the 🔊 symbol appears, a recording of what is shown within the spectrogram is available to download from the species library. In addition, QR codes are provided for selected examples, and these can be accessed immediately for listening purposes from most mobile devices.

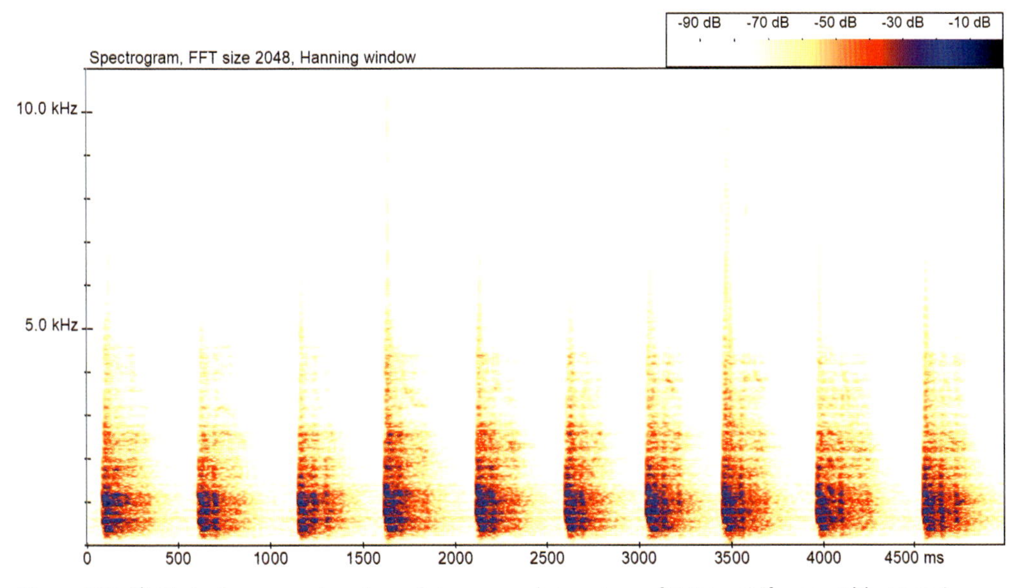

Figure 10.1 🔊 Hedgehog – a series of cough-type sounds (courtesy of G. Prince) (frame width: c.5 sec)

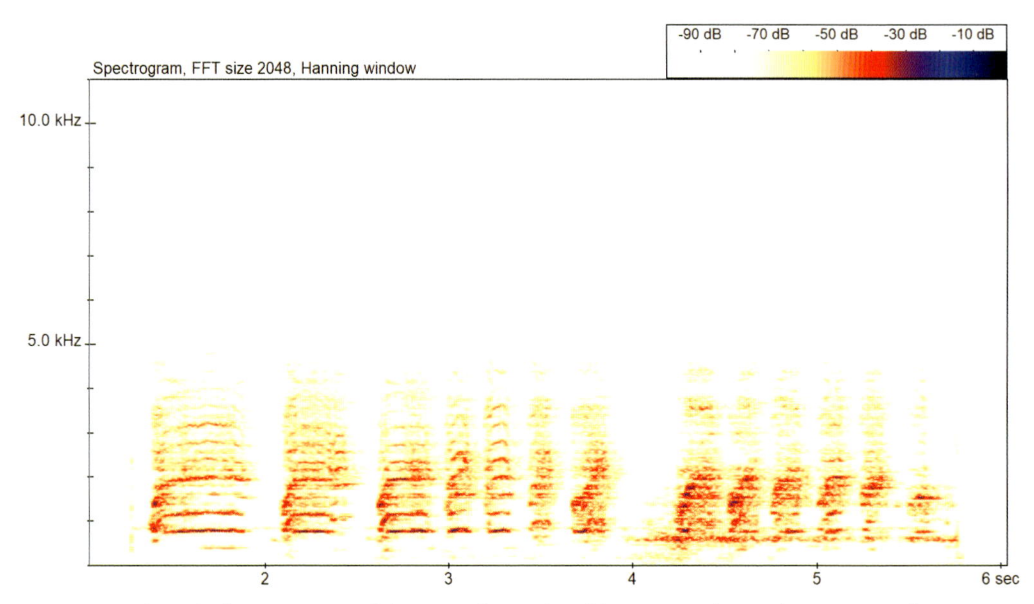

Figure 10.2 🔊 Hedgehog – a series of distress calls produced by a trapped juvenile (courtesy of G. Prince) (frame width: c.5 sec)

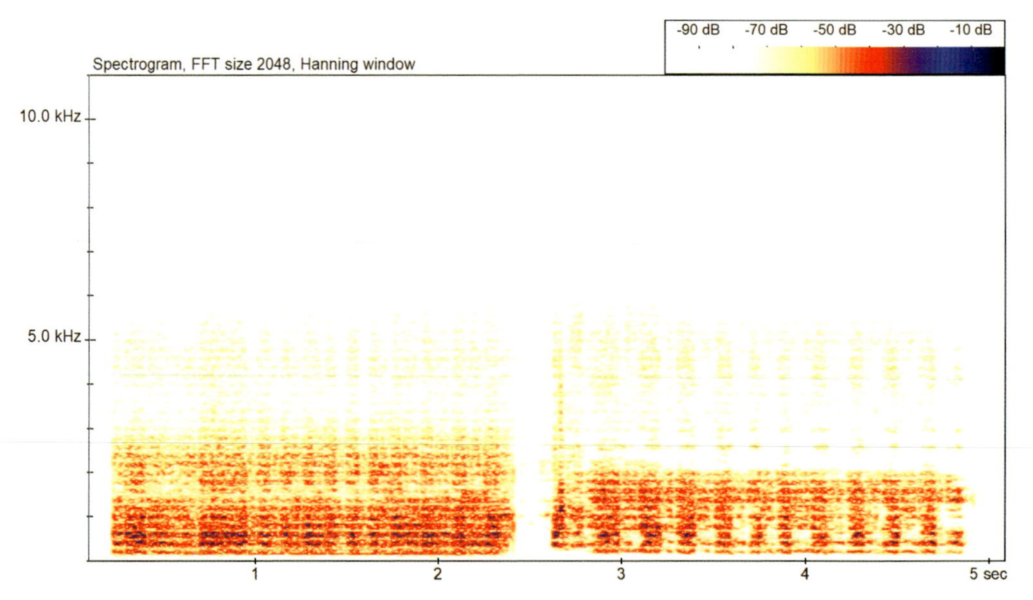

Figure 10.3 🔊 Hedgehog – a series of hissing sounds in response to a threat and/or agonistic encounter (courtesy of G. Prince) (frame width: c.5 sec)

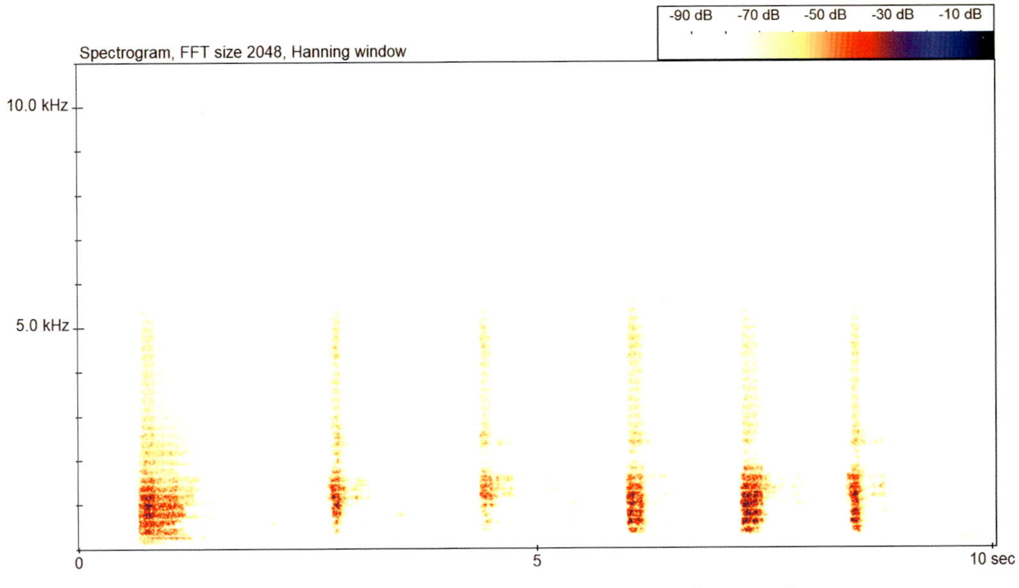

Figure 10.4 🔊 (QR) Hedgehog – a series of short challenge/warning calls during rival aggressive encounters (courtesy of G. Prince) (frame width: c.10 sec)

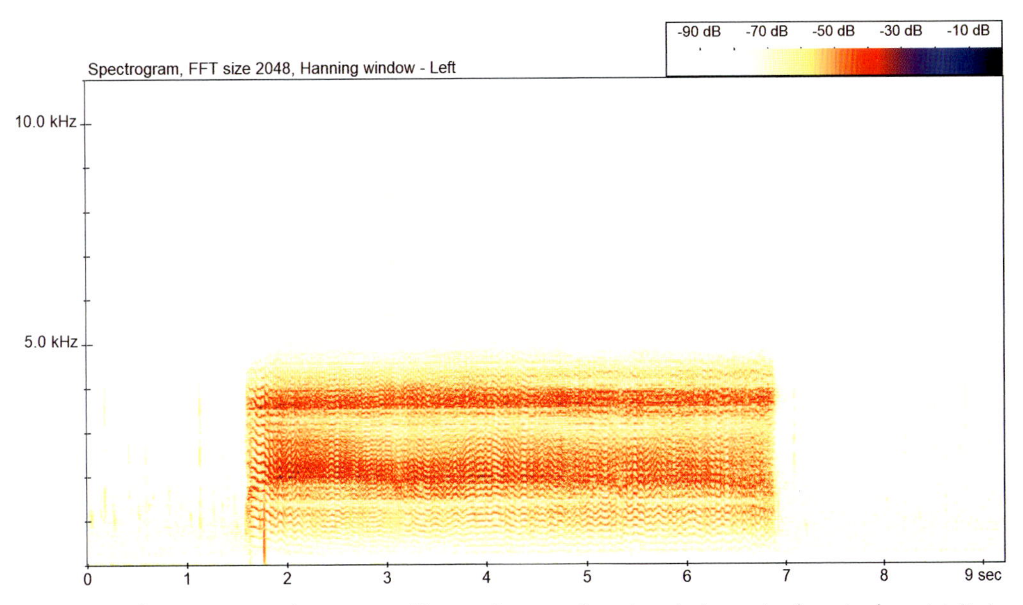

Figure 10.5 🔊 Hedgehog – a long scream-like warning sound produced when animal was in close vicinity to a badger (courtesy of B. Goodwin and A. Goodwin) (frame width: c.9 sec)

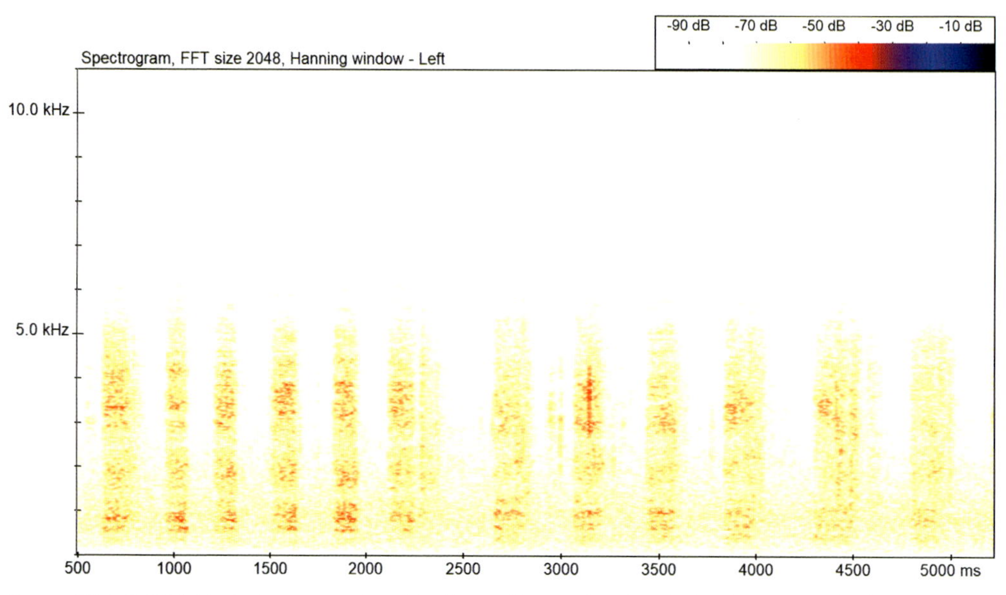

Figure 10.6 🔊 **(QR)** Hedgehog – a series of huffing calls associated with dominance display between rival animals (courtesy of A. Brand) (frame width: c.5 sec)

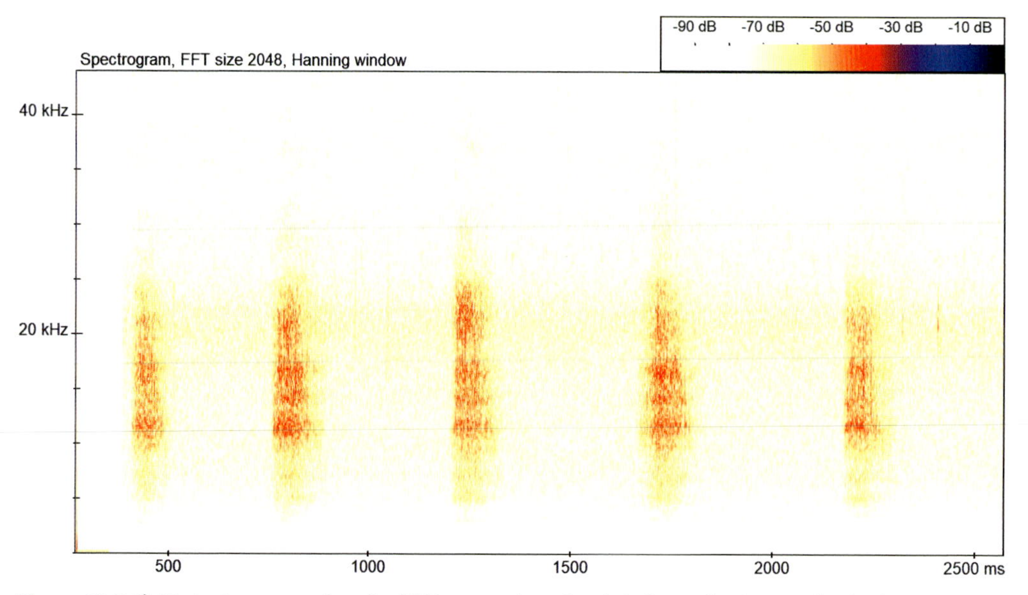

Figure 10.7 🔊 Hedgehog – a series of sniff-like sounds emitted during a dominance display between two animals (frequency scale: 0 to 44 kHz/frame width: c.2.5 sec)

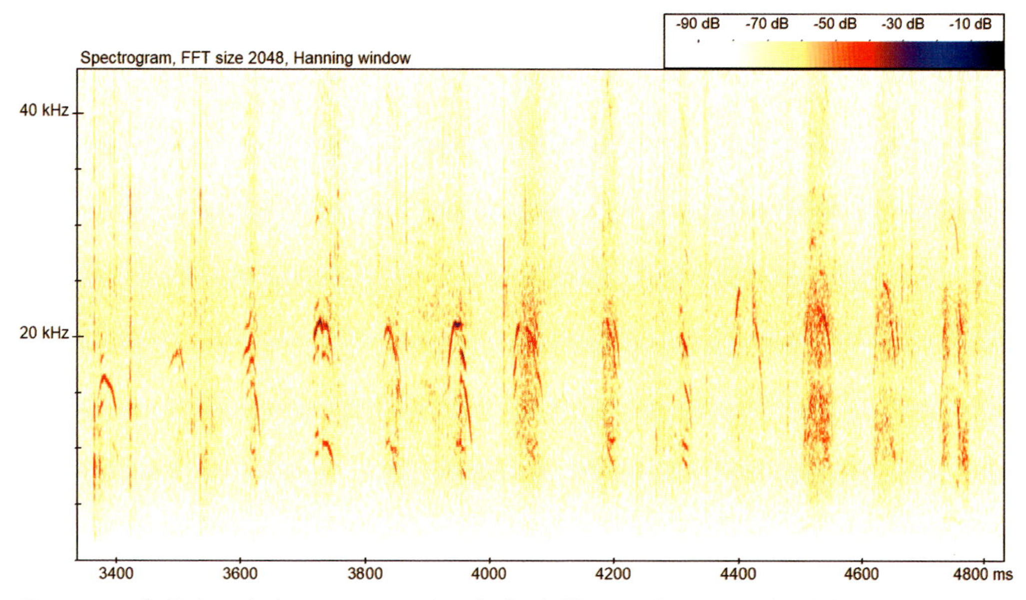

Figure 10.8 🔊 (QR) Hedgehog – a series of sniff/whistle-like sounds associated with foraging behaviour (frequency scale: 0 to 44 kHz/frame width: c.1.5 sec)

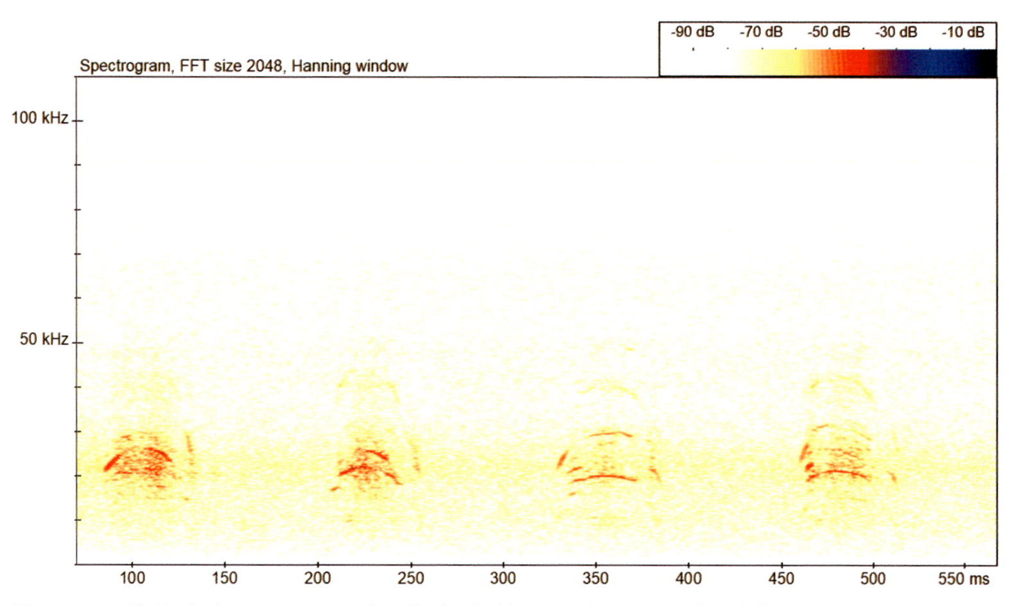

Figure 10.9 🔊 Hedgehog – a series of sniff/whistle-like sounds associated with foraging behaviour (frame width: c.0.5 sec)

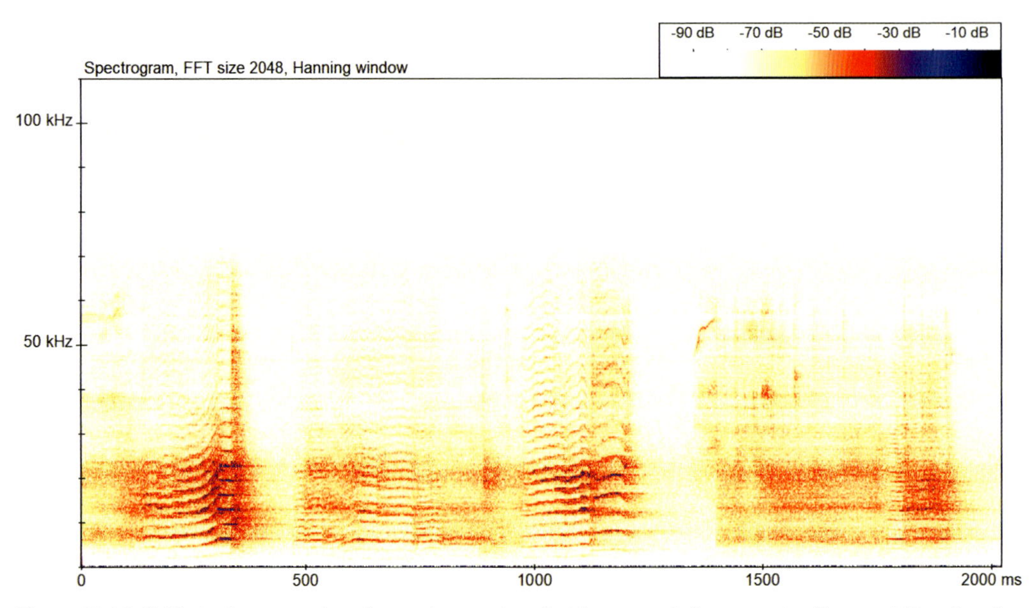

Figure 10.10 🔊 Hedgehog – a series of squeaks associated with an agonistic encounter (frame width: c.2 sec)

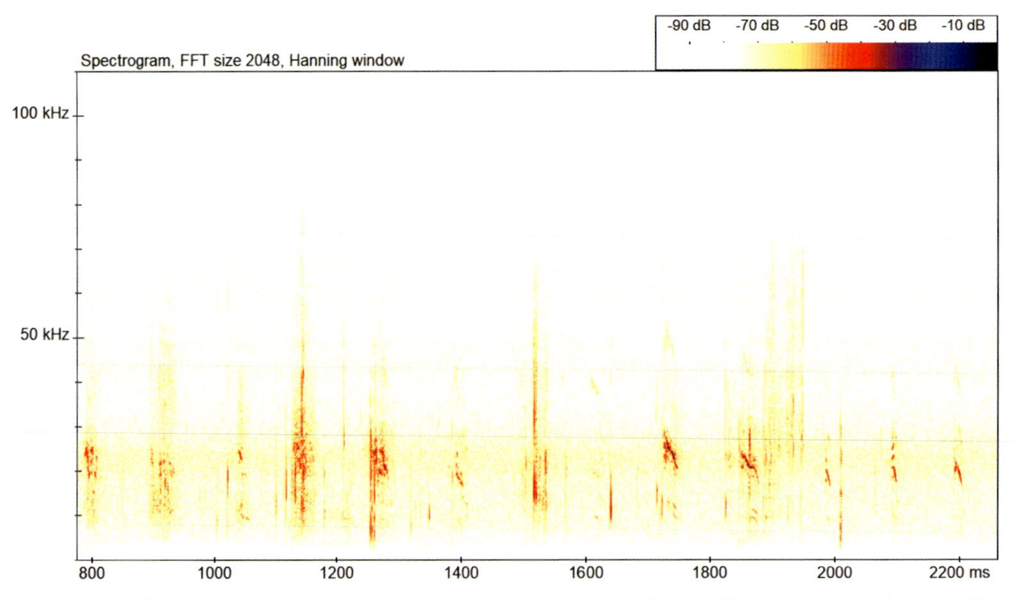

Figure 10.11 🔊 Hedgehog – a series of variable vocalisations recorded during foraging activity (frame width: c.1.5 sec)

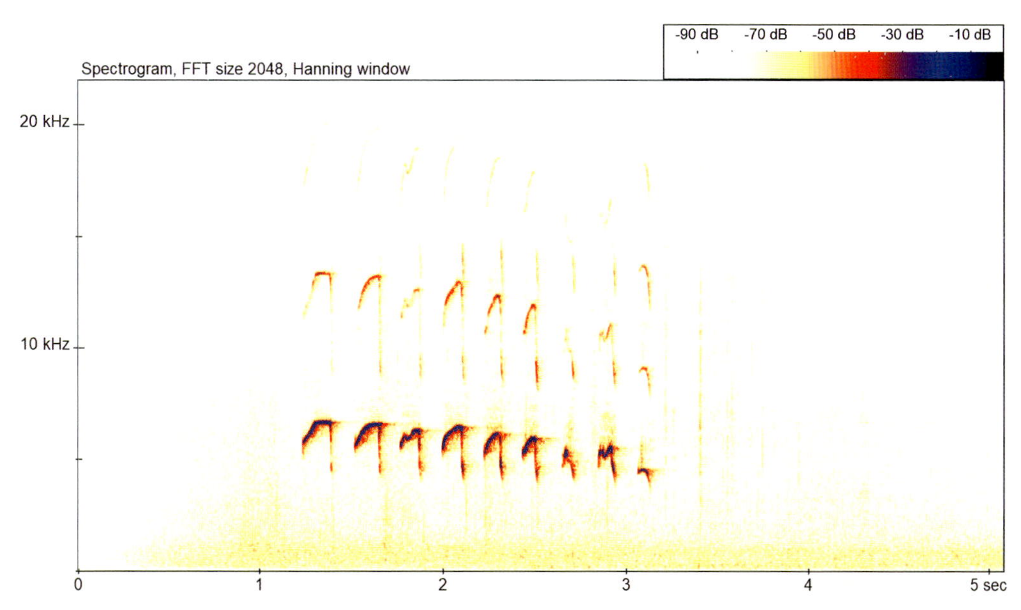

Figure 10.12 🔊 Hedgehog – a juvenile whilst accompanying mother before settling back into nesting area (courtesy of M. Ferguson) (frame width: c.5 sec)

QR codes relating to presented figures

Figure 10.4: Hedgehog	**Figure 10.6:** Hedgehog	**Figure 10.8:** Hedgehog (time expanded x10 for listening purposes)

Potential confusion between hedgehog and other species

Table 10.6 Summary of potential confusion between hedgehog and other species

	Potential confusion species
Confusion group	In the first instance, always consider habitat and distribution.
Other hedgehog species	Our only species of hedgehog, therefore not possible to confuse with other hedgehog species.
Species in other groups	Certain sounds may come across as being bird-like at times.

10.4 Mole vocalisations

In the following section, a brief summary of the key characteristics of mole will be described, along with details on the species-specific vocalisations and, where known, the behavioural function of these sounds.

10.4.1 Mole *Talpa europaea*

Length/Weight	Habitat preferences	Distribution
Males are bigger than females, with a head/body length of c.120 to 160 mm. Female body length ranges from c.115 to 145 mm. Males weigh 87 to 128 g, and females weigh 72 to 106 g.	Areas where soil depth allows for construction of tunnels Improved grassland Unimproved grassland Farmland Gardens Broadleaved woodland Moorland	Our only mole species, occurring throughout Britain, absent from Ireland.*
Diet		
Insectivorous, mostly feeding upon invertebrates found within soil, with earthworms featuring heavily.		

Daily activity pattern	Active throughout day and night, with periods of rest between active spells.
Seasonal activity pattern	Active throughout the year.
Mating activity pattern	Outside of mating season males and females are solitary. Mating takes place from as early as February through to May.
Other notes	Activity of moles in an area is usually easily established through the presence of distinctive molehills.

References
Gorman, 2008; *Mathews *et al.*, 2018; Mammal Society, 2022.

Acoustic behaviour including spectrogram examples

Mole are not regarded as being especially vocal, and are rarely heard due to their tendency to spend most of their life underground. Audible sounds have been encountered, including examples shown in Figures 10.13 and 10.14 relating to a single daytime encounter by David Mellor, when the animal was found on the surface and emitting these loud noises, although it could not be ascertained why it was doing so.

Figures 10.13 and 10.14 provide specific examples of spectrograms relating to this species, and where the 🔊 symbol appears, a recording of what is shown within the spectrogram is available to download from the species library. In addition, QR codes are provided for selected examples, and these can be accessed immediately for listening purposes from most mobile devices.

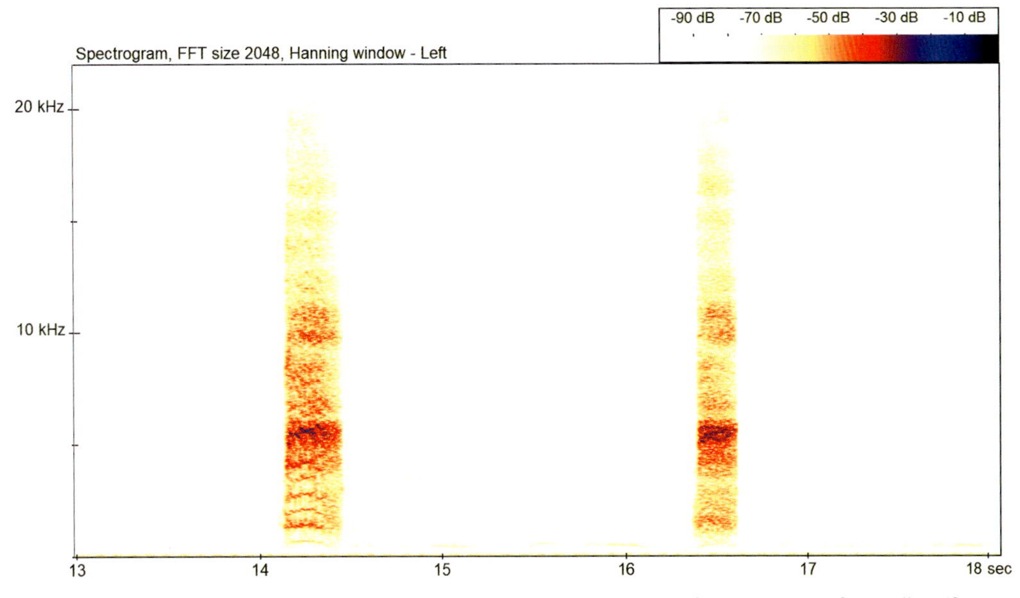

Figure 10.13 🔊 **(QR)** Mole – two calls produced by an animal on the surface (courtesy of D. Mellor) (frame width: c.5 sec)

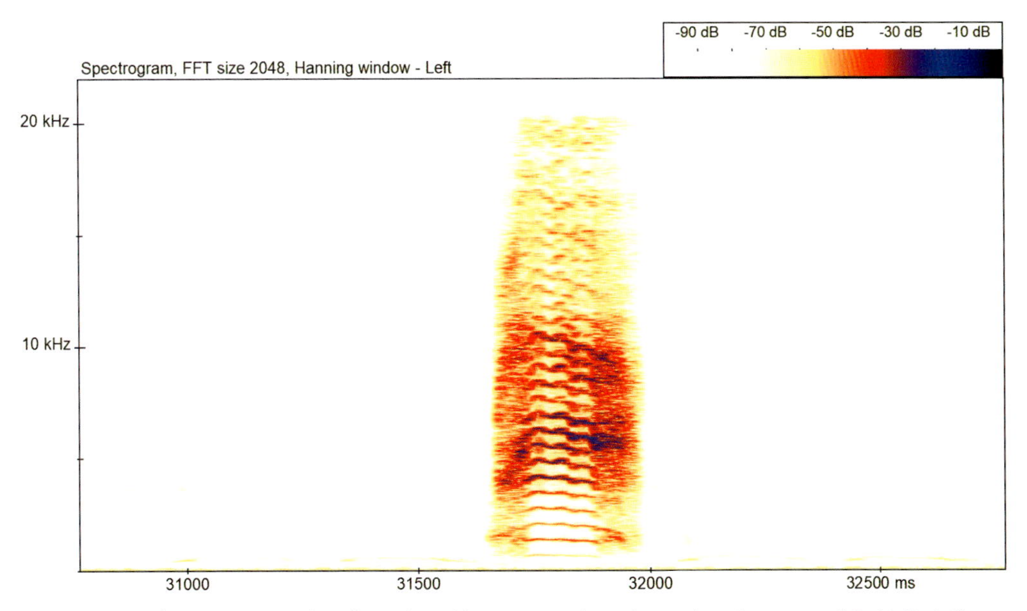

Figure 10.14 🔊 Mole – a single call produced by an animal on the surface (courtesy of D. Mellor) (frame width: c.2 sec)

QR codes relating to presented figures

Figure 10.13: Mole

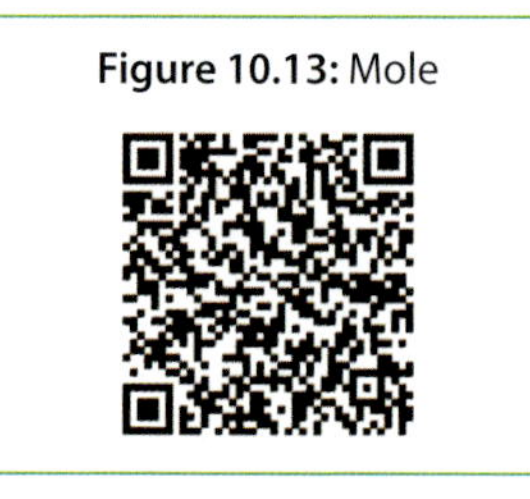

Potential confusion between mole and other species

Table 10.7 Summary of potential confusion between mole and other species

	Potential confusion species
Confusion group	In the first instance, always consider habitat and distribution.
Other mole species	This is the only species of mole within our area.
Species in other groups	Due to behaviour (i.e. predominantly living underground), unlikely to be recorded, and therefore unlikely to be confused with other species.

10.5 Shrew vocalisations

In the following section, a brief summary of the key characteristics of the six shrew species will be described, along with details on the species-specific vocalisations and, where known, the behavioural function of these sounds.

As well as using sound for communication (Simeonovska-Nikolova, 2004), it is thought that shrews may use acoustics for echo-orientation (Tomasi, 1979; Siemers *et al.*, 2009; Sanchez *et al.*, 2019). An interesting piece of research demonstrated that under laboratory conditions, common shrews *'used broadband echolocation pulses to locate protective cover'*, and to detect obstructions within tunnel systems (Forsman and Malmquist, 1988). These calls were described as being effective up to a distance of 200 mm. This study reinforced previous similar research on four North American shrew species (Gould *et al.*, 1964). Another study from 2009 which also used common shrew (and greater white-toothed shrew *Crocidura russula*) in its experiments demonstrated echolocation functioning in the behaviour of these species, describing shrew echolocation as not being as sophisticated as in bats, but stating that shrews use *'call reverberations for simple, close-range spatial orientation'* (Siemers *et al.*, 2009). In their discussion, the authors suggested that this mechanism is important for route planning, barrier avoidance and establishing escape routes. They also established that it was not likely that the echolocation served any purpose in hunting for prey. As such, the term 'echo-orientation' has been used to describe how shrews use sound to navigate within their environment (von Merten, 2011) as opposed to echolocation, used by bats to navigate and locate prey.

A further study which explored the differences in call structure of six different shrew species in Europe (Zsebok *et al.*, 2015) went on to suggest that these twittering calls may have a dual function (echo-orientation and conflict avoidance), but more studies in this respect were needed. This same study demonstrated that although, taking numerous measurements, it was possible, to a degree, to separate calls to species level (with c.66% accuracy), there was a wide range of overlap between the species studied. It is interesting that Siemers *et al.*, von Merten and Zsebok *et al.* all failed to find any evidence of the click calls described by earlier researchers, the suggestion being that modern technology that is now being used in research is finding calls and measuring their parameters more accurately than earlier researchers would have been able to.

An overview of shrew vocalisations for the three most widespread species of shrews present in mainland Britain (i.e. common shrew, pygmy shrew and water shrew) for the purpose of species identification was published by ourselves in *British Wildlife* (Newson *et al.*, 2020). Figure 10.15 shows the minimum frequencies produced by each of those species, and Figure 10.16 shows that same information as it relates to greater white-toothed shrew and lesser white-toothed shrew.

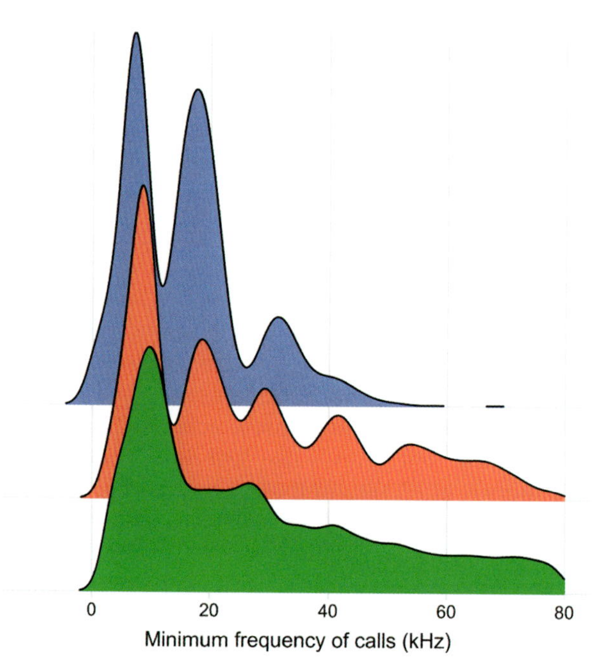

Figure 10.15 Minimum frequency of calls, with corresponding harmonic frequencies, for water shrew (blue), common shrew (red) and pygmy shrew (green) (updated from Newson *et al.*, 2020 with new material)

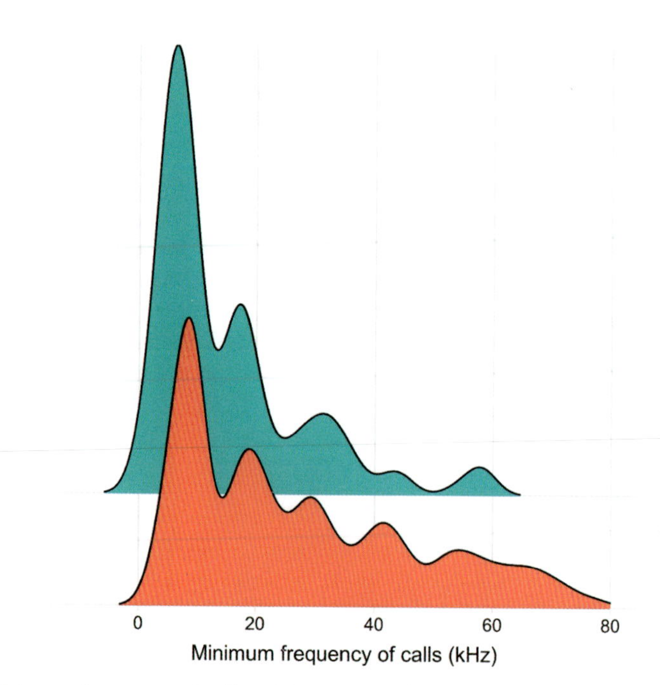

Figure 10.16 Minimum frequency of calls, with corresponding harmonic frequencies, for greater white-toothed shrew (red) and lesser white-toothed shrew (green).

10.5.1 Common shrew *Sorex araneus*

<table>
<tr><th>Length/Weight</th><th>Habitat preferences</th><th>Distribution</th></tr>
<tr><td>Males and females similar in size, with body length of 48 to 80 mm, and weight range of 5 to 14 g.</td><td rowspan="3">Unimproved grassland
Scrubland
Urban and gardens
Broadleaved woodland
Mixed woodland
Parkland
Hedgerows
Arable</td><td rowspan="3">Occurs throughout Britain (absent from Ireland), albeit scarcer in the far north of Scotland, and absent from many islands.*</td></tr>
<tr><th>Diet</th></tr>
<tr><td>Insectivorous, feeding across a range of invertebrates (e.g. earthworms, beetles, slugs, snails and spiders).</td></tr>
<tr><th>Daily activity pattern</th><td colspan="2">Active day and night, with short periods of rest.</td></tr>
<tr><th>Seasonal activity pattern</th><td colspan="2">Active throughout the year, although less so during winter months (does not hibernate).</td></tr>
<tr><th>Mating activity pattern</th><td colspan="2">Solitary and territorial outside of mating activities.
Promiscuous mating occurs between April and September.</td></tr>
<tr><th>Other notes</th><td colspan="2">Due to small size, difficult to encounter visually, but may be recorded using appropriately positioned camera traps.</td></tr>
<tr><td colspan="3">References
Churchfield and Searle, 2008a; *Mathews et al., 2018; Mammal Society, 2022.</td></tr>
</table>

Acoustic behaviour including spectrogram examples

For an overview of shrew vocalisations also refer to Section 10.5, preceding the species-specific accounts within this chapter.

Common shrew is rarely heard audibly, but produces identifiable sounds more frequently than most other small mammals. Some sounds are audible, at least for people with good hearing who are thus able to hear the lower-frequency part of the calls at close range. These high-pitched calls can be emitted when in contact (i.e. communication) with conspecifics and heterospecifics, or when in distress. More often they are heard during the breeding season (Churchfield and Searle, 2008a; von Merten, 2011).

Sounds produced have also been documented as serving an echo-orientation purpose, initially described as ultrasonic clicks in earlier studies, perhaps attributable to constraints in technology (von Merten, 2011). They have more recently been described as audible (sonic) twittering calls, variable in structure (Siemers *et al.*, 2009; von Merten, 2011; Zsebok *et al.*, 2015).

In the examples presented here we demonstrate how usually such calls are emitted in a series, with Figure 10.17 being a longer than typical encounter, albeit numerous emissions are to be expected. You should be aware, however, that occasionally a sequence of only a few calls may be recorded, and more rarely a single vocalisation. Figures 10.18

to 10.20 give you a progressively closer look at the call structure, and you can clearly see how the variance in frequency within a single call gives that warbling, vibrato effect so clearly heard when listening in time expansion (x10). Notice also how the maximum frequencies are reasonably stable throughout, and also the presence of harmonics. The second harmonic appears stronger in Figure 10.20, perhaps as a result of the microphone or recording specifications, whereby lower frequencies were not picked up as loudly, relative to higher frequencies.

The strength of second harmonics can also be seen in Figures 10.21 and 10.22, which act to emphasise a slight rising at the start, and falling at the end, of frequencies within a single call, but generally in most cases the frequency of calls remains fairly stable across a sequence. A further, more useful consideration for those based in Ireland is that of the three most widely distributed species (i.e. pygmy shrew, common shrew and water shrew), only pygmy shrew is present.

For most of our recordings of this species, the fundamental (first harmonic) is usually between 10 and 20 kHz. Very occasionally calls that do not align closely with what we would consider as typical for this species can be recorded, and as such, it is wise to always consider that examples beyond those shown may exist.

In addition to other shrew species as potential confusion species, of particular note among bird species within the region are the high-frequency 'mouse' calls of blue tit. Visually these calls can look very similar to common shrew (and Millet's shrew) in particular, but when played in x10 time expansion the warbling is slower than common and Millet's shrew.

Figures 10.17 to 10.22 provide specific examples of spectrograms relating to this species, and where the 🔊 symbol appears, a recording of what is shown within the spectrogram is available to download from the species library. In addition, QR codes are provided for selected examples, and these can be accessed immediately for listening purposes from most mobile devices.

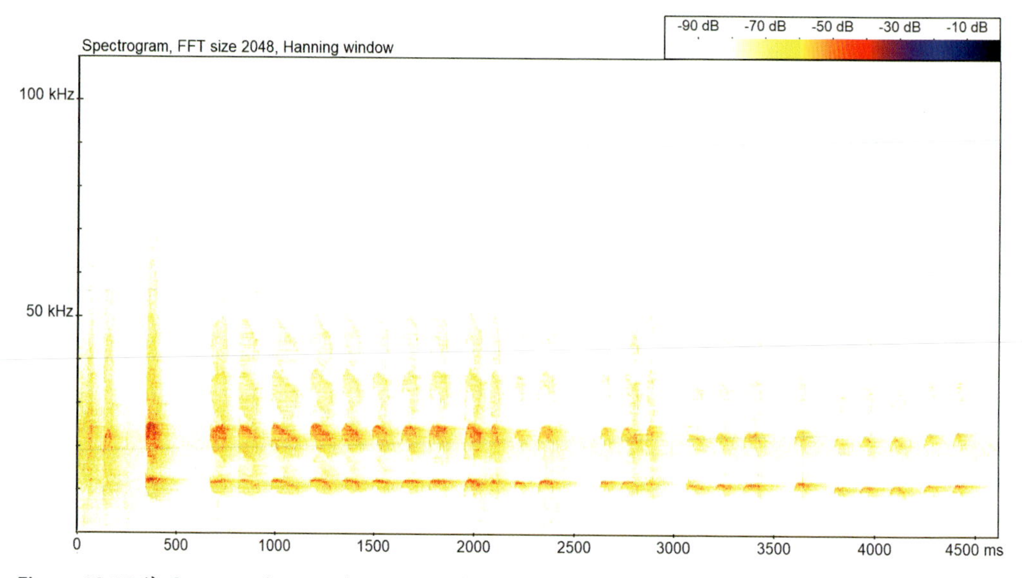

Figure 10.17 🔊 Common shrew – a long series of typical calls (frame width: c.5 sec)

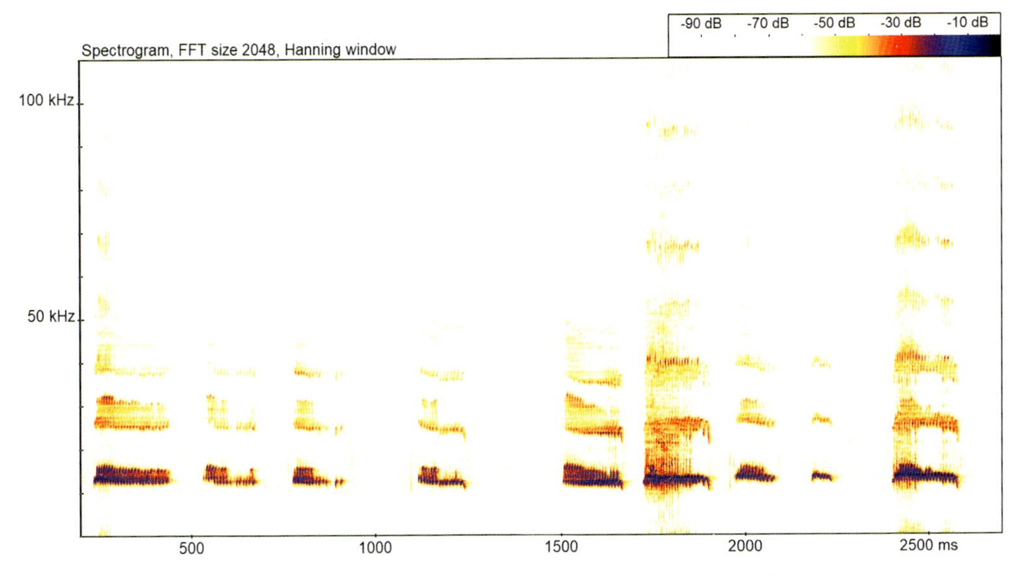

Figure 10.18 🔊 **(QR)** Common shrew – a section taken from a longer series of typical calls (courtesy of J. Wilson) (frame width: c.2.5 sec)

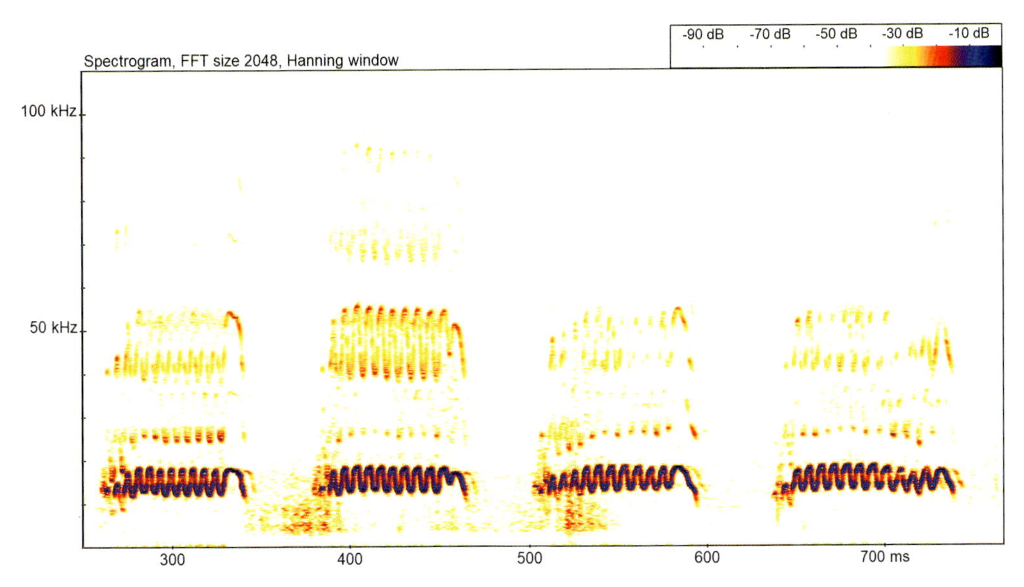

Figure 10.19 🔊 **(QR)** Common shrew – a series of four calls (courtesy of D. Galbraith) (frame width: c.0.5 sec)

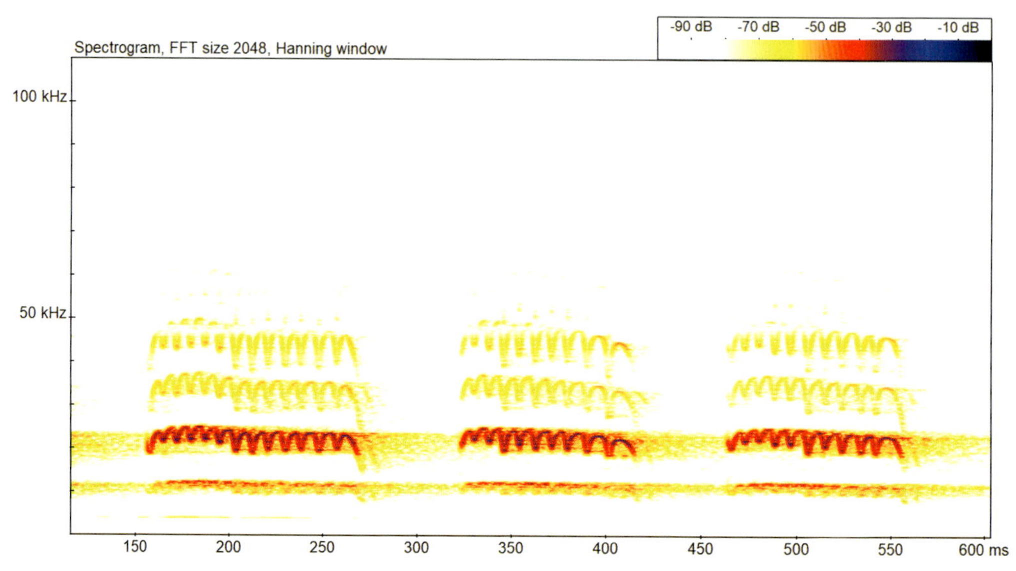

Figure 10.20 🔊 Common shrew – a series of three calls, showing more emphasis on the second harmonic (frame width: c.0.5 sec)

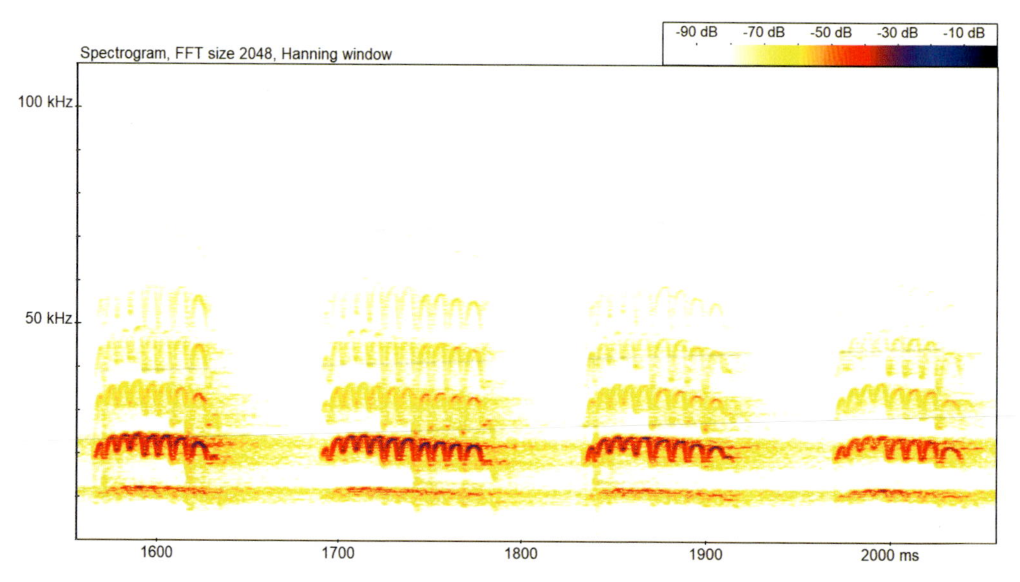

Figure 10.21 🔊 Common shrew – a series of four calls, with emphasis on second harmonic, showing slight rising in frequency at the start of each call (frame width: c.0.5 sec)

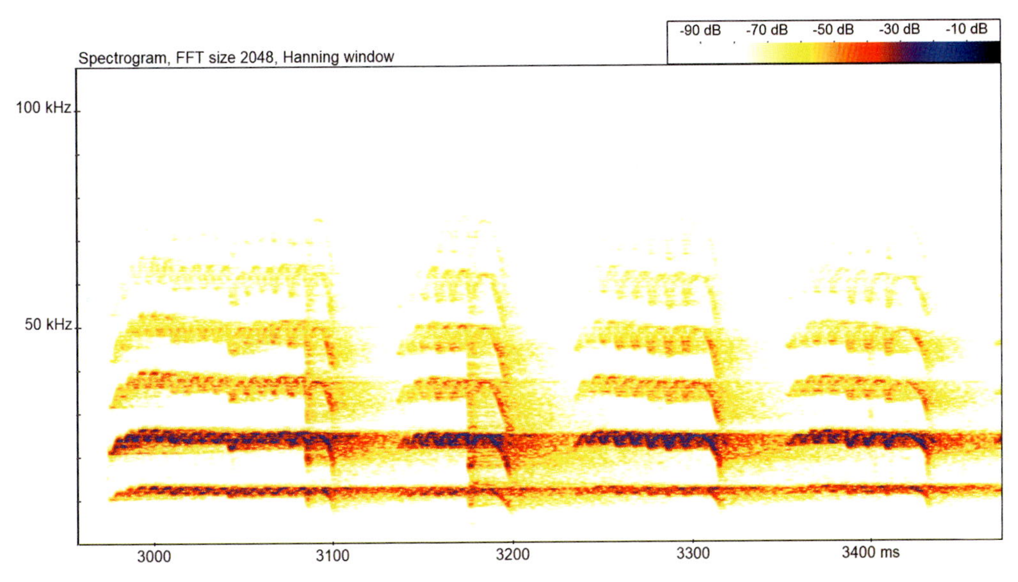

Figure 10.22 🔊 **(QR)** Common shrew – a series of four calls, with emphasis on second harmonic, showing slight rise in frequency at the start of each call, and a fall in frequency at the end (frame width: c.0.5 sec)

QR codes relating to presented figures

Figure 10.18: Common shrew call (time expanded x10 for listening purposes)	**Figure 10.19:** Common shrew call (time expanded x10 for listening purposes)	**Figure 10.22:** Common shrew call (time expanded x10 for listening purposes)

Potential confusion between common shrew and other species

Table 10.8 Summary of potential confusion between common shrew and other species

Confusion group	Potential confusion species
	In the first instance, always consider habitat and distribution.
Other shrew species	May get confused with Millet's shrew, but there is nowhere within the geographic range of this book where both species occur.
Species in other groups	Noctule bat social calls described as 'fast trills' often get mistaken for common shrews or vice versa. The 'mouse' call emitted by blue tit can sound similar.

10.5.2 Pygmy shrew *Sorex minutus*

Length/Weight	Habitat preferences	Distribution
Males and females similar in size, with head/body length of 40 to 60 mm. Weight range of 2.4 to 6.1 g.	Deciduous woodland Mixed woodland Grassland Gardens Parkland Arable	Occurs throughout Britain and Ireland.*
Diet		
Insectivorous, feeding across a range of invertebrates (e.g. beetles, spiders and woodlice).		

Daily activity pattern	Active day and night.
Seasonal activity pattern	Active throughout the year, but spends less time above ground during colder winter months (does not hibernate).
Mating activity pattern	Solitary and territorial outside of mating activities. Promiscuous mating occurs between April and October.
Other notes	Our smallest shrew species (and smallest terrestrial mammal over the geographic range of this book), and with the exception of the non-native greater white-toothed shrew, the only shrew species found in Ireland.

References
Churchfield and Searle, 2008c; *Lysaght and Marnell, 2016; *Mathews *et al.*, 2018.

Acoustic behaviour including spectrogram examples

For an overview of shrew vocalisations also refer to Section 10.5, preceding the species-specific accounts within this chapter.

Pygmy shrew are less vocal than common shrew, but known to emit a short audible 'chit' when alarmed or in contact with conspecifics (Churchfield and Searle, 2008c). Sounds produced have also been documented as serving an echo-orientation purpose, described as audible twittering calls that are variable in structure (von Merten, 2011; Zsebok *et al.*, 2015).

Typical pygmy shrew calls are different from all other shrew species in Britain and Ireland, in producing vocalisations that show a lot of variability in frequency, bandwidth and call structure within a sequence. A further, more useful consideration for those based in Ireland is that pygmy shrew is the only shrew species native to Ireland, and that although widespread and expanding its range, the calls of greater white-toothed shrew are distinctive from those of pygmy shrew.

In the examples presented here we demonstrate how calls are usually emitted in a series of often short bursts of activity as shown in Figure 10.23. In this example, the sequence commences rather harshly, is followed by descending FM calls, but also includes short FM bursts at the end of the sequence. When played in time expansion (x10), you typically do not get the same warbling that you expect to hear in common shrew. Also, relatively speaking, pygmy shrew produces calls covering a far greater bandwidth than common shrew.

Figures 10.24 and 10.25 show variations in call structure. Finally in Figures 10.26 and 10.27 we see examples where the calls rise steeply in frequency at the very start, before falling in frequency.

For most of our recordings for this species, the fundamental is usually between 10 and 20 kHz. Sometimes calls that do not align closely with what we would consider as being typical for this species can be recorded, and as such, it is a good idea to always consider that call structures beyond those shown may exist.

Figures 10.23 to 10.27 provide specific examples of spectrograms relating to this species, and where the 🔊 symbol appears, a recording of what is shown within the spectrogram is available to download from the species library. In addition, QR codes are provided for selected examples, and these can be accessed immediately for listening purposes from most mobile devices.

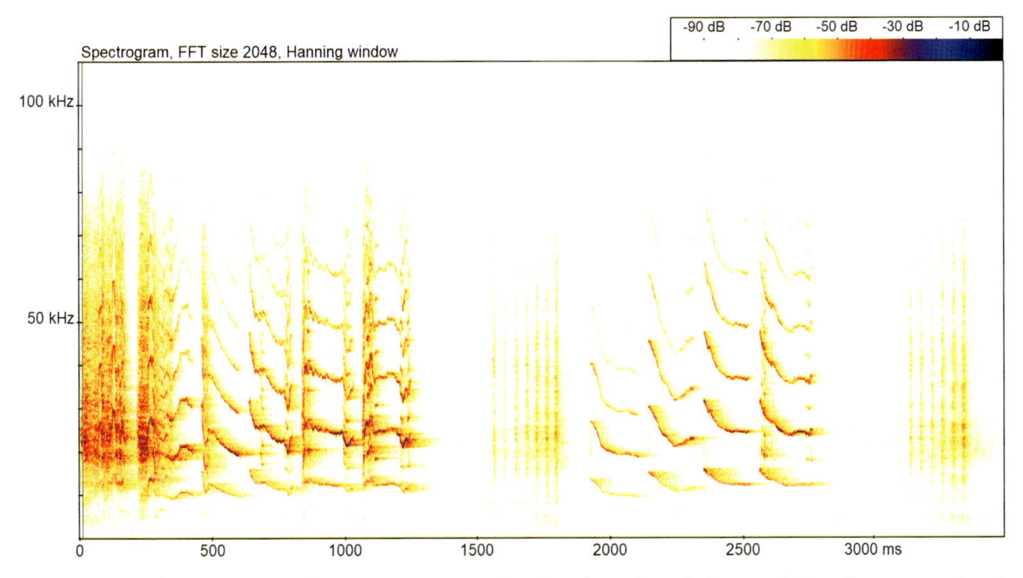

Figure 10.23 🔊 (QR) Pygmy shrew – a sequence showing three broad characteristics often encountered (frame width: c.3.5 sec)

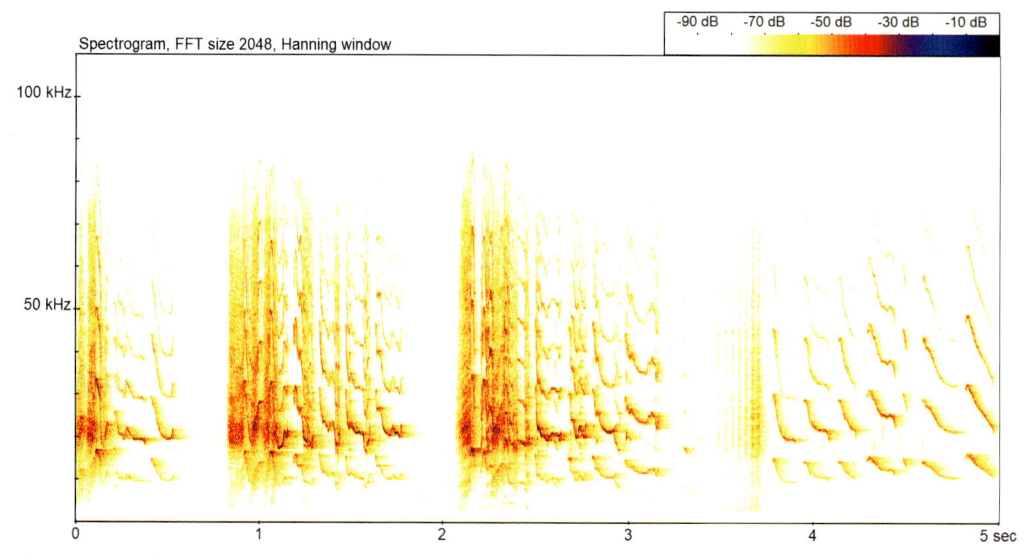

Figure 10.24 🔊 **(QR)** Pygmy shrew – a typical sequence of variable calls going through a relatively wide bandwidth of frequencies (frame width: c.5 sec)

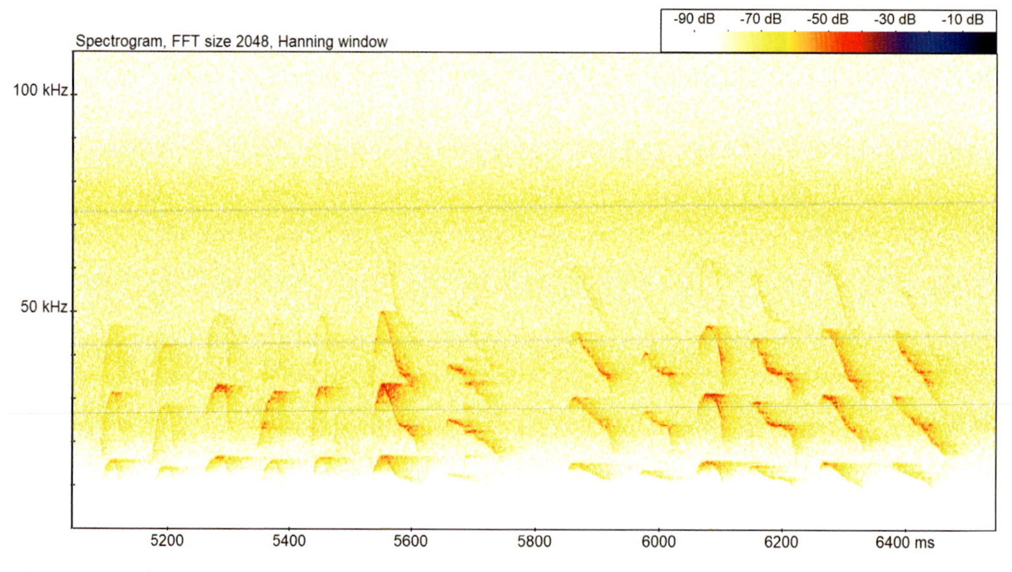

Figure 10.25 🔊 Pygmy shrew – a series of typical calls (courtesy of I. Kaergaard) (frame width: c.1.5 sec)

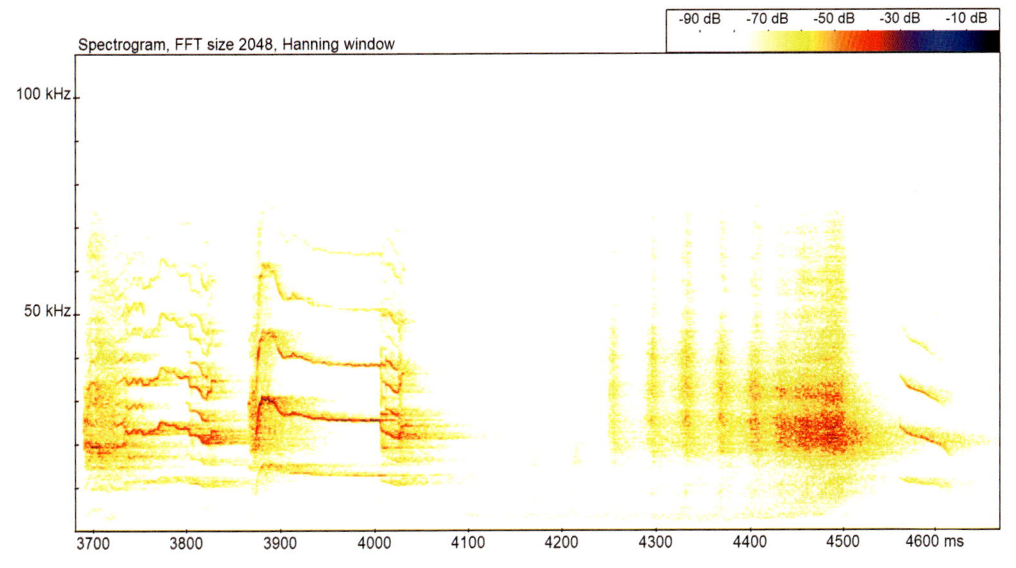

Figure 10.26 🔊 Pygmy shrew – a series of variable call structures, including an example of a call ascending sharply at the start, before dropping off and levelling out in frequency (frame width: c.1 sec)

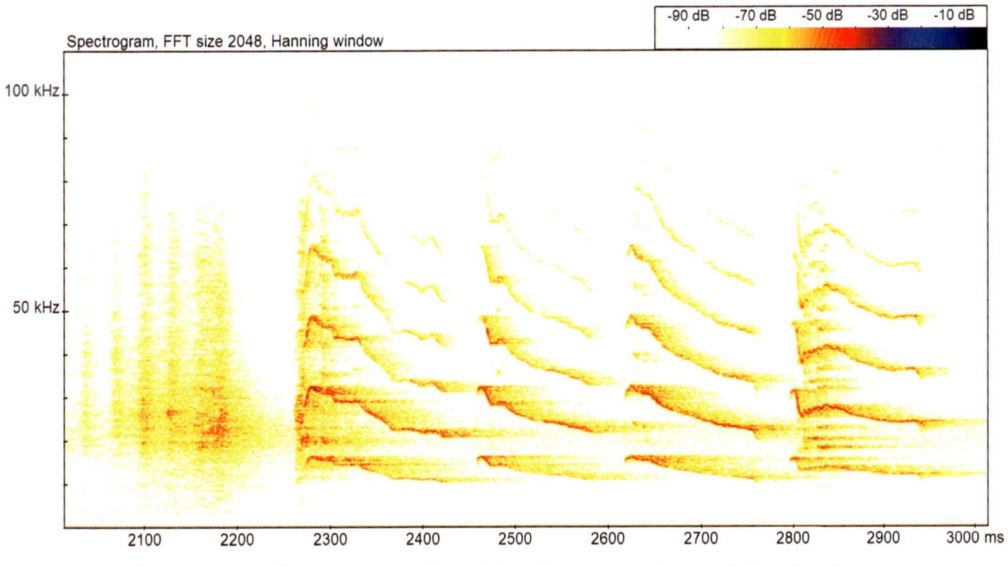

Figure 10.27 🔊 **(QR)** Pygmy shrew – a series of variable call structures (frame width: c.1 sec)

QR codes relating to presented figures

Figure 10.23: Pygmy shrew (time expanded x10 for listening purposes)

Figure 10.24: Pygmy shrew (time expanded x10 for listening purposes)

Figure 10.27: Pygmy shrew (time expanded x10 for listening purposes)

Potential confusion between pygmy shrew and other species

Table 10.9 Summary of potential confusion between pygmy shrew and other species

	Potential confusion species
Confusion group	In the first instance, always consider habitat and distribution.
Other shrew species	May be confused with water shrew.
Species in other groups	Unlikely to be confused with other small mammal species.

10.5.3 Millet's shrew *Sorex coronatus*

Length/Weight	Habitat preferences	Distribution
Similar to common shrew. Males and females similar in size, with body length of 48 to 80 mm, and weight range of 5 to 14 g.	Unimproved grassland Scrubland Urban and gardens Broadleaved woodland Mixed woodland Parkland Hedgerow Arable coastal Sand dunes Heath	Only found on the island of Jersey (Channel Islands), where common shrew is not present.
Diet		
Similar to common shrew. Insectivorous, feeding across a range of invertebrates (e.g. earthworms, beetles, slugs, snails and spiders).		
Daily activity pattern	Active day and night, with short periods of rest.	
Seasonal activity pattern	Active throughout the year, although less so during winter months (does not hibernate).	
Mating activity pattern	Solitary and territorial outside of mating activities. Promiscuous mating occurs between April and September.	
Other notes	Regarded as a sibling species to common shrew, and in many respects these species are very similar.	
References Churchfield and Searle, 2008b; McGowan and Gurnell, 2015.		

Acoustic behaviour including spectrogram examples

For an overview of shrew vocalisations also refer to Section 10.5, preceding the species-specific accounts within this chapter.

At the time of writing this, we have fewer examples of Millet's shrew vocalisations than any other shrew species. With this caveat, our current thinking is that the calls of Millet's shrew are extremely similar to common shrew, but with common shrew being absent from Jersey, there is no possibility for confusion within the geographic range of this book. Being one of only two shrew species on Jersey, where the second shrew species is lesser white-toothed shrew, which typically produces quite different calls, it should be reasonably easy to distinguish this species from other small mammals present on the island. For overseas readers wanting to distinguish Millet's shrew from common shrew, there is some evidence, although based on a small sample of Millet's shrew recordings, that the calls of Millet's shrew are a bit lower in frequency than common shrew, but the calls are otherwise very similar.

The examples we present to you here (Figures 10.28 to 10.30) are typical of the recordings we have encountered and appear to follow a similar call structure to those previously described within Section 10.5.1 (common shrew). Likewise, when listened to in time expansion you hear the same warbling vibrato you would expect if this were common shrew.

Figures 10.28 to 10.30 provide specific examples of spectrograms relating to this species, and where the 🔊 symbol appears, a recording of what is shown within the spectrogram is available to download from the species library. In addition, QR codes are provided for selected examples, and these can be accessed immediately for listening purposes from most mobile devices.

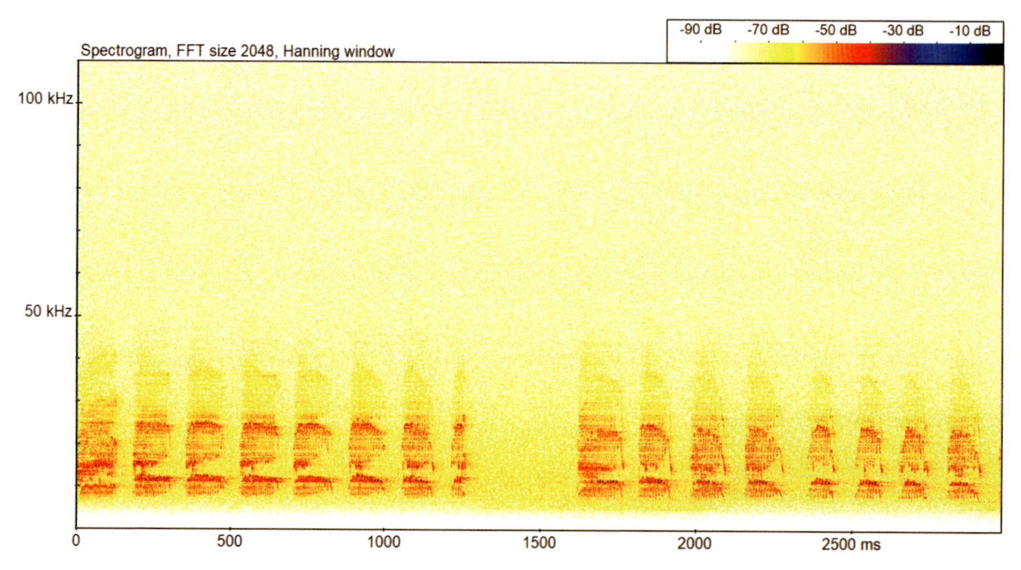

Figure 10.28 🔊 Millet's shrew – a series of typical calls, similar to those produced by common shrew (courtesy of L. Walsh) (frame width: c.3 sec)

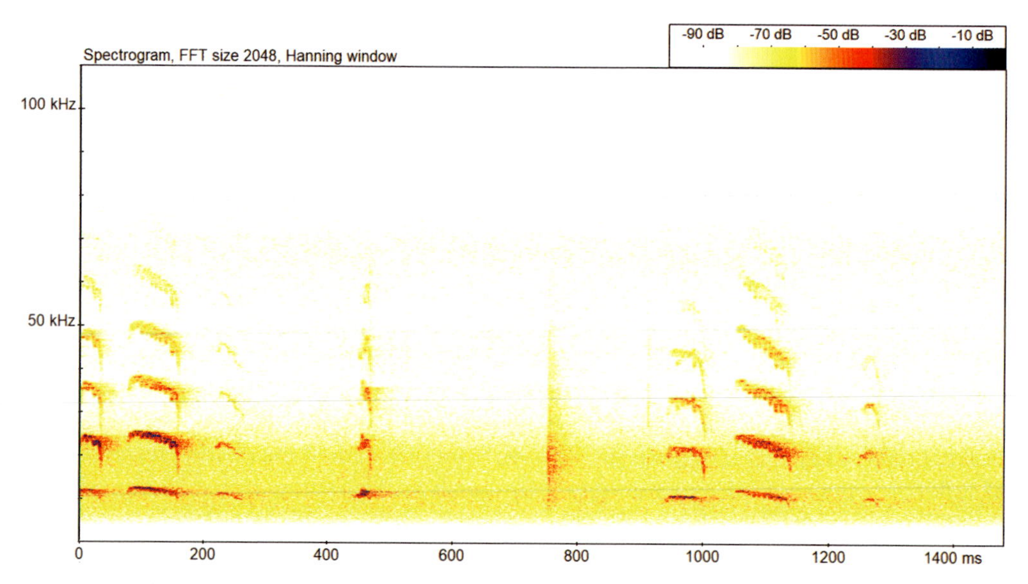

Figure 10.29 🔊 Millet's shrew – a series of typical calls, similar to those produced by common shrew (courtesy of L. Walsh) (frame width: c.1.5 sec)

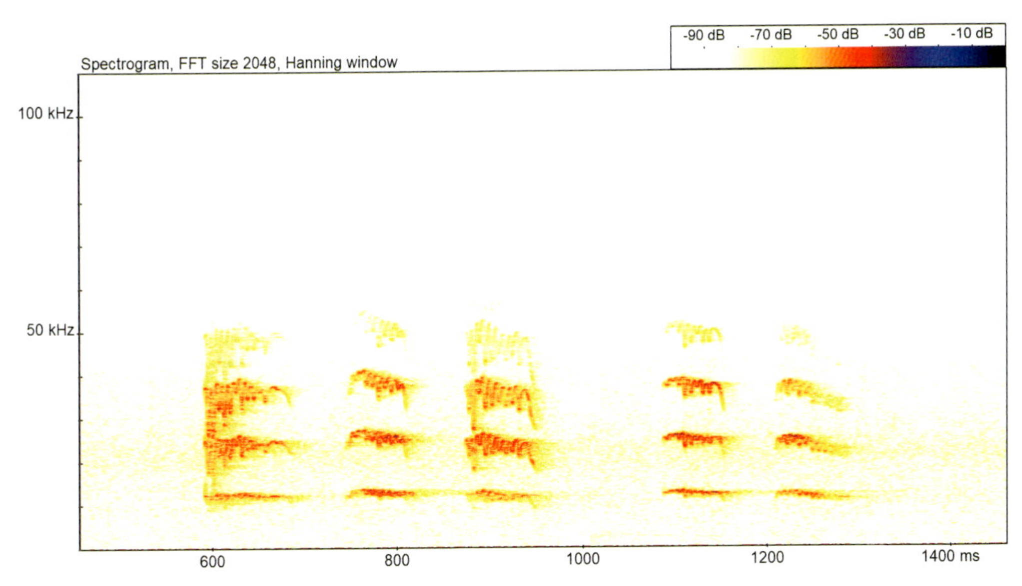

Figure 10.30 🔊 **(QR)** Millet's shrew – a series of typical calls, similar to those produced by common shrew (courtesy of L. Walsh) (frame width: c.1 sec)

QR codes relating to presented figures

Figure 10.30: Millet's shrew call (time expanded x10 for listening purposes)

Potential confusion between Millet's shrew and other species

Table 10.10 Summary of potential confusion between Millet's shrew and other species

	Potential confusion species
Confusion group	In the first instance, always consider habitat and distribution.
Other shrew species	May be confused with common shrew.
Species in other groups	Noctule bat social calls described as 'fast trills' often get mistaken for common shrews or vice versa. The 'mouse' call emitted by blue tit can sound similar.

10.5.4 Water shrew *Neomys fodiens*

Length/Weight	Habitat preferences	Distribution
Males and females similar in size, with head/body length of c.65 to 95 mm. Weight range of 12 to 18 g. **Diet** Insectivorous, feeding not only on aquatic invertebrates, but also terrestrial invertebrates, as well as frogs, newts and small fish.	Riparian habitat Rivers Streams Ponds Ditches Lakes Reedbeds Marshland Canals (Wet) Woodland	Occurring throughout Britain (absent from Ireland), albeit scarcer in northern Scotland, and absent from many islands.*

Daily activity pattern	Can be active day or night, but considered to be most active at night.
Seasonal activity pattern	Active throughout the year, but will spend less time on ground during colder winter months.
Mating activity pattern	Solitary and thought to be territorial outside of mating activities. Promiscuous mating occurs between April and September.
Other notes	Semi-aquatic.

References
Carter and Churchfield, 2006, *Churchfield, 2008a; *Mathews *et al.*, 2018; Mammal Society, 2022.

Acoustic behaviour including spectrogram examples

For an overview of shrew vocalisations also refer to Section 10.5, preceding the species-specific accounts within this chapter.

Water shrew calls within audible range may be heard by people with good hearing, although against a backdrop of running water (when present) may be hard to notice.

Their calls are heard more frequently during summer months, when it is noted for making audible squeaks and a rolling 'churr-churr' during threat or alarm behaviour (Churchfield, 2008a). Sounds produced have also been documented as serving an echo-orientation purpose, described as audible twittering calls, variable in structure (Zsebok *et al.*, 2015).

The minimum frequencies emitted by water shrew also tend to be noticeably lower (< 10 kHz) compared to common shrew and pygmy shrew, and with reference to the latter, water shrew call sequences appear to be less complex.

Compared with common shrew, there is not the same warbling when listened to slowed down (at x10 time expansion), at least across calls in a sequence, and instead the calls sound more mournful. Having said this, a degree of variation should be expected and separating water shrew from pygmy shrew could be difficult in some situations.

The examples we present to you here (Figures 10.31 to 10.33) are typical of the recordings we have encountered, with a slightly humped appearance, and with the minimum frequencies falling much further below 10 kHz than other shrew species, including common shrew. Also note that when zoomed in to view the calls, there is no impression of warbling, as you would see with common shrew; the call structures instead look relatively smooth.

Figures 10.34 and 10.35 differ because within each recording there are two animals present, and you can see areas where calls from each shrew overlap.

Figures 10.31 to 10.35 provide specific examples of spectrograms relating to this species, and where the 🔊 symbol appears, a recording of what is shown within the spectrogram is available to download from the species library. In addition, QR codes are provided for selected examples, and these can be accessed immediately for listening purposes from most mobile devices.

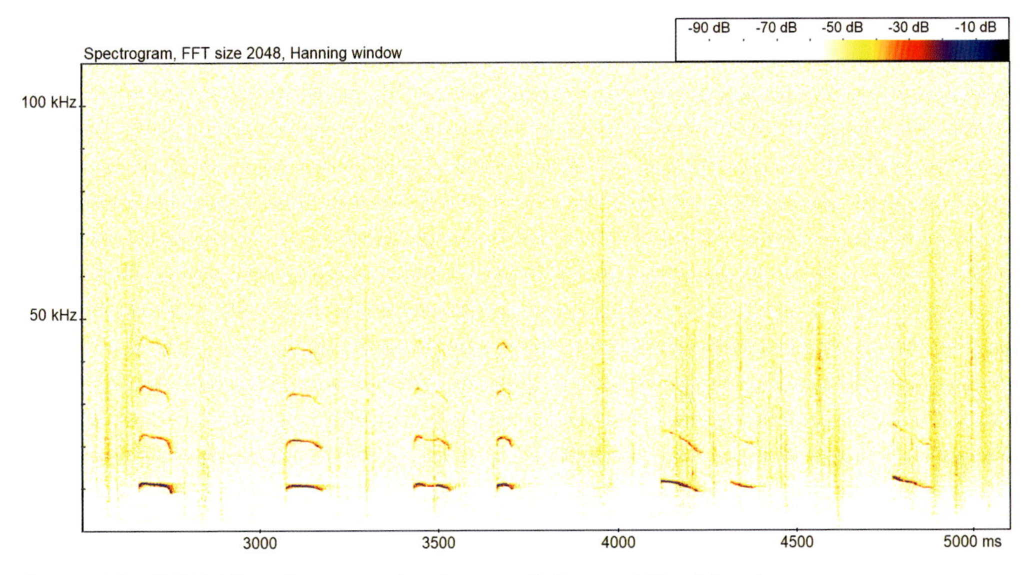

Figure 10.31 🔊 (QR) Water shrew – a series of seven calls (frame width: c.2.5 sec)

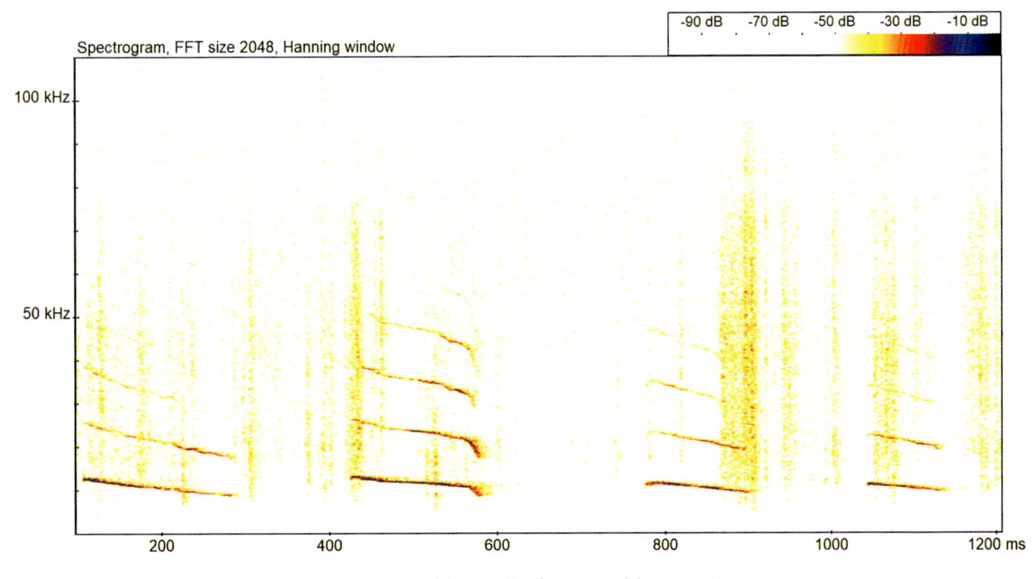

Figure 10.32 🔊 (QR) Water shrew – a series of four calls (frame width: c.1 sec)

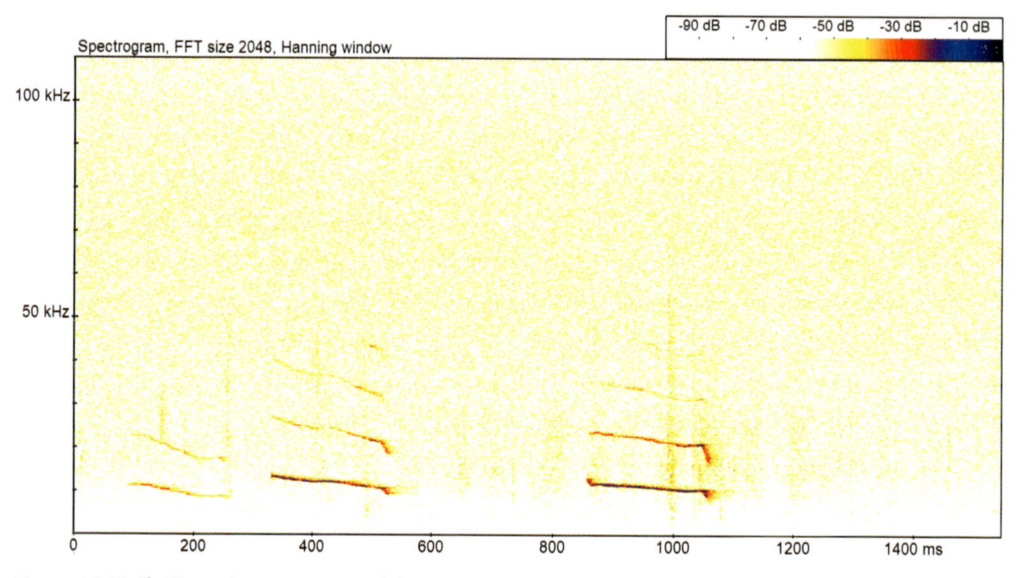

Figure 10.33 Water shrew – a series of three calls (frame width: c.1.5 sec)

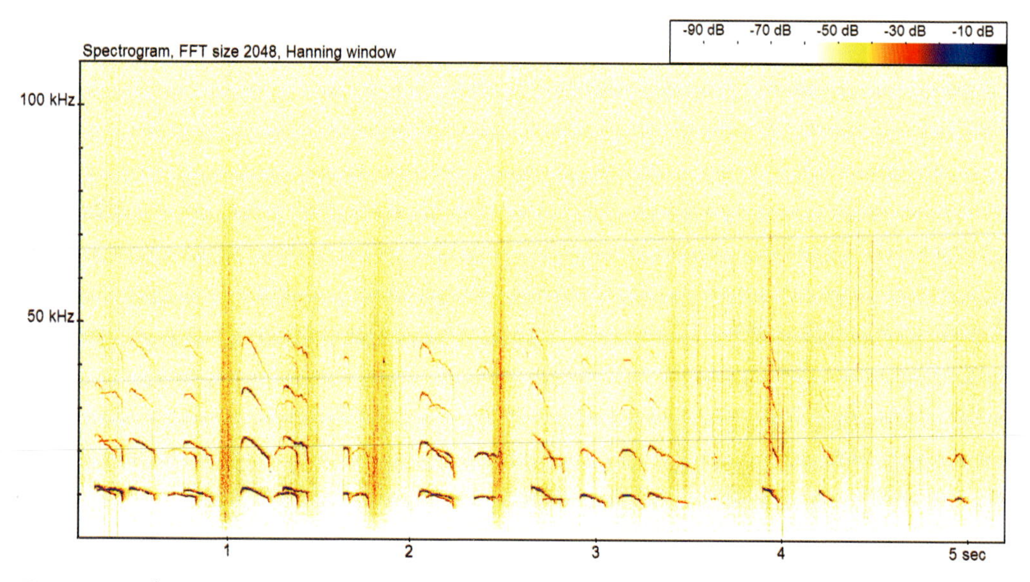

Figure 10.34 Water shrew – a longer sequence of events, with two animals present and calls sometimes overlapping (frame width: c.5 sec)

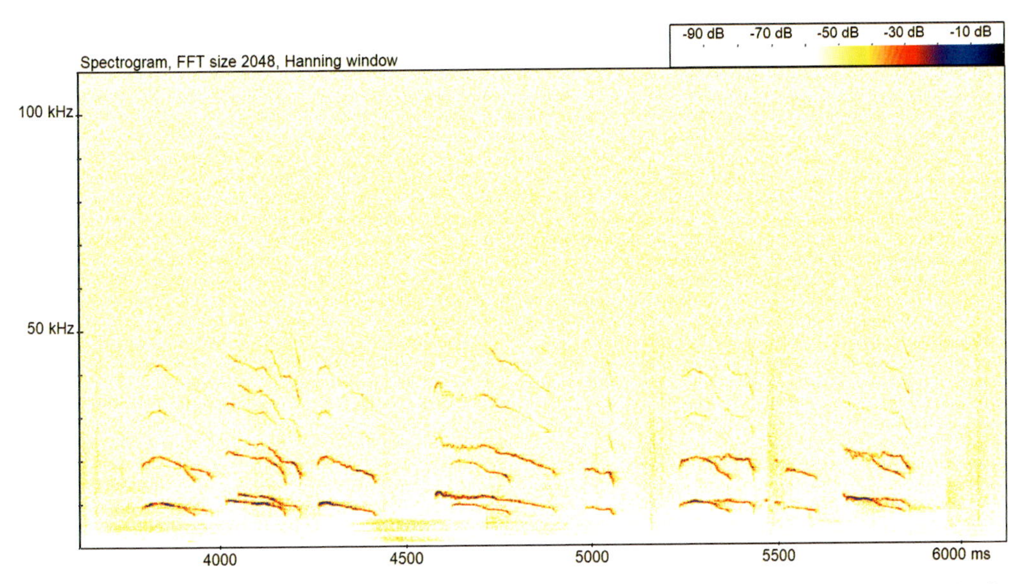

Figure 10.35 🔊 **(QR)** Water shrew – a longer sequence of events, with two animals present and calls sometimes overlapping (frame width: c.2.5 sec).

QR codes relating to presented figures

Potential confusion between water shrew and other species

Table 10.11 Summary of potential confusion between water shrew and other species

Confusion group	Potential confusion species
	In the first instance, always consider habitat and distribution.
Other shrew species	May be confused with pygmy shrew.
Species in other groups	Unlikely to be confused with other small mammals.

10.5.5 Lesser white-toothed shrew *Crocidura suaveolens*

Length/Weight	Habitat preferences	Distribution
Males and females similar in size, with head/body length of 50 to 75 mm. Weight range of 3 to 7 g. **Diet** Insectivorous, feeding across a range of invertebrates (e.g. beetles, flies, insect larvae and centipedes).	Deciduous woodland Mixed woodland Hedgerows Shoreland boulders and vegetation Coastal sand dunes, scrubland and heath	A non-native (naturalised), but possibly native shrew species confined within the range of this book to the Isles of Scilly, Jersey and Sark (Channel Islands).*

Daily activity pattern	Active day and night, but mostly overnight.
Seasonal activity pattern	Active throughout the year, but will spend less time above ground during colder winter months (does not hibernate).
Mating activity pattern	Solitary and territorial outside of mating activities. Promiscuous mating occurs between March and September.
Other notes	Its distribution within the geographical scope of this book does not overlap with greater white-toothed shrew.

References

*Churchfield and Temple, 2008; Mammal Society, 2022.

Acoustic behaviour including spectrogram examples

For an overview of shrew vocalisations also refer to Section 10.5, preceding the species-specific accounts within this chapter.

Lesser white-toothed shrew are rarely heard audibly, but may produce sounds when alarmed/threatened, audible to those of good hearing.

Due to the restricted distribution of lesser white-toothed shrew within the British Isles, it should be possible to confidently conclude its identification from acoustic recordings, at least in most situations. Particularly notable is that the fundamental (first harmonic) frequencies emitted by lesser white-toothed shrew, at c.20 kHz, tend to be much higher than all other shrew species, including greater white-toothed shrew. The distributions of lesser white-toothed shrew do not overlap with greater white-toothed shrew, pygmy shrew or water shrew within the geographic range of this book. As such, this aids confidence in identification from recordings.

For overseas readers wanting to distinguish pygmy shrew and water shrew from lesser white-toothed shrew, when recordings are played in x10 time expansion, you can expect to hear whistle-like calls with irregular and unpredictable changes in frequency within the sequence (i.e. not constant frequency), but not showing large changes in frequency as seen in pygmy shrew. These sound quite different from the strong warble or fluctuation in frequency that is heard in the calls of common shrew, and the calls are also much higher-pitched (higher minimum frequencies) than this species. Structurally, when comparing the calls of lesser white-toothed shrew against common shrew, if ever required to do so, the structure of common shrew vocalisations is usually much simpler due to their bandwidth being narrower, and ignoring the warble, frequencies of common shrew calls remain fairly consistent throughout an emission.

One important non-shrew confusion species to consider is brown rat (and black rat), which often emit calls of a similar structure and at similar frequencies. Accordingly careful consideration may be required in locations where both lesser white-toothed shrew and brown rat may be present. However, the irregular and unpredictable changes in frequency within the calls of lesser white-toothed shrew when played in x10 time expansion should help to exclude brown rat.

The examples we present to you here (Figures 10.36 to 10.38) are typical of the recordings we have encountered, and we especially wish to thank Darren Hart (Isles of Scilly Wildlife Trust) for providing us with these.

Figures 10.36 to 10.38 provide specific examples of spectrograms relating to this species, and where the 🔊 symbol appears, a recording of what is shown within the spectrogram is available to download from the species library. In addition, QR codes are provided for selected examples, and these can be accessed immediately for listening purposes from most mobile devices.

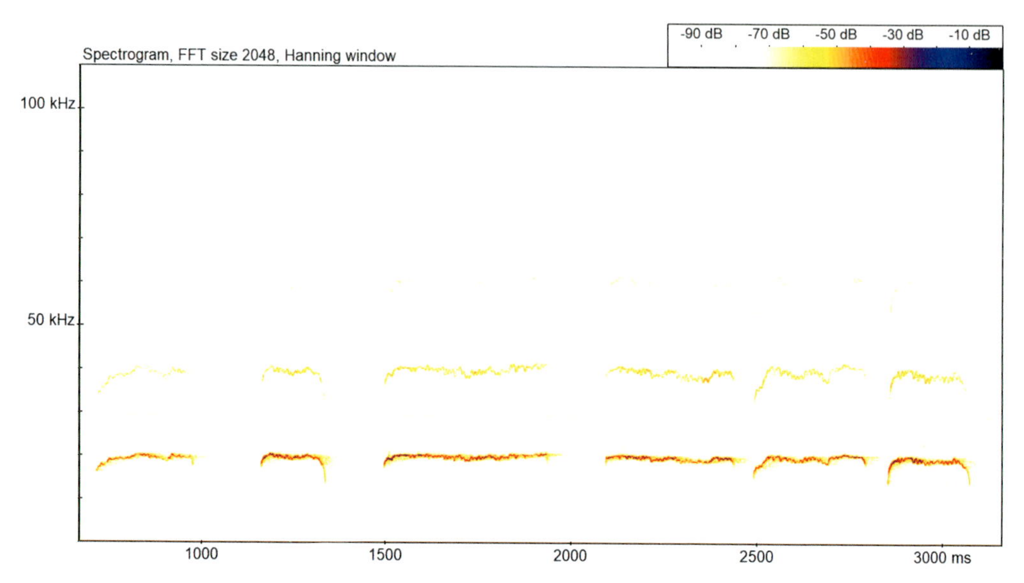

Figure 10.36 🔊 **(QR)** Lesser white-toothed shrew – a series of six calls produced by a single individual (courtesy of D. Hart, Isles of Scilly Wildlife Trust) (frame width: c.2.5 sec)

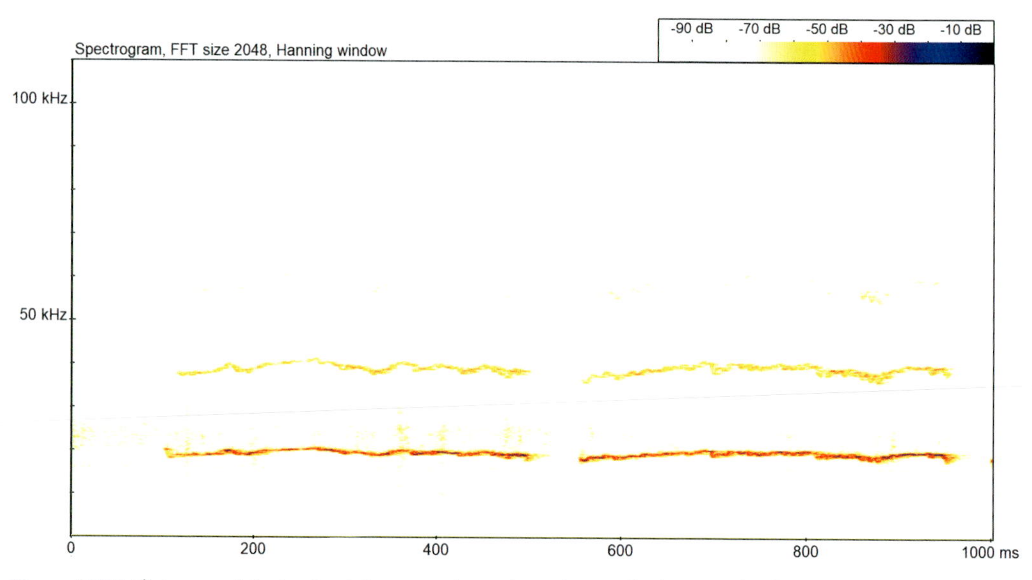

Figure 10.37 🔊 Lesser white-toothed shrew – a closer view of two calls showing the fluctuation/warble in the calls (courtesy of D. Hart, Isles of Scilly Wildlife Trust) (frame width: c.1 sec)

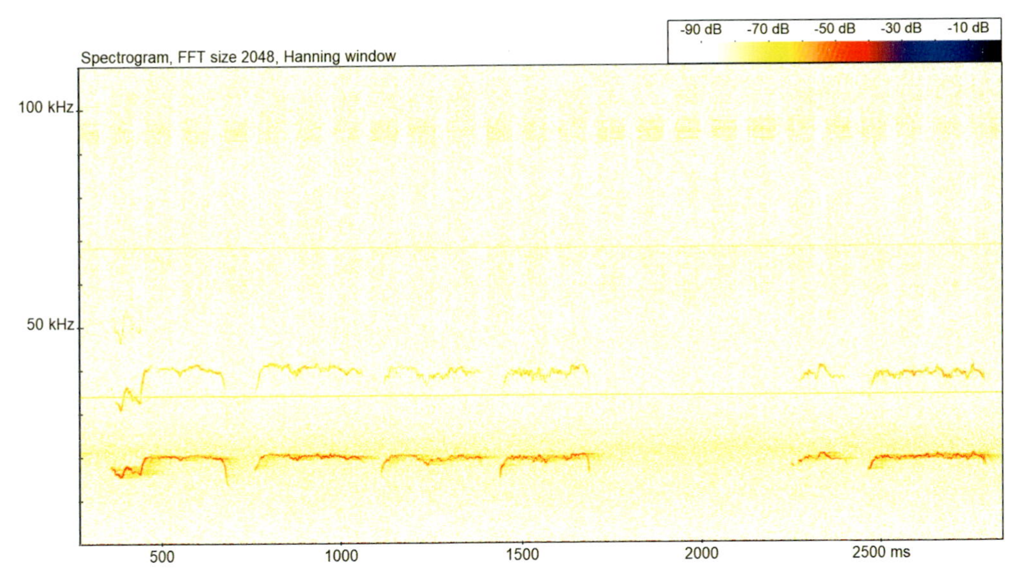

Figure 10.38 Lesser white-toothed shrew – showing variation in call durations and shape of calls (courtesy of D. Hart, Isles of Scilly Wildlife Trust) (frame width: c.2.5 sec)

QR codes relating to presented figures

Figure 10.36: Lesser white-toothed shrew call (time expanded x10 for listening purposes)

Potential confusion between lesser white-toothed shrew and other species

Table 10.12 Summary of potential confusion between lesser white-toothed shrew and other species

Confusion group	Potential confusion species
	In the first instance, always consider habitat and distribution.
Other shrew species	Unlikely to be confused with other shrew species.
Species in other groups	May be confused with brown rat/black rat 22 kHz distress calls.

10.5.6 Greater white-toothed shrew *Crocidura russula*

<table>
<tr><th>Length/Weight</th><th>Habitat preferences</th><th>Distribution</th></tr>
<tr><td>Males and females similar in size, with head/body length of 60 to 90 mm.
Weight range of 11 to 14.5 g.</td><td rowspan="3">Deciduous woodland
Mixed woodland
Grassland
Gardens
Parkland
Arable</td><td rowspan="3">Non-native (naturalised), but possibly native species in Guernsey, Alderney and Herm (Channel Islands). Non-native in the Republic of Ireland, and more recently recorded in Sunderland (north-east England) (2021).* </td></tr>
<tr><th>Diet</th></tr>
<tr><td>Insectivorous, feeding across a range of invertebrates (e.g. spiders, woodlice, snails, slugs, earthworms and centipedes). Occasionally known to prey upon small rodents, amphibians and lizards.</td></tr>
<tr><th>Daily activity pattern</th><td colspan="2">Active day and night.</td></tr>
<tr><th>Seasonal activity pattern</th><td colspan="2">Active throughout the year.</td></tr>
<tr><th>Mating activity pattern</th><td colspan="2">Considered to be less solitary and territorial than other shrew species, but more territorial during mating season.
Monogamous relationships occur, with mating season from February to October.</td></tr>
<tr><th>Other notes</th><td colspan="2">Its distribution within the geographical scope of this book does not overlap with lesser white-toothed shrew.</td></tr>
</table>

References
Churchfield, 2008b; *Lysaght and Marnell, 2016; *Mammal Society, 2022.

Acoustic behaviour including spectrogram examples

For an overview of shrew vocalisations also refer to Section 10.5, preceding the species-specific accounts within this chapter.

Whilst the frequency of greater white-toothed shrew calls are similar to common shrew, the sound when played in time expansion (x10) is normally quite different. In particular, similar to the closely related lesser white-toothed shrew, the whistle-like calls of greater white-toothed shrew show irregular and unpredicted changes in frequency within the calls (i.e. not constant frequency), and the calls in a sequence often sound like they have quite an abrupt end when played. These sound quite different from the strong warbling but otherwise quite predictable calls of common shrew.

Although there may be a variety of call structures (e.g. Figure 10.39), the calls of greater white-toothed shrew often include hump-shaped examples (Figures 10.40 to 10.42), occurring at c.10 kHz or above. Separation from lesser white-toothed shrew does not appear to be a problem, due to the difference in dominant frequencies (higher at c.20 kHz for lesser white-toothed shrew), but within the geographic scope of this book, there is not thought to be any overlap in the distribution between greater and lesser white-toothed shrew. Occasionally within our recordings, and more so than with other shrew species, we have seen the emission of a harsher sound in amongst the more typical vocalisations, as seen in Figure 10.43 at c.400 ms, 900 ms, 1,200 ms and 1,400 ms.

Figures 10.39 to 10.43 provide specific examples of spectrograms relating to this species, and where the 🔊 symbol appears, a recording of what is shown within the spectrogram is available to download from the species library. In addition, QR codes are provided for selected examples, and these can be accessed immediately for listening purposes from most mobile devices.

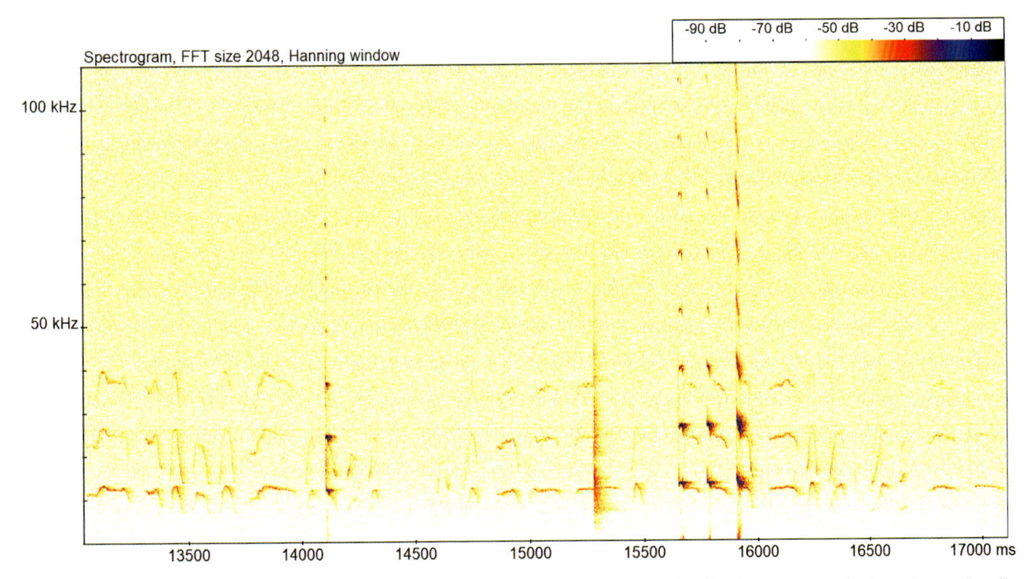

Figure 10.39 🔊 (QR) Greater white-toothed shrew – a long series of calls showing typical variety of call structures (frame width: c.4 sec)

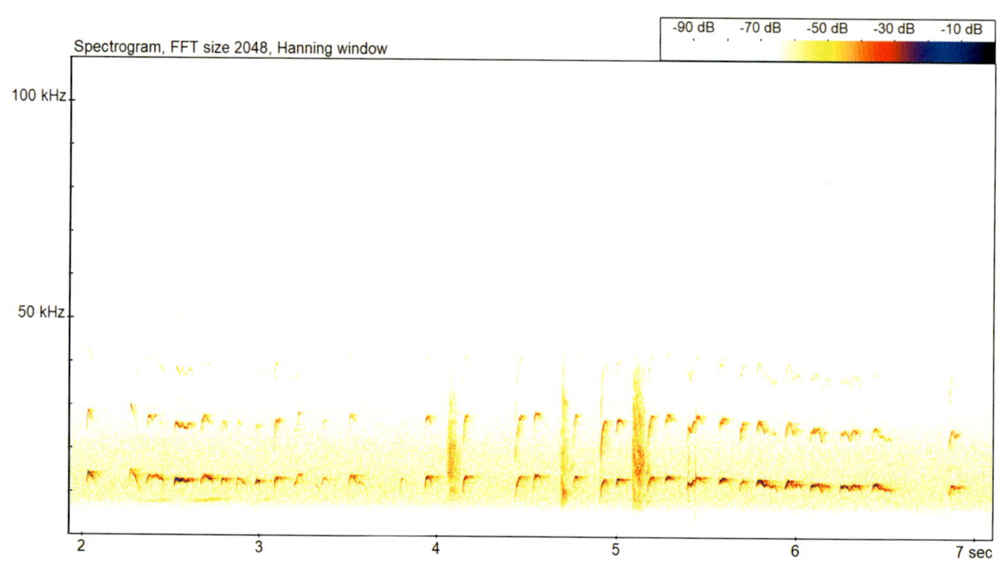

Figure 10.40 (QR) Greater white-toothed shrew – a long series of mostly hump-shaped typical calls (frame width: c.5 sec)

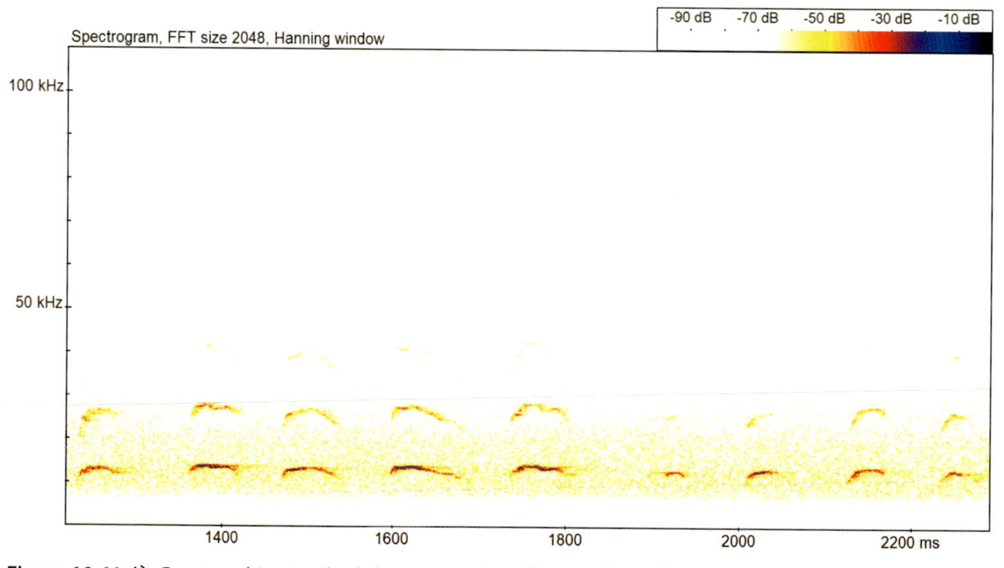

Figure 10.41 Greater white-toothed shrew – a series of hump-shaped typical calls (frame width: c.1 sec)

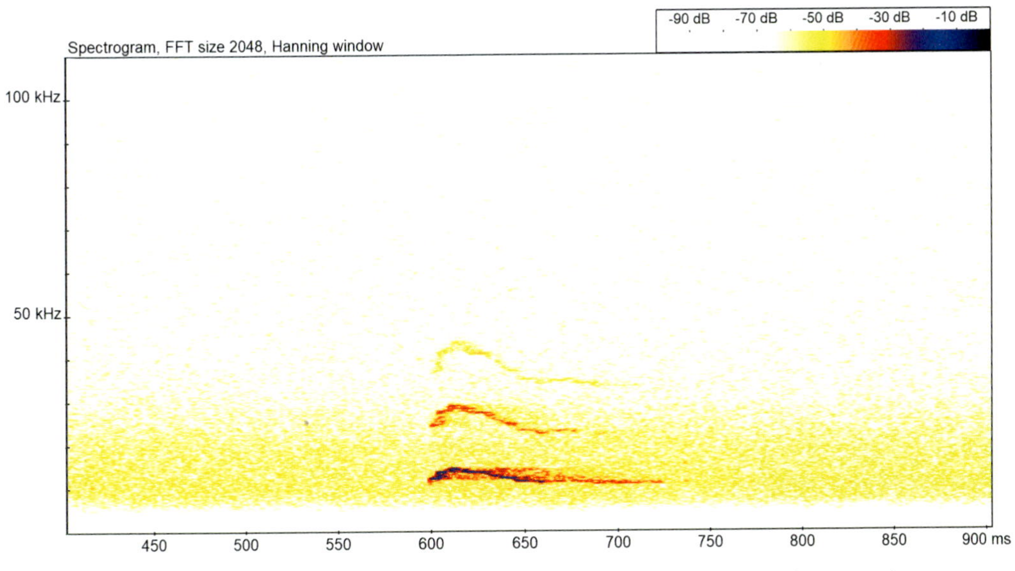

Figure 10.42 Greater white-toothed shrew – a single hump-shaped call (frame width: c.0.5 sec)

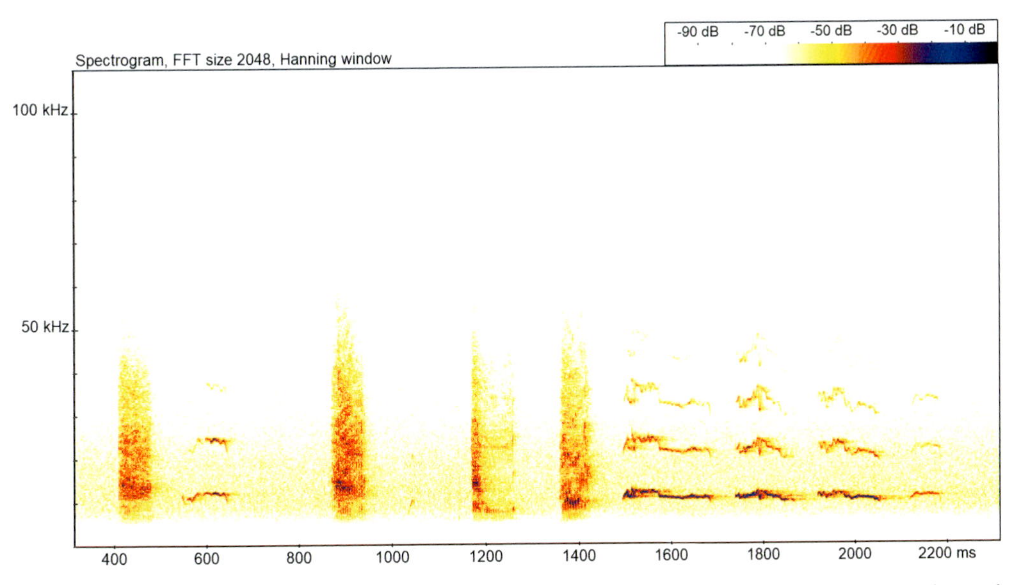

Figure 10.43 Greater white-toothed shrew – a series of typical calls, but with additional (four) harsher and more broadband emissions (frame width: c.2 sec)

QR codes relating to presented figures

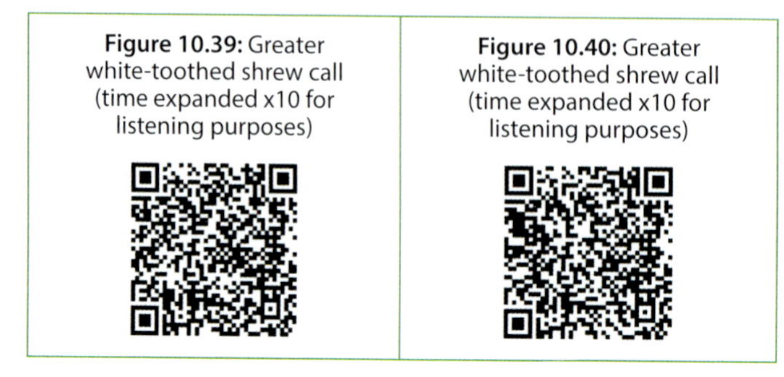

Figure 10.39: Greater white-toothed shrew call (time expanded x10 for listening purposes)

Figure 10.40: Greater white-toothed shrew call (time expanded x10 for listening purposes)

Potential confusion between greater white-toothed shrew and other species

Table 10.13 Summary of potential confusion between greater white-toothed shrew and other species

	Potential confusion species
Confusion group	In the first instance, always consider habitat and distribution.
Other shrew species	Unlikely to be confused with other shrew species.
Species in other groups	Unlikely to be confused with other small mammals.

10.6 Potential application of bioacoustics for monitoring insectivores

The potential for application of bioacoustics for monitoring moles and hedgehogs is probably quite limited, in that moles tend to be fairly self-evident when present in any case, and hedgehogs can be fairly conspicuous when active. At most, vocalisations made by hedgehog may provide additional information to support behavioural observations.

Shrews can be detected using an ultrasonic microphone (e.g. bat detector) that is positioned up to five metres away from the animal, without reducing the likelihood of detection, with this distance increasing to up to 14 metres when the microphone is placed at height. This means that data on the presence of shrew species can be successfully collected incidentally during routine static bat detector surveys, where bat detectors are typically deployed at height to record bats. This may be particularly valuable for shrews, where there are mortality risks associated with live trapping and handling.

The use of acoustic surveys could be used in combination with live trapping, where the trapping involves the use of Longworth traps that are fitted with shrew holes that catch most small mammal species but are specifically designed to allow shrews to escape. This is likely to be more appealing for volunteers and for use in citizen science small mammal monitoring projects, particularly since live trapping of shrews can only be carried out under licence. This approach would require less training and experience in small mammal trapping, and therefore has the potential to increase volunteer participation. It could also reduce the survey effort required to check traps if shrews are not likely to be caught, and it should avoid mortality risks associated with the live trapping and handling of shrews. In addition, by eliminating the need to capture shrews, the sensitivity of the trap treadle could be reduced to avoid triggering by shrews, thus increasing the potential capture incidence of other small mammal target species.

Similarly, there are clear opportunities to use acoustics to monitor the distribution and spread of invasive shrew species: the greater white-toothed shrew in northern England for example, where this species may adversely impact on native populations of pygmy shrew, as has been observed previously in the Republic of Ireland.

There are many valuable conservation-led reasons for understanding, as widely as possible, how to identify the presence of shrews (or potentially their absence) in order to take appropriately focused action which may benefit conservation. Table 10.14 gives some examples of where acoustic studies could be deployed, either stand-alone or in conjunction with other survey techniques, in order to gather useful data.

Table 10.14 Usefulness of acoustic studies – shrew examples

Species	Potential applications/benefits
All species	An opportunity to add considerably more data where information on any specific species is lacking (e.g. less well-covered parts of Britain and Ireland) or out of date. This may be particularly cost-effective if bat surveys are already being carried out in these areas, and the recordings can be mined for small mammals (e.g. using the BTO Acoustic Pipeline for a first analysis).
All species	A wider awareness of acoustics for this group means there is greater chance of someone realising that something unknown or unexpected is occurring in their area (e.g. presence of a native species or new invasive species, previously thought to be absent).
All species	Additional information to show health of local populations.
Greater white-toothed shrew **Lesser white-toothed shrew** **Millet's shrew**	Monitoring local populations to help confirm species presence at a local level, or for describing species distributions at a larger scale.

APPENDIX I

Glossary

Algorithm	A set of rules or a procedure that is to be followed in calculations or other problem-solving operations, especially by computer software.
Amplitude	The volume at which a sound is produced.
Attenuation	The reduction in amplitude of a sound as it travels through air. Higher frequencies usually attenuate more quickly than lower frequencies of the same intensity.
Audible	A sound that is within the normal hearing range of a human (< 20 kHz).
Bandwidth	A measure of the width of a range of frequencies, or the difference between minimum and maximum frequencies.
Broadband	A call that travels through a wide frequency range.
Component	A distinct single call separated in time from other calls.
Conspecific	Between animals of the same species.
Constant frequency (CF)	Frequency remains stable over time.
Decibel (dB)	Unit used to measure the intensity (loudness) of sound.
FM	*See* Frequency modulation.
fmax	Maximum or start frequency. The highest frequency emitted during a call.
FmaxE	Frequency of maximum energy, or peak frequency. The frequency level within a sound containing the most energy.
fmin	Minimum or end frequency. The lowest frequency emitted during a call.
Frequency	The word used to describe the measured sound level in terms of hertz or kilohertz. For example, 'the frequency produced is at 50 kHz'.
Frequency of maximum energy	*See* FmaxE.
Frequency modulation (FM)	Frequency changes over time.
Full spectrum	A recording format used by some ultrasonic devices/bat detectors (e.g. Batlogger M, EchoMeter Touch, Anabat Swift) whereby sounds of all frequencies are recorded accurately and in real time. The ultrasonic sounds cannot be heard by the surveyor, unless the bat detector has a function for creating simultaneous audible noise (e.g. auto-heterodyning). Full spectrum recordings are beneficial for sound analysis, as all sound within the parameters of the detector's capabilities are recorded, including any harmonics produced and/or sound from other sources occurring at the same time.
Fundamental	The first harmonic (H^1) created by any incidence of sound occurring.
Harmonics	Layers of sound occurring on top of each other, in multiples of the fundamental (lowest) noise, which is also called the first harmonic (H^1). Additional harmonics above the first (or fundamental) are described as second (H^2), third (H^3), etc.

Hertz (Hz)	A unit of frequency indicating one cycle per second. *Also see* Kilohertz (kHz).
Heterospecific	Between animals of different species.
Inaudible	A sound that is not within the normal hearing range of a human.
Infrasound	Noise occurring at a frequency below which a human with perfect hearing would be expected to hear (typically < 20 Hz).
Inter-component interval	The period occurring from the start of one component (i.e. distinct sound) to the start of the next component produced by the same source.
Kilohertz (kHz)	A unit of frequency indicating 1,000 cycles per second. Also see Hertz (Hz).
Millisecond (ms)	A unit of time that is equal to a thousandth of a second. Also see second (sec).
Narrowband	A call that travels through a narrow frequency range.
Omnidirectional	Receiving sound from all directions, as opposed to from one specific direction.
Oscillogram	A graph showing amplitude against time.
Peak frequency	*See* FmaxE.
Power spectrum	A graph showing amplitude against frequency for a recorded sound.
Quasi-constant frequency (QCF)	Nearly constant frequency.
Second (sec)	The typical base unit for measuring time. *Also see* millisecond (ms).
Sequence	A group of component calls closely associated with each other. An enumerated collection of calls that make up a vocalisation with a recognisable pattern.
Signal to noise ratio (SNR)	'Signal' is the noise from the target sound source, whereas 'noise' refers to background sound. In general, the louder the signal, relative to the noise, the better the recording.
SNR	*See* signal to noise ratio.
Sonic	Sound that is audible based on normal human hearing range.
Spectrogram	A graph showing sound as a visual representation of frequency against time.
Syllable	Each differing aspect, in terms of structure, within a single call component.
Time expansion	A process for recording sound whereby the original noise is slowed down by a factor (usually ×10), meaning that the duration of the event is expanded by the same factor, and the frequency is reduced, in order to become audible. Because all of the original sound is retained (albeit slowed down) time expanded recordings are beneficial for sound analysis, as all sound within the parameters of the detector's capabilities are recorded, including any harmonics produced and/or sound from other sources occurring at the same time.
Ultrasound	Noise occurring at a frequency beyond which a human with perfect hearing would be expected to hear, typically characterised as sound > 20 kHz.

APPENDIX II

Case Studies

In this appendix we provide three case studies, where we consider the opportunities that acoustics can offer for improving our understanding of particular species of mammal.

Case Study 1: Acoustic (ultrasonic) identification of hazel dormouse

The hazel dormouse was chosen as a target species to look at the efficacy of acoustic methods for monitoring small mammal species. The hazel dormouse is a good study species because they produce a range of ultrasonic vocalisations at the low frequency end of the ultrasonic spectrum (c.30 kHz or below), which combined with a higher amplitude than many other small mammal species (e.g. mice and voles), means that the calls of hazel dormice can be detected from further away; up to about five metres away from a bat detector (Newson and Pearce, 2021). Furthermore, they are a priority species of conservation importance, where monitoring by traditional methods (e.g. nest boxes/ nest tubes) can only be undertaken by appropriately licensed personnel. Successful application of acoustic methods could therefore open up opportunities for greater participation by amateur naturalists to survey for hazel dormouse, including away from established monitoring sites, with possibilities for increasing our understanding of their distribution and abundance.

Prior to carrying out acoustic surveys for any mammal species, it is important to consider how much survey effort is needed, and to target surveys according to an appropriate time of year, and time of day or night. In the case of hazel dormouse, which hibernate for up to six months of the year, surveys are best undertaken during their active period from April to October inclusive. Additionally, because they are most active at night (a strategy to avoid predation), recording should be carried out between dusk and dawn. Given that small mammals vocalise to communicate between conspecifics, we might expect the calling rate of hazel dormouse to be higher during courtship, mating and parental–neonatal interactions, and for there to be seasonal patterns in call frequency.

In a previous study (Newson and Pearce, 2021), we monitored a captive pair of hazel dormice at the British Wildlife Centre in 2021, from sunset to sunrise, between late April and the end of October. From this, manual analysis of the data identified 20,680 recordings (five-second .wav files) that contained hazel dormouse calls.

Looking at the seasonal and nightly patterns in vocal activity, the pair of captive hazel dormice were found to be vocal across the whole period and were recorded on up to 10% of nights between May and mid-October, with a peak in activity during the second half of August and early September (Figure A1, left-hand plot). These data suggest that acoustic surveys for hazel dormouse could be carried out at any time within their active period (April to October) but that surveys during August and September may increase the chances of detection. Nevertheless, detections were infrequent, and the main period of acoustic activity was during the middle of the night across all months (Figure A1, middle and right-hand plots). As such, walked dusk transect surveys (catering for the first one to two hours after sunset), which are typically carried out for bats, are unlikely to be an effective method for monitoring hazel dormice. Instead, to maximise their detection, static unmanned recording devices would be preferable.

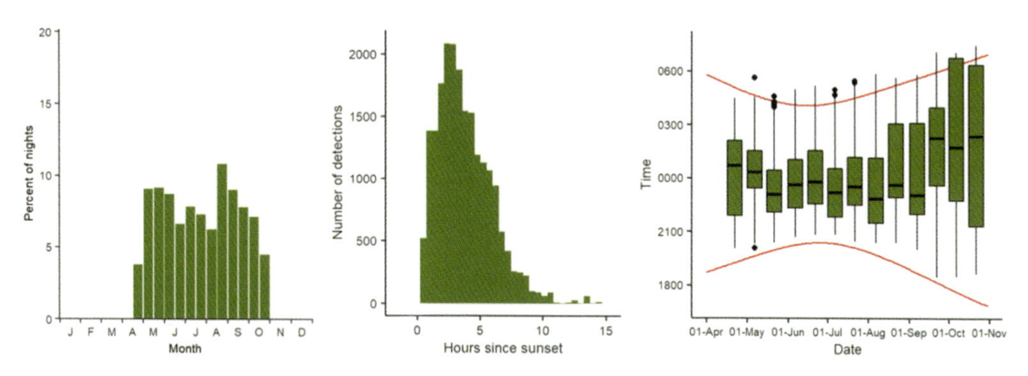

Figure A1 Seasonal and nightly activity of a pair of captive hazel dormice. The left-hand plot shows the percentage of nights on which the hazel dormouse was detected every half-month through the season, providing the periods of main activity for this species. The middle plot shows the overall spread of recordings with respect to sunset time, calculated over the whole season. The right-hand plot shows the spread of recordings with respect to sunset and sunrise times (red lines) summarised for each half-month through the season. For this last seasonal plot, the individual boxplots show quartiles (lower, median and upper) with lines extending to 1.5 times the interquartile range, and small dots show outliers.

A closer look at the data found that dormouse calls were often recorded in clusters of consecutive recordings at the same time of night, suggesting periods when this species was vocal. If we look at the data in relation to the average number of vocal events per night for each half-month (Figure A2), calling was greatest between mid-May and mid-August, with peaks in the second half of June and July, and a notable decline at the end of the survey period.

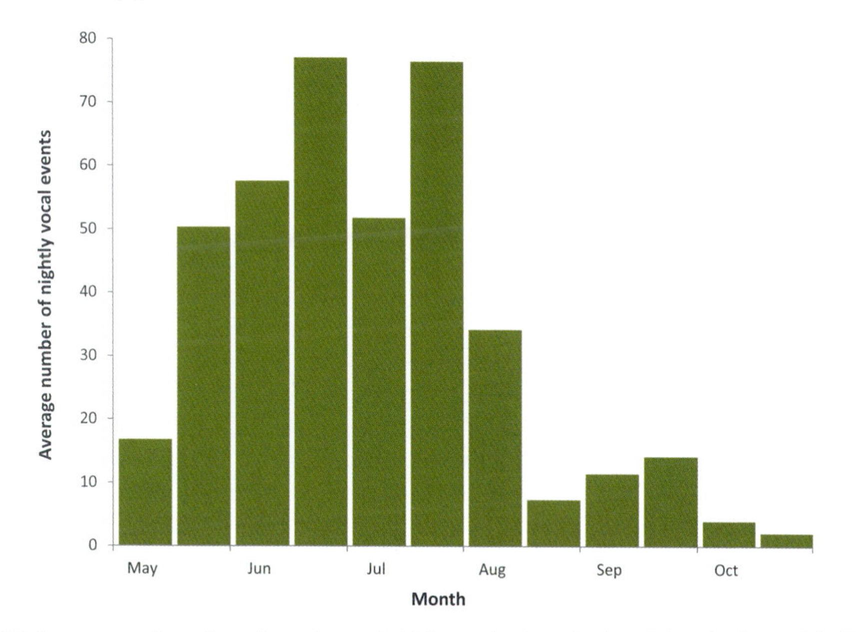

Figure A2 Average number of vocal events per night for a pair of captive hazel dormice for each half-month of the survey period

Considering the above results together, it may be most cost-effective to survey hazel dormice acoustically between about the second half of June and the first half of September, using static detectors that are programmed to be operational from sunset to sunrise.

Following the collection of data from a captive pair of hazel dormice, a further study was undertaken to examine whether the calling rates observed in captive animals are transferable to animals in the field. Our chosen study site was Bradfield Woods National Nature Reserve in Suffolk where hazel dormice (34 adults) were reintroduced in 2006. The population has been comprehensively monitored ever since, via monthly box checks (200 boxes) and the microchipping of all animals found. As such, Bradfield Woods provides a unique opportunity to compare the level of acoustic detections against estimates of population size.

Acoustic recorders were deployed at four locations in Bradfield Woods between the beginning of May and the end of October 2022. Three of the monitoring sites (L1 to L3) were within areas where good numbers of hazel dormouse were known to be present (i.e. woodland plots that supported nest boxes that are regularly checked and where hazel dormouse have been consistently recorded). The fourth monitoring site (L4) was located within a non-intervention plot, where no nest boxes have been installed, an area that has never been surveyed for dormice. The purpose for including plot L4 was to examine whether any correlations between the size of the hazel dormouse populations (quantified from microchipped animals) and the number of acoustic detections associated with plots L1 to L3 could be applied to areas outside these to estimate the size of the dormouse population within an unmonitored woodland area (i.e. in our case, plot L4).

Acoustic detectors were deployed in pairs, with one (Song Meter Mini Bat) located within the scrub/coppice layer at approximately 1.2 m above ground level, and the other (Song Meter 4 Bat) at four to five metres above ground level, within the lower tree canopy. For the higher detector, the ultrasonic microphone was connected to a five-metre cable, and the recorder itself was fixed close to ground level, so that batteries and SD cards could be changed without needing to use ladders, and to minimise disturbance. The settings of the recorders were the same as those used for the captive dormouse study (Newson and Pearce, 2021). The reason for deploying acoustic recorders at two height levels was to determine whether the detection rates of hazel dormice are influenced by the woodland structure, and also whether the seasonal movements between these woodland layers has an influence on the chance of detecting hazel dormice. For example, during the spring when temperatures fluctuate, animals typically move between their ground hibernation nests and the lower understorey shrub layer. Come early summer when insects and nectar sources provide important food resources, dormice will move up into the canopy level, and then in late summer, when trees and shrubs are in fruit, they move lower into the shrub level again to exploit these fruit and seed resources.

The data was analysed using the BTO Acoustic Pipeline to extract all .wav files that contained acoustic contacts, which includes bats, bush-crickets and small mammals. This dataset was then verified manually to ensure all hazel dormouse calls within the data were identified. A total of 3,015 sound files contained hazel dormouse calls, with all recordings from plots L1 to L3 (i.e. no dormice were detected from plot L4). The survey effort here, of using six bat detectors over a six-month period, is significantly greater than used in the captive study; the total output of all six machines equated to about 15% of the number of recordings from a single bat detector recording a captive pair of hazel dormice.

Under field conditions, hazel dormouse was recorded throughout the survey period, and similar to the captive data, there was a peak in activity during August and early September, which was much more pronounced in the field data. The percentage of nights when hazel dormouse was recorded during this period was 7–8% (Figure A3, left-hand plot), which is comparable to results for the captive animals during the same period

(6–10% of nights; Figure A1, left-hand plot). Similar to the captive study, the main period of vocal activity was in the middle of the night, three to four hours after sunset (Figure A3, middle plot), and again this was consistent throughout the survey period (Figure A3, right-hand plot).

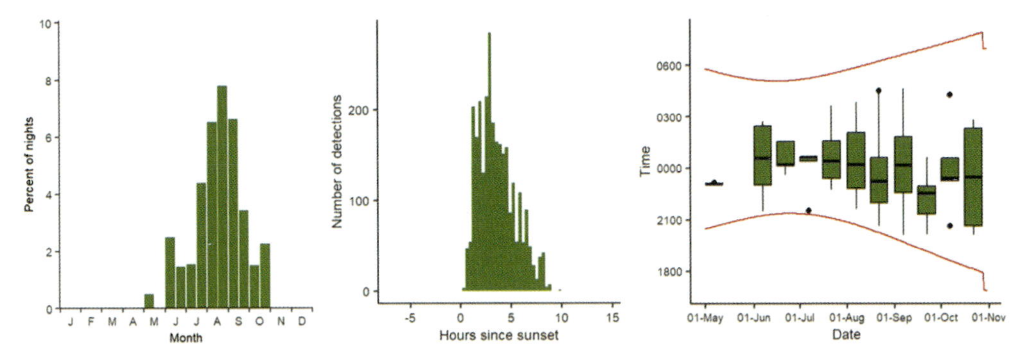

Figure A3 Evaluation of seasonal and nightly vocal activity of hazel dormouse in Bradfield Woods, Suffolk.

The left-hand plot shows the percentage of nights on which the hazel dormouse was detected every half-month through the season, showing the periods of main activity for this species. The middle plot shows the overall spread of recordings with respect to sunset time, calculated over the whole season. The right-hand plot shows the spread of recordings with respect to sunset and sunrise times (red lines) summarised for each half-month through the season. For this last seasonal plot, the individual boxplots show quartiles (lower, median and upper) with lines extending to 1.5 times the interquartile range, and small dots show outliers.

We further examined hazel dormouse vocal activity in terms of the number of vocal events per night and the duration of these events (Figure A4). The average number of vocal events detected under field conditions was less than 10% of the number of detections from captive animals (Figure A2) and unlike the captive study, the highest calling rate was in August and September. The average duration of vocal events varied considerably, from five-second contacts up to more than four minutes in duration. Longer periods of calling were reported in August and early September, when the number of vocal events was also at its highest. The height of the detector deployment did not appear to impact on detection rates.

These data suggest that acoustic surveys for hazel dormice, although possible throughout their active period (May–October), may be most cost-effective if carried out during August and September. It demonstrates that although data under controlled captive conditions can provide valuable information, which can be used to guide the design of field surveys, this needs to be followed by field surveys to ground truth how best to target survey effort according to patterns of activity. This will be influenced by seasonal and habitat variations between study sites. The main factors that likely contribute to differences in seasonal acoustic activity between captive and wild hazel dormice will be the absence of competition for a mate and food resources under captive conditions.

Further acoustic work on the hazel dormouse population at Bradfield Woods is planned, and it is hoped that with a larger sample of data, it will be possible to determine whether the number of acoustic contacts correlates with the number of animals present in each of the study plots. Anecdotally, Simone Bullion, who has been monitoring and pit-tagging

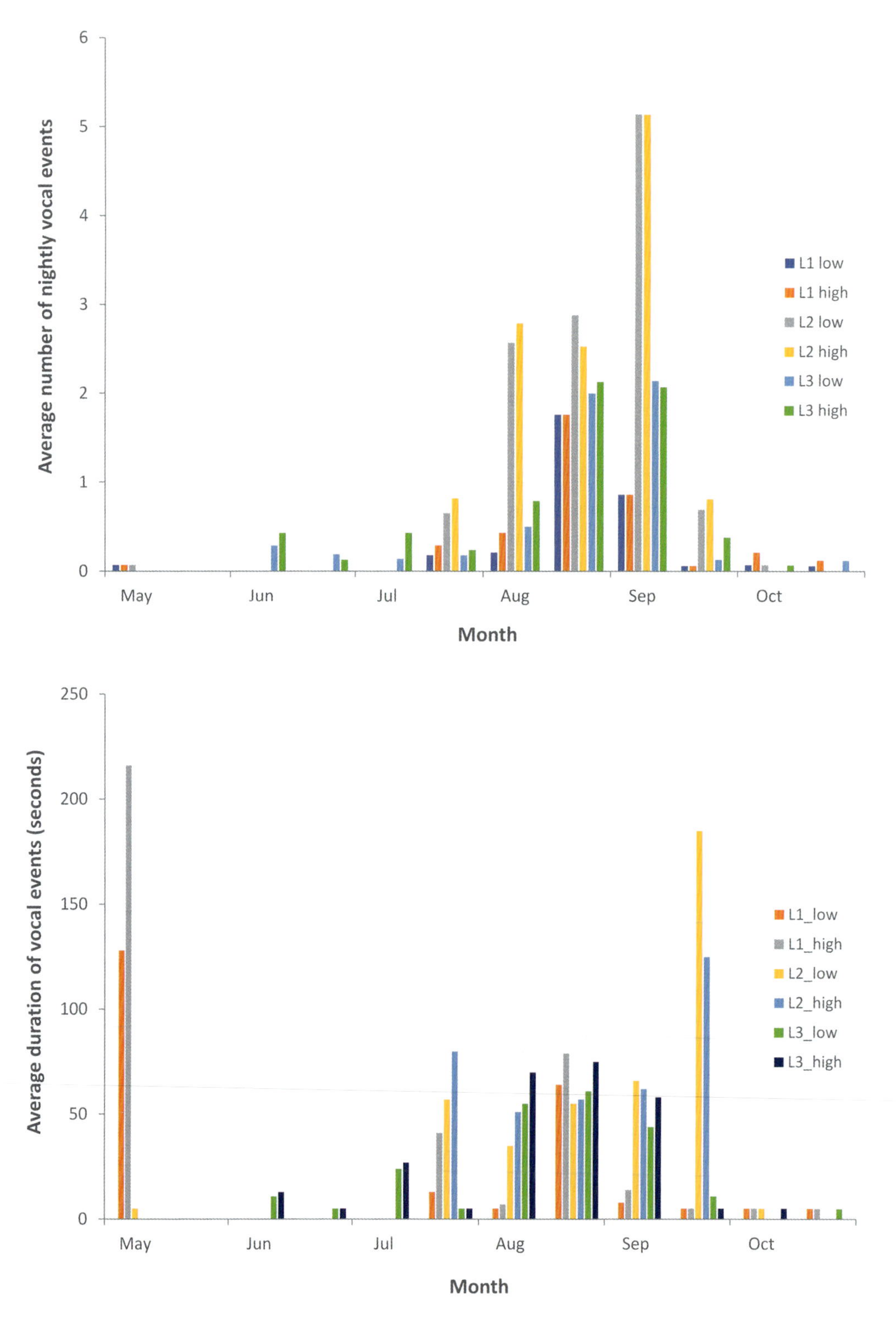

Figure A4 Vocal activity – average number of nightly vocal events (top) and duration of vocal events (bottom) of hazel dormouse detected every half-month in Bradfield Woods, Suffolk.

the dormouse population at Bradfield Woods since 2006, confirmed that plot L2, where acoustic activity was highest, *'has always been a reliable area for dormice in boxes'*. If correlations are identified from these further studies, this could open up the possibility for acoustic surveys to be used for estimating the size of hazel dormouse populations within areas of woodland. For example, they could be used to supplement data collected from nest box schemes where pit-tagging is not carried out, as well as providing population estimates for woodland sites where dormouse box schemes are not established.

In addition to hazel dormouse, the calls of wood mouse, yellow-necked mouse, common shrew, pygmy shrew and bank vole were also recorded during the Bradfield Wood study. Over the six-month survey period, 567 recordings contained *Apodemus* calls, 595 recordings included the calls of *Sorex* species, whilst bank vole calls were only represented by five sound files. Given that other small mammal species are likely to be more abundant than the hazel dormouse, these data clearly demonstrate the lower detection rate of these species compared to hazel dormouse. Similar to hazel dormouse, a greater number of acoustic contacts for *Apodemus* and *Sorex* species were reported during August and September (Figure A5, top).

A study on the application of acoustic methods to monitor small mammal community composition more generally was previously carried by the authors in 2020 (Newson and Pearce, 2021). The typical activity patterns (i.e. diurnal, nocturnal and/or crepuscular) of a species will influence the time period over which vocalisations are most likely to be recorded, and in that previous study the acoustic recorders were active throughout the 24-hour period. A comparison of small mammal acoustic detections during the day and night found that 95% of *Apodemus* contacts were at night, whereas voles and shrews were recorded more often during the day, with only 21% and 27% of the total recordings of these species being at night.

The decision to set the recording period from sunset to sunrise at Bradfield Woods was made to coincide with the times during which we knew, from captive individuals, that hazel dormouse were most active. In addition to the typically low detection rates of vole species, the absence of daytime recording in the Bradfield Wood study likely accounts for the very low number of vocalisations recorded rather than an absence of voles within the area. Similarly, the number of shrew contacts would have likely been higher if the recorders had been set to record over a 24-hour period.

The 2020 study also looked at the influence of recorder deployment height on detection rates of small mammals. For example, whether a species is predominantly ground-dwelling or arboreal might influence the chosen height of the microphone. In that study, similar to the study here in Bradfield Woods, the detectors were paired; one was set close to ground level (c.0.3 to 0.5 m height) and the other at height (c.2.5 to 3 m). The study found that recording closer to the ground increased the chances of recording *Apodemus* and vole species, whereas there was no evidence that deploying microphones low down (at between 0.3 and 0.5 m) increased the chances of recording *Sorex* species – if anything, the converse was true, with shrews being detected more often (almost twice as often) by recorders deployed at height.

Although detectors were not deployed at ground level in Bradfield Woods (since hazel dormouse is an arboreal species), the variation in the deployment height of the detectors (1.2 m and 4–5 m) used in this study provides a further opportunity to investigate the influence of height on the detection distance of small mammals (see Figure A5, bottom).

Successful detection of hazel dormouse did not appear to be influenced by the height of the detector, although vocal activity was slightly more pronounced from recorders

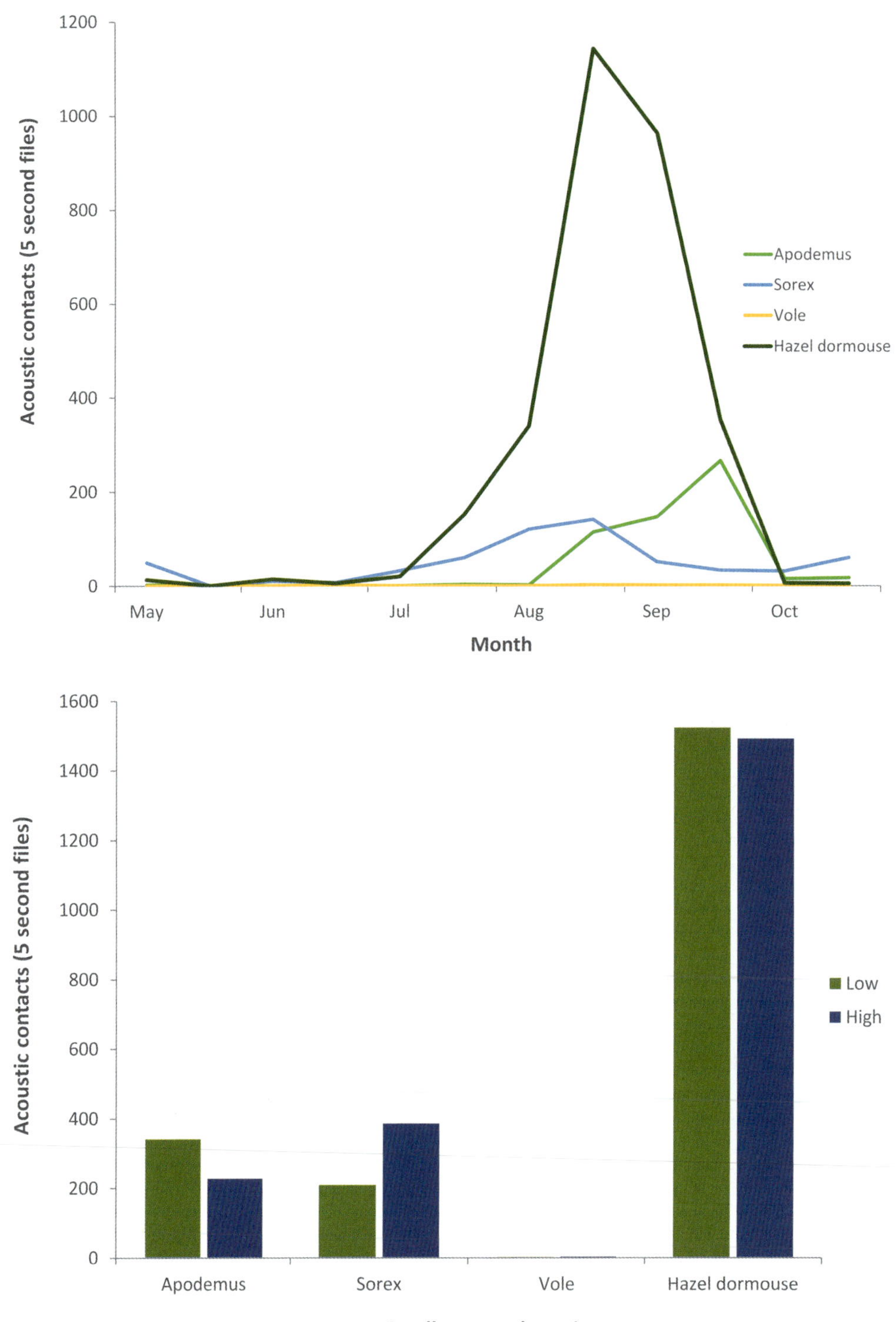

Figure A5 Total number of acoustic contacts for small mammal species recorded during each half-month (top) and the influence of height on detection rates for small mammals (bottom), recorded at Bradfield Woods, April to October 2022

deployed at height (i.e. closer to the tree canopy) during some months, notably June to early August, when animals are more likely to be utilising food resources associated with the tree canopy (Figure A4). The apparent absence of influence of height could in part be due to the high amplitude and lower frequency of vocalisations made by hazel dormouse, whereby even the lower recorders were able to detect calls made by animals from higher up within the tree canopy. Comparatively, and consistent with the 2020 study, *Apodemus* species were detected more often by the lower detectors whereas shrews were recorded more often by recorders deployed at height.

These initial studies indicate that acoustic surveys provide a useful tool for detecting the presence of small mammals, with the key advantage of surveys being non-invasive. With the development of algorithms such as those provided through the BTO Acoustic Pipeline to assist with the extraction and identification of small mammal calls, acoustic monitoring of this species group could be carried out over a wider area (i.e. outside of traditional monitoring sites) to improve our understanding of the distribution and possibly relative abundance of these species. Further investigation of deployment methods should in time enable the establishment of species-specific survey protocols that would enhance the efficacy of acoustic methods for monitoring small mammals.

Case Study 2: Acoustic (audible) identification of edible dormouse

Edible dormouse is the target species in this study. Edible dormouse is a non-native species and their current range falls mostly within 35 km of Tring, Buckinghamshire (England), where they were first introduced in 1902. Although a slow-spreading species, the Mammal Society population estimate suggests that their numbers have doubled between 1995 and 2018 (Mathews *et al.*, 2018). They are an invasive pest species causing damage to forestry interests and potential harm to native species (e.g. hole-nesting birds, bats, hazel dormouse) via competition for food resources and nest sites. When they occur close to human settlements, they frequently make use of gardens and buildings and may cause significant damage. Data from local management of populations by culling suggests they now cover almost the entire Chilterns woodland area bounded by the M25 motorway, M40 motorway and Chilterns Scarp, with current verified populations/records beyond, including north of Reading, eastwards towards London, south towards Slough, west Essex and Ascot in south-east Berkshire (Mammal Society, 2021).

Unlike most other small mammals, edible dormice vocalise predominantly in the audible range. They are a long-lived arboreal, nocturnal rodent which, similar to hazel dormouse, has a short period of activity, hibernating underground for seven (or more) months of the year. They are most vocal between June and August, when breeding takes place, and vocalisations are most often reported at dusk. Current monitoring of edible dormouse populations appears to be restricted to a single nest box project and data obtained from local authority and Defra reports on culling licences. If this species can be identified reliably from their calls, there is the potential for acoustic surveys to be used to monitor changes in their distribution.

This case study happened by chance when, in mid-June 2022, during a bat roost count at Penham Green in Hertfordshire, edible dormouse were heard vocalising from a nearby treeline, and some sample recordings were made using the iPhone 'Voice Memo' app. One of the surveyors, Jo Sutherland, said she regularly hears edible dormouse along the boundary treeline at the back of her property in south-east Buckinghamshire and kindly agreed to the deployment of acoustic recorders in her garden, in an attempt to try and

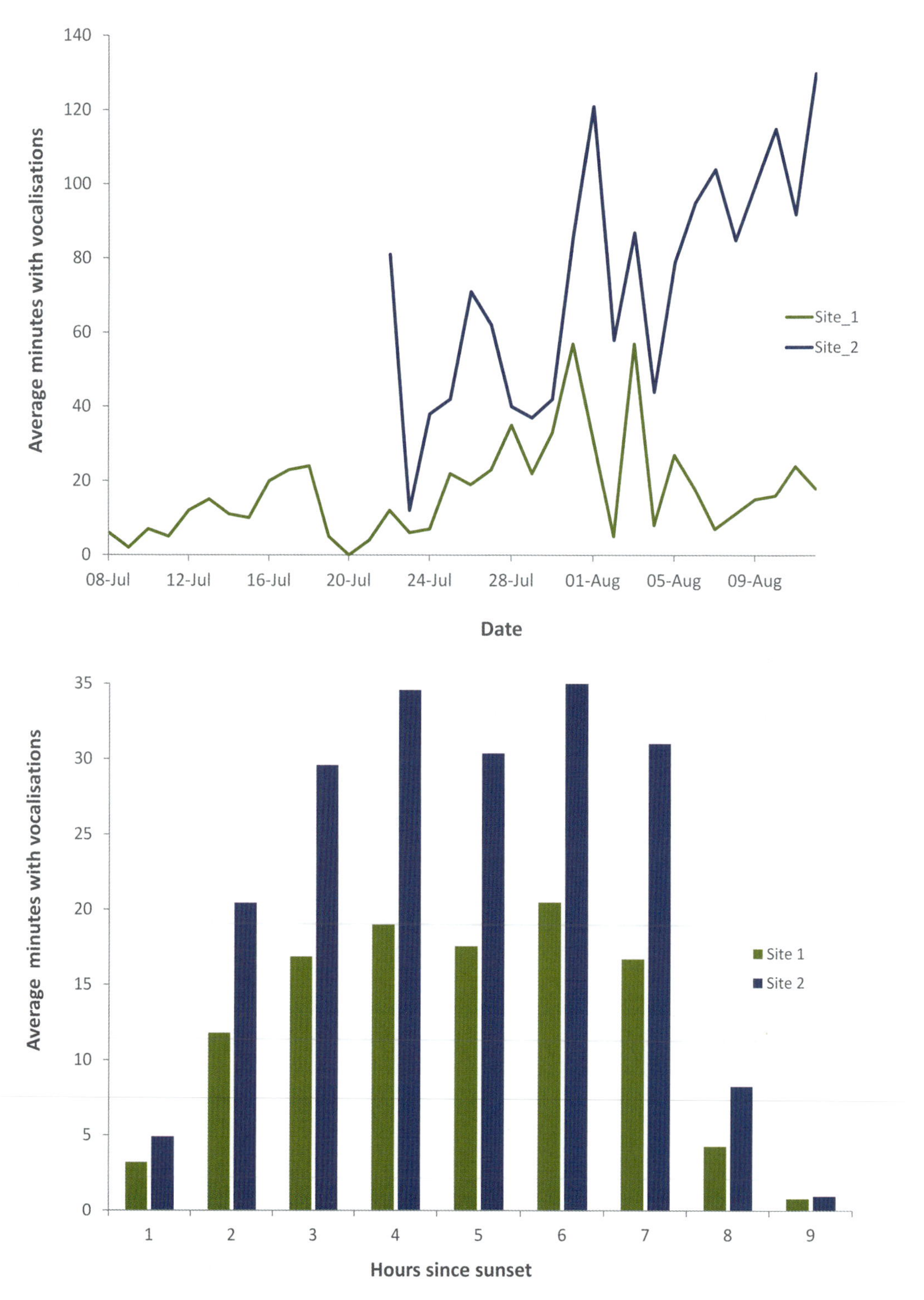

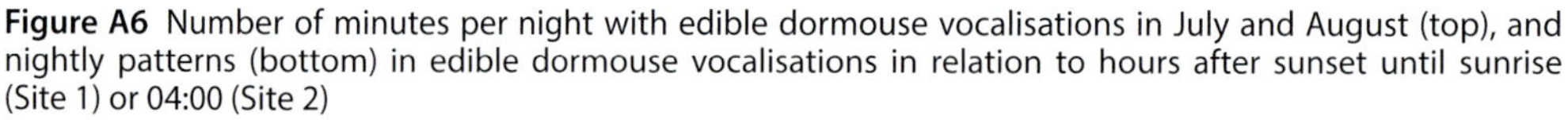

Figure A6 Number of minutes per night with edible dormouse vocalisations in July and August (top), and nightly patterns (bottom) in edible dormouse vocalisations in relation to hours after sunset until sunrise (Site 1) or 04:00 (Site 2)

record the repertoire of edible dormouse vocalisations for this book (Site 1). Recorders were deployed in early July until 12 August 2022.

Coincidentally, a few weeks later, we were contacted by Hilary Phillips, who was enquiring about the feasibility of using acoustics for monitoring edible dormice. Hilary regularly hears edible dormice in woodland to the rear of her garden (also located in south-east Buckinghamshire), and she agreed to the deployment of recorders (Site 2) from 22 July until 12 August 2022.

Song Meter Mini Bat (Wildlife Acoustics Inc.) with acoustic microphones were used in the study, attached to two-metre-high tripods. The recorders were set to continuously record all sounds up to 45 kHz from 22:00 hours until sunrise (Site 1), and until 04:00 (Site 2). Due to the residential character of the survey locations, the start time was set between 40 to 90 minutes after sunset to minimise the likelihood of recording people using the adjacent gardens. The length of sound files was set to 60 seconds. In the absence of a classifier at time of writing, to automatically detect and identify likely calls of edible dormouse, the recordings were reviewed manually to identify edible dormouse vocal activity. The development of classifiers for detection and identification of edible dormouse calls within continuous recordings are in development (BTO Acoustic Pipeline) and will be useful for improving the efficiency of this survey method.

The two study sites offered an opportunity to monitor edible dormouse using acoustic methods for two habitat types: (1) along a treeline that affords connectivity between woodlands, and (2) a woodland that is known to support a breeding population of edible dormice. The sites were located within 5.5 km of each other and lie within 15 km of Tring, Buckinghamshire.

With a survey effort of 36 nights (Site 1) and 22 nights (Site 2) respectively, a total of 617 and 1,621 one-minute recordings contained edible dormouse calls from each site. The level of vocal activity was notably lower along the treeline (Site 1) compared to the woodland habitat (Site 2) (Figure A6). There were however similarities in the pattern of vocal activity at the two sites, with an increase in call rate noted during early August. Vocal activity remained high during August at the woodland (Site 2), whereas there was greater variability and a decline in edible dormouse activity associated with the treeline (Site 1). Nevertheless, with the exception of a single night, edible dormouse calls were recorded on all nights during the survey period, and vocalisations were recorded throughout the night, with peaks in activity during the middle of the night (three to seven hours after sunset) (Figure A6, bottom). Since no data was collected before 22:00 it is not possible to determine the level of activity close to dusk, but anecdotal evidence suggests that activity within the first hour after sunset is likely to be higher than reported in this study. Nevertheless, these data suggest that the middle of the night may be the optimum period for detecting edible dormice.

This preliminary study provides evidence for the viability of using acoustic methods to detect and identify edible dormouse. The peak in vocal activity in the middle of the night suggests that static surveys may be more effective than manned surveys at dusk. Nevertheless, a large volume of data was collected during this short survey period and the time taken to extract calls from the recordings was significant. The development of classifiers for detection and identification of edible dormouse calls within continuous recordings would be useful to improve the efficiency of this survey method. As we found, care is needed to avoid confusing edible dormouse calls with bird species, notably tawny owl begging calls, so classifiers that also consider the identification of nocturnal bird species may aid accurate identification.

Case Study 3: Water vole combining trail camera and acoustics

Trail cameras have been applied in several preliminary studies (carried out by the authors) to explore the efficacy of acoustic methods for monitoring a range of mammal species in Britain and Ireland. Here we present the results of one such study, where water vole are the target species. Water vole are a priority species of conservation importance in the UK, with reintroductions being a key method for restoring populations that have suffered declines. If water vole can be reliably detected using acoustic methods in the field, it may provide a non-invasive and complementary approach to the more traditional survey protocols.

Bob Reed gave us access to Sawbridgeworth Marsh Nature Reserve, which supports a good water vole population. Two small mammal trail camera boxes, baited with apple and seed mix, were deployed approximately 20 m apart along one bank of a well-vegetated drain with field evidence of water vole activity (latrines, runways, burrows, feeding signs). Browning Strike Force HD Pro X cameras (with close-focus lenses) were used to record video files (20-second duration; 5-second capture delay). A Song Meter Mini Bat (Wildlife Acoustics Inc.) was used to gather acoustic data, set to continuously record all sound between 16 and 192 kHz, and deployed midway between the two mammal boxes at c.0.75 m above ground level. The survey period ran for two weeks (27 July to 9 August 2021). Technical issues prevented the collection of sound data on 30 July and no trail camera footage was gathered from Camera 2 on 4 August. The equipment was checked daily and bait was replenished as required.

Small mammals were recorded in 4,752 video files. Of these files, water voles accounted for 34%. Other small mammal species picked up by the trail cameras included bank vole, brown rat, common shrew and pygmy shrew. Water vole visited the mammal boxes most often during the early part of the survey period, but activity was rapidly replaced by brown rat from day four onwards (Figure A7, top). Brown rat were clearly attracted to the bait, often monopolising the mammal boxes for extended periods (more than 30 minutes at a time). This demonstrates how baiting alters animal behaviour, potentially to the detriment of other species, creating a bias in survey data towards the more dominant species.

With the exception of bank vole, all other small mammal species detected by trail cameras were also recorded acoustically. A total of 645 sound files (5-second .wav files) contained small mammal calls and of these files, water vole vocalisations represented less than 5%. The quantity of small mammal acoustic data increased through the survey period as more brown rats (in particular juveniles) occupied the area, and brown rat accounted for more than 90% of the total small mammal recordings (Figure A7, bottom). For the shrew species, detection by trail camera and acoustic methods was comparable whereby these species accounted for 2% and 1% respectively. The low number of water vole calls, and lack of success in recording bank vole vocalisations, demonstrates the difficulty of recording vole species acoustically (compared to other small mammals), but provides initial evidence for water vole being more vocal than other vole species.

The trail camera data found that water voles were attracted to the bait, but they rarely fed within the mammal box, preferring to gather food and return it to their burrows to cache or feed. Additionally, with the exception of two occasions when two voles unintentionally entered the mammal box at the same time, voles visited the mammal boxes singly. By comparison, brown rat fed within or close to the box (still in view of the camera) sometimes for prolonged periods, and on 10% of occasions, they were seen to feed in small groups of two to four animals, typically consisting of juveniles with and without adults. Differences in sociality between voles and rats will influence the

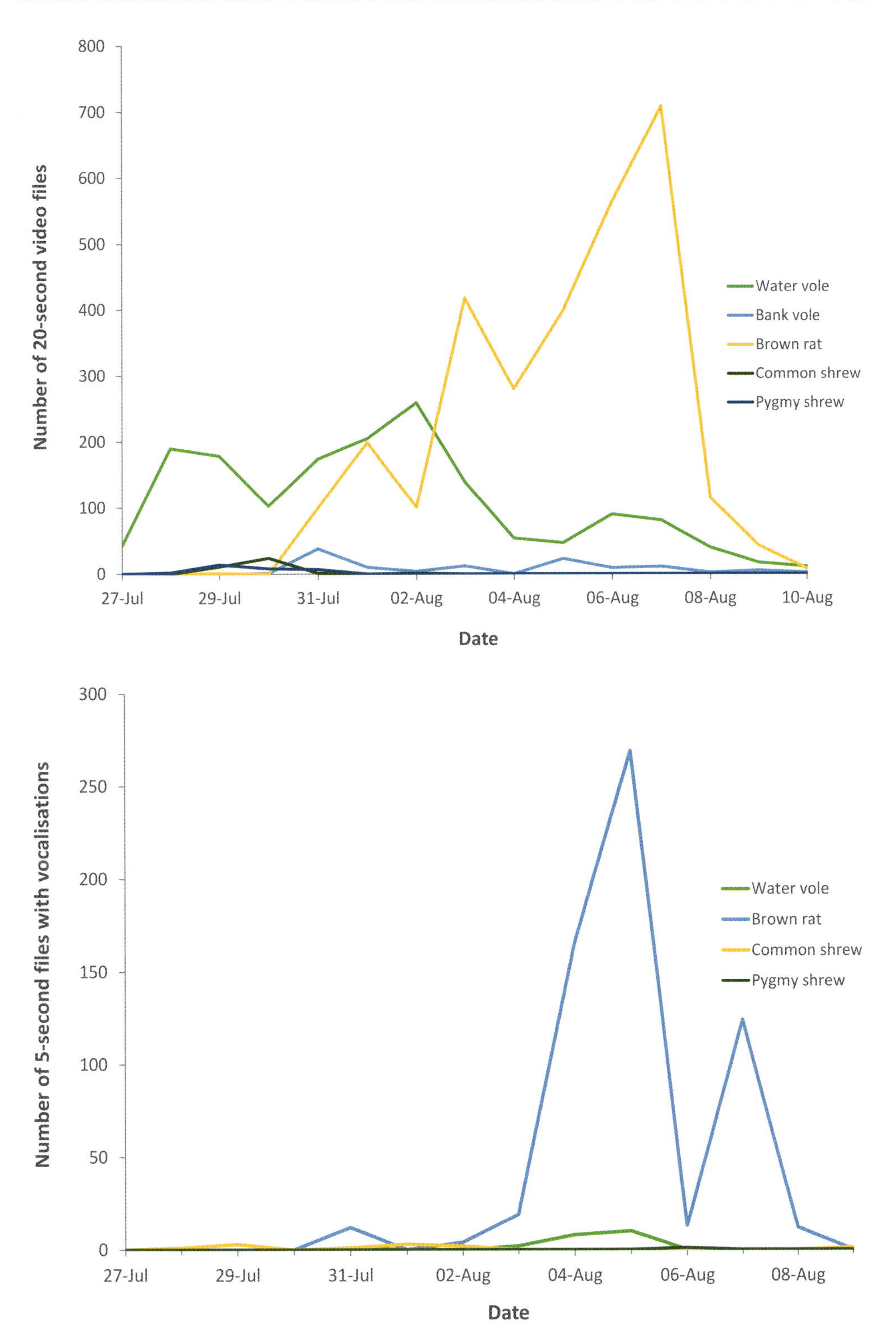

Figure A7 Comparison of mammal activity recorded during the survey period based on trail camera (top) and acoustic data (bottom)

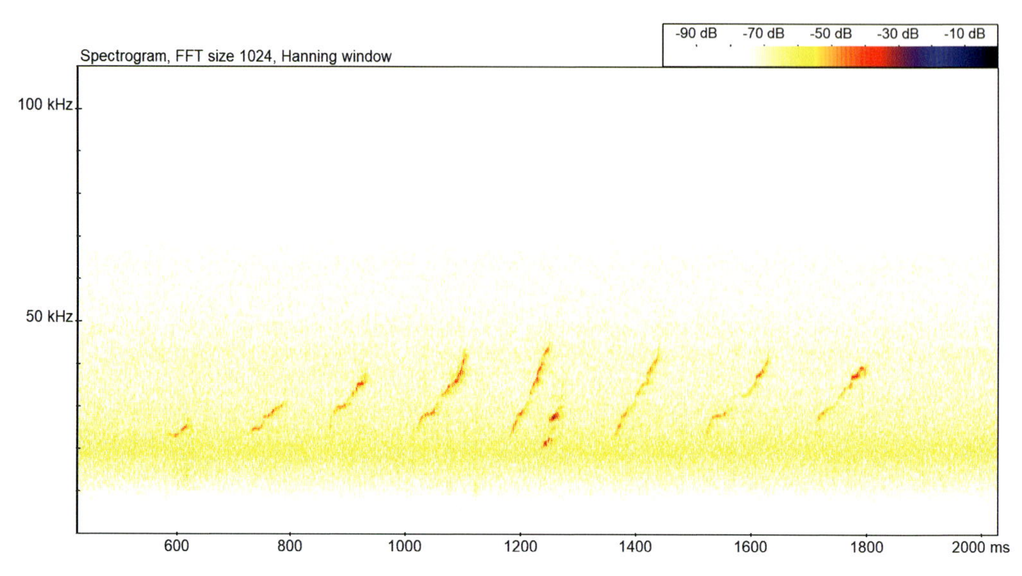

Figure A8 Short-duration water vole call sequence

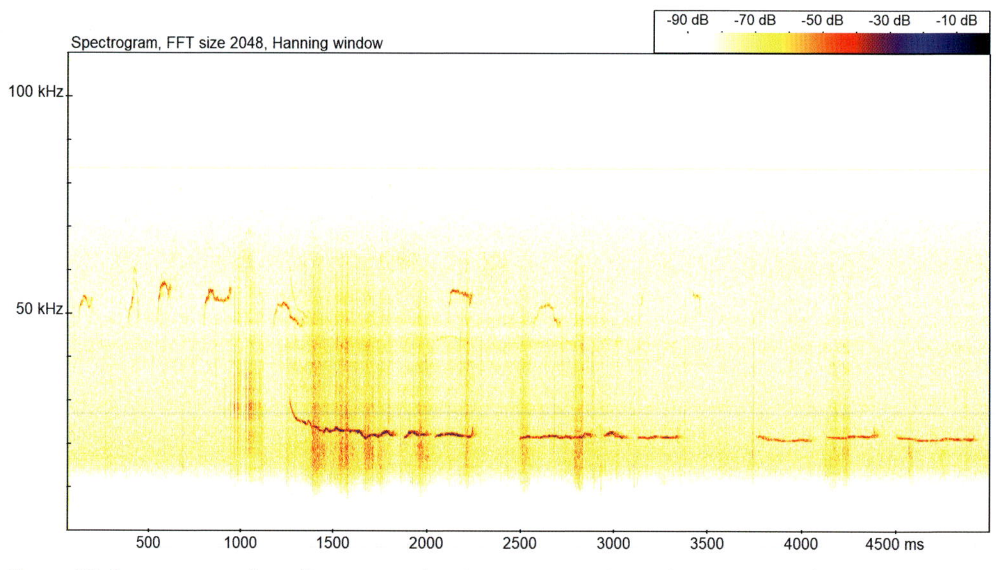

Figure A9 Brown rat complex call sequence, showing non-aggressive and agonistic vocalisations.

frequency of vocal interactions, which will in part account for the low detectability of water voles, compared to brown rat. Furthermore, calls of water voles were typically only given over a short duration (Figure A8), whereas more complex sequences of brown rat calls, regularly made up of non-aggressive and agonistic vocalisations, over consecutive sound files, were often recorded (Figure A9).

Figure A10 (top) shows the daily time period of small mammal activity recorded by trail cameras. Water voles (and bank vole) visited mammal traps at all times throughout a 24-hour time period, and frequently during the day. Comparatively, brown rat visited mammal boxes more often during the night. In contrast, regardless of the dominance of

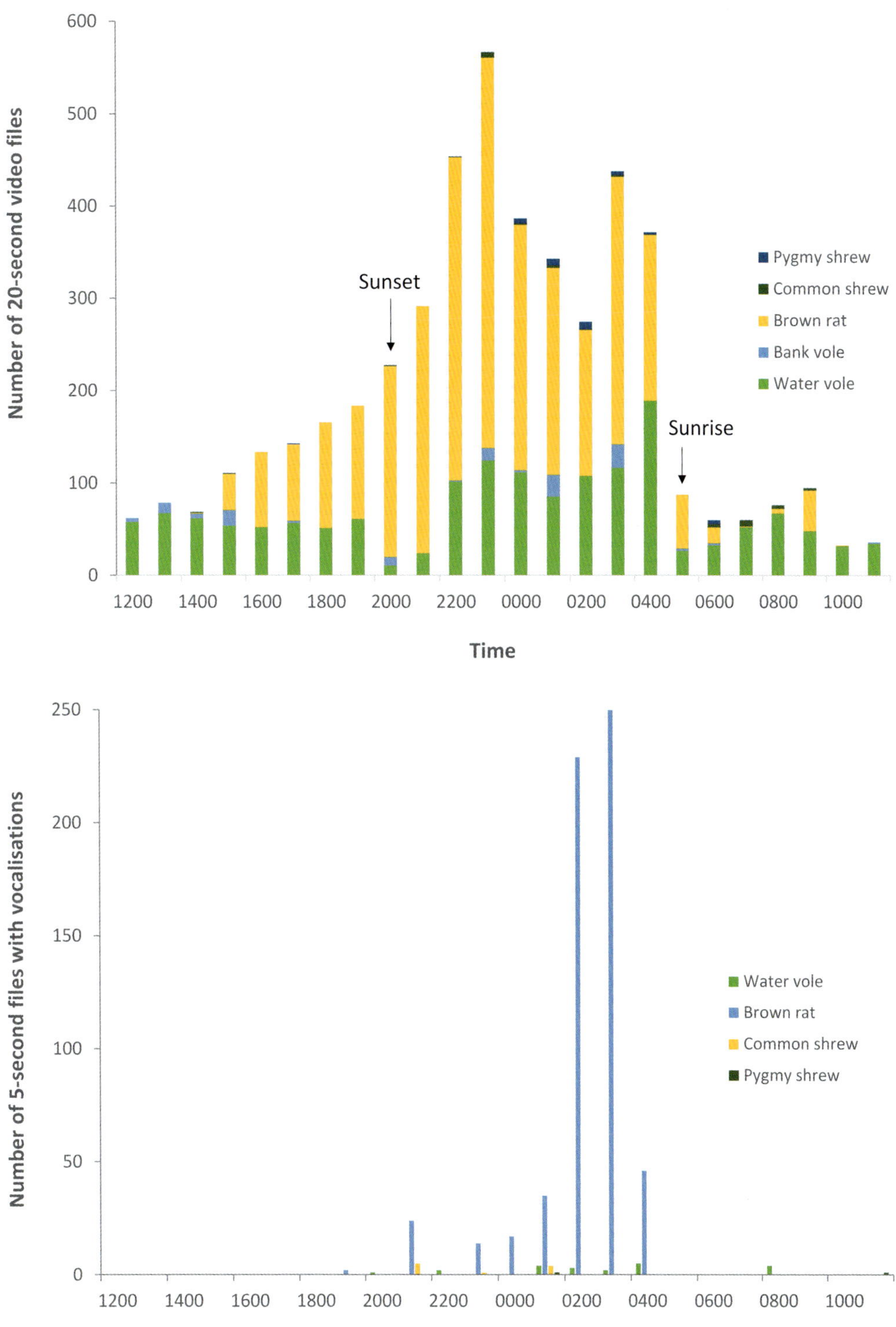

Figure A10 Mammal detections by trail cameras (top) and acoustic recorders (bottom) in relation to time of day.

brown rat calls, all small mammals were principally detected acoustically during the night; see Figure A10 (bottom). Small mammals will be more susceptible to predation during the daylight hours and the trail camera footage showed that animals were more cautious during the day, visiting mammal boxes only sporadically and for shorter periods of time. It is likely small mammals remain underground or within dense vegetation during the day, and thus out of range of the microphones. Furthermore, there are likely to be fewer vocalisations during the day, in order to avoid exposure to predators.

The water vole sound files in Chapter 9 were all obtained from recordings made from captive animals. This short study shows that water vole vocalisations can be recorded and identified using acoustic methods in the field, and more effectively than for other vole species, such as bank vole. As with captive water voles, the call rate in the field is low, and a large volume of acoustic data would need to be collected for a small return. The use of classifiers to automatically detect and identify water vole calls, such as the BTO Acoustic Pipeline, is therefore essential.

The study also demonstrates the value of combining acoustic methods with trail camera data, at the least to help guide the development of an appropriate survey protocol, since video footage will provide important behavioural information. The use of bait in this study likely impacted negatively upon vole habitat use, because of the rapid increase in brown rat numbers. Additional studies without the use of bait and over a prolonged period would be beneficial. In particular, additional work to look at the nightly and seasonal patterns in water vole vocalisations would be useful to inform the design of a species-specific survey protocol, ideally to reduce the overall quantity of data whilst ensuring a representative sample of water vole acoustic data is still collected. For example, if vocalisations are most likely to be recorded during the hours of darkness, it may be efficient to tailor surveys to a shorter nightly period (e.g. from sunset to sunrise ± 1 hour).

References

Abass, G. (2021). How we can save our endangered red squirrels using AI. *The Optimist*. Available online: https://www.standard.co.uk/optimist/sustainable/how-we-can-save-endangered-red-squirrels-using-ai-b959343.html

Aebischer, N. J., Davey, P. D. and Kingdon, N. G. (2011). *National Gamebag Census: Mammal Trends to 2009*. Game & Wildlife Conservation Trust, Fordingbridge. http://www.gwct.org.uk/ngcmammals

Agranat, I. (2014). *Detecting Bats with Ultrasonic Microphones. Understanding the Effects of Microphone Variance and Placement on Detection Rates*. Wildlife Acoustics, Inc. 10 November 2014. https://www.wildcare.eu/media/wysiwyg/pdfs/UltrasonicMicrophones.pdf

Ancillotto, L., Sozio, G., Mortelliti, A. and Russo, D. (2014). Ultrasonic communication in Gliridae (Rodentia): the hazel dormouse (*Muscardinus avellanarius*) as a case study. *Bioacoustics* 23 (2): 129–41.

Ancillotto, L., Mori, E., Sozio, G., Solano, E., Bertolino, S. and Russo, D. (2017). A novel approach to field identification of cryptic *Apodemus* wood mice: calls differ more than morphology. *Mammal Review* 47 (1): 6–10.

Anderson, J. W. (1954). The production of ultrasonic sounds by laboratory rats and other mammals. *Science* 119: 808–809. https://doi.org/10.1126/science.119.3101.808

Baker, P. A. and Harris, S. (2008). Ch. 9 (pp. 407–22), in Harris, S. and Yalden, D. W. (eds), *Mammals of the British Isles: Handbook*, 4th edition. Mammal Society, London.

Baldwin, M. (2022). Wildlife Online – an educational website, produced by Marc Baldwin, about British wildlife: https://www.wildlifeonline.me.uk

Balharry, E., Jefferies, D. J. and Birks, J. D. S. (2008). Ch. 9 (pp. 447–55), in Harris, S. and Yalden, D. W. (eds), *Mammals of the British Isles: Handbook*, 4th edition. Mammal Society, London.

Barataud, M. (2015). *Acoustic Ecology of European Bats: Species Identification, Study of their Habitats and Foraging Behaviour*. Biotope Editions, Publications Scientifiques du Museum.

Barré, K., Le Viol, I., Julliard, R., Pauwels, J., Newson, S. E., Julien, J-F., Claireau, F., Kerbiriou, C. and Bas, Y. (2019). Accounting for automated identification errors in acoustic surveys. *Methods in Ecology and Evolution* 10(8): 1171–88. https://doi.org/10.1111/2041-210X.13198

Barrett-Hamilton, G. E. H., Hinton, M. A. C. and Wilson, E. A. (1910). *A History of British Mammals*, Vol. 2, part XI, pp. 227–28. Gurney and Jackson, London. https://doi.org/10.5962/bhl.title.55827

Bednářová, R., Hrouzková-Knotková, E., Burda, H., Sedláček, F. and Šumbera, R. (2012). Vocalizations of the giant mole-rat (*Fukomys mechowii*), a subterranean rodent with the richest vocal repertoire. *Bioacoustics iFirst* 1–21. https://doi.org/10.1080/09524622.2012.712749

Berry, R. J., Tattersall, F. H. and Hurst, J. (2008). Ch. 5 (pp. 141–49), in Harris, S. and Yalden, D. W. (eds), *Mammals of the British Isles: Handbook*, 4th edition. Mammal Society, London.

Birks, J. (2017). *Pine Martens*. The British Natural History Collection. Whittet Books.

Birks, J. D. S. and Kitchener, A. C. (2008). Ch. 9 (pp. 476–85), in Harris, S. and Yalden, D. W. (eds), *Mammals of the British Isles: Handbook*, 4th edition. Mammal Society, London.

Brandt, C., Malmkvist, J., Nielsen, R. L., Brande-Lavridsen, N. and Surlykke, A. (2013). Development of vocalization and hearing in American mink (*Neovison vison*). *Journal of Experimental Biology* 216: 3542–50. https://doi.org/10.1242/jeb.080226

Briggs, B. and King, D. (1998). *The Bat Detective: A Field Guide for Bat Detection*. Batbox Ltd.

Bright, P. W. and Morris, P. A. (2008). Ch. 5 (pp. 76–81), in Harris, S. and Yalden, D. W. (eds), *Mammals of the British Isles: Handbook*, 4th edition. Mammal Society, London.

Bright, P., Morris, P. and Mitchell-Jones, T. (2006). *The Dormouse Conservation Handbook*, 2nd edition. English Nature, Peterborough.

Brudzynski, S. M. (2009). Communication of adult rats by ultrasonic vocalization: biological, socio-biological, and neuroscience approaches. *Institute for Laboratory Animal Research (ILAR) Journal* 50 (1): 43–50.

Buffenstein, R. (1996). Ecophysiological responses to a subterranean habitat; a Bathyergid perspective. *Mammalia* 60 (4): 591–605. https://doi.org/10.1515/mamm.1996.60.4.591

Bullion, S. and Looser, A. (2019). *Guidance for Using Hazel Dormouse Footprint Tunnels*. Suffolk Wildlife Trust (June 2019).

Burgdorf, J., Kroes, R. A., Moskal, J. R., Pfaus, J. G., Brudzynski, S. M. and Panksepp, J. (2008). Ultrasonic vocalizations of rats (*Rattus norvegicus*) during mating, play, and aggression: behavioral concomitants, relationship to reward, and self-administration of playback. *Journal of Comparative Psychology* 122 (4): 357–67.

Campbell-Palmer, R., Gow, D., Campbell, R., Dickinson, H., Girling, S. J., Gurnell, J., Halley, D., Jones, S., Lisle, S., Parker, H., Schwab, G. and Rosell, F. (2016). *The Eurasian Beaver Handbook: Ecology and Management of Castor fiber*. Pelagic Publishing, Exeter.

Carter, P. and Churchfield, S. (2006). *The Water Shrew Handbook*. Mammal Society.

Castillo, D. (2019). Who is present? Individuality in the call structure of the Eurasian Otter (*Lutra lutra*) whistle. Master's thesis, University of Zurich. Available online at: http://www.parcs.ch/wpz/pdf_public/2021/41672_20210225_180019_masterarbeit_dominik_del_castillo_2019.pdf

Chapman, N. G. (2008). Ch. 11 (p. 564), in Harris, S. and Yalden, D. W. (eds), *Mammals of the British Isles: Handbook*, 4th edition. Mammal Society, London.

Charlton, B. D., Newman, C., Macdonald, D. W. and Buesching, C. D. (2020). Male European badger churrs: insights into call function and motivational basis. *Mammalian Biology* 100: 429–38. https://doi.org/10.1007/s42991-020-00033-x

Christian, S. F. (1993). Behavioural ecology of the Eurasian badger (*Meles meles*): space use, territoriality and social behaviour. PhD thesis, University of Sussex, UK.

Churchfield, S. (2008a). Ch. 7 (pp. 271–75), in Harris, S. and Yalden, D. W. (eds), *Mammals of the British Isles: Handbook*, 4th edition. Mammal Society, London.

Churchfield, S. (2008b). Ch. 7 (pp. 280–83), in Harris, S. and Yalden, D. W. (eds), *Mammals of the British Isles: Handbook*, 4th edition. Mammal Society, London.

Churchfield, S. and Searle, J. B. (2008a). Ch. 7 (pp. 257–65), in Harris, S. and Yalden, D. W. (eds), *Mammals of the British Isles: Handbook*, 4th edition. Mammal Society, London.

Churchfield, S. and Searle, J. B. (2008b). Ch. 7 (pp. 265–66), in Harris, S. and Yalden, D. W. (eds), *Mammals of the British Isles: Handbook*, 4th edition. Mammal Society, London.

Churchfield, S. and Searle, J. B. (2008c). Ch. 7 (pp. 267–71), in Harris, S. and Yalden, D. W. (eds), *Mammals of the British Isles: Handbook*, 4th edition. Mammal Society, London.

Churchfield, S. and Temple, R. K. (2008). Ch. 7 (pp. 276–80), in Harris, S. and Yalden, D. W. (eds), *Mammals of the British Isles: Handbook*, 4th edition. Mammal Society, London.

Churchfield, S., Barber, J. and Quinn, C. (2000). A new survey method for Water Shrews (*Neomys fodiens*) using baited tubes. *Mammal Review* 30 (3–4): 249–54. https://doi.org/10.1046/j.1365-2907.2000.00074.x

Ciechanowski, M. and Sachanowicz, K. (2014). Fat dormouse *Glis glis* (Rodentia: Gliridae) in Albania: synopsis of distributional records with notes on habitat use. *Acta Zoologica Bulgarica* 66 (1): 39–42.

Ciucci, P., Catullo, G. and Boitani, L. (2009). Pitfalls in using counts of roaring stags to index red deer (*Cervus elaphus*) population size. *Wildlife Research* 36 (2): 126–33. https://doi.org/10.1071/WR07121

Clausen, K. C., Malmkvist, J. and Surlykke, A. (2008). Ultrasonic vocalisations of kits during maternal kit-retrieval in farmed mink, *Mustela vison*. *Applied Animal Behaviour Science* 114: 582–92.

Coffey, K. R., Marx, R. G. and Neumaier, J. F. (2019). DeepSqueak: a deep learning-based system for detection and analysis of ultrasonic vocalizations. *Neuropsychopharmacology* 44: 859–68. https://doi.org/10.1038/s41386-018-0303-6

Cole, M., Kitchener, A. C. and Yalden, D. W. (2008). Ch. 5 (pp. 72–76), in Harris, S. and Yalden, D. W. (eds), *Mammals of the British Isles: Handbook*, 4th edition. Mammal Society, London.

Cooke, A. S. and Farrell, L. (2008). Ch. 11 (p. 617), in Harris, S. and Yalden, D. W. (eds), *Mammals of the British Isles: Handbook*, 4th edition. Mammal Society, London.

Cowan, D. P. and Hartley, F. G. (2008). Ch. 6 (pp. 201–10), in Harris, S. and Yalden, D. W. (eds), *Mammals of the British Isles: Handbook*, 4th edition. Mammal Society, London.

Crawley, D., Coomber, F., Kubasiewicz, L., Harrower, C., Evans, P., Waggitt, J., Smith, B. and Mathews, F. (2020). *Atlas of the Mammals of Great Britain and Northern Ireland (The Mammal Society)*. Pelagic Publishing, Exeter.

Cresswell, W. (2023). Bioacoustics as a tool for red squirrel conservation. Countryside Job Service. Available online: https://www.countryside-jobs.com/article/2023-01-18-bioacoustics-as-a-tool-for-red-squirrel-conservation

Delahay, R., Wilson, G., Harris, S. and Macdonald, D. W. (2008). Ch. 9 (pp. 425–36), in Harris, S. and Yalden, D. W. (eds), *Mammals of the British Isles: Handbook*, 4th edition. Mammal Society, London.

Di Cerbo A. R. and Biancardi, C. M. (2012). Monitoring small and arboreal mammals by camera traps: effectiveness and applications. *Acta Theriologica* 58: 279–83. https://doi.org/10.1007/s13364-012-0122-9

Diggins, C. A. (2021). Behaviors associated with vocal communication of squirrels. *Ecosphere* 12(6). https://doi.org/10.1002/ecs2.3572

Douhard, M., Bonenfant, C., Gaillard, J-M., Hamann, J-L., Garel, M., Michallet, J. and Klein, F. (2013). Roaring counts are not suitable for the monitoring of red deer *Cervus elaphus* population abundance. *Wildlife Biology* 19: 94–101. https://doi.org/10.2981/12-037

Dunstone, N. and Macdonald, D. W. (2008). Ch. 9 (pp. 485–94), in Harris, S. and Yalden, D. W. (eds), *Mammals of the British Isles: Handbook*, 4th edition. Mammal Society, London.

Elmeros, M., Madsen, A. B. and Berthelsen, J. P. (2003). Monitoring of reintroduced beavers (*Castor fiber*) in Denmark. *Lutra* 46 (2): 153–62.

ENETWILD consortium, Podgórski, T., Acevedo, P., Apollonio, M., Berezowska-Cnota, T., Bevilacqua, C., Blanco, J. A., Borowik, T., Garrote, G., Huber, D., Keuling, O., Kowalczyk, R., Mitchler, B., Michler, F. U., Olszańska, A., Scandura, M., Schmidt, K., Selva, N., Sergiel, A., Stoyanov, S., Vada, R. and Vicente, J. (2020). *Guidance on Estimation of Abundance and Density of Wild Carnivore Population: Methods, Challenges, Possibilities*. https://doi.org/10.2903/sp.efsa.2020.EN-1947

Fellowes, M. D. E., Acquaah-Harrison, K., Angeoletto, F., Santos, J. W. M. C., da Silva Leandro, D., Rocha, E. A., Pirie, T. J. and Thomas, R. L. (2020). Map-A-Mole: greenspace area influences the presence and abundance of the European Mole *Talpa europaea* in urban habitats. *Animals (Basel)* 10 (6): 1097. https://doi.org/10.3390/ani10061097

Ferreira, L. S., Damo, J. S., Sábato, V., Baumgarten, J. E., Rodrigues, F. H. G. and Sousa-Lima, R. (2019). Using playbacks to monitor and investigate the behaviour of wild maned wolves. *Bioacoustics*. https://doi.org/10.1080/09524622.2019.1691655

Flowerdew, J. R. and Tattersall, F. H. (2008). Ch. 5 (pp. 125–37), in Harris, S. and Yalden, D. W. (eds), *Mammals of the British Isles: Handbook*, 4th edition. Mammal Society, London.

Forsman, K. A. and Malmquist, M. G. (1988). Evidence for echolocation in the common shrew, *Sorex araneus*. *Journal of Zoology* 216 (4): 655–62.

Galbraith, M. and Gaywood, C. (2008). The proposed trial re-introduction of European beaver: The Giardia issue. *Journal of The Royal Environmental Health Institute of Scotland* 14 (4): 12–13. Available online: https://www.researchgate.net/publication/322369999_The_proposed_trial_re-introduction_of_European_beaver_The_Giardia_issue.

Garcia, M., Gingras, B., Bowling, D. L., Herbst, C. T., Boeckle, M., Locatelli, Y. and Fitch, W. T. (2016). Structural classification of Wild Boar (*Sus scrofa*) vocalizations. *Ethology* 122 (4): 329–42. https://doi.org/10.1111/eth.12472

Gilbert, F. F. (1965). Analysis of basic vocalizations of the Ranch Mink. *Journal of Mammalogy* 50 (3): 625–27. https://doi.org/10.2307/1378797

Glass, D., Scott, D. M., Donoher, D. and Overall, A. D. J. (2015). The origins of the County Kildare Dormouse. *Biology and Environment: Proceedings of the Royal Irish Academy* 115B (1): 11–16. https://doi.org/10.3318/bioe.2015.02

Gnoli, C. and Prigioni, C. (1995). Preliminary study on the acoustic communication of captive otters (*Lutra lutra*). *Hystrix* 7 (1–2): 289–96.

Gordon, S. (2015). An investigation into the variation of the vocalisations emitted by wild populations of the Eurasian otter (*Lutra lutra*) at a holt in Fife, Scotland. Honours thesis (40061004), Napier University, Edinburgh (Supervisor: Dr P. White).

Gorman, M. L. (2008). Ch. 7 (pp. 250–55), in Harris, S. and Yalden, D. W. (eds), *Mammals of the British Isles: Handbook*, 4th edition. Mammal Society, London.

Gorman, M. L. and Reynolds, P. (2008). Ch. 5 (pp. 107–10), in Harris, S. and Yalden, D. W. (eds), *Mammals of the British Isles: Handbook*, 4th edition. Mammal Society, London.

Gould, E., Negus, N. C. and Novick, A. (1964). Evidence for echolocation in shrews. *Journal of Experimental Zoology* 156: 19–38.

Goulding, M., Kitchener, A. C. and Yalden, D. W. (2008). Ch. 11 (p. 561), in Harris, S. and Yalden, D. W. (eds), *Mammals of the British Isles: Handbook*, 4th edition. Mammal Society, London.

Green, J., Green, R. and Jefferies, D. J. (1984). A radio tracking survey of otters *Lutra lutra* on a Perthshire river system. *Lutra* 27: 85–145.

Gurnell, J. and Flowerdew, J. R. (2019). *Live Trapping Small Mammals: A Practical Guide*. Mammal Society, London.

Gurnell, J., Kenward, R. E., Pepper, H. and Lurz, P. W. W. (2008a). Ch. 5 (pp. 66–72), in Harris, S. and Yalden, D. W. (eds), *Mammals of the British Isles: Handbook*, 4th edition. Mammal Society, London.

Gurnell, J., Lurz, P. W. W. and Halliwell, E. C. (2008b). Ch. 5 (pp. 57–66), in Harris, S. and Yalden, D. W. (eds), *Mammals of the British Isles: Handbook*, 4th edition. Mammal Society, London.

Gurnell, J., Lurz, P. and Wauters, L. (2012). *Squirrels*. Mammal Society.

Gyger, M. and Schenk, F. (1983). Semiotical approach to the ultrasonic vocalization in the wood mouse *Apodemus sylvaticus* L. *Behaviour* 84 (3): 244–57.

Gyger, M. and Schenk, F. (1984). Ultrasonic vocalization in the wood mouse *Apodemus sylvaticus* L. *Acta Zoologica Fennica* 171: 97–99.

Halls, S. A. (1981). The influence of olfactory stimuli on ultrasonic calling in murid and cricetid rodents. PhD thesis, University of London.

Hammerschmidt, K., Radyushkin, K., Ehrenreich, H. and Fischer, J. (2012). The structure and usage of female and male mouse ultrasonic vocalizations reveal only minor differences. *PLoS ONE* 7 (7). https://doi.org/10.1371/journal.pone.0041133

Harrington, L., Birks, J., Chanin, P. and Tansley, D. (2020). Current status of American mink *Neovison vison* in Great Britain: a review of the evidence for a population decline. *Mammal Review* 50 (2): 157–69. https://doi.org/10.1111/mam.12184

Harris, S. (2022). Invasions, plagues and conservation – the history of ship rats in Britain and Ireland. *British Wildlife* 34: 157–67.

Heffner, H. (1980). Hearing in Glires: Domestic rabbit, cotton rat, feral house mouse, and kangaroo rat. *Journal of the Acoustical Society of America* 68 (6): 1584–99.

Heffner, R. S. and Heffner, H. E. (1985). Hearing in mammals: the least weasel. *Journal of Mammalogy* 66 (4): 745–55. https://doi.org/10.2307/1380801

Heffner, R. S., Heffner, H. E., Contos, C. and Kearns, D. (1994). Hearing in prairie dogs: transition between surface and subterranean rodents. *Hearing Research* 73: 185–89. https://doi.org/10.1016/0378-5955(94)90233-X

Heffner, R. S., Koay, G. and Heffner, H. E. (2020). Hearing and sound localization in Cottontail rabbits, *Sylvilagus floridanus*. *Journal of Comparative Physiology A* 206: 543–52. https://doi.org/10.1007/s00359-020-01424-8

Henrich, M., Niederlechner, S., Kroschel, M., Thoma, S., Dormann, C. F., Hartig, F. and Heurich, M. (2020). The influence of camera trap flash type on the behavioural reactions and trapping rates of red deer and roe deer. *Remote Sensing in Ecology and Conservation*. https://doi.org/10.1002/rse2.150

Hewison, A. J. M. and Staines, B. W. (2008). Ch. 11 (pp. 611–12), in Harris, S. and Yalden, D. W. (eds), *Mammals of the British Isles: Handbook*, 4th edition. Mammal Society, London.

Hodgdon, H. E. and Larson, J. S. (1973). Some sexual differences in behaviour within a colony of marked beavers (*Castor canadensis*). *Animal Behaviour* 21: 147–52. https://doi.org/10.1016/S0003-3472(73)80052-1

Hoffmeyer, I. and Sales, G. D. (1977). Ultrasonic behaviour of *Apodemus sylvaticus* and *Apodemus flavicollis*. *Oikos* 29 (1): 67–77.

Holley, A. J. F. (1992). Studies on the biology of the brown hare (*Lepus europaeus*) with particular reference to behaviour. Thesis, Durham University. Available at Durham E-Theses Online: http://etheses.dur.ac.uk/6135/

Hutterer, R. and Peters, G. (2001). The vocal repertoire of *Graphiurus parvus*, and comparisons with other species of dormice. *Trakya University Journal of Scientific Research* Series B, 2 (2): 69–74.

Iason, G. R., Hulbert, I. A. R., Hewson, R. and Dingerkus, K. (2008). Ch. 6 (pp. 220–28), in Harris, S. and Yalden, D. W. (eds), *Mammals of the British Isles: Handbook*, 4th edition. Mammal Society, London.

Iesari, V., Catorci, A., Scocco, P., Bieber, C. and Fusani, L. (2017). Vocal behaviour of the edible dormouse (*Glis glis*) during the mating season. Master's degree in applied geobotany, experimental thesis produced by V. Iesari. School of Biosciences and Veterinary Medicine, University of Camerino.

Jefferies, D. J. and Woodroffe, G. L. (2008). Ch. 9 (pp. 437–47), in Harris, S. and Yalden, D. W. (eds), *Mammals of the British Isles: Handbook*, 4th edition. Mammal Society, London.

Jennings, N. (2008). Ch. 6 (pp. 210–20), in Harris, S. and Yalden, D. W. (eds), *Mammals of the British Isles: Handbook*, 4th edition. Mammal Society, London.

Juškaitis, R. and Büchner, S. (2013). *The Hazel Dormouse,* Muscardinus avellanarius. Verlags KG Wolf (formerly Westarp Wissenschaften).

Kapusta, J. and Sales, G. D. (2009). Male–female interactions and ultrasonic vocalization in three sympatric species of voles during conspecific and heterospecific encounters. *Behaviour* 146 (7): 939–62.

Kapusta, J., Sales, G. D. and Czuchnowski, R. (2007). Aggression and vocalization of three sympatric vole species during conspecific and heterospecific same-sex encounters. *Behaviour* 144 (3): 283–305.

Kim, E. J., Kim, E. S., Covey, E. and Kim, J. J. (2010). Social transmission of fear in rats: the role of 22 kHz ultrasonic distress vocalization. *PLoS One* 5 (12). https://doi.org/10.1371/journal.pone.0015077

Kitchener, A. C. and Daniels, M. J. (2008). Ch. 9 (pp. 397–406), in Harris, S. and Yalden, D. W. (eds), *Mammals of the British Isles: Handbook*, 4th edition. Mammal Society, London.

Kopij, G. (2014). Distribution and abundance of the Red Squirrel *Sciurus vulgaris* in an urbanised environment. *Acta Musei Silesiae, Scientiae Naturales* 63: 255–62. https://doi.org/10.2478/cszma-2014-0022

Kruger, M. C., Sabourin, C. J., Levine, A. T. and Lomber, S. G. (2021). Ultrasonic hearing in cats and other terrestrial mammals. *Acoustics Today* 17 (1): 18. https://doi.org/10.1121/AT.2021.17.1.18

Kruuk, H. (2006). *Otters – Ecology, Behaviour and Conservation*. Oxford University Press.

Lahvis, G. P., Alleva, E. and Scattoni, M. L. (2011). Translating mouse vocalizations: prosody and frequency modulation. *Genes, Brain, and Behavior* 10 (1): 4–16.

Lambin, X. (2008). Ch. 5 (pp. 99–107), in Harris, S. and Yalden, D. W. (eds), *Mammals of the British Isles: Handbook*, 4th edition. Mammal Society, London.

Langbein, J., Chapman, N. G. and Putman, R. J. (2008). Ch. 11 (p. 595), in Harris, S. and Yalden, D. W. (eds), *Mammals of the British Isles: Handbook*, 4th edition. Mammal Society, London.

Libera, M. D., Passilongo, D. and Reby, D. (2015). Acoustics of male rutting roars in the endangered population of Mesola red deer *Cervus elaphus italicus*. *Mammalian Biology* 80: 395–400. https://doi.org/10.1016/j.mambio.2015.05.001

Lishak, R. S. (1982a). Vocalizations of nestling grey squirrels. *Journal of Mammalogy* 63: 446–52.

Lishak, R. S. (1982b). Grey squirrel mating calls: a spectrographic and ontogenic analysis. *Journal of Mammalogy* 63: 1–3.

Lishak, R. S. (1984). Alarm vocalizations of adult grey squirrels. *Journal of Mammalogy* 65: 1–4.

Littlewood, N. A., Hancock, M. H., Newey, S., Shackelford, G. and Toney, R. (2021). Use of a novel camera trapping approach to measure small mammal responses to peatland restoration. *European Journal of Wildlife Research* 67: 12. https://doi.org/10.1007/s10344-020-01449-z

Long, A. M., Moore, N. P. and Hayden, T. J. (1998). Vocalizations in red deer (*Cervus elaphus*), sika deer (*Cervus nippon*), and red × sika hybrids. *Journal of Zoology* 244 (1): 123–34. https://doi.org/10.1111/j.1469-7998.1998.tb00014.x

Lupanova, A. S. and Egorova, M. A. (2015). Vocalization of sex partners in the house mouse (*Mus musculus*). *Journal of Evolutionary Biochemistry and Physiology* 51 (4): 324–31.

Lysaght, L. and Marnell, F. (2016). *Atlas of Mammals in Ireland 2010–2015*. National Biodiversity Data Centre, Waterford.

Macdonald, D. and Barrett, P. (1995). *Collins Field Guide: Mammals of Britain and Europe*. HarperCollins, London.

Malkemper, E. P., Mason, M. J. and Burda, H. (2020). Functional anatomy of the middle and inner ears of the red fox, in comparison to domestic dogs and cats. *Journal of Anatomy* 236 (6): 980–95. https://doi.org/10.1111/joa.13159

Mammal Society (2021). Edible dormouse (*Glis glis*) by Roger Trout – Invasive Species Week. Mammal Society. https://www.mammal.org.uk/2021/05/edible-dormouse-glis-glis-by-roger-trout-invasive-species-week/

Mammal Society (2022). Full Mammal Species Hub. Mammal Society. Website with supporting factsheets: https://www.mammal.org.uk/species-hub/full-species-hub/discover-mammals/

Marchlewska-Koj, A. (2000). Olfactory and ultrasonic communication in bank voles. *Polish Journal of Ecology* 48: 11–20.

Marsh, A. C. W. and Montgomery, W. I. (2008). Ch. 5 (pp. 137–41), in Harris, S. and Yalden, D. W. (eds), *Mammals of the British Isles: Handbook*, 4th edition. Mammal Society, London.

Mathews, F., Kubasiewicz, L. M., Gurnell, J., Harrower, C. A., McDonald, R. A. and Shore, R. F. (2018). *A Review of the Population and Conservation Status of British Mammals. A Report by the Mammal Society under Contract to Natural England, Natural Resources Wales and Scottish Natural Heritage*. Natural England, Peterborough.

Mauri, L., Luschi, P. and Apollonio, M. (1994). Vocal repertoire of fallow deer (*Dama dama*). *Bollettino di zoologia* 61. https://doi.org/10.1080/11250009409356035

McDonald, R. and Harris, S. (2006). *Stoats and Weasels*. Mammal Society.

McDonald, R. A. and King, C. M. (2008a). Ch. 9 (pp. 455–67), in Harris, S. and Yalden, D. W. (eds), *Mammals of the British Isles: Handbook*, 4th edition. Mammal Society, London.

McDonald, R. A. and King, C. M. (2008b). Ch. 9 (pp. 467–76), in Harris, S. and Yalden, D. W. (eds), *Mammals of the British Isles: Handbook*, 4th edition. Mammal Society, London.

McGowan, D. and Gurnell, J. (2014). *Small Mammal Survey*. Jersey 2014. Available online: https://www.gov.je/SiteCollectionDocuments/Government%20and%20administration/R%20Small%20Mammal%20Survey%20Jersey%202014%2020150729%20DM.pdf

McRae, T. R. (2020). A review of squirrel alarm-calling behavior: what we know and what we do not know about how predator attributes affect alarm calls. *Animal Behavior and Cognition* 7 (2): 168–91. https://doi.org/10.26451/abc.07.02.11.2020

McRae, T. R. and Green, S. M. (2014). Joint tail and vocal alarm signals of gray squirrels (*Sciurus carolinensis*). *Behaviour* 151 (10): 1433–52. https://doi.org/10.1163/1568539X-00003194

McRae, T. R. and Green, S. M. (2015). Gray squirrel alarm call composition differs in response to simulated aerial versus terrestrial predator attacks. *Ethology Ecology & Evolution*. https://doi.org/10.1080/03949370.2015.1087433

Meek, P. D., Ballard, G.-A., Fleming, P. J. S., Schaefer, M., Williams, W., Falzon, G. (2014). Camera traps can be heard and seen by animals. *PLoS ONE* 9 (10). https://doi.org/10.1371/journal.pone.0110832

Meek, P. D., Ballard, G.-A. and Fleming, P. J. S. (2015). The pitfalls of wildlife camera trapping as a survey tool in Australia. *Australian Mammalogy* 37: 13–22. https://doi.org/10.1071/AM14023

Meek, P., Ballard, G., Fleming, P. and Falzon, G. (2016). Are we getting the full picture? Animal responses to camera traps and implications for predator studies. *Biology and Evolution* 6 (10): 3218–25. https://doi.org/10.1002/ece3.2111

Melcore, I., Ferrari, G. and Bertolino, S. (2020). Footprint tunnels are effective for detecting dormouse species. *Mammal Review* 50: 226–30. https://doi.org/10.1111/mam.12199

Middleton, N. E. (2020). *Is That a Bat? A Guide to Non-Bat Sounds Encountered During Bat Surveys*. Pelagic Publishing, Exeter.

Middleton, N. E., Froud, A., French, K. (2022). *Social Calls of the Bats of Britain and Ireland*, 2nd edition. Pelagic Publishing, Exeter.

Millman, R. (2021). Saving red squirrels with AI and cloud computing. ITPro. Available online: https://www.itpro.co.uk/cloud/cloud-computing/361509/saving-red-squirrels-with-ai-and-cloud-computing

Miska-Schramm, A., Kapusta, J. and Kruczek, M. (2018). Copper influence on bank vole's (*Myodes glareolus*) sexual behaviour. *Ecotoxicology* 27 (3): 385–93.

Morris, P. (2011). *The Hedgehog*. Mammal Society.

Morris, P. A. (2008). Ch. 5 (pp. 81–85), in Harris, S. and Yalden, D. W. (eds), *Mammals of the British Isles: Handbook*, 4th edition. Mammal Society, London.

Morris, P. A. and Reeve, N. J. (2008). Ch. 7 (pp. 241–49), in Harris, S. and Yalden, D. W. (eds), *Mammals of the British Isles: Handbook*, 4th edition. Mammal Society, London.

Morris, C., O'Reilly, C., Turner, P., Halliwell, L., O'Meara, D. and Sheerin, D. (2013). A novel non-invasive method for detecting the harvest mouse (*Micromys minutus*). Mammal Society, *Mammal News* Summer 2013, Issue 166: 22.

Mortelliti, A. and Boitani, L. (2008). Inferring red squirrel (*Sciurus vulgaris*) absence with hair tubes surveys: a sampling protocol. *European Journal of Wildlife Research* 54: 353–56. https://doi.org/10.1007/s10344-007-0135-x

Mullan, S. and Saunders, R. (2019). Ch. 7: European Rabbits (*Oryctolagus cuniculus*), in Yeates, J. (ed.), *Companion Animal Care and Welfare: The UFAW Companion Animal Handbook*, 1st edition. John Wiley & Sons.

Müller-Schwarze, D. (2011). *The Beaver: Its Life and Impact*, 2nd edition. Cornell University Press.

Narins, P. M., Stoeger, A. S., O'Connell-Rodwell, C. (2016). Infrasonic and seismic communication in the vertebrates with special emphasis on the Afrotheria: an update and future directions, in Suthers, R., Fitch, W., Fay, R. and Popper, A. (eds), *Vertebrate Sound Production and Acoustic Communication: Springer Handbook of Auditory Research*, Vol. 53. Springer, Cham. https://doi.org/10.1007/978-3-319-27721-9_7

Newson, S. E. and Pearce, H. (2022). The potential for acoustics as a conservation tool for monitoring small terrestrial mammals. *JNCC Report* No. 708, JNCC, Peterborough.

Newson, S. E., Middleton, N. and Pearce, H. (2020). The acoustic identification of small terrestrial mammals in Britain. *British Wildlife* 32: 186–94.

Newson, S. E., Allez, S. L., Coule, E. K., Guille, A. W., Henney, J. M., Higgins, L., McLellan, G. D., Simmons, M. C., Sweet, L., Whitelegg, D. and Atkinson, P. W. (2023). Bailiwick Bat Survey: 2022 season report. *BTO Research Report* 750. BTO, Thetford.

Newton-Fisher, N., Harris, S., White, P. and Jones, G. (1993). Structure and function of red fox *Vulpes vulpes* vocalisations. *Bioacoustics* 5: 1–31. https://doi.org/10.1080/09524622.1993.9753228

Nicastro, N. (2004). Perceptual and acoustic evidence for species-level differences in meow vocalizations by domestic cats (*Felis catus*) and African wild cats (*Felis silvestris lybica*). *Journal of Comparative Psychology* 118 (3): 287–96. https://doi.org/10.1037/0735-7036.118.3.287

Osipova, O. and Rutovskaya, M. V. (2000). Information transmission in bank voles by odor and acoustic signals (signalling communication). *Polish Journal of Ecology* 48: 21–36.

Owens, J. L., Olsen, M., Fontaine, A., Kloth, C., Kershenbaum, A. and Waller, S. (2017). Visual classification of feral cat *Felis silvestris catus* vocalizations. *Current Zoology* 63 (3): 331–39. https://doi.org/10.1093/cz/zox013

Papin, M., Pichenot, J., Guérold, F. and Germain, E. (2018). Acoustic localization at large scales: a promising method for grey wolf monitoring. *Frontiers in Zoology* 15 (11). https://doi.org/10.1186/s12983-018-0260-2

Passilongo, D., Mattioli, L., Bassi, E., Szabó, L. and Apollonio, M. (2015). Visualizing sound: counting wolves by using a spectral view of the chorus howling. *Frontiers in Zoology* 12 (22). https://doi.org/10.1186/s12983-015-0114-0

Persinger, M. A. (2013). Infrasound, human health, and adaptation: an integrative overview of recondite hazards in a complex environment. *Natural Hazards* 70: 501–25. https://doi.org/10.1007/s11069-013-0827-3

Pocock, M. J. O. and Jennings, N. (2006). Use of hair tubes to survey for shrews: new methods for identification and quantification of abundance. *Mammal Review* 36 (4): 299–308. https://doi.org/10.1111/j.1365-2907.2006.00092.x

Pollock, M. M., Lewallen, G. M., Woodruff, K., Jordan, C. E. and Castro, J. M. (2017). *The Beaver Restoration Guidebook: Working with Beaver to Restore Streams, Wetlands, and Floodplains. Version 2.0.* United States Fish and Wildlife Service, Portland, Oregon. Available online at: https://www.fws.gov/oregonfwo/promo.cfm?id=177175812

Powell, R. A. and Zielinski, W. J. (1989). Mink response to ultrasound in the range emitted by prey. *Journal of Mammalogy* 70 (3): 637–38. https://doi.org/10.2307/1381439

Putman, R. J. (2008). Ch. 11 (p. 592), in Harris, S. and Yalden, D. W. (eds), *Mammals of the British Isles: Handbook*, 4th edition. Mammal Society, London.

Quy, R. J. and Macdonald, D. W. (2008). Ch. 5 (pp. 149–55), in Harris, S. and Yalden, D. W. (eds), *Mammals of the British Isles: Handbook*, 4th edition. Mammal Society, London.

Ražen, N., Kuralt, Ž., Fležar, U., Bartol, M., Černe, R., Kos, I., Krofel, M., Luštrik, R., Majić Skrbinšek, A. and Potočnik, H. (2020). Citizen science contribution to national wolf population monitoring: what have we learned? *European Journal of Wildlife Research* 66 (46). https://doi.org/10.1007/s10344-020-01383-0

Reby, D. and McComb, K. (2003). Anatomical constraints generate honesty: acoustic cues to age and weight in the roars of red deer stags. *Animal Behaviour* 65: 519–30. https://doi.org/10.1006/anbe.2003.2078

Reby, D., Cargnelutti, B., Joachim, J. and Aulagnier, S. (1998a). Spectral acoustic structure of barking in roe deer (*Capreolus capreolus*). Sex-, age- and individual-related variations. *Comptes Rendus de l'Académie des Sciences* III (4): 271–79. https://doi.org/10.1016/s0764-4469(99)80063-8

Reby, D., Hewison, A. J. M., Cargnelutti, B., Angibault, J.-M. and Vincent, J.-P. (1998b). Use of vocalizations to estimate population size of roe deer. *Journal Of Wildlife Management* 62 (4): 1342–48. https://doi.org/10.2307/3802000

Reby, D., Hewison, M., Izquierdo, M. and Pépin, D. (2001). Red deer (*Cervus elaphus*) hinds discriminate between the roars of their current harem-holder stag and those of neighbouring stags. *Ethology* 107: 951–59. https://doi.org/10.1046/j.1439-0310.2001.00732.x

Rehnus, M., Wehrle, M. and Obrist, M. K. (2019). Vocalisation in the mountain hare: calls of a mostly silent species. *European Journal of Wildlife Research* 65: 95. https://doi.org/10.1007/s10344-019-1331-1

Rödel, H. G., Landmann, C., Starkloff, A., Kunc, H. P. and Hudson, R. (2013). Absentee mothering – not so absent? Responses of European rabbit (*Oryctolagus cuniculus*) mothers to pup distress calls. *Ethology* 119: 1024–33. https://doi.org/10.1111/eth.12149

Rodolfi, G. (1994). Dormice *Glis glis* activity and hazelnut consumption. *Acta Theriologica* 39 (2): 215–20.

Roper, T. J. (2010). *Badger*. HarperCollins, London.

Rossi, I., Mauri, L., Laficara, S. and Appollonio, M. (2002). Barking in roe deer (*Capreolus capreolus*): seasonal trends and possible functions. *Hystrix* 13 (1–2): 13–18. https://doi.org/10.4404/hystrix-13.1-2-4181

Russ, J. (2012). *British Bat Calls: A Guide to Species Identification*. Pelagic Publishing, Exeter.

Russ, J. (2021). *Bat Calls of Britain and Europe: A Guide to Species Identification*. Pelagic Publishing, Exeter.

Sales, G. D. (2010). Ultrasonic calls of wild and wild-type rodents (pp. 77–88), in Brudzynski, S. M. (ed.), *Handbook of Mammalian Vocalization An Integrative Neuroscience Approach*. Academic Press, London.

Sales, G. D. and Pye, J. D. (1974). *Ultrasonic Communication by Animals*. Chapman & Hall, London.

Sample, G. (2006). *Collins Field Guide to Wildlife Sounds in Britain and Ireland*. HarperCollins, London.

Sanchez, L., Ohdachi, S. D., Kawahara, A., Echenique-Diaz, L. M., Maruyama, S. and Kawata, M. (2019). Acoustic emissions of *Sorex unguiculatus* (Mammalia: Soricidae): assessing the echo-based orientation hypothesis. *Ecology and Evolution* 9: 2629–39. https://doi.org/10.1002/ece3.4930

Schötz, S., van de Weijer, J. and Eklund, R. (2017). *Phonetic Characteristics of Domestic Cat Vocalisations.* Proceedings of 1st Intl Workshop on Vocal Interactivity in-and-between Humans, Animals and Robots (VIHAR), Skvöde, Sweden, 25–26 August 2017.

Schuh, D., Hoy S. T. and Selzer, D. (2004). Vocalization of rabbit pups in the mother–young relationship. Proceedings of 8th World Rabbit Congress, 7–10 September 2004, Puebla, Mexico.

Shore, R. F. and Hare, E. J. (2008). Ch. 5 (pp. 88–99), in Harris, S. and Yalden, D. W. (eds), *Mammals of the British Isles: Handbook,* 4th edition. Mammal Society, London.

Siemers, B. M., Schauermann, G., Turni, H. and von Merton, S. (2009). Why do shrews twitter? Communication or simple echo-based orientation. *Biology Letters* 5: 593–96. https://doi.org/10.1098/rsbl.2009.0378

Simeonovska-Nikolova, D. M. (2004). Vocal communication in the bicoloured white-toothed shrew *Crocidura leucodon. Acta Theriologica* 49 (2): 157–65.

Staines, B. W., Langbein, J. and Burkitt, T. D. (2008). Ch. 11 (p. 581), in Harris, S. and Yalden, D. W. (eds), *Mammals of the British Isles: Handbook,* 4th edition. Mammal Society, London.

Stoddart, D. M. and Sales, G. D. (1985). The olfactory and acoustic biology of wood mice, yellow-necked mice and bank voles. *Symposia of the Zoological Society of London* 55: 117–39.

Strachan, R. (1999). *Water Voles.* British Natural History Series, Vol. 27. Whittet Books, London.

Studd, E. K., Derbyshire, R. E., Menzies, A. K., Simms, J. F., Humphries, M. M., Murray, D. L. and Boutin, S. (2021). The purr-fect catch: using accelerometers and audio recorders to document kill rates and hunting behaviour of a small prey specialist. *Methods in Ecology and Evolution* 12 (4): 1–11.

Takahashi, N., Kashino, M. and Hironaka, N. (2010). Structure of rat ultrasonic vocalizations and its relevance to behavior. *PLoS One* 5 (11): https://doi.org/10.1371/journal.pone.0014115

Tapper, S. and Yalden, D. (2010). *The Brown Hare.* Mammal Society.

Thomas, J. A. and Jalili, M. S. (2004). Review of echolocation in insectivores and rodents (pp. 547–63), in Thomas, J. A., Moss, J. F. and Vater, M. (eds), *Echolocation in Bats and Dolphins.* University of Chicago Press, Chicago.

Thomas, E. and Wilson, E. (2018). *Guidance for Surveying Hedgehogs.* Hedgehog Street. Available online: https://www.hedgehogstreet.org/wp-content/uploads/2018/06/Guidance-for-surveying-hedgehogs.pdf

Thomsen, L. R., Campbell, R. D. and Rosell, F. (2007). Tool-use in a display behaviour by Eurasian beavers (*Castor fiber*). *Animal Cognition* 10: 477–82. https://doi.org/10.1007/s10071-007-0075-6

Tomasi, T. E. (1979). Echolocation by the Short-Tailed Shrew *Blarina brevicauda. Journal of Mammalogy* 60 (4): 751–759. http://www.jstor.org/stable/1380190

Trout, R. C. and Harris, S. (2008). Ch. 5 (pp. 117–25), in Harris, S. and Yalden, D. W. (eds), *Mammals of the British Isles: Handbook,* 4th edition. Mammal Society, London.

Tullo, E., Ponzetta, M. P., Trunfio, C., Gardoni, D., Ferrari, S. and Guarino, M. (2015). Acoustic analysis of some characteristics of red deer roaring. *Italian Journal of Animal Science* 14: 383–88. https://doi.org/10.4081/ijas.2015.3773

Twigg, G. I., Buckle, A. P. and Bullock, D. J. (2008). Ch. 5 (pp. 155–58), in Harris, S. and Yalden, D. W. (eds), *Mammals of the British Isles: Handbook,* 4th edition. Mammal Society, London.

Vannoni, E. and McElligott, A. G. (2007). Individual acoustic variation in fallow deer (*Dama dama*) common and harsh groans: a source-filter theory perspective. *Ethology* 113: 223–34. https://doi.org/10.1111/j.1439-0310.2006.01323.x

Vannoni, E. and McElligott, A. G. (2008). Low frequency groans indicate larger and more dominant fallow deer (*Dama dama*) males. *PLoS ONE* 3(9). https://doi.org/10.1371/journal.pone.0003113

von Merten, S. (2011). Spatial exploration and acoustic orientation in shrews: a comparative approach on the three sympatric species *Sorex araneus, Sorex minutus* and *Crocidura leucodon.* Dissertation, Tübingen.

von Merten, S., Hoier, S., Pfeifle, C. and Tautz, D. (2014). A role for ultrasonic vocalisation in social communication and divergence of natural populations of the house mouse (*Mus musculus domesticus*). *PLoS ONE* 9 (5). https://doi.org/10.1371/journal.pone.0097244.

Wegge, P., Pokheral, C. P. and Jnawali, S. R. (2004). Effects of trapping effort and trap shyness on estimates of tiger abundance from camera trap studies. *Animal Conservation* 7: 251–56. https://doi.org/10.1017/S1367943004001441

Wembridge, D., Johnson, G., Al-Fulaij, N. and Langton, S. (2022). *The State of Britain's Hedgehogs 2022*. British Hedgehog Preservation Society. Website: https://www.britishhedgehogs.org.uk/new-state-of-britains-hedgehog-report-issued-today/

Wilsson, L. (1971). Observations and experiments on the ethology of the European beaver (*Castor fiber* L.). *Viltrevy* 8 (3): 1–266.

Wöhr, M. (2018). Ultrasonic communication in rats: appetitive 50 kHz ultrasonic vocalizations as social contact calls. *Behavioral Ecology and Sociobiology* 72 (1): 14.

Wöhr, M. and Schwarting, R. K. W. (2013). Affective communication in rodents: ultrasonic vocalizations as a tool for research on emotion and motivation. *Cell and Tissue Research* 354 (1): 81–97.

Wong, J., Stewart, P. D. and Macdonald, D. W. (1999). Vocal repertoire in the European Badger (*Meles meles*): structure, context and function. *Journal of Mammalogy* 80 (2): 570–88. https://doi.org/10.2307/1383302

Woodroffe, G. (2007). *The Otter*. Mammal Society.

Woodroffe, G. L., Lambin, X. and Strachan, R. (2008). Ch. 5 (pp. 110–17), in Harris, S. and Yalden, D. W. (eds), *Mammals of the British Isles: Handbook*, 4th edition. Mammal Society, London.

Woods, M. (2010). *The Badger*, 2nd edition. Mammal Society.

Yahner, R. H. (1980). Barking in a primitive ungulate, *Muntiacus reevesi*: function and adaptiveness. *The American Naturalist* 116 (2): 157–77.

Yalden, D. W. and Morris, P. A. (1993). *The Analysis of Owl Pellets*. Occasional Publication No. 13. The Mammal Society, London.

Yeon, S. C., Kim, Y. K., Park, S. J., Lee, S. S., Lee, S. Y., Suh, E. H., Houpt, K. A., Chang, H. H., Lee, H. C., Yang, B. G. and Lee, H. J. (2011). Differences between vocalization evoked by social stimuli in feral cats and house cats. *Behavioural Processes* 87 (2): 183–89. https://doi.org/10.1016/j.beproc.2011.03.003

Zala, S. M., Nicolakis, D., Marconi, M. A., Noll, A., Ruf, T., Balazs, P. and Penn, D. J. (2020). Primed to vocalize: wild-derived male house mice increase vocalization rate and diversity after a previous encounter with a female. *PLoS ONE* 15 (12). https://doi.org/10.1371/journal.pone.0242959

Zsebok, S., Czaban, D., Farkas, J., Siemers, B. M. and von Merten, S. (2015). Acoustic species identification of shrews: twittering calls for monitoring. *Ecological Informatics* 27: 1–10.

Index